International Lithosphere Program
Publication No. 0104

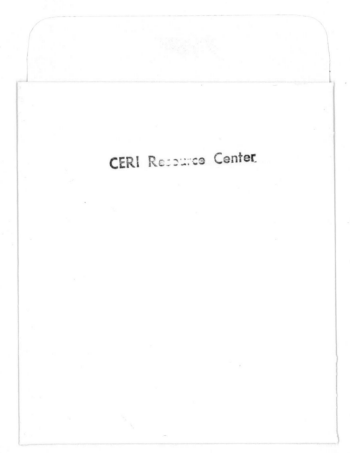

Plateau Uplift

The Rhenish Shield – A Case History

Edited by
K. Fuchs   K. von Gehlen   H. Mälzer
H. Murawski   A. Semmel

With 185 Figures

Springer-Verlag
Berlin   Heidelberg   New York   Tokyo   1983

Professor Dr. Karl Fuchs
Geophysikalisches Institut, Universität Karlsruhe,
Hertzstraße 16, D-7500 Karlsruhe 21, FRG

Professor Dr. Kurt von Gehlen
Institut für Geochemie, Petrologie und Lagerstättenkunde, Universität Frankfurt,
Postfach 111932, D-6000 Frankfurt 11, FRG

Professor Dr. Hermann Mälzer
Geodätisches Institut, Universität Karlsruhe,
Kaiserstraße 12, D-7500 Karlsruhe 1, FRG

Professor Dr. Hans Murawski
Geologisch-Paläontologisches Institut, Universität Frankfurt,
Senckenberganlage 32-34, D-6000 Frankfurt, FRG

Professor Dr. Arno Semmel
Institut für Physische Geographie, Fachbereich Geowissenschaften,
Senckenberganlage 36, D-6000 Frankfurt, FRG

ISBN 3-540-12577-9 Springer-Verlag Berlin Heidelberg New York Tokyo
ISBN 0-387-12577-9 Springer-Verlag New York Tokyo Heidelberg Berlin

Library of Congress Cataloging in Publication Data. Main entry under title: Plateau uplift. 1. Geology-Germany (West)-Rheinisches Schiefergebirge. I. Fuchs, K. QE269.P56 1983 554.3'43 83-10447

This work is subject to copyright. All rights are reserved, whether the whole or part of the material is concerned, specifically those of translation, reprinting, re-use of illustrations, broadcasting, reproduction by photocopying machine or similar means, and storage in data banks. Under § 54 of the German Copyright Law, where copies are made for other than private use, a fee is payable to "Verwertungsgesellschaft Wort", Munich.

© by Springer-Verlag Berlin Heidelberg 1983
Printed in Germany

The use of registered names, trademarks, etc. in this publication does not imply, even in the absence of a specific statement, that such names are exempt from the relevant protective laws and regulations and therefore free for general use.

Offsetprinting: Beltz Offsetdruck, Hemsbach/Bergstraße.
Bookbinding: J. Schäffer OHG, Grünstadt.
2131/3130-543210

Jürgen Henning Illies
14. 3. 1924 – 2. 8. 1982

Dedication

In the early 1970's, geological discussions centred around the plate tectonics theory, and thus questions regarding epeirogenesis seemed to be of less relevance. In order to redress the balance, the Deutsche Forschungsgemeinschaft (German Research Society) readily accepted the proposal of a number of German geoscientists to investigate the causes of vertical uplift in central Europe using, as their example, the well-studied Rhenish Massif. After preparatory discussions within the National Committee for the International Geodynamics Project and the Senate Commission for Joint Research in Earth Sciences, the Senate of the German Research Society recommended the initiation, at the start of 1976, of the priority programme *Vertical movements and their causes using, as example, the Rhenish Massif*. The aim of this programme was to study questions relat-d to the driving mechanism(s) for the uplift and to approach their solution using modern geoscientific techniques.

The preparations were guided by Professor Dr. Henning Illies, Karlsruhe, who became, subsequently, coordinator of the programme. Within the 6 years of support, from 1976 to 1982, by his prolific ideas and everlasting personal effort, Henning Illies - supported by a coordinating committee - was able to bring together geodesists, geologists, geomorphologists, geophysicists, and petrologists from universities and state offices and to stimulate them towards the final goal. From the very beginning the German groups sought cooperation with geoscientists in neighboring countries. This cooperation culminated in 1979 in a large-scale European experiment to explore the earth's crust along a 600 km long explosion-seismic profile between Göttingen and the French town Romilly sur Seine in which 116 geoscientists from nine countries participated.

Unfortunately, Henning Illies did not see the completion of the priority programme, having died after a long, serious illness on 2nd August 1982, two days after the official closing date. This final volume for which he worked and to which his last energies were committed is dedicated to his memory.

A final colloquium was held on 7-8 October 1982 in Aachen where more than 130 geoscientists from all countries participated and where the essential results were presented to the scientific community in abbreviated form. It can be stated that one can no longer define a uniform uplift of the Rhenish Massif. Rather, there exists episodic uplift with several centres and different rates of uplift. A number of questions remain unsolved and new ones are being raised which will be pursued in future projects. This final volume is being submitted as the German contribution to the International Lithosphere Programme on the occasion of the XVIII General Assembly of the International Union of Geodesy and Geophysics at Hamburg in August 1983.

My sincere thanks are due to all collaborators in the priority programme who have contributed to the successful completion of these joint investigations. At this moment special thanks are also due to the referees and coordinators who have devoted themselves unselfishly to bringing the project to a good end.

Eugen Seibold

Präsident
Deutsche Forschungsgemeinschaft

Preface

This volume offers an account of the scientific outcome of a Priority Programme (Schwerpunktprogramm) which the Deutsche Forschungsgemeinschaft (DFG; German Research Society) promoted during 6 years from 1976 to 1982. In the understanding of the DFG, a Priority Programme involves the financing and coordination of research efforts of a group of investigators, possibly from several institutions, and is intended to concentrate on one particular topic and/or on one major area of interest over a period of, as a rule, 5 years.

Discussions on the feasibility of a major programme on vertical movements started in 1971. The tentative programme was published under the title *Terrestrial Vertical Movements* in the DFG's planning document *Grauer Plan IV: 1972-1974*; p. 152. Deliberations within the National Committee for the International Geodynamics Project and in the Senate Commission for Joint Research in Earth Sciences (Geokommission) set out the scope of the project and envisaged an investigation of the Rheinische Schild (i.e., the Rhenish Massif east and west of the River Rhine). Seventeen geoscientists participated in a round-table discussion on the subject *Vertical movements and their causes as exemplified in the case of the Rheinische Schild*, which took place at Bonn-Bad Godesberg on 8th July 1975. At this meeting the scientific objectives came more clearly into focus.

A Programme Committee formulated the proposal to be put to the Senate of the DFG. Given favorable response there, the scientists involved were encouraged to submit formal applications. Professor Dr. Henning Illies, Karlsruhe, was appointed Coordinator. A Coordinating Committee chosen to support him consisted of the following geoscientists:
K. Fuchs, Karlsruhe; K. von Gehlen, Frankfurt; H. Mälzer, Karlsruhe; H. Murawski, Frankfurt; and A. Semmel, Frankfurt.

After the application had been scrutinized by a group of reviewers and had been granted funding from the main budget of the DFG, 27 working groups were free to proceed as from 1st August 1976. Fifty to eighty geoscientists involved in the Priority Programme met annually for a late autumn colloquium (1976 and 1977 in Königstein/Taunus; 1978-1981 in Neustadt an der Weinstraße) to exchange experiences, to report results and to engage in discussions. A high level of cooperative endeavour evolved in the course of time and it soon became evident that it was an advantage to have chosen a study area within the Federal Republic of Germany. Familiar exposures could now be re-examined with new perspectives and studied by new methods. On three occasions during the years 1978 to 1980, the participants gathered together in the field to discuss, on site, problems which remained open. Areas visited included the Westerwald, Limburg and Neuwied Basin, Hunsrück, Eifel, Hohes Venn, and the Lower Rhenish Embayment. In addition, a number of smaller meetings of working groups took place during the closing stages of the project in order to evaluate the results obtained in the field. Requests for future funding and written reports arrived regularly in the hands of an assessment group each spring. The Coordinating Committee usually met twice a year in order to provide advice and direction, especially at points where different shades of opinions had developed. By this means a highly efficient system of quality control was maintained. During the 6 years of the Priority Programme a total of around 50 individual projects received financial support of a total of ca. 7.9 million DM. The various projects participating in the Priority Programme represented several geosciene disciplines in these proportions: Geology, 18 projects; Mineralogy/Geochemistry, 8; Solid Earth Geophysics, 11; Geodesy, 2; Geomorphology, 10.

During the Priority Programme much importance was attached to the support of young scientists and to the cooperation with other countries. Thirtyfive to fourthy young scientists were funded in the Priority Programme. A number of scientists from other countries cooperated for set terms of time within German working groups. Current reports of the work were given to meetings at home and abroad and links were forged with working groups active in other countries.

The objectives of the Priority Programme fit very well into the scope of the International Lithosphere program. Especially ILP Working Groups 1, 5 and 9, as well as Coordinating Committees 4 and 5 are strongly related to the results of the Priority Programme.

The results of the various investigations were presented at a final colloquium at the Geological Institute of the Rheinisch-Westfälische Technische Hochschule, Aachen, on 7th and 8th October 1982.

It is quite evident that the progress of the investigations and the results obtained have justified the decision to launch this Priority Programme. Even if the objectives have not in every case been fully achieved, it is proper to say that the overall result is a significant advance in the investigation of vertical crustal movements and their causes. A set of fertile new questions has emerged, to which future workers may address themselves. The successful completion of the programme was achieved only because of the excellent spirit of interdisciplinary cooperation, which all involved helped to maintain. Grateful thanks are expressed to each and every one of the contributors to the success of this Priority Programme.

D. Maronde
Deutsche Forschungsgemeinschaft
Bonn

Acknowledgments

This book contains the results of a 6 years Priority Research Programme established by the Deutsche Forschungsgemeinschaft (German Research Society) on the recommendation of its Senatskommission für Geowissenschaftliche Gemeinschaftsforschung. Also in the name of all active participants of this programme we want to thank for the initiation and for continuous support by the Deutsche Forschungsgemeinschaft.

We are deeply indebted to Henning Illies, former coordinator of this Priority Research Programme who contributed immensely to its success from its beginning to his death on 2nd August 1982 by stimulation, encouragement, and new ideas given to all participants.

The editors thank all contributors for their cooperation in the preparation of the manuscripts and also Springer-Verlag Heidelberg for the excellent printing. Our special thanks are due to Dr. Mechie (Geophysikalisches Institut Karlsruhe) and Springer-Verlag for their help in improving the English of the manuscripts.

We owe our thanks to Dr. D. Maronde for his excellent and knowledgeable administration of the Priority Programme.

K. Fuchs, Karlsruhe
K. v. Gehlen, Frankfurt
H. Mälzer, Karlsruhe
H. Murawski, Frankfurt
H. Semmel, Frankfurt

Contents

1 *Plateau Uplift of the Rhenish Massif - Introductory Remarks*
  J.H. Illies and K. Fuchs 1

2 *Regional Tectonic Setting and Geological Structure of the Rhenish Massif*
  H. Murawski, H.J. Albers, P. Bender, H.-P. Berners, St. Dürr,
  R. Huckriede, G. Kauffmann, G. Kowalczyk, P. Meiburg,
  R. Müller, A. Muller, S. Ritzkowski, K. Schwab, A. Semmel,
  K. Stapf, R. Walter, K.-P. Winter, and H. Zankl 9

3 *Pre-Quaternary Uplift in the Central Part of the Rhenish Massif*
  W. Meyer, H.J. Albers, H.P. Berners, K.v.Gehlen, D. Glatthaar, W. Löhnertz, K.H. Pfeffer, A. Schnütgen, K. Wienecke, and H. Zakosek 39

4 *Plateau Uplift During Pleistocene Time* 47

   4.1 Plateau Uplift During Pleistocene Time - Preface
      A. Semmel 48

   4.2 The Early Pleistocene Terraces of the Upper Middle Rhine
      and Its Southern Foreland - Questions Concerning Their
      Tectonic Interpretation
      A. Semmel 49

   4.3 Distribution and Dimension of Young Tectonics in
      the Neuwied Basin and the Lower Middle Rhine
      E. Bibus 55

   4.4 The Rhine Valley Between the Neuwied Basin and the
      Lower Rhenish Embayment
      K. Brunnacker and W. Boenigk 62

   4.5 The Tectonic Position of the Lower Mosel Block in
      Relation to the Tertiary and Old Pleistocene Sediments
      E. Bibus 73

   4.6 Cenozoic Deposits of the Eifel-Hunsrück Area Along the
      Mosel River and Their Tectonic Implications
      J. Negendank 78

4.7	Neotectonic Movements at the Southern and Western Boundary of the Hunsrück Mountains (Southwestern Part of the Rhenish Massif) L. Zöller	89
4.8	The Lower Pleistocene Terraces of the Lahn River Between Dietz (Limburg Basin) and Laurenburg (Lower Lahn) W. Andres and H. Sewering	93
4.9	Quaternary Tectonics and River Terraces at the Eastern Margin of the Rhenish Massif K.H. Müller and S. Lipps	98
4.10	The Late Tertiary-Quaternary Tectonics of the Palaeozoic of the Northern Eifel R. Müller	102
4.11	The Quaternary Destruction of an Older, Tertiary Topography Between the Sambre and Ourthe Rivers (Southern Ardennes) H.P. Berners	108

5 *Volcanic Activity* 111

5.1	Distribution of Volcanic Activity in Space and Time H.J. Lippolt	112
5.2	Tertiary Volcanism of the Hocheifel Area H.-G. Huckenholz	121
5.3	Volcanism in the Southern Part of the Hocheifel E. Bussmann and V. Lorenz	129
5.4	Tertiary Volcanism in the Siebengebirge Mountains K. Vieten	131
5.5	Tertiary Volcanism in the Westerwald Mountains K. von Gehlen and W. Forkel	133
5.6	Tertiary Volcanism in the Northern Hessian Depression K.H. Wedepohl	134
5.7	The Quaternary Eifel Volcanic Fields H.-U. Schmincke, V. Lorenz, and H.A. Seck	139
5.8	Carbon Dioxide in the Rhenish Massif H. Puchelt	152
5.9	Eocene to Recent Volcanism Within the Rhenish Massif and the Northern Hessian Depression - Summary H.A. Seck	153

6 *Present-Day Features of the Rhenish Massif* 163

6.1	Height Changes in the Rhenish Massif: Determination and Analysis H. Mälzer, G. Hein, and K. Zippelt	164

6.2	Stress Field and Strain Release in the Rhenish Massif H. Baumann and J.H. Illies	177
6.3	General Pattern of Seismotectonic Dislocation and the Earthquake-Generating Stress Field in Central Europe Between the Alps and the North Sea L. Ahorner, B. Baier, and K.-P. Bonjer	187
6.4	Historical Seismicity and Present-Day Microearthquake Activity of the Rhenish Massif, Central Europe L. Ahorner	198
6.5	Microearthquake Activity Near the Southern Border of the Rhenish Massif B. Baier and J. Wernig	222
6.6	Geothermal Investigations in the Rhenish Massif R. Haenel	228
6.7	The Gravity Field of the Rhenish Massif W.R. Jacoby, H. Joachimi, and C. Gerstenecker	247

7 *Crust and Mantle Structure, Physical Properties and Composition* 259

7.1	The Long-Range Seismic Refraction Experiment in the Rhenish Massif J. Mechie, C. Prodehl, and K. Fuchs	260
7.2	Combined Seismic Reflection-Refraction Investigations in the Rhenish Massif and Their Relation to Recent Tectonic Movements R. Meissner, M. Springer, H. Murawski, H. Bartelsen, E.R. Flüh, and H. Dürschner	276
7.3	Electrical Conductivity Structure of the Crust and Upper Mantle Beneath the Rhenish Massif H. Jödicke, J. Untiedt, W. Olgemann, L. Schulte, and V. Wagenitz	288
7.4	The Evolution of the Hercynian Crust - Some Implications to the Uplift Problem of the Rhenish Massif P. Giese	303
7.5	Large-Scale Mantle Heterogeneity Beneath the Rhenish Massif and Its Vicinity from Teleseismic P-Residuals Measurements S. Raikes and K.-P. Bonjer	315
7.6	Crustal Xenoliths in Tertiary Volcanics from the Northern Hessian Depression K. Mengel and K.H. Wedepohl	332
7.7	Crustal Xenoliths and Their Evidence for Crustal Structure Underneath the Eifel Volcanic District G. Voll	336

7.8 Mantle Xenoliths in the Rhenish Massif and the
    Northern Hessian Depression
    H.A. Seck and K.H. Wedepohl 343

7.9 Relation of Geophysical and Petrological Models of
    Upper Mantle Structure of the Rhenish Massif
    K. Fuchs and K.H. Wedepohl 352

8 *Attempts to Model Plateau Uplift* 365

8.1 Gravity Anomaly and Density Distribution of the Rhenish
    Massif
    J. Drisler and W.R. Jacoby 366

8.2 Uplift, Volcanism and Tectonics: Evidence for Mantle
    Diapirs at the Rhenish Massif
    H.J. Neugebauer, W.-D. Woidt, and H. Wallner 381

9 *Epilogue: Mode and Mechanism of Rhenish Plateau Uplift*
    K. Fuchs, K. von Gehlen, H. Mälzer, H. Murawski, and
    A. Semmel 405

Contributors

You will find the addresses at the beginning of the respective contributions

Ahorner, L. 187,198
Albers, H.J. 9,39
Andres, W. 93

Baier, B. 187,222,315
Bartelsen, H. 276
Baumann, H. 177
Bender, P. 9
Berners, H.-P. 9,39,108
Bibus, E. 55,73
Boenigk, W. 62
Bonjer, K.-P. 187,315
Brunnacker, K. 62
Bussmann, E. 129

Drisler, J. 366
Dürr, St. 9
Dürschner, H. 276

Flüh, E.R. 276
Forkel, W. 133
Fuchs, K. 1,260,352,405

Gehlen, K.v. 39,133,405
Gerstenecker, C. 247
Giese, P. 303
Glatthaar, D. 39

Haenel, R. 228
Hein, G. 164
Huckenholz, H.-G. 121
Huckriede, R. 9

Illies, J.H. 1,177

Jacoby, W.R. 247,366
Joachimi, H. 247
Jödicke, H. 288

Kaufmann, G. 9
Kowalczyk, G. 9

Lippolt, H.J. 112
Lipps, S. 98
Löhnertz, W. 39
Lorenz, V. 129,139

Mälzer, H. 164,405
Mechie, J. 260
Meiburg, P. 9

Meissner, R. 276
Meyer, W. 39
Mengel, K. 332
Müller, K.H. 98
Müller, R. 9,102
Muller, A. 9
Murawski, H. 9,276,405

Negendank, J. 78
Neugebauer, H.J. 381

Olgemann, W. 288

Pfeffer, K.H. 39
Prodehl, C. 260
Puchelt, H. 152

Raikes, S. 315
Ritzkowski, S. 9

Schmincke, H.-U. 139
Schnütgen, A. 39
Schulte, L. 288
Schwab, K. 9
Seck, H.A. 139,153,343
Semmel, A. 9,48,49,405
Sewering, H. 93
Springer, M. 276
Stapf, K. 9

Untiedt, J. 288

Vieten, K. 131
Voll, G. 336

Wagenitz, V. 288
Wallner, H. 381
Walter, R. 9
Wedepohl, K.H. 134,332,343,352
Wernig, J. 222
Wienecke, K. 39
Winter, K.-P. 9
Woidt, W.-D. 381

Zakosek, H. 39
Zankl, H. 9
Zippelt, K. 164
Zöller, L. 89

1 Plateau-Uplift of the Rhenish Massif – Introductory Remarks*

J. H. Illies† and K. Fuchs[1]

Abstract

The plateau uplift of the Rhenish Massif in the Federal Republic of Germany was the object of a multi-disciplinary research program which started in 1976. The program comprised geological, geophysical, geodetic, petrological, geochemical, geomorphologic, and age-dating surveys together with attempts to model the uplift numerically. This introduction summarises the problems of plateau uplift of the Rhenish Massif as they appeared at the beginning and during the starting phase of the research program. Special emphasis is placed on the following subjects: the Rhenish Massif in the foreland of the Alpine collision front and its relation to the tectonics in the Upper Rhine Graben and the graben of the Lower Rhenish Embayment; shear heating vs. hot spot hypothesis; interaction between uplift, rifting, volcanism and mantle heterogeneities, asymmetries of plateau uplift west and east of the Rhine.

---

[1] Geophysikalisches Institut, Universität Karlsruhe, Hertzstraße 16, D-7500 Karlsruhe 21, Fed. Rep. of Germany

* Contribution No. 253, Geophysical Institute Karlsruhe

Epeirogeny

Epeirogeny (Gilbert, 1890) – the process of broad, gentle warping – was already a standard term of classical geology towards the end of the last century. Since then, this process has been quantified by mapped isopachs and isobases. Facies and fauna manifest the exogen reactions and the development with time of epeirogenic processes. Geodetic measurements reveal recent movements.

During the Upper Mantle Project and the Geodynamics Project interesting models for the cause of epeirogeny were proposed mainly for the morphology of the ocean floor. These are based on the interrelation of heat flow, density and isostatic adjustment. The topography of the mid-ocean ridges and the subsidence of the lithosphere towards the ocean basins was explained plausibly.

The new International Lithosphere Project places special emphasis on the continental lithosphere. Here additional boundary conditions, given by complex geological structures and their history have to be taken into account. In a review article, Menard (1973) pointed out a number of possibilities for epeirogenic movements.

The investigation of epeirogeny has been advanced strongly by the study of recent crustal movements. A map of recent vertical movements of Switzerland (Gubler et al., 1981) shows that the rate of uplift culminates at the Simplon with 1.7 mm a^{-1}. Here, it is most likely the isostatic response of a continuing crustal thickening at the Alpine colli-

---

Plateau Uplift, ed. by K. Fuchs et al.
© Springer-Verlag Berlin Heidelberg 1983

sion front. While the Alps possess a crustal root with a thickness of up to about 50 km, the crustal arching of the Black Forest and Vosges is supported by a closed mantle doming underneath the Rhine Graben culminating at a depth of 24-25 km (Edel et al., 1975). The isobases of the post-Eocene uplift of the graben shoulders parallel, by and large, the contours of the Moho (Illies, 1974). Morphological studies indicate that the Black Forest and Vosges have risen since the Upper Pliocene by at most 800 m. However, the recent variations of elevations are negative (Mälzer and Schlemmer, 1975), which was not expected. Further, the valleys of the rivers emptying into the Upper Rhine Graben area are filled with young gravel of thickness partially in excess of 40 m. Here an inversion of the direction of vertical movements has started recently.

Subsidence, as an inversion of vertical movement, dominates the structural history of the North Sea depression (Ziegler, 1975). The central graben, most active in the Jurassic and Cretaceous, was flanked originally by strongly elevated shoulders. In the Bathonium (Upper Dogger), the rifting was connected with volcanic and plutonic activity. In the Palaeocene, graben tectonics ceased here leaving an extinct rift valley. The narrow graben was replaced by a wide through which included the former shoulders. The inversion caused the graben shoulders to subside, perhaps because the heat generator at depth had vanished. A review of the research on the plateau uplift of the Rhenish Massif up to 1978 has been presented by Illies et al. (1979).

The Rhenish Massif in the Foreland of the Alpine Collision Front

The Rhenish Massif, wedged in between the Upper Rhine Graben and the Lower Rhenish Embayment is also part of the West European Rift belt. After a multiphase activity of eruption in the Tertiary, the volcanic activity started again in the Pleistocene, and the last eruption took place only 1000 years ago (Erlenkeuser et al., 1972). Contrary to this young magmatic history it was surprising that the heat flow determined in the drill hole Ochtendung (Kappelmeyer, 1977), has only a normal value. At least the upper part of the crust did not show any abnormal thermal behaviour.

In the framework of plate tectonics the Rhenish Massif is an unstable block (Fig. 1) in the perimeter of the Alpine collision front. In the vicinity of this collision front an intraplate tectonics was active with different intensity. Also, at present the block is mobile. Ahorner et al. (1972) deduced NW-SE directed compressive stress trajectories from the Alps to the Lower Rhenish Embayment from earthquake fault plane solutions. In situ stress determinations have shown that the tectonic stresses culminate in the Central Alps and diminish abruptly towards the northern foreland (Illies and Greiner, 1976, 1979). Middle Europe consolidated during the Variscan orogeny to form a rigid abutment relative to the Alpine orogeny. It opposes a rapid stress release by tectonic deformation. Only the Upper Rhine Graben forms a zone of weakness. The bundle of lineaments, active as an extension graben, strikes in the direction of the sinistral shear component of the recent stress field. Therefore, it seems to be predestined to recast the Alpine stresses by strain release as active tectonics and seismicity. Müllerried (1921) noticed that originally normal faults at the rim of the Rhine Graben became strike-slip faults. This observation was confirmed independently by other methods. Fault plane solutions of earthquakes show that the Upper Rhine Graben reacts as a sinistral shear track (Ahorner, 1975; Bonjer, 1981). Ahorner (1975) estimated the seismic slip rate at approximately 0.05

Fig. 1. The shear motion of the Upper Rhine Graben terminates at the southern rim of the Taunus mountains. The Rhenish Massif is thrown into a tensional process which also reactivates the fan-shaped graben of the Lower Rhenish Embayment. The process was facilitated because the central graben of the North Sea acted as a crash-zone (Greiner and Illies, 1977)

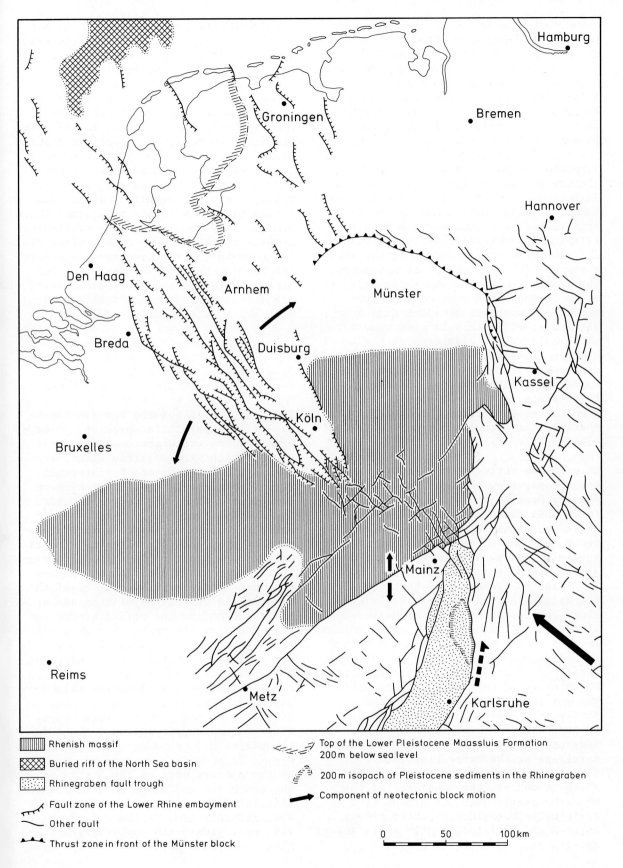

Fig. 1

mm a^{-1} (= 50 m Ma^{-1}). At the northern end of the Upper Rhine Graben neotectonic activity is most pronounced (Schneider and Schneider, 1975; Schwarz, 1976). Also, seismicity at the southern border of the Taunus mountains near Mainz and Groß Gerau is stronger than in the middle part of the Upper Rhine Graben.

From the southern rim of the Taunus through the depression of Idstein and Limburg to the mountains of the Westerwald a grid of faults, partially still active, traverses the Rhenish Massif (Stengel-Rutkowski, 1976). The studies in the Neuwied basin and along the Mosel river should track the young movements. The fault plane solutions indicate a trend of the horizontal component of maximum compression of about N135°E (Ahorner, 1975). This is also true for the considerable seismicity of the Lower Rhenish Embayment. Here, Ahorner (1962) deduced an increase of the dilatation rates towards the NW from an analysis of Pleistocene rupture motions. A corresponding fan opening of the fault system continues into the Netherlands until the city of Utrecht.

Between the active tectonics of the Alps and the passive foreland block of the Rhenish Massif, the Upper Rhine Graben forms a mediating element. Since the Pliocene it has acted as a sinistral shear zone which has remained seismotectonically active until the present day. At the southern rim of the Taunus mountains the shear motion is blocked, and the orientation of the originally tensional graben of the Hessian Depression is not suited to carry this shear motion further to the north. Instead, a belt of earthquake epicentres and mild neotectonics traverses the Rhenish Massif from Frankfurt to Bonn in the direction of the maximum horizontal compressive stress axis. Near Bonn, active graben tectonics begins again, however, now under tension corresponding to the new direction of the axis. Like the front flap of a lederhosen protects the fly of the trousers, the mountains of the Rhenish Massif cover the hidden rift by the plastic behaviour of their finely layered and foliated, mostly argillaceous rocks.

Thus, only near the surface does the active rift-belt appear to be interrupted by the Rhenish Massif. The formation of discrete ruptures requires a brittle behaviour of the rock masses. This is encountered in the Saxothuringian and Moldanubian basement of the Rhenish Massif. The basement of the Lower Rhenish Embayment, as known from the mountains of the Hohes Venn, was deformed during the Caledonian orogeny and subsequently folded during the Variscan orogeny resulting in an additional stiffening. Both areas, north and south of the Rhenish Massif, reacted competently by rupture deformation. The 4- to 10-km-thick sedimentary layers of the Rhenish Massif, however, behaved incompetently in the same stress field, i.e., they were deformed ductilly with only a few exceptions.

Shear Heating vs. Hot Spot

The young uplift of the Rhenish Massif is not an unparalleled process. In other areas of the world there are microplates which were uplifted strongly during the Pliocene and Pleistocene. During the International Symposium on Recent Crustal Movements in Palo Alto, Calif. (Whitten et al., 1979), the Sierra Nevada was visited during an excursion program. This block is located between the Great valley of California in the west and the Basin and Range Province in the east, and it consists of granitic intrusions with an age of about 130 Ma and the relics of Palaeozoic schists. The mountain ranges, which are about 600 km long and 100 km wide, were still a peneplain in the Miocene; the uplift started in the Pliocene. In the Pleistocene the block was tilted westwards, its eastern rim reaching 4400 m above sea level. Rivers and glaciers have strongly furrowed the old peneplain

At the eastern edge of the Sierra Nevada, the Basin and Range Province begins with the Owens river rift valley which subsided only towards the end of the Pliocene. Widespread Quaternary volcanism took place here, with the youngest eruptions of the Mono Craters occurring around 1200 B.C. So far, there is no indication for a mantle diapir below the Sierra Nevada. The heat flow is 16.74 mWm^{-2}, the smallest value in the western U.S.A. How is this compatible with the strong uplift?

The Sierra Nevada is framed by two active wrench faults, the San Andreas fault and the conjugate Garlock fault. The San Andreas fault is recognized today as an active plate boundary. In fact, active tectonics and seismicity continue east for another 1000 km to the Wasatch fault near Salt Lake City. In this active mosaic of foreland blocks of the western USA the Sierra Nevada forms a rigid microplate. The Colorado plateau, bordered by the Wasatch fault and the active Rio Grande rift valley, is in a similar position.

At the Palo Alto Symposium several mechanisms for the uplift of the Sierra Nevada and the Colorado plateau were discussed. Schubert and Yuen (1978) and other authors pointed out, as a possible relationship between the horizontal tectonics of these blocks in the Quaternary, the occurrence of shear heating at the lithosphere-asthenosphere boundary and consequent thermal expansion.

Since the notion of the an "Eifel hot spot" (e.g., Duncan et al., 1972) has lost much of its attractiveness since new radiometric age data has become available (e.g., Cantarel and Lippolt, 1977), it seems worthwhile to consider, also, the model of a transient shear heating for the Rhenish Massif, especially since here, widespread volcanism and young uplift are in apparent contradiction to the indication of a somewhat reduced heatflow ranging between 54.42 and 71.16 mWm^{-2} (Haenel, 1976).

The changes during the Pliocene of the stress pattern and motion at its southern rim, caused the Rhenish Massif to react by two different kinematic responses. Firstly the block as a whole yielded by an anticlockwise rotation. This is supported by extensional faulting at and parallel to the southern rim of the Rhenish Massif. At its southeastern margin corresponding dextral shear displacements are observed (Illies and Greiner, 1976). Secondly, the massif yielded by internal brittle fracturing and deformation caused by a wedging of the apex of the block east of the Rhine Graben into the southern rim of the Rhenish Massif. Although the strain rates at the Rhine Graben of about 0.05 mm a^{-1} (Ahorner, 1975) are smaller by orders of magnitude compared to those at the San Andreas fault (12-15 mm a^{-1}), and although the total displacement parallel to the axis of the Rhine Graben is not more than 300 m since Upper Pliocene/ Pleistocene times, nevertheless it should be explored whether the young motions of the Rhenish Massif, both vertical and horizontal, can be explained simply by using a static model or whether a dynamic model with horizontal block movements provides a better description.

Interaction Between Uplift, Rifting, and Volcanism

After the formation of the main river terraces of the Rhine and Mosel, uplift of the Rhenish Massif of at least 150 m accompanied by widespread volcanism in the West Eifel and the Neuwied basin. During the same period, considerable displacements took place in the areas of subsidence such as the Upper Rhine Graben, and the Lower Rhenish Embayment. Near the end of the Pleistocene the tectonic activity started again. In the Upper Rhine Graben region the present-day geodetic height changes exceed the average rate since the beginning of the Pleistocene by a factor of ten. The different heights of the lowermost terrace indicate that the recent rate of uplift has been continuing for at least the past 10,000 years. In the Lower Rhenish Embayment some faults form considerable steps within Pleistocene terrace surfaces. Here, the recent height changes are also a multiple of the estimated average rates in the Pleistocene. The Rhenish Massif reacted with spectacular volcanism evidenced by the Maar-craters of the Eifel (Laacher See). The northwest Eifel continues its fast uplift at the present day.

The Rhenish Massif responded promptly with uplift and volcanism, to intensified rifting of the graben segments in the north and south. The horizontal tectonics of the frame produced vertical tectonics in the Rhenish Massif as its most obvious result. The young uplift of the Rhenish Massif took place episodically. Is there any relation to the phases of young rifting in the Upper Rhine Graben or the Lower Rhenish Embayment and to the timing of the Quaternary volcanism?

Stress distribution and seismotectonics are both compatible with the model of a tensional graben, prevented by material behaviour, traversing the Rhenish Massif. In contrast to the relatively uniform rates of Quaternary uplift obtained from the deformation of terraces, the pattern of recent vertical motion is rather dispersed. Obviously, the uniform long term uplift, when resolved into small time slices, is a process changing in space and time.

To understand the young patterns of motion and their relation to the various boundary conditions, it becomes necessary to consider also the Tertiary history of the massif. It is important to know the Tertiary base of reference to realize to what extent the Quaternary uplift of the Rhenish Massif followed old contours.

Stress Field, Volcanism, and Mantle Heterogeneities

From seismotectonics and geodetic measurements it is realized that the fan-shaped crustal spreading of the Lower Rhenish Embayment (Ahorner, 1962, 1975) is continuing until present. Obviously, the process affects the Rhenish Massif. The pattern of young mobile lineaments reveals a latent brittle structure of the basement. The Quaternary volcanism of the West Eifel, of the Neuwied basin, and of the mountain Rodder Berg near the city of Mehlem used this weakening structure for its ascent. This volcanism is connected preferably with NW-SE striking lineaments (Schmincke et al., 1983; Fig. 1), i.e., it follows the trend of the trajectories of the active stress field. Typically, no detectable surface fracturing was connected with it. One can imagine that fluid and volatile phases of the magma only took advantage of the access provided by the prevailing stress field. Fissures opening parallel to the direction of maximum compressive stress during hydraulic fracturing may serve as a parallel to understand the ascent of magma in the present stress field. The continuation of dilatation in the Lower Rhenish Embayment, young tensional faults, seismicity, and Quaternary volcanism in the Rhenish Massif give the impression that the West European graben system is about to activate its missing link.

The important question is whether the upwelling of mantle material was the primary process or whether it was a passive reaction to the openings of fissures in the crustal stress field. How does the crustal uplift picture fit into the pattern of occurrence of Quaternary magmatism and seismicity? At the beginning of the multi-disciplinary research program it was not known whether the uplift contours indicated by morphological, geological, and geodetic investigations have a counterpart in the upper mantle. Therefore, deep seismic and magnetotelluric sounding had to be carried out. Studies of gravity and heat flow contributed to the spectrum of data which should constrain acceptable models of plateau uplift of the Rhenish Massif.

Horizontal tensional strain causes vertical crustal thinning and a corresponding doming of mantle material. Is the uplift of the Rhenish Massif related to this phenomenon? Does the doming of the Rhenish Massif correspond to the arching of the shoulders of the Upper Rhine Graben? The explosion seismic and magnetotelluric experiments combined with studies of teleseismic travel-time anomalies have tried to give a detailed answer. The investigation of Quaternary volcanism and its mantle xenoliths are important indications of the related crust-mantle interaction.

Uprising mantle material, possibly an asthenolith, should be accompanied by an increased heat flow. At the beginning of the research program, it was thought that the Rhenish Massif appeared to have cool to normal temperatures. Possibly the incompetent behaviour of the thick sediments has prevented the development of deep fissures with hydrothermal convection, in contrast to the situation in the Upper Rhine Graben. Tensional processes in the brittle fracture regime of the crust develop characteristic reactions in the subcrustal lithosphere and asthenosphere. The unloaded substratum responds by phase transformation, thermal expansion, partial melting, and fractionation. Is the body of anomalous material in the upper mantle responsible for the uplift and is it the carrier of the source magma driving the volcanism in the Rhenish Massif? Here, petrological and geophysical data are strongly interrelated.

The relationship between uplift, volcanism, and mantle diapirs in a tensional stress field does not occur everywhere. McKenzie (1978) and others have shown recently from a number of examples that crustal stretching leads to crustal thinning with consequent isostatic subsidence. The Pannonian and North Sea basins are typical examples. Why, therefore, is tensional tectonics causing uplift of the Rhenish Massif, while in other regions it is causing subsidence? What is the role of the asthenolith in the Rhenish Massif? Is it the reaction of the mantle to changes in the stress field, and does it replace the notion of a wandering hot spot (Duncan et al., 1972)?

Asymmetries of Plateau Uplift West and East of the Rhine

Geophysical investigations of the Rhenish Massif have drawn attention to an asymmetry of the lithosphere west and east of the Rhine. A volume of low velocity material was discovered by Raikes (1980) centred under the volcanic region west of the Rhine. The explosion seismic experiment revealed a similar asymmetry for the crust and uppermost mantle. Further, the Quaternary volcanism is restricted to the part of the Rhenish Massif west of the Rhine. Geodetic height changes are strongest in the west as well. If the western part of the Rhenish Massif is geophysically the active part, then we should expect even higher rates of uplift during the Pleistocene.

East of the Rhine, the volcanism in the Westerwald mountains is older; the formerly active mantle seems to be extinct. Physiographically, the Rhenish Massif appears to be lifted up uniformly on both sides of the river Rhine. However, does this also mean that the uplift occurred on both sides at the same time?

References

Ahorner, L., 1962. Untersuchungen zur quartären Bruchtektonik der Niederrheinischen Bucht. Eiszeitalter Gegw., 13:24-105.

Ahorner, L., 1972. Erdbebenchronik für die Rheinlande 1964-1970. Decheniana, 125 (1/2):259-283.

Ahorner, L., 1975. Present-day stress field and seismotectonic block movements along major fault zones in Central Europe. In: Pavoni, N., Green, R. (eds.) Recent crustal movements. Tectonophysics, 29:233-249.

Ahorner, L., Murawski, H., and Schneider, G., 1972. Seismotektonische Traverse von der Nordsee bis zum Apenin. Geol. Rundsch., Stuttgart, 61:915-942.

Bonjer, K.P., 1981. The seismicity of the Upper Rhinegraben - a continental rift system. In: Petrovski, J., Allen, C.R. (eds.) Proc Int. Res. Conf. Intra-Cont. Earthqu., Sept. 17-21, 1979, p.107-115.

Cantarel, P. and Lippolt, H.J., 1977. Alter und Abfolge des Vulkanismus der Hocheifel. Neues Jahrb. Geol. Paläontol. Mk., 1977:600-612.

Duncan, R.A., Petersen, N., and Hargraves, R.B., 1972. Mantle plumes, movement of the European plate and polar wandering. Nature (London), 239:82-86.

Edel, J.B., Fuchs, K., Gelbke, C., and Prodehl, C., 1975. Deep structure of the southern Rhinegraben area from seismic refraction investigations. Z. Geophys., 41:333-356.

Erlenkeuser, H., Frechen, I., Straka, H., and Willkomm, H., 1972. Das Alter einiger Eifelmaare nach neuen petrologischen, pollenanalytischen und Radiokarbon-Untersuchungen. Decheniana, 125:113-129.

Gilbert, C.K., 1890. Lake Bonnerville. Monogr. 1, U.S. Geol. Surv. Washington, 438 pp.

Greiner, G. and Illies, J.H., 1977. Central Europe: Active or residual tectonic stresses. Pageoph., 115:11-26.

Gubler, E., Kahle, H.-G., Klingele, E., Mueller, St., and Oliver, R., 1981. Recent crustal movements in Switzerland and their geophysical interpretations. Tectonophysics, 71:125-152.

Haenel, R., 1976. Die Bedeutung der terrestrischen Wärmestromdichte für die Geodynamik. Geol. Rundsch., 65: 797-809.

Illies, J.H., 1974. Intra-Plattentektonik in Mitteleuropa und der Rheingraben. Oberrhein. Geol. Abh., 23:1-24.

Illies, J.H. and Greiner, G., 1976. Regionales Stress-Feld und Neotektonik in Mitteleuropa. Oberrhein. Geol. Abh., 25:1-40.

Illies, J.H. and Greiner, G., 1979. Holocene movements and state of stress in the Rhinegraben rift system. In: Whitten, C.A., Green, R., Meade, B.K. (eds.) Tectonophysics, 52:349-359.

Illies, J.H., Prodehl, C., Schmincke, H.-U., and Semmel, A., 1979. The Quaternary uplift of the Rhenish Shield in Germany. Tectonophysics, 61:197-225.

Kappelmeyer, O., 1977. Erkundung des Temperaturfeldes in der Eifel mit einer Forschungsbohrung Ochtendung. BGR-Bericht Nr. ET 4213, BGR Hannover, 21 pp.

Mälzer, H. and Schlemmer, H., 1975. Geodetic measurements and recent crustal movements in the southern Upper Rhinegraben. In: Pavoni, N., Green, R. (eds.) Recent crustal movements. Tectonophysics, 29:275-282.

McKenzie, D., 1978. Some remarks on the development of sedimentary basins. Earth Planet. Sci. Lett., 40:25-32.

Menard, H.W., 1973. Epeirogeny and plate tectonics. EOS, 54:1244-1260.

Müllerried, F., 1921. Klüfte, Harnische und Tektonik der Dinkelberge und des Basler Taféljuras. Verk. Naturhist.-Med. Verein Heidelberg, 15:1-46.

Raikes, S., 1980. Teleseismic evidence for velocity heterogeneity beneath the Rhenish Massif. J. Geophys., 48: 80-83.

Schneider, E.F. and Schneider, H., 1975. Synsedimentäre Bruchtektonik im Pleistozän des Oberrheintal-Grabens zwischen Speyer, Worms, Hardt und Odenwald. Münster. Forsch. Geol. Paläontol., 36:81-126.

Schuber, G. and Yuen, D.A., 1978. Shear heating instability in the earth's upper mantle. Tectonophysics, 50: 197-205.

Schwarz, E., 1976. Präzisionsnivellement und rezente Krustenbewegung dargestellt am nördlichen Oberrheingraben. Z. Vermess., 101:14-25.

Stengel-Rutkowski, W., 1976. Idsteiner Senke im Limburger Becken im Licht neuer Bohrergebnisse und Aufschlüsse (Rheinisches Schiefergebirge). Geol. Jahrb. Hessen, 104:183-224.

Whitten, C.A., Green, R., and Meade, B.K., 1979. Recent crustal movements, 1977. Proc. 6. Int. Symp. Recent Crustal Movements. Stanford Univ., Palo Alto, July 25-30, 1977. Tectonophysics, 52 (1-4):603pp.

Ziegler, P.A., 1975. Geological evolution of the North Sea and its tectonic framework. Am. Assoc. Petrol. Geol. Bull., 59:1073.

2 Regional Tectonic Setting and Geological Structure of the Rhenish Massif

H. Murawski[1], H. J. Albers[2], P. Bender[3], H.-P. Berners[4], St. Dürr[3], R. Huckriede[3], G. Kauffmann[3], G. Kowalczyk[1], P. Meiburg[5], R. Müller[10], A. Muller[4], S. Ritzkowski[6], K. Schwab[7], A. Semmel[8], K. Stapf[9], R. Walter[4], K.-P. Winter[1], and H. Zankl[3]

Abstract

This chapter contains an outline of the geological history of the Rhenish Massif and the development of its margins and surrounding areas. Section 2.2 gives some main aspects of the development of the Variscan geosyncline and orogeny and a short description of the pre-Variscan geological history of this region. Since the end of the Variscan orogeny many different epeirogenetic processes have occurred here (Sect. 2.3). During these long geological times uplift and subsidence changed in the different parts of the Massif and in the surrounding areas.

The Rhenish Massif cannot be seen as an isolated complex; its history is often connected with the development of the surrounding areas. The maps and profiles of this chapter presents some of the main proofs for this fact. On the other hand, the development of the western and the eastern part of the Rhenish Massif differs in many cases. It is impossible to understand many symptoms of the latest development of the Massif without knowledge of its pre-Tertiary geological history.

---

1 Geologisch-Paläontologisches Institut der J.W. Goethe-Universität, Senckenberganlage 32-34, D-6000 Frankfurt am Main 1, Fed. Rep. of Germany

2 Landesamt für Ökologie, Landschaftsentwicklung und Forstplanung, D-4350 Recklinghausen, Fed. Rep. of Germany

3 Institut für Geologie und Paläontologie, Fachbereich Geowissenschaften, Philipps-Universität, Lahnberge, D-3550 Marburg 1, Fed. Rep. of Germany

4 Lehrgebiet Allgemeine und Historische Geologie, Geologisches Institut der Rheinisch-Westfälischen Technischen Hochschule Aachen, Wüllnerstr. 2, D-5100 Aachen, Fed. Rep. of Germany

5 Geologisch-Paläontologisches Institut der Technischen Hochschule Darmstadt, Schnittspahnstr. 9, D-6100 Darmstadt, Fed. Rep. of Germany - z.Zt. Universidad Autonoma de Nuevo Leon, Instituto de Geologia, Apartado Postal 104, 67 700 Linares, N.L. Mexico

6 Geologisch-Paläontologisches Institut und Museum der Universität, Goldschmidt-Str. 3, D-3400 Göttingen, Fed. Rep. of Germany

7 Geologisches Institut der Technischen Universität, Adolf Römerstr. 2 A, D-3392 Clausthal-Zellerfeld, Fed. Rep. of Germany

8 Institut für Physische Geographie der J.W. Goethe-Universität, Senckenberganlage 36, D-6000 Frankfurt am Main 1, Fed. Rep. of Germany

9 Institut für Geowissenschaften, Johann Gutenberg-Universität, Saarstr. 21, D-6500 Mainz, Fed. Rep. of Germany

10 Lehrstuhl für Geologie und Paläontologie und Geologisches Institut der Rheinisch-Westfälischen Technischen Hochschule Aachen, Wüllnerstr. 2, D-5100 Aachen, Fed. Rep. of Germany

---

Plateau Uplift, ed. by K. Fuchs et al.
© Springer-Verlag Berlin Heidelberg 1983

2.1 Introductory Note (H. Murawski)

The Rhenish Massif described in this book includes the Ardennes, the Eifel, the Rhenish Massif East of the Rhine River and their margins, which disappear outwards below series of Upper to Post-Variscan age (Fig. 1). The geological situation and the development of these margins is very complicated. Section 2.3 will give some help in understanding structure and genesis of these margins.

The term Variscan will often be used here. It is therefore necessary to explain something of the history of this term. The name Variscan is derived from the Roman name for the German town of Hof, Eastern Bavaria: curia Variscorum/Varistorum. The famous Austrian geologist Eduard Suess introduced the term Variscan Mountains into the literature in 1888. This regional term includes only the mountains between the Ardennes and the Vosges in the West to the Sudetes in the East. In the first quarter

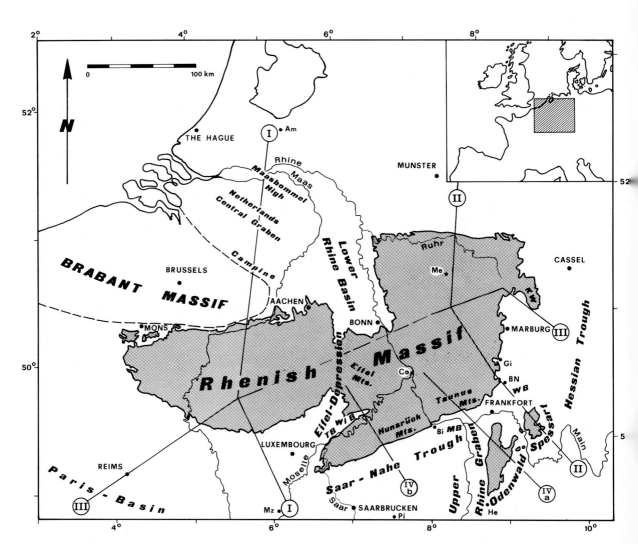

Fig. 1. Outline map of the Rhenish Massif. The numbers I to IVb designate the profile lines of Figs. 12-17 (Sect. 2.3). Am Ammersfoort; BN Bad Nauheim; Bi Bingen; Co Coblence; Gi Giessen; He Heidelberg; KW Kellerwald; MB Mayence Basin; Me Meschede; Mz Metz; Pi Pirmasens; TB Trier Basin; WB Wetterau Basin; WiB Wittlich Basin

of this century the well-known German geologist Hans Stille changed the sense of the term Variscan from regionally descriptive to chronological for tectonic events within a given time span. He applied this term not only to the Variscan Mountains in the sense of Suess, but – in his own words – for mountain building (better: orogeny) in the late Palaeozoic throughout the world. However, in 1892 the French geologist Marcel Bertrand used the term Hercynian instead of Variscan. Therefore, in the French literature the names appear: plissements hercyniens (= Variscan folds) and chaîne hercynienne (= Variscan Mountains). In the English literature Hercynian Orogeny is the same as Variscan Orogeny in the German literature. There is another complication with regard to the term Hercynian strike, which in German literature means often a strike direction SE-NW or ESE-WNW. Moreover, the term herzynische Fazies is used in German literature for a special Devonian facies type in the Rhenohercynian geosyncline.

2.2 Variscan and Pre-Variscan History (H. Murawski)

The Variscan

In contrast to Rheinischer Schild (= Rhenish Shield) in the sense of H. Cloos (1939) the Rhenish Massif described in this book has an easterly long axis (Fig. 1). To a high degree it owes this shape to the Variscan tectonic history. The strike of the fold axes of the main anticlines and synclines (Fig. 2) and the whole structure of this part of the Variscan Mountains, the Rhenohercynian produced this elongated form. Over a long distance the Rhenohercynian is bordered in the South by the Saar-Nahe-Trough and its continuation to the East. This trough has a fill of Variscan Molasse of Upper Carboniferous and Permian age and is a special Molasse basin in the interior of the Variscan Mountains. The long axis of this trough strikes ENE-WSW. The northern boundary of the Rhenohercynian is the Subvariscan Foredeep (Subvaristische Vortiefe), which is also elongated ENE-WSW. In the western part of the Rhenohercynian the pre-Palaeozoic and Palaeozoic of the Ardennes dip below the Mesozoic cover without tectonic boundaries. These Mesozoic rock sequences are the northern extrusions of the Paris Basin (Fig. 1). The eastern boundary of the Rhenish Massif as a part of the Rhenohercynian is the Hessian Trough, also called Hessian Depression (Hessische Senke) elsewhere in this volume, which strikes north-south. The fill of this trough includes rock series of Uppermost Palaeozoic and Mesozoic to Cenozoic. Farther northeast – on the eastern border of the Hessian Trough – the Rhenohercynian is again exposed in the Harz Mountains.

One of the most important structures of the Rhenish Massif is a trough-like zone, the Eifel Depression (Eifeler Nord-Süd-Zone) situated in the western Eifel Mountains. This structure, which is only part of a broader system with North-South strike, plays a special role not only in Mesozoic time (see Sect. 2.3), but also in the geological history of the Rhenohercynian. West of this structure, in the Ardennes and Hohes Venn, rock series of pre-Devonian and Devonian age are exposed in some massifs, which suffered a low-grade metamorphism. These rocks are exposed over long distances in the cores of the Variscan anticlines (e.g., Massif of Stavelot/Hohes Venn, Massif of Rocroi, Massif of Serpont, Massif of Givonne). After Kramm (1982) wide areas of the pre-Devonian and Lower Devonian rocks in the southern flank of the Stavelot-Venn Massif suffered a low-grade metamorphism (350°-$450^\circ C$). There is no doubt that a post-Devonian - Variscan metamorphism exists here. On the other hand Kramm could find no irrefutable proofs for a Caledonian metamorphism in this region. It is possible that in other parts of the Ardennes arguments for a Caledonian metamorphism beside the Variscan metamorphism are given. In the course of Kramm's investigations it was discovered that the geothermal gradient during the metamorphism in the southern flank of the Stavelot-Venn Massif was about 60° C km^{-1}. In this region the Devonian and the Lower Carboniferous must have had a thickness of 6000-7500 m. The grade of metamorphism, the calculated geothermal gradient and the estimated rock thickness are thus in good accordance.

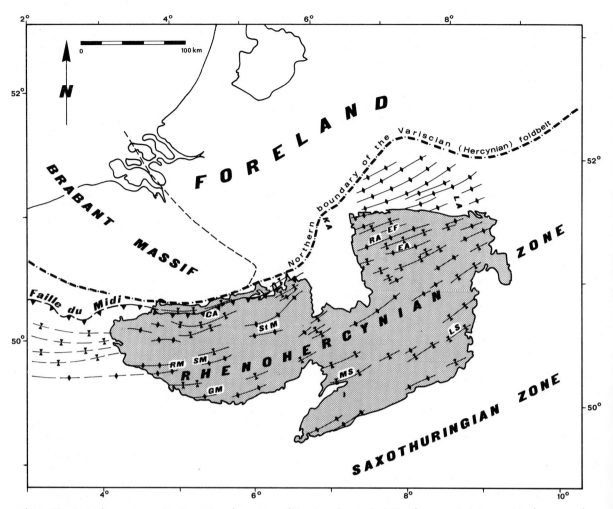

Fig. 2. Outline map of the Rhenish Massif showing the Variscan fold tectonics; main anticlines and synclines. CA Condroz anticline; EA Ebbe anticline; EF Ennepe fault; GM Givonne Massif; KA Krefeld arch; LA Lippstadt Arch; LS Lahn syncline; MS Mosel syncline; RA Remscheid anticline; RM Rocroi Massif; SM Serpont Massif; StM Stavelot/Venn Massif

East of the massifs of the Ardennes a special uplift zone elongated in NNW-SSE direction was found by seismic studies and boreholes in the northern part of the Lower Rhine Basin (also called Lower Rhenish Embayment elsewhere in this volume): the Krefeld Arch (Krefelder Gewölbe; Fig. 2); below a relatively thin cover of Cretaceous and younger sediments there is Lower Carboniferous, Devonian, Cambro-Ordovician (+ Silurian ?) and at great depth crystalline basement (Plein et al., 1982). In contrast to the Cambro-Ordovician of the Stavelot Massif and other massifs of the Ardennes, these series are more or less unfolded here. It seems clear today that the Krefeld Arch and the Lippstadt Arch (Lippstädter Gewölbe, Fig. 2) to the East are of Variscan origin. A magnetic anomaly in the southern part of the Krefeld Arch was interpreted as the effect of a magnetic body at depth, perhaps a basic pluton, of post-Asturian to pre-Upper Permian age. In contrast to the situation in the Ardennes, exposed pre-Devonian in the cores of Variscan anticlines of the eastern Rhenish Massif, east of the Eifel Depression, is only known at a few points and in small extent (e.g., Remscheid Anticline, Ebbe Anticline). The rock series of these locations are unmetamorphosed and fossiliferous. There the pre-Devonian to Devonian boundary is uncertain in its nature, so it is unclear if there is a distinct unconformity or not. The so-called Vordevon (= pre-Devonian) of the southern part of the Taunus Mountains could not yet be accurately dated. It is clear that there are differences between these and other rocks of the Taunus Mountains in terms of rock type and their low-grade metamorphism missing in Devonian rocks of the Taunus Mountains. The relatively different model ages of 394 and 304 Ma from some rocks of the Southern Taunus place the metamorphism in the Variscan only. So we have to think that the educts (original materials) must be of pre-Devonian age. In contrast, rock series of low-grade metamorphism are exposed in the southern part of the Hunsrück Mountains (Soonwald), which had also been considered pre-Devonian formerly; but they are Devonian rocks, which suffered Variscan metamorphism. In this region a more or less gradational change occurs from the region with low-grade metamorphism in the south to very low-grade metamorphism in the north.

The Variscan west of the Eifel Depression is characterized in its northern part by a large overthrust zone ("Faille du Midi", Fig. 2). The movements on this overthrust were in northwesterly direction toward the Subvariscan Foredeep (Molasse basin). It was possible to prove in some areas (e.g., Southern Belgium) that the foredeep was overrun by these overthrusts. A comparable situation of overthrusting is unknown from the northern border of the eastern part of the Rhenish Massif, east of the Eifel Depression. There, the rock series of the foredeep are folded with intensity increasing to the north. There are many different faults, also reversed faults, but overthrust tectonics in the sense of the Faille du Midi is not found. Recently the question has been considered whether the great Ennepe Fault (Ennepe-Störung) in the northeastern part of the Rhenish Massif could be a listric fault, possible with overthrust character at depth. The cause of the conspicuous overthrust tectonics in the western part of the Rhenish Massif must be the Brabant Massif situated North of the Ardennes. This Massif was deformed in Caledonian times. It served as a buttress against the folding of the Variscan orogeny in the Ardennes. Such a foreland situation is unknown in the northeastern part of the Rhenish Massif east of the Eifel Depression.

The name Rhenohercynian (Rhenoherzynikum) was created by Kossmat (1927). He divided the Variscan Mountains of Central Europe into different zones, a zonation which today is essentially accepted. Going from North to South we pass the foreland and the foredeep (Subvariscan Foredeep), the latter with extensive coal deposits of Upper Carboniferous age. The next zone is the Rhenohercynian (Fig. 2). It consists of folded series of Devonian and Lower Carboniferous. It includes the Rhenish Massif and the Harz Mountains. The combination of the name Rhine and "Harz" gives the name Rhenohercynian. The next zone to the south, named Saxothuringian, consists of series from Cambrian to Lower Carboniferous age. The name was given because the mountains of Saxony and Thuringia contain these series in a typical formation. The northwestern part of this zone contains rock series which have suffered regional metamorphism: the Mid-German Crystalline Rise (Mitteldeutsche Schwelle in the sense of Brinkmann). The original materials of its metamorphic rocks appear to be more or less of the same age as the sedimentary rocks of the unmetamorphosed Saxothuringian. The next zone to the south, composed essentially of different basement series, including granites and other plutonites, is the Moldanubian (Moldanubikum), from the rivers Moldau in Czechoslovakia and Danube in Germany and Austria, which flow through this zone.

It is evident that the name Rheinisches Schiefergebirge (= Rhenish Slate Mountains) was chosen according to the predominant role of the cleavage in this area. It is clear that this cleavage is concentrated more or less in rock types with special petrographical properties as in fine-grained clastic deposits. Cleavage is much rarer in graywackes and sandstones. It is more or less absent in the reef limestone massifs. Folding and cleavage are connected in genesis and time. Thus folds and cleavage of the

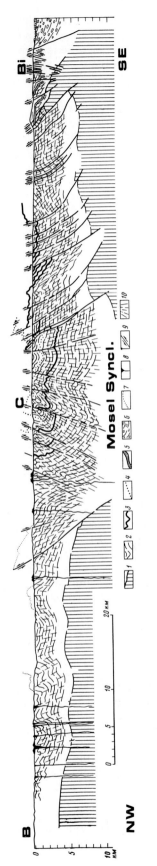

Fig. 3. Geological cross section along the Rhine valley from Bingen (Bi) over Coblence (C) to Bonn (B). (Simplified after Meyer and Stets, 1975, Table 1.) 1 Crystalline basement; 2 Lower Devonian; 3 boundary Siegenian/Emsian; 4 Ems quarzite; 5 Middle Devonian; 6 Lower Permian (Rotliegendes); 7 Tertiary (Mayence Basin); 8 Volcanoes (Tertiary more or less); 9 normal faults and reverse faults; 10 cleavage

Rheno-Ardennes show a predominant northwestern vergence. This can be seen very well in the deeply scoured Rhine valley between Bingen and Bonn (Fig. 3). There are two exceptions to the vergence in the middle and the southern part of this cross-section. In the large syncline of the Mosel (Moselmulde, middle part of the section) a vast fan of cleavage and vergence exists which diverges downwards. The southernmost part of the section is characterized by a steep vergence to the south combined with distinct imbrication structures. Both the Mosel Syncline and the southern imbrication zone demonstrate special tectonic activities in these areas. Figure 4

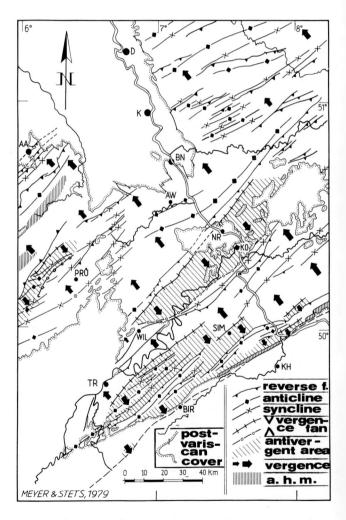

Fig. 4. Structural map of the central part of the Rhenish Massif. (After Meyer and Stets, 1980, p.728, Fig. 2)

gives more information about the strike of folds and vergence of a large-scale area of the Rhenish Massif. It shows very clearly the predominance of the northerly vergence of this region. Only few areas are antivergent in relation to the general vergence to northwest, e.g., the northern flank of the Mosel Syncline (between WTL, NR and KO of Fig. 4 and farther east) and a small zone adjoining the southern boundary of the Rhenish Massif. As is seen in Fig. 4, the Eifel Depression did not destroy the general North-West vergence, so it is evident that this tendency is the result of the existence of an extensive stress field in Variscan time (see also the cross-sections of the eastern Rhenish Massif in Weber, 1981, Fig. 3). On the other hand, it is difficult to follow other tectonic structures such as the large anticlines, synclines, great reverse faults and overthrusts across the Eastern Eifel and the Eifel Depression from east to west. This characterizes the importance of this north-south element. It is not impossible that the Eastern Eifel and the Eifel Depression represent something like a great north-south hinge, which compensated the different movements of the blocks in the east and west.

The different regions of the Rheno-Ardennes generally show extended and well-recognizable synclines and anticlines (Fig. 2). Some authors tried to find the value of tectonic confinement in the Rhenish Massif with the help of index numbers (e.g., Wunderlich; Breddin). Quite apart from the fact that these results, based on geometrical reconstructions, must be seen as minimum values only, it can be stated that in some regions of the Rhenohercynian great reverse faults, imbricate structures and, in addition, overthrusts are indicated. The best example of overthrust tectonics is demonstrated in the region of the northern boundary of the Ardennes. Miners in Northern France and Belgium have long known the fact that older rock series (e.g., Lower Carboniferous and Devonian) cover the coal-bearing Upper Carboniferous in the Subvariscan Foredeep along a tectonic contact. It was evident that the cover came from the south, so they gave the name Faille du Midi to this great fault or overthrust system. In the beginning of the 1970's Breddin pointed to the possibility that the Aachen Anticline may be the eastern prolongation of the Faille du Midi. The Aachen Anticline is not a simple anticlinal structure. It has a strong northwest vergence and many reverse faults produce an imbrication structure. Breddin (1973) concluded from the results of intensive tectonic investigation in the Ardennes region that a northwest-moving nappe must have existed in Variscan times. New data on this problem were gained by deep seismic reflection in the course of the project (DFG-funded) *Geotraverse Rhenoherzynikum* (Meissner et al., 1981).

A very good seismic reflector was found at a depth of about 4 km below the Hohes Venn along a profile of 30 km length, starting east of Aachen and crossing the Hohes Venn in NW-SE direction. The deep reflector ascends from 4 km below the Hohes Venn to the northwest and seems to join the Aachen Anticline (Fig. 5). This is in best agreement with results obtained by Belgian and Dutch geologists and geophysicists in Southern Belgium (Bless et al., 1980). There is no doubt that the northwestern part of the Ardennes must have been transported in northwestern direction towards the Brabant Massif. Some geologists assume that this northwest transport comprised the whole Ardennes. There is also no doubt that the Faille du Midi is an immense overthrust system. It is very probable that the clear reflector below the Hohes Venn is part of the overthrust system. It would be of great interest to learn more about the character and the age of the autochthonous below the nappe. The geophysical data give insufficient informations. An answer will possibly come from a deep borehole only. Belgian geologists believe that the Dinant Syncline has overthrust the autochthonous Namur Syncline in Variscan time, based on the results of intensive geological and geophysical research (Bless et al., 1980). It must be clarified in future whether the situation in the Hohes Venn is similar. In this connection it is of great interest that investigations of seismicity in the Hohes Venn by Ahorner (this volume) show

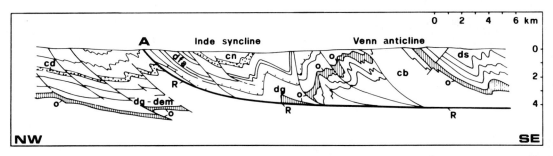

Fig. 5. Geological cross-section through the Hohes Venn and its foreland. (After Meissner et al., 1981, Fig. 4). R seismic reflector; cb Cambrian; o Ordovician; Devonian: dg (Gedinnian); ds (Siegenian); dem (Emsian), dfa (Famennian). Carboniferous: cd (Dinantian), cn (Namurian). A Aachen anticline

that there are earthquake hypocentres at about the depth of the good reflector; it is remarkable that the focal mechanisms have a strong horizontal component. Thus, depth and stress direction are in agreement with the Variscan overthrust tectonics.

Beside the nappes in the northwestern part of the Rhenohercynian, indication for overthrust tectonics seems to be given also in other parts of the Massif. Some authors (e.g., Weber, 1981; Engel et al., 1982/1983) believe that rock series near Giessen in southeastern part of the Massif should be tectonic slices at the base of a Giessen Nappe. They assume that this nappe may be derived from the northwestern margin of the Mid-German Crystalline Rise south of the Rhenohercynian. Small slices of Silurian rocks in the Kellerwald Mountains (eastern part of the Massif) are possibly allochthonous too. Others, e.g., Quade et al. (1981), point out that overthrusting may have played a greater role in the southern part of the Rhenish Massif than thought in the last decades. Overthrust tectonics in the northern flank of the Lahn Syncline shows folding, imbricating and overthrusting across distances of at least 5 km. The vergence is to the northwest. But these authors think that these nappes can not come as far as from the southernmost border of the Rhenish Massif. This is in discussion between Quade and Franke, but undoubtedly the discussion on nappes and overthrusts in the Rhenish Massif now going on will continue in the future.

Besides the problems of the type of tectonics in the Rhenish Massif, new thoughts about the age of tectonics also emerged recently. Formerly geologists thought that the age of tectonics within the Variscan era might differ in the southern and northern parts of the Massif, related to different tectonic phases. Today some authors suggest that folding tectonics may occur like an orogenetic wave in the sense of Wunderlich from south to north, rejecting a sharp separation into distinct tectonic phases in this Massif between south and north. Accordingly in the eastern part of the Rhenish Massif, the orogenetic wave should have started in the southeast part of the Massif, propagating in northwesterly direction (Franke et al., 1978). This seems to be paralleled by the fact that the K/Ar ages, produced by the weak metamorphism (connected in time with the Variscan tectonics) diminish from south to north in the Massif. So an age of metamorphism of about 325 Ma is found in the Taunus Mountains, but the metamorphism in the northern Sauerland is younger than 305 Ma (Ahrendt et al., 1978). This would mean that the orogenetic wave needed 20-30 Ma to make its way from the present Taunus Mountains to the Ruhr region. Connected with this orogenetic wave is also an advance of the flysch troughs from southeast to northwest (Franke et al., 1978). This is in harmony with the concept of a synorogenetic origin of the flysch sedimentation. It is a difficult question whether this concept also holds in the region west of the Eifel Depression. It is, however, certain that the overthrusting in the

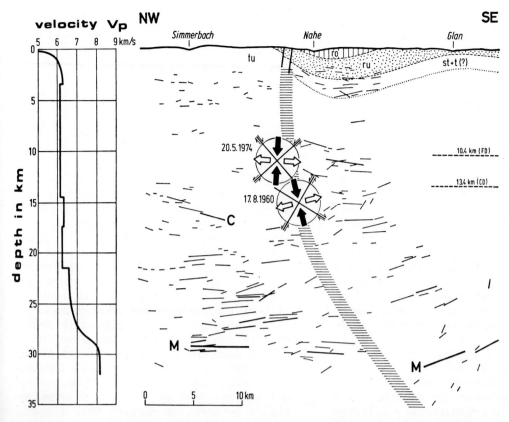

Fig. 6. Cross-section through the region on both sides of the Hunsrück-Südrand-Störung with two hypocentres of earthquakes. (After Ahorner and Murawski, 1975, p.71, Fig. 5). Lines main reflectors of deep reflection seismic investigations. M Moho (= Crust/Mantle boundary); C Conrad discontinuity (= boundary Lower/Upper Crust); Hatched probable trace of the fault zone; The velocity/depth diagram on the left after Meissner et al. (1974). tu Lower Devonian; st + t(?): Carboniferous and Devonian (?); ru Lower Rotliegendes; ro Upper Rotliegendes (Rotliegendes = Lower Permian). Right FD Förtsch discontinuity; CD Conrad discontinuity

northern part of the Ardennes is younger than the overridden Upper Carboniferous. On the other hand, it is clear that this tectonic activity is contemporanous and cogenetic to the folding with distinct vergence to the northwest. This means that the tectonic activity here is of Uppermost Carboniferous age. It is not fully clear whether there may have been succeeding later movements. In contrast to the northern part post-Lower Carboniferous tectonics (possibly Sudetic phase) occurred in the southern part of the Rhenish Massif (e.g., in the Southern Hunsrück).

The southern margin of the Rhenish Massif (Hunsrück and Taunus Mountains) east of the Eifel Depression is of tectonic nature. A distinct fault zone is proven to exist along the southern border of the Hunsrück Mountains: Hunsrück southern border fault. Results of deep seismic reflection (DFG project *Geotraverse Rhenoherzynikum*) demonstrate concave listric faults, curving to the south (Fig. 6). The measurements on faults at the Earth surface immediately give vertical or dip overturned to the south. On the surface of these faults the series of Upper Carboniferous and Permian have subsided several thousand meters. So the Saar Nahe Basin became a very deep trough, the Saar Nahe Trough (Fig. 1). Today this large ancient fault system is not active as a whole, but earthquakes do occur here and there, mainly at intersections with SE-NW striking faults (Ahorner and Murawski, 1975). Earth tidal tilt measurements were carried out along a profile crossing the Hunsrück southern border fault (M. Bonatz, Bonn; C. Gerstenecker, Darmstadt, and J. Zschau, Kiel). The profile consists of five borehole stations and four stations in mines located at different distances on both sides of the fault zone. The observed amplitude factors and phase lags are strongly disturbed. The amount of disturbance depends on the distance of the station from the fault zone. The investigation is continuing. A detailed final report about the project will be published later separately from this book. The Hunsrück southern border fault can be recognized on satellite photographs, but its appearance is not so predominant as on geological maps, where it is recorded from Bingen/Rhine to the region of the upper Saar river. In the western part of this system, however, it has not the clear and homogenous form as seen in the eastern part near the Rhine river. In the intersection area between the Saar-Nahe Trough and the Eifel Depression, e.g., in the area of the upper Mosel river, geological maps and aerial photographs show traces of the fault zone clearly. They bisect the Mesozoic and thus demonstrate post-Jurassic movements. These younger movements are less pronounced than the Upper Carboniferous and Permian ones (Schunck, 1979). Far west, the fault zone becomes more and more blurred.

For the recent southern border of the Ardennes this fault is without importance, for the southern margin of the Ardennes is clearly an erosional feature. If the fault zone has a prolongation to the west, it must be below the thick cover of the Paris Basin (see Sect. 2.3).

According to results of investigations of K.P. Winter (unpublished) the prolongation of the Hunsrück-Südrand-Störung East of the Rhine along the southern border of the Taunus Mountains can be observed in LANDSAT-1 and aerial photographs. But according to these aerial photographs, it has not the form of a closed fault, but is a bundle of faults. It is evident that the recent picture is also influenced by young tectonic activity in the northern part of the Upper Rhine Graben. The prolongation of the fault zone farther East in the region of the Hessian Trough is much less clear than in the intersection zone produced by the southern part of the Eifel Depression and the Saar Nahe Trough. In the Hessian Trough several tectonic events occurred since the Variscan (see Sect. 2.3, Fig. 15). Thick Tertiary volcanics cover vast regions and in some areas the salt layers of the Upper Permian (Zechstein) complicate the clear identification of the older tectonic pattern. Nevertheless LANDSAT-1 photographs show in this region some lineations in accordance with those of the Variscan time, although it seems to be difficult to identify them as prolongations of the southern border faults of the Rhenish Massif. Uplifts composed of Variscan rock series in some areas of the Hessian Trough demonstrate that the Variscan rock series continue in strike deep in the Hessian Trough below a thick cover of younger sediments, and surface again in the Harz Mountains. But we note similarities and dissimilarities between the Rhenish Massif and the Harz Mountains in their rock series.

In contrast to the Harz Mountains and many other parts of the Variscan chain, no granites or other plutonites of Variscan age are exposed in the Rhenish Massif today. In the Massif of Stavelot (Ardennes) several small intrusions of subvolcanic rocks in quartz-dioritic to tonalitic composition have been found. They are restricted to pre-Devonian rocks and probably belong to the Caledonian Cycle. The tonalite of Lammersdorf (Hohes Venn) is the most famous of them. Ignoring this we are left with one of the important questions whether there are no plutons in this part of the Variscan or whether the level of erosion has not yet reached the plutonic stockwork here. The only geological indication for possible plutonism of Variscan age in the subsurface of the Rhenish Massif is ore veins (lead-zinc ores and siderites), if one accepts the idea that such ore production has been connected with plutonic activity. On the other hand, there are some ore deposits in the Rhenish Massif which have no connection with a synorogenic or late orogenic plutonism, having originated in the geosynclinal phase as submarine deposits, e.g., iron-ore deposits of the Lahn-Dill region. These ores show a strong relationship with submarine lavas of Variscan age. The baryte-pyrite deposits of Meggen in the northeastern part of the Rhenish Massif occur as volcano-sedimentary sediments in a special basin of the Rhenohercynian geosyncline. The submarine volcanism of the geosynclinical stage of this region shows a typical evolution in the part east of the Eifel Depression. The oolithic iron ores of Lower to Middle Devonian age in the Eifel Mountains are of sedimentary origin. The origin of the iron is unknown.

The Rhenohercynian trough, predating the Variscan orogeny, was not a simple sedimentation basin. It was bordered in the north by branches of the Old Red Continent and in the south by the Mid-German Crystalline Rise, but its bottom morphology changed greatly with time of the geosynclinal development. There was no stability for any special basin in this trough. The axes of these special basins changed their positions, so that the thickness of the different rock series and their facies vary during the geosynclinal stage. It should be remembered that the sediment thickness must not be an indicator for the depth of the sea in every case. It is only an indication about the subsidence rate of the basin and of adequate sediment supply. Beside the morphology of the basin bottom, produced by subsidence and uplift of some parts, big limestone reefs (mainly in the Middle Devonian) and masses of submarine basaltic lava flows produced additional morphology. The geosynclinal volcanism of the Middle and Upper Devonian (Lower Carboniferous partly) is known from the eastern part of the Rhenish Massif only. It is absent in the contemporaneous rock series in the Eifel Depression (Eifel-Kalkmulden) and in the region to the west. Winter (1965) located in this region small layers of bentonites and tuffs in the Devonian and in 1981 conducted tephrostratigraphical correlations in the Eifel Mountains and Ardennes. The derivation of these volcanic ashes, strongly altered by later crystallization, is quite unclear.

With the help of palaeogeographical maps Meyer and Stets (1980, Figs. 3-8) tried to reconstruct the history of sedimentation for the Lower and Middle Devonian in the western and central Rhenish Massif. It is possible to demonstrate change in location of the different special sedimentation basins with the aid of isopaches and characteristical facies data. Certainly the axes of the different special basins have changed their location in time; in the region of the Moselle Syncline the Lower and Middle Devonian sediments assume a cumulative maximum thickness of 12 km (Meyer and Stets, 1980, Fig. 9). Additionally, investigations in the Eifel Depression provide clear evidence for facies boundaries with north-south direction in the Devonian (Winter, 1965, and in: Meyer et al., 1977). This clearly marks the importance of the north-south direction in this special part of the Massif even at the geosynclinal stage.

Investigations by the Department of Geology and Palaeontology of the University of Göttingen (special program DFG, SFB 48) of the eastern part of the Rhenish Massif have contributed largely to the sedimentation history in the Rhenohercynian trough. It is always dependent on the vertical movements of the sea bottom and the supply of sediments. The palaeogeographical maps constructed in this program allow us to define the change of location of the shelf and the foreshelf basin for the time of Middle Devonian, Upper Devonian and Lower Carboniferous (Franke et al., 1978, Figs. 3, 4 and 6). These results demonstrate that this region was a mobile shelf. This mobility caused considerable shifts of facies and changes of facies properties in different parts of the trough. On the other hand, the investigations demonstrated some regularities, e.g., three phases in the sedimentation history of this part of the trough caused by vertical movements of the sea bottom, but also of the land regions from which the sediments were derived. In the *first phase* (Lower and Middle Devonian with increasing activity toward Lower Carboniferous) neritic clastics came from the Old Red Continent in the north, while the sea floor subsided continuously. The effect was an enormous accumulation of sediments in a widespread neritic shelf area. Franke et al. (1978) regard these sediments as a Caledonian Molasse, for they were produced during denudation and erosion of parts of the Caledonides. It is, however, probable that such a debris did not come from the central part of the Caledonides. Uplifts of Caledonian rock series north of the Rhenohercynian trough seem to have been the source areas. The sediments of the Caledonian Molasse in the early Devonian reached even the southern part of the trough. Subsequently this type of sedimentation, somewhat diminished, retreated to the north. In the *second phase* (Middle and Upper Devonian and Lowermost Carboniferous) subsidence and sedimentation were less intensive than in phase 1. The neritic sedimentation occurred only in the northwestern part. The other regions show pelagic sedimentation in wide areas. One of the characteristics was the growth of large limestone reefs on submarine swells. Submarine volcanism (basaltic lavas, pyroclastics, subordinate keratophyrs) was at maximum at this time. In the *third phase* clastic sedimentation, essentially of turbidites, came from the southern border of the basin, the Mid-German Crystalline Ridge. This Variscan Flysch phase began in the Devonian, increased in the Upper Devonian and peaked in the Lower and Middle Carboniferous (Namurian). It is a synorogenetic sedimentation predominantly of graywackes and is followed by the Molasse sedimentation without a sharp break. The flysch front in Carboniferous time runs from southeast to northwest. The migration of the flysch front is dependent on orogenetic processes, for folding also seems to move from southeast to northwest like an orogenetic wave.

The flysch phase has a further speciality in the Rhenish Massif. There is a remarkable change of facies along the northern trough from east to west. In the eastern part of the Rhenish Massif the Lower Carboniferous shows the so-called Culm facies, a more or less clastic sedimentation. In the western part, starting east of Düsseldorf, the facies of the Lower Carboniferous changes to the Kohlenkalk facies, a limestone-rich sedimentation. This is caused by different bathymetry of the geosyncline of this time here and the fact that clastic materials could not reach the western part of the northern flysch trough.

The Molasse sedimentation occurred in a trough at the northern margin of the Rhenish Massif and in the Saar Nahe Trough South of the Rhenohercynian. There is a remarkable difference between these two regions. The Subvariscan foredeep in the North is of paralic type. A widespread shelf area follows to North especially in the eastern part. The Saar Nahe Trough is not of paralic type. It is a special Molasse basin in the inner part of the Variscan Mountains. Another difference between these two basins is that while the coal-bearing series in the northern basin start in the Middle Carboniferous (Namurian) and end in the Upper to Uppermost Westfalian, the coal seams in the Saar Nahe Trough begin in Westfalian B. The last small seams are known from the Lower Permian here.

The Pre-Variscan and the Origin of the Rhenohercynian Geosyncline

The geological history described so far is the Variscan one. The sedimentation in the special Rhenish geosyncline occurred in the Devonian and Lower Carboniferous and, in the Subvariscan foredeep, in the Upper Carboniferous. It is of interest to learn something about the pre-Devonian development of this region. In several areas of the Rhenish Massif rock series are exposed, which are of pre-Devonian age, as mentioned (Fig. 7). The best examples are in the Ardennes and the Brabant Massif. It is possible to reconstruct the palaeography of the pre-Devonian in these regions. The outcrops of pre-Devonian in the eastern part of the Rhenish Massif are less and it is much more difficult to interpret the pre-Devonian for this area. Data from Ordovician and Silurian rocks exist from exposures in the cores of the Remscheid Anticline and the Ebbe Anticline.

There is a difficulty in the reconstructions everywhere because of a strong tectonic deformation by folding, imbri-

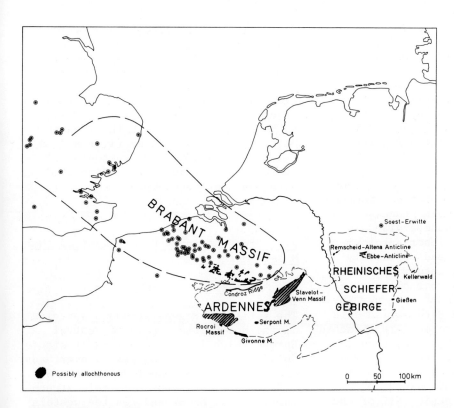

Fig. 7. Outcrops and borings with Lower Palaeozoic rocks in the Brabant Massif and the Rhenish Massif. (After Walter, 1980, p. 13, Fig. 1)

cating and overthrusting in the Variscan orogeny. To include these changes Walter (1980, Figs. 3-7) tried to draw palaeogeographical maps including the region from western England to the eastern border of the Rhenish Massif. The basis of these maps are data on sedimentology and facies. They give the time from late Precambrian/Lower Cambrian to the Silurian (Llandovery-Ludlow). To demonstrate the difficulties of palinspastic reconstructions in this region, Walter gives an interesting example. If one tries to correlate the facies area of the Anticline of Condroz (Ardennes) and of the Ebbe Anticline (eastern part of the Rhenish Massif) for Ordovician time, it is evident that the facies area of the Stavelot-Venn Massif (Ardennes), represented by the Salm (= Lower Ordovician), today is situated at least 20-30 km too far north. Walter assumed that this may have been caused by exceptionally intensive overthrusting to the north. This is in agreement with the results of deep seismic reflection and investigations by Dutch and Belgian geologists that indicated strong Variscan overthrust tectonics in the northwestern part of the Ardennes. On the other hand, this confirms our opinion that the geological situation west and east of the Eifel Depression (in the widest sense) is very different.

The Rhenish Massif has a history of sedimentation and tectonics evidently belonging to the Caledonian cycle. But this region seems not to be a special geosyncline restricted to northwestern and central Europe. Rather the sedimentation basin depended on subsiding processes of different blocks of the Precambrian basement. On the other hand, the tectonics is not accompanied by high-grade metamorphism or by plutonism of any importance. There is no noticeable closed Molasse stage following the orogenic processes, if one disregards the sediments of the first phase of the Variscan geosyncline, which were termed Caledonian Molasse by the Göttingen group. But this is a special type of Molasse, which is not confined to a basin such as a foredeep in the traditional sense.

It seems clear that the rock series of the Variscan Rhenohercynian geosyncline overlie continental basement, perhaps a thinned continental crust, and there is no indication of oceanic crust underlying the marine Palaeozoic sediments. On the other hand, it is impossible to prove the existence of a real subduction zone of oceanic lithosphere in the Variscan belt. We want to emphasize this particularly in contrast to Burrett (1972), who proposed two subduction zones between the Rhenohercynian and the Saxothuringian. Real indications for the existence of a subduction zone in this region have never been found. The possibility that there may have been no oceanic plate subduction tectonics prompted new models of the orogenic processes which produced the Rhenohercynian orogenetic belt. The most important former tectonic concept, advanced especially in the first half our century by Stille, was based on the ideas that all continents grow by accretion of orogens. The oldest part would be a Precambrian core and then the various orogens are attached one after another to the core, thus resulting in a continent with zonal orogenic pattern. Today much more complicated ideas are being discussed on the development of the orogenic belt of the Variscides. Some propose that there is no indication for any autonomous plate tectonics in Central Europe at Variscan time, rather this is probably an example of intra-plate tectonics (e.g., Matthews, 1978). It may be influenced or initiated by movements of megaplates far outside of the region. Another concept for the geosynclinal development and for the orogeny is that of mantle diapirs. This idea may be seen in contrast to plate-tectonics, if it takes aim at fixism in tectonics (e.g., Krebs and Wachendorf, 1973). The most recent genetic hypothesis for the Rhenohercynian was proposed by Weber (1978) and extended by Behr (1978) to the whole Variscan of Central Europe. Weber starts with the premise that the Rhenohercynian developed on the continental crust, and therefore a simple plate-tectonics model is inapplicable here. He assumed that folding and related tectonics in the Rhenohercynian crust was caused by underflow movements (subfluence) in southerly direction. According to Weber (1981) subfluence may be regarded as the reaction of continental crust to subduction of the underlying lithospheric mantle. Such subduction may be more or less the same process like this of the Ampferer subduction (A-subduction). Such an A-subduction should have occurred in the region of the Mid-German Crystalline Rise if we follow this hypothesis. This model was elaborated in the course of the geological investigations of the Göttingen group (Engel et al., 1982/1983). It is to be noted here that in the Rhenish Massif west of the Rhine river and in the Mosel Syncline some tectonic features cannot be fully interpreted with the subfluence model.

2.3 Pre-Quaternary Development of the Massif Margins

The Margins of the Massif

The Variscan orogeny produced a region in the northwestern part of Central Europe which is characterized by strong folding, cleavage, and partly overthrust tectonics: the Rhenohercynian (see Sect. 2.2). The axes of the main fold structures strike NE-SW and, in the western part of the Ardennes, E-W (Fig. 2). The long axis of the Rhenish Massif is more or less parallel to these structures, but in some areas the boundaries are distinctly different in direction.

The northwestern part of the *northern margin* of the Massif is parallel to the northern boundary between the Variscan rock series of the Ardennes including its foredeep and the Brabant Massif in the north. The Brabant Massif was deformed in the Caledonian orogeny. The basement is covered there by unfolded Devonian and Carboniferous and a thin cover of Mesozoic and Cenozoic rocks. In contrast, the folded Upper Carboniferous of the northeastern flank of the Massif dips below the Cretaceous cover of the Münsterland Basin. This means that the northwestern part of the margin follows the tectonic structures, while the northeastern part represents an erosion line only. Between these two parts a special tectonic structure enters the Rhenohercynian Massif: the Lower Rhine Basin (Niederrheinische Bucht). The Variscan basement has subsided and the basin is filled by a thick rock sequence of Tertiary and Quaternary age. In the interior of the

Lower Rhine Basin there is a dome-like structure: the Krefeld Arch (Krefelder Gewölbe); see Sect. 2.2, Fig. 2. This structure is definitely of pre-Permian age and is a conservative element in the history of the region; e.g., not only during the Variscan orogeny but also in the Lower and Upper Cretaceous and in the Tertiary.

The *southern margin* of the Massif also shows variable forms. In the southwest, west of the Eifel Depression, the folded complex of the Ardennes dips gently without any noticeable tectonic evidence such as faults below the cover series of the Paris Basin. The dip of the cover is $2°-5°$ toward the Paris Basin in Luxembourg and about $2°$ toward the west or southwest in Belgium. The boundary between the Rhenohercynian and the Saxothuringian is here unknown. If this boundary exists in the region, it must be below the cover of the Paris Basin. In contrast, the southern margin east of the Eifel Depression follows the Saar-Nahe Trough, an important structure which contains the boundary between the Rhenohercynian in the north and the Saxothuringian in the south (see Sect. 2.2). This deep trough is filled with sediments of Variscan Molasse. It is interesting that in the Upper Carboniferous the origin of the sediments was from north (Hunsrück Mts.) and from south, while in the early Lower Permian (Unter-Rotliegendes) it may have come mainly from the south. It seems that in late Lower Permian (Ober-Rotliegendes) a change occurred in this system, for the grain size of the sediments now diminished from north to south in this basin. The southern source area of the trough sediments is unknown today since it is covered by thick Mesozoic sediments. Between the two parts of the southern margin of the Rhenish Massif the Eifel Depression (Eifeler Nord-Süd-Zone) crosses the Rhenohercynian (see Sect. 2.2). We shall show later that this special structure played an important role in the post-Variscan history of the Rhenish Massif.

The *eastern margin* of the Rhenish Massif is independent of the strike directions of the Variscan fold systems. Here the folded Variscan dips below the Mesozoic and Cenozoic fill of a great north-south structure, the Hessian Trough (Hessische Senke). The recent geological appearance shows that the border is not homogeneous. At many places it is an erosional line, at others faults are to be found. It is sure that at least in the northern part of the Hessian Trough the strongest subsidence of the Variscan basement is located many kilometers east of the current western border of the trough. The complicated geological pattern of this Trough was generated by different processes in the geological history of this structure (Murawski, 1960). While the Hessian Trough is the eastern margin of the Rhenish Massif, the Rhenohercynian continues at depth below the Hessian Trough to the Harz Mountains.

This short description of the contours of the Rhenish Massif shows that the recent pattern was produced by many processes affecting the Massif and its environment. It is therefore necessary to discuss the important stages of the history of the Massif and to assess the interaction between the geological events in the Massif region and those in the surroundings. In the following a brief description is given of the post-Variscan development of these regions.

The Development of the Margins and Forelands of the Massif

Permian

Since the beginning of the Permian many regions along the periphery of the Rhenish Massif show directions in the morphology of newly produced basins different from the Variscan NE-SW direction. In the Lower Permian (Rotliegendes) the northern border of the region with erosion and denudation was located far north (Fig. 8). But there are great differences between the development in the western and the eastern parts of this region demonstrating the effect of important elements. In the western forelands of the Ardennes a vast platform produced by fluvial and aeolian sedimentation spread out far to the north. Volcanites are unknown here, while in the eastern area volcanic lavas are well known east of the Emsland in the Lower

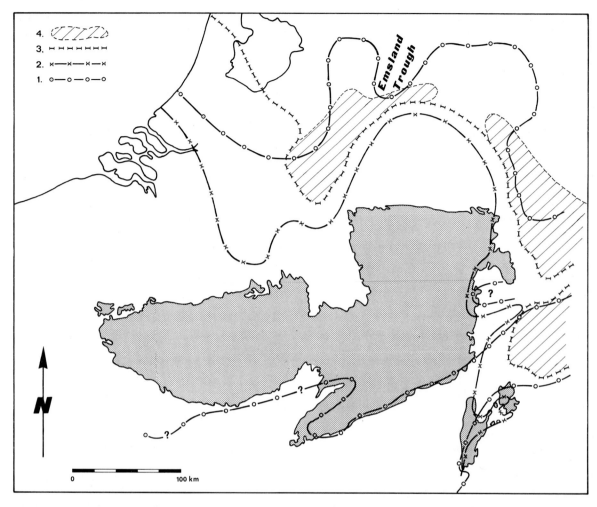

Fig. 8. Palaeogeographical outline map of the Permian. 1 Boundary denudation area: sedimentation area in Lower Permian (Rotliegendes); 2 coast line of the Lowermost Upper Permian (Zechstein 1; Z1); 3 boundary of the "Z1-Salinar" (salt-bearing rock sequences); 4 salt basins Z1. *Question mark* line unsure; *Dotted* Rhenish Massif, Odenwald and Spessart Mts.

Permian (Rotliegendes) basin of northern Germany. Between these two areas a special basin, the Emsland Trough (Emsland-Trog) with a NNE-SSW long axis shows stronger subsidence. This Trough seems to be the northern prolongation of the Eifel Depression (Murawski, 1964). Farther east the northern part of the Hessian Trough and its prolongation to the north are characterized by curving of the borderline of the Lower Permian to the south (Fig. 8). The Lower Permian south of the Rhenish Massif follows the Variscan direction (SW-NE). This can be seen in special basins such as that of Wittlich/Mosel and particularly the Saar-Nahe Trough which follows the southern boundary of the Hunsrück and Taunus Mountains. But it seems that some N-S elements were also active at this time in the Upper part of the Lower Permian, e.g., some N-S faults between the Odenwald and Spessart Mountains. This faulting allowed a local extension of the sedimentation in these areas (Kowalczyk, 1982).

The tendency to produce structures as basins and swells with northerly direction increased in the Upper Permian (Zechstein). As seen in Fig. 8, the coast lines north of the Rhenish Massif at this time changed their positions from those of the Lower Permian. The coastline of the lowermost Zechstein (Z1) moved to the northern border of the Brabant Massif in the west and to the northern edge of the recent Cretaceous Münsterland Basin in the east. While the southern border of the evaporite basin of Z1 is situated farther north than the coast line of Z1 in the west, they are parallel and close together in the east. This difference is accentuated by the N-S structure of the Emsland Basin, which is documented by the coast line and salt basin border of Z1 curving southward. This structure shows the close relationship to the Eifel Depression better than in Lower Permian (Murawski, 1964). A similar situation is evident to the east in the area of the Hessian Trough. This subsiding basin allowed the "Zechstein"-sea to advance temporarily until the present region of the Upper Rhine and was clearly independent of the Variscan SW-NE directions. This is accentuated by the fact that no marine "Zechstein" sediments occur in the former Variscan Molasse basin south of the Taunus and Hunsrück Mountains in the Saar-Nahe Trough. On the other hand the former SW-NE directions persist in the Upper Permian in special cases, e.g., in the Spessart Mountains where the different facies zones and the general coast line are dominated by the former Variscan structures (Kowalczyk et al., 1978).

Triassic and Lower Jurassic

The development of the structural or palaeographical patterns of the Permian time continued in the Triassic and Lower Jurassic (Fig. 9). Vast regions of the Ardennes and the Brabant Massif and of the eastern part of the Rhenish Massif were denudated during this time. The Eifel Depression between the eastern and western Rhenish Massif became a continuous subsiding basin. It permitted a passage to be established between the North German-Netherland Basin in the north and the sedimentation basins of eastern France and southern Germany. The Hessian Trough had a similar function east of the Rhenish Massif. The isopaches of the Lower Triassic (Buntsandstein) trend in a SSW-NNE direction preferentially, but in some cases the old Variscan direction is preserved in the form of minor swells.

In the Eifel Depression the Lower Triassic (Buntsandstein) and the Upper Triassic (Keuper) sediments are terrestrial. But it seems to be clear that in the Middle Triassic (Muschelkalk) and Lower Jurassic (Liassic) this structure was a marine branch like the Hessian Trough, connecting epicontinental sea basins in the north and the south.

In the platform north of the Brabant Massif there is little temporal change in subsiding, but it seems that the earliest movements in the Netherlands Central Graben system occurred in Lower Triassic time; no Upper Triassic evaporites are known. In the region of the East Netherland platform uplift occurred in the Liassic. On the other hand the southern limit of Upper Triassic salt deposition and the 500 m isopache of the Liassic are clearly connected to the northern border of the Münsterland Basin (Fig. 9).

Upper Jurassic and Cretaceous

During the Middle and Upper Jurassic the palaeography changed distinctly. The former N-S structures such as those of the Eifel Depression or the Hessian Trough lost their predominance. A vast continental mass originated consisting of the Brabant Massif, the Ardennes, the Eastern Rhenish Massif and the present Harz Mountains (Fig. 10). North of this continental mass, the WNW-ESE direction dominated tectonics and morphology. The development of the areas west and east of the former N-S structures (Emsland Trough and Eifel Depression) were more or less parallel. North of the Brabant Massif the system of the Netherlands Central Graben grew into a subsiding basin with its long axis in NW-SE direction in the Upper Jurassic and Lower Cretaceous. This basin is divided into two special basins by an uplift or swell. North of the present Münsterland Basin which remained part of the Rhenish Massif in the Upper Jurassic and Lower Cretaceous, a large subsiding basin developed during this time

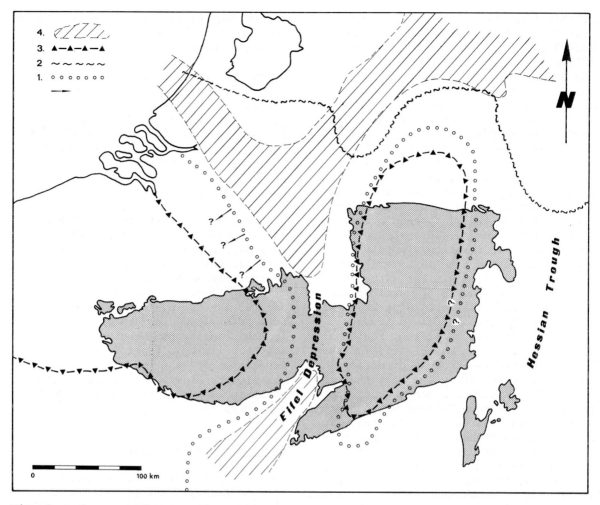

Fig. 9. Palaeographical outline map of the Lower Triassic and the Lower Jurassic. *1* Boundary denudation area: sedimentation area in the Lower Triassic (Buntsandstein); *2* southern border of salt-bearing series of the Lower Triassic (Röt-Salinar); *3* coast line of the Lower Jurassic (Liassic); *4* Liassic main sedimentation basins (thickness > 500 m). *Arrows* a possibly wider primary extension of the Buntsandstein basin; *Question marks* distant extension unknown; *Dotted* Rhenish Massif, Odenwald and Spessart Mountains

(Lower Saxony basin). It has a WNW-ESE long axis. After a time of expansion of the Luxembourg and Lorraine Basins in the Lower and Middle Triassic, the development of the Paris Basin began in the Upper Triassic south of the Ardennes. The slight relief at the southern margin of the Ardennes and in the area south of it supported the formation of the Paris Basin (Fig. 9). The phase of strongest subsidence of the basin and widest expansion into the surrounding massifs was in Middle and Upper Jurassic. On the eastern flank of the present Upper Rhine Graben a wide basin (Südwestdeutsches Becken, Südwestdeutsche Großscholle) had existed since the beginning of the Mesozoic. The sedimentation in this basin ended totally at the boundary Jurassic/Cretaceous and by uplift this region, together with the Rheno-Ardennes, became a shield-like continental mass (= Rheinischer Schild in the sense of Cloos, 1939).

The geological development in parts of the northern margin of the continental mass (Ardennes, Rhenish Massif, Harz

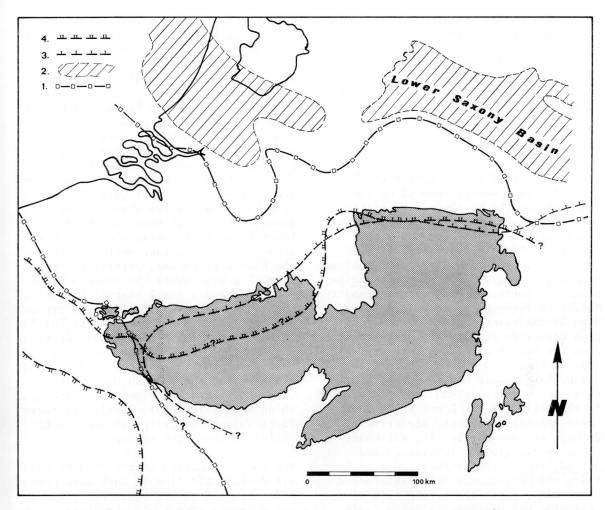

Fig. 10. Palaeogeographical outline map of the Cretaceous. 1 Coast line of the boundary Uppermost Jurassic/Lowermost Cretaceous; 2 main Cretaceous basins; 3 coast line of the Cenomanian; 4 coast line of the Campanian. *Question mark* distant extension unknown or unsure; *Dotted* Rhenish Massif, Odenwald and Spessart Mountains

Mountains) was different in the Cretaceous. The submerge of the northern foreland of the Brabant Massif began in the Uppermost Lower Cretaceous and the beginning of the Upper Cretaceous in a gradual manner. On the other hand, the Cenomanian sea gained ground swiftly in the Münsterland Basin and the northern parts of the Eastern Rhenish Massif (Fig. 10). In the western part of the Rhenish Massif the transgression became more and more intensive during the Upper Cretaceous. The whole Brabant Massif and its westerly extension and large parts of the Ardennes were transgressed by the sea in the uppermost Santonian, the Campanian and Maastrichtian. For the Upper Maastrichtian the sea had a regressive tendency (Albers and Felder, 1979). In contrast to this, the sea moved back very slowly to the north in the Northeastern Rhenish Massif. In the Hessian Trough (except in its northernmost part) and in Southern Germany outside the Alps (except in a small region north of Regensburg/Danube), no marine Cretaceous sediments were deposited. On the other hand, the Paris basin remained a sedimentation area. The transgressions of the sea into this basin came mainly from the Southeast and partly from the Northwest. In the Cenomanian and the younger series the course of the coast lines is often uncertain in the northern part of the basin, i.e., the southern border of the Ardennes (Mégnien, 1980). It seems that the extension of the Cretaceous sea and the sediments thickness in the Paris Basin is less than that of the Jurassic.

Tertiary

The third period of the uplift of the Rhenish Massif was influenced by the development of the southern part of the North Sea Basin (Fig. 11). Earlier, the Eifel Depression had been a dominant structure which tectonically affected the northern foreland, e.g., the Emsland Trough (see Figs. 8 and 9). On the other hand, the northern flank of the Brabant Massif influenced the coast line direction or the basin borders over a long time span (see Figs. 8-10). This changed from the beginning of the Tertiary (Fig. 11). It seems that the northern flank of the Brabant Massif had no further influence. A new active SE-NW system developed in the region of the Lower Rhine Basin Lower Rhenish Embayment (Niederrheinische Bucht). As seen in Fig. 11, all coast lines reacted to the subsidence and curved here to the southeast during Tertiary and Pleistocene (see Pleistocene coast line, Fig. 11). During this time tectonics was manifest in SE-NW fault strikes and today active tectonics is demonstrated by recent seismic activity in the area (Ahorner, 1983). The N-S directions of the Eifel Depression clearly lost intensity and the SE-NW direction has become the dominant role, but the N-S pattern of the Tertiary Eifel volcanoes shows that the N-S direction was not fully inactive. On the other hand, the Tertiary Siebengebirge volcanoes, south of Bonn, follows the SE-NW direction, which is so prominent in the Rhine Valley and the Lower Rhine Basin.

West of this zone, in the region of the Brabant Massif and the Ardennes, the Tertiary coast lines have more or less an E-W direction and it can be seen that the Oligocene sea covered the Brabant Massif and large parts of the Ardennes (Albers and Felder, 1981). To the east, the coast lines surround the eastern Rhenish Massif north of the Münsterland Basin, and there is a distinct influence of the Hessian Trough. This is most evident in the Oligocene, specially the Middle Oligocene. On the other hand, there is a remarkable difference between the Hessian Trough and the Lower Rhine Basin. In the Hessian Trough the Middle Oligocene coast line pass the trough from north to south and this way the North Sea had a small connection with the Mediterranean across the Hessian Trough, the Upper Rhine Graben, and the Rhone Graben. The Upper Oligocene coast line has a regressive phase in this trough. This is in contrast to the Lower Rhine Basin where the Upper Oligocene invaded farther southeast than the Middle Oligocene, perhaps in the form of a bay toward the Eastern Eifel mountains and a small channel following the recent Rhine Valley (see the arrows of Fig. 11 and Meyer et al. this Vol.). It seems that this trend continues to the Lower Miocene (Hemmoorian), for there are some indications that the sea transgressed the Rhenish Massif in a very small branch following the present Rhine Valley (Martini, 1981). All these facts suggest differing tectonic activities in these two regions.

The southern margin of the Rhenish Massif during this time is much more complicated than in the Upper Paleozoic and Mesozoic. The fault structures, active in the Upper Carboniferous and Permian (Hunsrück Southborder Fault, see Sect. 2.2), are only locally active. The former more or less continuous fault zone along the southern Hunsrück border is dislocated by young active N-S and NW-SE faults (Bausch, 1980, unpublished). The Upper Rhine Graben and the Mayence Basin (Mainzer Becken) reach the southern border of the Rhenish Massif and produce a very complicated tectonic picture. The complications can be proven very clearly in the pattern of the photo lineations. There are many directions originating in the young tectonics as that of the Upper Rhine Graben and others which persisted from old Variscan structures. Some directions interfere between these two (Winter, 1981, unpublished).

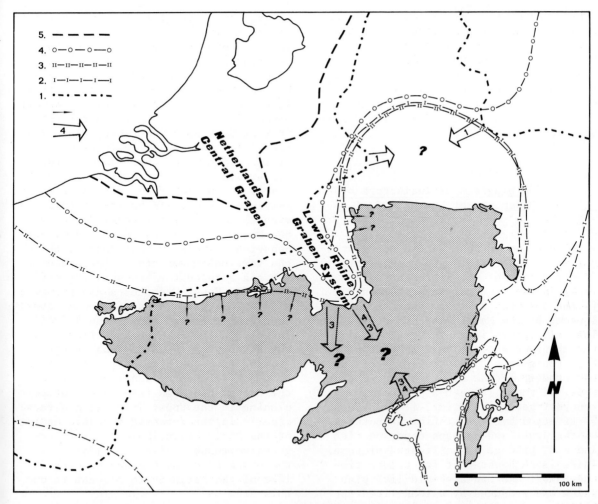

Fig. 11. Palaeogeographical outline map of the Tertiary and the Pleistocene.
1 Coast line of the Lower Palaeocene (Thanetian); 2 coast line of the Middle Oligocene (Rupelian); 3 coast line of the Upper Oligocene (Chattian); 4 coast line of the Lower Miocene (Hemmorian); 5 coast line of the Pleistocene. *Small arrows* a possibly wider extension of the Oligocene basin; *large arrows with embossed number* sure transgression directions, reconstructions based upon the discovery of isolated marine or brackish rock samples in some exposures of this region. *Number* geological age; *Question mark* distant extension unknown; *Dotted* Rhenish Massif, Odenwald and Spessart Mountains

In the area where the NE-SW striking structures of the Saar-Nahe Trough and the N-S structures of the Eifel Depression meet, the N-S structures distinctly influence the tectonic pattern of the Saar-Nahe Trough by faulting and flexuring. Between the Jurassic and Quaternary the southern branch of the Eifel Depression (Trierer Bucht, Lothringer Senke) subsided more strongly than the Saar-Nahe Trough. There is no proof for the Hunsrück Southborder fault (Hunsrück-Südrandstörung) to have been active everywhere during the Tertiary and Quaternary (Schunck, 1979). The western part of the Ardennes may have been submerged in the Palaeocene and it seems certain that from time to time large areas of the Ardennes were covered by the Oligocene sea. A clear connection with the Paris Basin during this time is not evident at the southern border of the Ardennes. Thus, many different epeirogenic processes have occurred in the region of the Rhenish Massif since the Variscan. But there was a distinct change in the Tertiary and Pleistocene. In contrast to the Mesozoic, strong volcanic activity occurs, additionally the epeirogeny, and the SE-NW direction in faulting and subsidence became predominant in some parts of the Rhenish Massif and its northern foreland.

Facies-Time Diagrams as Indicators of Vertical Movements of the Massif

It must be noted that we did not distinguish between certain and probable reconstructions of the different coast lines in the figures. Nevertheless, it is the best way to learn that the development of the Massif margins during geological history is an integral of differing processes. We must understand that there is not *one* simple massif as the product of *one* single building process. If there is such a process in youngest geological time, it doubtlessly is affected by pre-existing structures of the crust and perhaps upper mantle and must produce a complicated picture. There is no doubt that for a long time the Massif has had a shield-like plan which was produced by a long-lived uplift process with many oscillations. This is demonstrated by Figs. 12-14, 16 and 17, showing the oscillations of the Massif in transgressions and regressions from the Permian to Quaternary. This includes facies control in different times and regions. Figure 1 (Sect. 2.2) shows the location of the different facies-time diagrams. They are not real cross-sections as geological profiles. They should only clarify the facies of the different times and regions of the Massif margins. Parallel with the description of the facies situation in the margins and the forelands of the Massif, the facies-time diagrams show the general situation of the central part of the Massif schematically (terms: Variscan Mountain Chain, Rhenish Mass, Rhenish Massif in all diagrams).

Diagram I - I (Fig. 12)

This diagram presents the situation of the Ardennes and their forelands (Fig. 12 see page 32/33). It includes the Brabant Massif and the Netherlands Basin in the north as well as the northeastern border of the Paris Basin in the south. It runs from Amersfoort/Netherland (N) to Metz/Lorraine (S). At first glance it is evident that the two forelands differ distinctly in their geological history. The history of the northern margin of the Massif has to be combined with three facts: the development of the North Sea in the widest sense, the behaviour of the Brabant Massif, and the vertical movements of the Ardennes. The sea entered the Ardennes three times: Jurassic, Upper Cretaceous and Lower to Middle Tertiary. The situation of the southern margin of the Ardennes is different. In the Permian and Triassic the history of the Saar-Nahe Trough (coming from the NE) and the extension of the Eifel Depression (coming from the North) had some influence, but beginning in the Upper Triassic and very clearly in the Jurassic the development of the Paris Basin indicated a transgressive tendency (Voisin, 1981). On the other hand the Cretaceous coast line of the Paris Basin did not reach this area.

Diagram II - II (Fig. 13)

This diagram, more or less parallel to diagram I - I (Fig. 12), demonstrates the situation in the eastern part of the Rhenish Massif (Fig. 13 see page 32/33). Coming from Lower Saxony in the north, crossing the Münsterland Basin and leaving the Rhenish Massif near Bad Nauheim, then across the Wetterau Basin it reaches the Spessart Mountains. The northern boundary of the Variscan orogenic belt (Sect. 2.2, Fig. 2) is the boundary for all transgressions of the sea from Permian to Lower Cretaceous. This is documented by the limit between Lower Saxony Basin and Münsterland Bight of our diagram. Only the Upper Cretaceous sea transgressed, and reached the northern areas of the Rhenish Massif south of the Ruhr River (Fig. 10). Later on the coast lines of the Uppermost

Cretaceous and Tertiary return to the outer border of the Münsterland Basin (Fig. 11). This is similar to the situation in Permian to Lower Cretaceous (Figs. 8-10). All this is remarkably different from diagram I - I (Fig. 15) in the Ardennes and even the behaviour of the Upper Cretaceous coast lines is not totally congruent between the Ardennes and the eastern part of the Rhenish Massif (Fig. 10). The difference between diagram I - I and II - II is also striking in the southern part. Here diagram II - II reaches the Spessart Mountains, where Variscan metamorphic rocks are exposed: the metamorphic northern belt of the Saxothuringian (see Sect. 2.2). The geological history of this region, including the Wetterau between Spessart Mountains and the Rhenish Massif, is quite different from that of the northern part of the diagram II - II. There is no evidence for Jurassic and Cretaceous deposits (Fig. 10). The Triassic is similar to that of the Hessian Trough and Southern Germany. The Permian more or less corresponds to the Saar-Nahe Trough. On the other hand, fault tectonics of the northern branch of the Upper Rhine Graben and the southern parts of the Hessian Trough influenced the situation in the Tertiary. It is evident that the situation in the northern and southern part of the diagram is different from that of diagram I - I. But it is also evident that these variations are caused by the particular geological developments in the different areas.

Diagram III - III (Fig. 14, see p. 34/35)

Diagram III - III runs from the Champagne to the Hessian Trough, presenting the situation of the southwestern border of the Ardennes and that of the eastern boundary of the Rhenish Massif. The northern boundary of Paris Basin reached the southwestern flank of the Ardennes (Figs. 9 and 10) so that in the Jurassic the sea transgressed far north. On the other hand the Cretaceous of the Paris Basin did not reach the recent southern boundary of the Ardennes. Quite different from this is the geological history of the eastern branch of the Rhenish Massif. While the Jurassic and Cretaceous are predominant in the Champagne they are less important in this part of the Hessian Trough at the eastern boundary of the Rhenish Massif. In other parts of the Hessian Trough the Lower and Middle Jurassic is existent, but the Upper Jurassic and Cretaceous are absent. Figure 15 (see page 36) after Meiburg (1982) gives a schematic diagram of the tectonic situation of the southern border of the Lower Saxony Basin to the southern part of the Hessian Trough for the time from Permian to Tertiary. There is good harmony between all areas from Permian to Lower Jurassic. But the Lower Saxony Basin and the Hessian Trough differ from Middle Jurassic to Lower Cretaceous. From Upper Cretaceous to Tertiary the vertical movements are similar in all regions. Another remarkable difference between east and west of this diagram is the strong Tertiary volcanism as a typical phase of the Hessian Trough and the important role of the Hessian Trough in Middle Oligocene as a marine connecting passage between north and south (Fig. 11). All this is absent in the western part of the diagram, for its history is strongly determined by the development of Paris Basin.

(Caption to Fig. 12, figure see p. 32/33)
Facies-time diagram I-I (see Fig. 1, Sect. 2.2). *Dark* marine or brackish; *light* fluvio-lacustrial; *open circles* conglomerates; *dotted* sands or sandstones; *hatched* clay; *brick* carbonates. *Brick and hatches*: marl; *hooklets convergent downwards*: volcanics. *Hooklets divergent downwards*: salt-bearing series. *Lare Arrows* direction of vertical movements; *Question mark* distant extension unknown or boundary unsure

(Caption to Fig. 13, figure see p. 32/33)
Facies-time diagram II-II (see Fig. 1, Sect. 2.2). Symbols see Fig. 12. *Black triangles*: sediments from inland ice glaciers

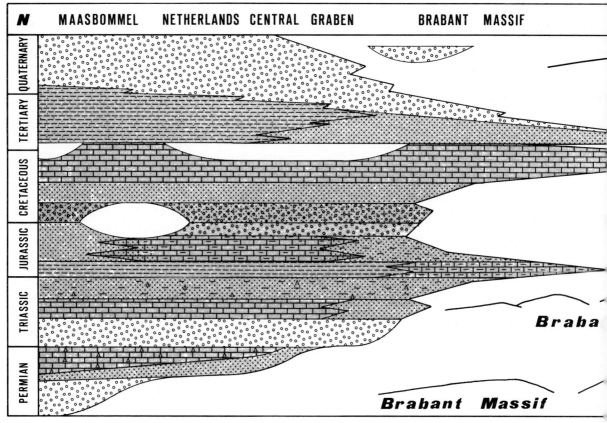

Fig. 12 (Figure caption see page 31)

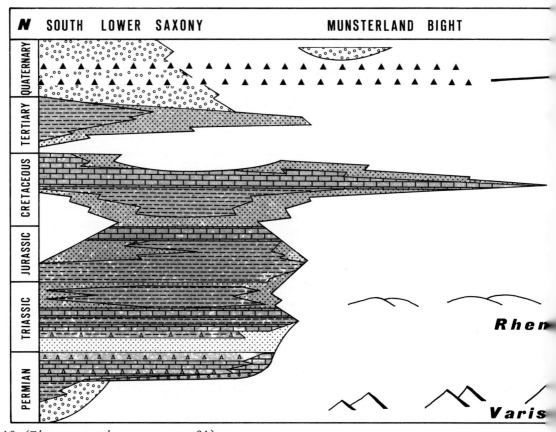

Fig. 13 (Figure caption see page 31)

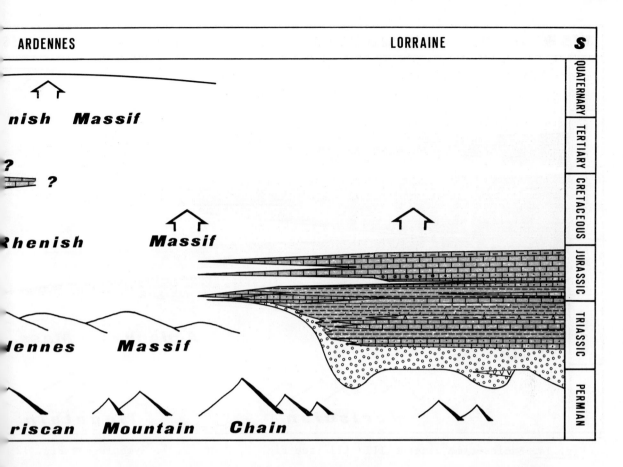
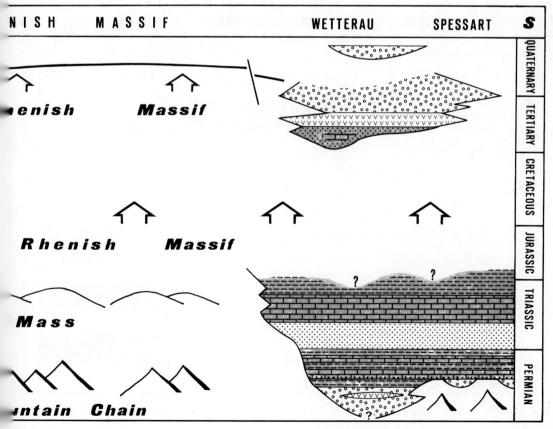

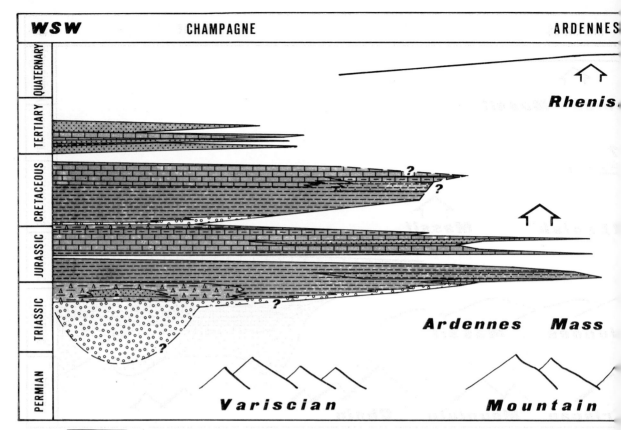

Fig. 14. Facies-time diagram III - III (see Fig. 1, Sect. 2.2). Symbols see Fig. 12

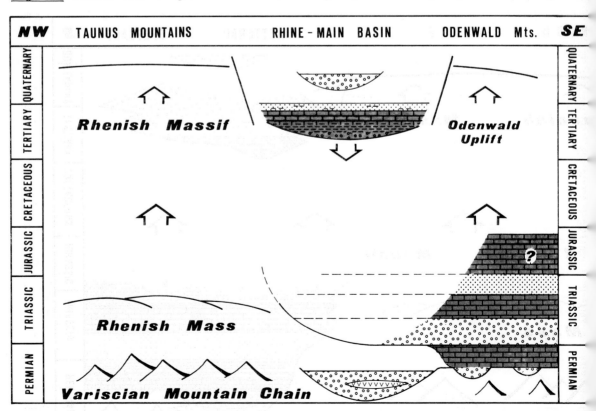

Fig. 16. Facies-time diagram IVa (see Fig. 1, Sect. 2.2). Symbols see Fig. 12

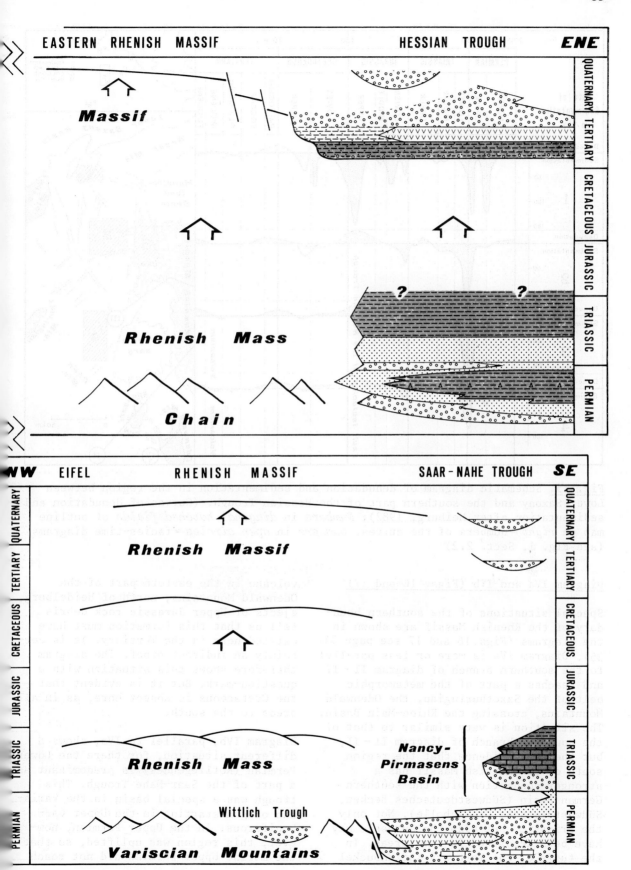

Fig. 17. Facies-time diagram IVb (see Fig. 1, Sect. 2.2). Symbols see Fig. 12

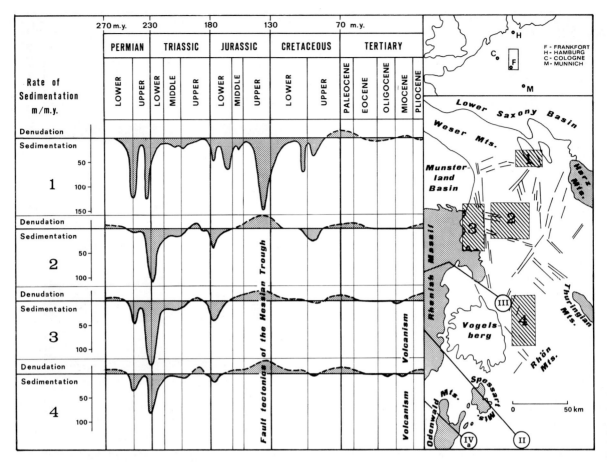

Fig. 15. Schematic diagram of denudation and sedimentation in the region between Lower Saxony and the southern part of the Hessian Trough. (Curves of denudation and sedimentation after Meiburg, 1982). Numbers in *diagonal hatched fields* of outline map on *right* = numbers of the curves. Numbers in *open circles* = facies-time diagrams (see Fig. 1, Sect. 2.2)

Diagram IVa and IVb (Figs. 16 and 17)

Special situations of the southern boundary of the Rhenish Massif are shown in the diagrams (Figs. 16 and 17 see page 34/35). Diagram IVa is more or less parallel to the southern branch of diagram II - II and reaches a part of the metamorphic belt of the Saxothuringian, the Odenwald Mountains, crossing the Rhine-Main Basin. The situation is very similar to that of the southern branch of diagram II-II, but there is evidence that the region south of the Rhenish Massif has a stronger connection with the southern German Basin (Südwestdeutsches Becken, Südwestdeutsche Großscholle). Not only the Triassic, but also the Jurassic may have been deposited in some areas. In the tuffs of the Tertiary Katzenbuckel volcano in the eastern part of the Odenwald Mountains, north of Heidelberg, ejects of Upper Jurassic rock debris tell us that this formation must have existed here in the Tertiary. It is certainly an indirect proof. The diagram therefore shows this situation with a question-mark. But it is evident that the Cretaceous is absent here, as in all areas to the south.

Diagram IVb, parallel to IVa, shows a different situation, for there the Lower Permian (Rotliegendes) is predominant as a part of the Saar-Nahe Trough. This trough was a special basin in the Variscan orogen starting in the Upper Carboniferous. In the Upper Permian, however, this region was uplifted, so that the marine Upper Permian did not reach the area. It is unclear where the northern boundary of the Triassic from the Nancy-Pirmasens Basin and the southern Palatine lies. This Triassic basin is located south of the Saar-Nahe Trough. In our diagram Jurassic and Cretaceous are absent, marine Tertiary also. The coast line of the marine, brackish or lacustrine Tertiary lies farther east, at the western boundary of the Mayence Basin (Mainzer Becken). It seems that the region intersected by our diagram was a denudation area in Tertiary times.

In this brief summary of the geological history of the margins of the Rhenish Massif over the long time span from Permian to Tertiary, its complexity is evident as a result of interactions between the Rhenish Massif and the surrounding areas and also of interactions between the different parts of the Massif. The following chapters present the younger history of the Rhenish Massif itself and the causes of its vertical movements.

Acknowledgements. Colleagues of the authors kindly provided information and data relating to the maps and profiles. For critical discussions we wish to extend our thanks to Prof. Jacoby (Frankfurt), and for the graphical representation of the figures to Mr. Laschet (Aachen).

References (Sect. 2.2)

Ahorner, L. and Murawski, H., 1975. Erdbebentätigkeit und geologischer Werdegang der Hunsrück-Südrand-Störung. Z. Dtsch. Geol. Ges., 126:63-82.

Ahrendt, H., Hunsiker, J.C., and Weber, K., 1978. K/Ar-Altersbestimmungen an schwach metamorphen Gesteinen des Rheinischen Schiefergebirges. Z. Dtsch. Geol. Ges., 129:229-247.

Behr, H.J., 1978. Subfluenz-Prozesse im Grundgebirgs-Stockwerk Mitteleuropas. Z. Dtsch. Geol. Ges., 129:283-318.

Bless, M.J.M., Bouckaert, J., and Paproth, E., 1980. Paleogeography of Upper Westphalian deposits in NW Europe with reference to the Westphalian C North of the mobil Variscan belt. Meded. Rijks Geol. Dienst, N.S., 28:101-147.

Breddin, H., 1973. Tiefentektonik und Deckenbau im Massiv von Stavelot-Venn (Ardennen und Rheinisches Schiefergebirge). Geol. Mitt., 12:81-130.

Burett, C.J., 1972. Plate tectonics and the Hercynian Orogeny. Nature (London), 239:155-157.

Cloos, H., 1939. Hebung - Spaltung - Vulkanismus. Geol. Rundsch., 30:401-525.

Engel, W., Franke, W., and Langenstrassen, F., 1982/1983. Palaeozoic Sedimentation in the Northern Branch of the Mid-European Variscides - Essay of an Interpretation. In: Martin, H., Eder, F.W. (eds.) Intracontinental fold belts in case studies in the Variscan Belt of Europe and the Damara Belt in Namibia (in press).

Franke, W., Eder, W., and Langenstrassen, F., 1978. Main aspects of geosynclinal sedimentation in the Rhenohercynian zone. Z. Dtsch. Geol. Ges., 129: 201-216.

Kossmat, F., 1927. Gliederung des varistischen Gebirgsbaues. Abh. Sächs. Geol. Landesamt, 1, 39 p.

Kramm, U., 1982. Die Metamorphose des Venn-Stavelot-Massivs, nordwestliches Rheinisches Schiefergebirge: Grad, Alter und Ursache. Decheniana, 135: 121-178.

Krebs, W., and Wachendorf, H., 1973. Proterozoic-Palaeozoic geosynclinal and orogenic evolution of Central Europe. Geol. Soc. Am. Bull., 84: 2611-2630.

Matthews, S.C., 1978. Caledonian connexions of Variscan tectonism. Z. Dtsch. Geol. Ges., 129:423-428.

Meissner, R., Bartelsen, H., and Murawski, H., 1981. Thin-skinned tectonics in the northern Rhenish Massif, Germany. Nature (London), 290:399-401.

Meissner, R. and Vetter, U., 1974. The northern end of the Rhinegraben due to some geophysical measurements. In: Illies, H.J., Fuchs, K. (ed.). Approaches to Taphrogenesis:236-243.

Meyer, W. and Stets, J., 1975. Das Rheinprofil zwischen Bonn und Bingen. Z. Dtsch. Geol. Ges., 126:15-29.

Meyer, W. and Stets, J., 1980. Zur Paläogeographie von Unter- und Mitteldevon im westlichen und zentralen Rheinischen Schiefergebirge. Z. Dtsch. Geol. Ges., 131:725-751.

Meyer, W., Stoltidis, I., and Winter, J., 1977. Geologische Exkursion in den Raum Weyer - Schuld - Heyroth - Niederehe - Üxheim - Ahütte. Decheniana, 130:322-334.

Plein, E., Dörholt, W., and Greiner, G., 1982. Das Krefelder Gewölbe in der Niederrheinischen Bucht, Teil einer großen Horizontalverschiebungszone? In: Reiche, E. (ed.) Krefelder und Lippstädter Gewölbe. Fortschr. Geol. Rheinland Westfalen, 30:15-30.

Quade, H., Nyk, R., and Walde, R., 1981. Überschiebungstektonik in der Eisenerzlagerstätte Fortuna bei Berghausen/Dill (Rheinisches Schiefergebirge). Z. Dtsch. Geol. Ges., 132:29-41.

Schunck, K., 1979. Der Kreuzungsbereich Eifeler Nord-Süd-Zone und Saar-Nahe-Senke. Luftbildgeologische Analyse eines Schollenmosaiks. Diss. Frankfurt/Main (Eigenverlag).

Walter, R., 1980. Lower Paleozoic Paleography of the Brabant Massif and its Southern Adjoining Areas. Meded. Rijks Geol. Dienst, 32-2:14-25.

Weber, K., 1978. Das Bewegungsbild im Rhenoherzynikum. Abbild einer varistischen Subfluenz. Z. Dtsch. Geol. Ges., 129:249-281.

Weber, K., 1981. The structure development of the Rheinisches Schiefergebirge. In: Zwart, H.J., Dornsiepen, U.F. (eds.) The Variscan orogene in Europe. - Geol. Mijnbouw, 60:149-159.

Winter, J., 1965. Das Givetium der Gerolsteiner Mulde (Eifel). Fortschr. Geol. Rheinland Westfalen, 9:277-322.

Winter, J., 1981. Exakte tephrostratigraphische Korrelation mit morphologisch differenzierten Zirkonpopulationen (Grenzbereich Unter/Mitteldevon, Eifel-Ardennen). Neues Jahrb. Geol. Paläontol. Abh., 162:97-136.

References (Sect. 2.3)

Ahorner, L., 1983. Historical Seismicity and present-day micro-earthquake activity of the Rhenish Massif, Southern Europe (this vol.).

Albers, H.J. and Felder, W.M., 1979. Litho-, Biostratigraphie und Palökologie der Oberkreide und des Alttertiärs (präobersanton - Dan/Paläozän) von Aachen-Südlimburg. Aspekte der Kreide Europas. IUGS (A), 6:47-84.

Albers, H.J. and Felder, W.M., 1981. Feuersteingerölle im Oligomiozän der Niederrheinischen Bucht als Ergebnis mariner Abrasion und Carbonatlösungsphasen auf der Kreide-Tafel von Aachen-Südlimburg. Fortschr. Geol. Rheinland Westfalen. 29:469-488.

Cloos, H., 1939. Hebung - Spaltung - Vulkanismus. Geol. Rundsch., 30:401-525.

Kowalczyk, G., 1982. Das Rotliegende zwischen Taunus und Spessart. Geol. Abh. Hessen 84 (in press).

Kowalczyk, G., Murawski, H., and Prüfert, A., 1978. Die paläogeographische und strukturelle Entwicklung im Südteil der Hessischen Senke und ihrer Randgebiete seit dem Perm. Jahresber. Mitt. Oberrhein. Geol. Verein, N.F., 60:181-205.

Martini, E., 1981. Sciaenides (Pisces) aus dem Basisbereich der Hydrobien-Schichten des Oberrheingrabens, des Mainzer Beckens und des Hanauer Beckens (Miozän). Senckenbergiana Lethaea, 62:93-123.

Mégnien, C., 1980. Synthèse géologique du Bassin de Paris. Mem. BRGM, Paris, 101 (vol. 1) and 102 (vol. 2).

Meiburg, P., 1982. Saxonische Tektonik und Schollenkinematik am Ostrand des Rheinischen Massivs. Geotekton. Forsch., 62:(I-II), 267 p.

Murawski, H., 1960. Das Zeitproblem bei der Tektogenese eines Großgrabensystems. Ein taphrogenetischer Vergleich zwischen Hessischer Senke und Oberrheingraben. Notizbl. Hess. Landesamt Bodenforsch., 88:294-342.

Murawski, H., 1964. Die Nord-Süd-Zone der Eifel und ihre nördliche Fortsetzung. Publ. Serv. Géol. Luxembourg, 14:285-308.

Schunck, K., 1979. Der Kreuzungsbereich Eifeler Nord-Süd-Zone und Saar-Nahe-Senke. Diss. Univ. Frankfurt/Main (Eigenverlag).

Voisin, L., 1981. Analyse geomorphologique d'une région type: l'Ardenne occidentale. Serv. Repr. Thèses Univ. Lille III, t. II:499-883.

3 Pre-Quaternary Uplift in the Central Part of the Rhenish Massif

W. Meyer[1], H. J. Albers[2], H. P. Berners[3], K. v. Gehlen[4], D. Glatthaar[5], W. Löhnertz[6], K. H. Pfeffer[7], A. Schnütgen[7], K. Wienecke[1], and H. Zakosek[8]

Abstract

During the Mesozoic, the Rhenish Massif was not covered any more by the sea since Lower Jurassic times, except for the northern and northwestern margins which were affected by short ingressions during the Upper Cretaceous. The first uplifting movements took place at the end of the Cretaceous and in the early Tertiary in the northwestern part of the massif. But in spite of these movements the surface of the entire Massif did not rise much above sea level during the Lower Tertiary. A short marine ingression covered large areas of the western Rhenish Massif in the Middle/Upper Oligocene. The main uplift started at the end of the Oligocene and continued up to the present time. Towards the end of the Miocene the uplift accelerated. The Rhine valley exists since the Middle Miocene.

Introduction

Tertiary sediments as indicators of vertical movements of the Rhenish Massif are almost only preserved near its margins. Therefore, relict etchplains (Rumpfflächen) had to be used as reference levels in the central parts of the Massif, in addition to small patches of Tertiary sediments.

The term Rumpfflächen is used here for planation surfaces with hills and tors developed under near-tropical Mesozoic-Tertiary weathering conditions. The plains cut across the underlying rocks almost independent of the geological structure (Büdel, 1982).

There are often several levels of stepped etchplains (Rumpftreppen). They are connected by gently inclined ramp-like slopes with slopes of more than 2°, or

---

1 Geologisches Institut, Universität Bonn, Nußallee 8, D-5300 Bonn 1, Fed. Rep. of Germany

2 Landesanstalt für Ökologie, Landschaftsentwicklung und Forstplanung, Castroper Str. 312-314, D-4350 Recklinghausen, Fed. Rep. of Germany

3 Institut für Geologie und Paläontologie, RWTH Aachen, Wüllnerstr. 2, D-5100 Aachen, Fed. Rep. of Germany

4 Institut für Geochemie, Petrologie und Lagerstättenkunde, Universität Frankfurt, D-6000 Frankfurt 11, Fed. Rep. of Germany

5 Geographisches Institut, Ruhr-Universität Bochum, D-4630 Bochum 1, Fed. Rep. of Germany

6 Ripsdorf-Johannesweg 3, D-5378 Blankenheim, Fed. Rep. of Germany

7 Geographisches Institut, Universität Köln, Albertus-Magnus-Platz, D-5000 Köln 41, Fed. Rep. of Germany

8 Institut für Bodenkunde, Universität Bonn, Nußallee 13, D-5300 Bonn, Fed. Rep. of Germany

in distinct steps (etchplain escarpments), with the formation of triangular reentrants, etchplain strips and intramontane plains within higher rumpfflächen, and the occurrence of inselbergs (Bremer and Pfeffer, 1978).

The landforms of the Central European Mittelgebirge are characterized by Pleistocene valleys and interfluves. The tops of the interfluves are surfaces with dimensions of more than 75×75 m and less than $2°$ inclination. The uniform level of these extended plains on the tops of the interfluves, only surmounted by single round hills, as well as a partial cover of decomposition residues permits this interpretation.

Pre-Tertiary Development

In the Eifel region, Variscan tectonic processes, together with the differing resistance of the exposed rocks, have been essential in the formation of post-Paleozoic relief and paleogeography. However, the amount of tectonically influenced relief was much more important than local petrographical inhomogeneities.

Mesozoic history starts with the working-out of a N-S trending Variscan fold axis depression zone as a channel (with some islands above Devonian reef complexes). The thickness of the Middle Buntsandstein sediments ingressing from the south may reach more than 200 m in regions of syn-sedimentary subsidence in the north (Mechernich Depression) and south (northern Bitburg Depression). During the Upper Buntsandstein, the sediments progressively overlapped former margins.

With the marine character increasing towards the Upper Triassic, the western shoreline moved towards the Ardennian continent (Lucius, 1948), and the eastern parts of the Rhenish Massif sank below sea level. In the section Bitburg-Mechernich, syn-sedimentary Mesozoic subsidence reached 250 m, with Liassic sediments as the last stratigraphic members.

Between the Liassic and Santonian, the Rhenish Massif then rose above sea level but not more than 200 m. The consequence was an erosion of large parts of the Triassic sedimentary cover before the Upper Cretaceous or at least pre-Eocene (Neuwied, Antweilergraben) or pre-Oligocene (Arensberg volcano). The present isobases of the Buntsandstein in the Eifel Depression now show a maximum difference of 800 m in the section Bitburg-Mechernich without major faults, which was mostly acquired during this time span.

In the northwestern part of the massif the flat pre-Santonian surface has an inclination of not more than 4% (about $2°$) to the NW due to post-Cretaceous displacements without any fracture. Here all kinds of clastic rocks (Paleozoic sandstones, shales, etc.) are equally cut by the truncation surfaces, with the partial exception of Cambrian quartzites of greater thickness which locally rise above the surfaces. Paleozoic calcareous rocks are widely karstified. Several phases of karstification can be distinguished since Cretaceous times in this part of the massif (Albers and Felder, 1981).

In the northwestern area paleosol sediments are the oldest Cretaceous sediments (Upper Santonian or older, Albers and Felder, 1979, Murawski et al., 1983) which were formed by the erosion of weathered zones in central parts of the massif.

Shallow fluviatile channels were cut into the etchplains. Later these terrestrial sediments were partly eroded and the etchplains were slightly flattened by marine abrasion (Lower Campanian and younger).

First uplifts can be recorded for the Lower Campanian. Very important uplifting (160 m) took place during Upper Maastrichtian time. Then the central and northwestern parts of the massif lost their pre-Maastrichtian sedimentary cover and parts of the weathered zone. Here, wide areas of the etchplains were once more subjected to marine abrasion. The Upper Maastrichtian sea reached central areas of at least the western part of the massif (Albers and Felder, 1979; Altmeyer, 1982).

The Formation of Grey Loams

In the geological and pedological literature on the Rhenish Massif, grey loams (Graulehme, paleo-plastosols, fossil plastosols) are mainly interpreted as pre-Quaternary warm-age weathering products. They have been used as stratigraphic markers (mostly assumed to represent pre-Oligocene age). However, grey loams on Lower Devonian non-carbonate sediments could also be products of rock decomposition by hydrothermal ascending solutions, even in connection with young volcanism. Decomposition of rocks has taken place in the Rhenish Massif since Jurassic times.

In an attempt to solve this important problem, five locations in the Eifel (Daun, Satzvey and Bonn) where hydrothermal decomposition seems to have taken place, were sampled and analyzed chemically and mineralogically. The zones of decomposition are from a few cm up to several m wide and typically developed along strike and dip of steeply inclined sediments. They reach down to more than 20 m below the surface. The zones of decomposition do not follow all fissures (as would be expected for weathering from above) but are restricted to closely delimited areas.

The zones of decomposition are characterized by clayey, plastic material which was bleached by loss of Fe and Mn oxides. The oxides are often accumulated on both sides of the bleached zones in less or not decomposed rocks.

Clay minerals are illites (80%-100%) and kaolinite in proportions identical in the decomposition products and the country rock. Often secondary hematite occurs within the zones of decomposition.

In pre-Quaternary soils, however, a transformation to boluslike, kaolinitic warm-age loams seems to have taken place; these have lost cations (particularly Fe, Mn, Ca, Mg, K).

As both possibilities of grey loam formation seem to have been active, interpretation as fossil soils and stratigraphic markers should only be made after special investigation.

Uplift during Paleogene Times

The southern Ardennes region (Belgium, France) is an area of sedimentation during the Eocene. It covers the Rocroi, Croix Scaille, Haute Lesse and St. Hubert massifs, as well as parts of the southern margin of the Dinant basin (Voisin, 1981). During the Oligocene, the area of sedimentation widened northward and then covered the Dinant basin and the region of the Condroz and Stavelot massifs. In the Upper Miocene (post Bolderien), strong uplift started at the southern margin of the Ardennes and moved from here to the east as well as to the north.

In the northwestern part of the Rhenish Massif, no evidence is known for a regression between the uppermost Maastrichtian and the Paleocene. A hiatus in the lowermost Paleocene is probably caused by marine erosion. The sea seems to have covered at least the northwestern parts of the Massif until the uppermost Lower or lowermost Middle Paleocene, maybe also more central parts. Because of lack of sediments in the Rhenish Massif from the Upper Paleocene to the Lower Eocene, no conclusions can be drawn regarding climatic conditions for this time.

In the Stavelot-Massif pre-Oligocene sediments and paleosols are completely restricted to the northwestern margin. In the central parts they have been eroded before the Oligocene or by the Oligocene transgression. In this region etchplains were finally remodeled by Oligocene submarine erosion. Remains of dissolution of Cretaceous carbonates (Feuerstein-Eluvium, argile à silex) are the products of Mio-Pliocene and Pleistocene karstification, which began with the Miocene regression (Albers and Felder, 1981).

Several levels of etchplains can be reconstructed in a northwest-southeast trending zone from the Hohes Venn to the Mosel. North of the Mosel the lowest etchplain has an elevation of 380-400 m. 30 km further north the dominating etchplain level is 520-540 m. North of a line Prüm-Jünkerath, etchplains of about 580-610 m characterize the landscape. Only ridges of quartzite up to 690 m tower above the etchplains. The

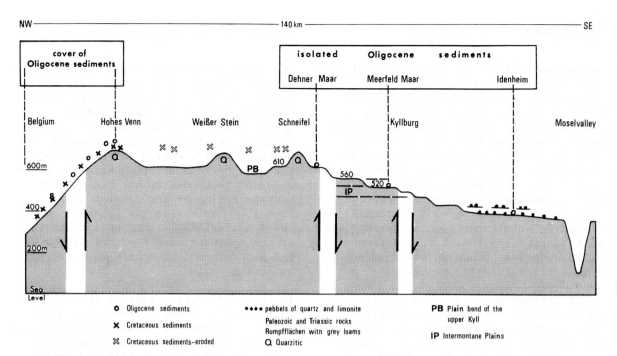

Fig. 1. Morphological section through the northwestern Rhenish Massif (Pfeffer, unpublished)

etchplain which borders the Mosel shows a gentle inclination towards the Mosel valley. Pebbles of quartz and limonite and Pliocene and Quaternary loams are found on top of it. A pre-Oligocene weathering zone and Eocene sediments can be found near Binsfeld. Exposures show lower zones of deep, clayey weathering to grey loams. The transition between the etchplains occurs ramp-like or in steps with intermediate levels (etchplain escarpments). Intramontane plains are developed within the higher relief where the bedrock is limestone, pelitic rock or sandstone. Geomorphology suggests a post-Triassic to Oligocene development of several levels (stepped etchplain) with differences in relief of more than 300 m. But traces of Oligocene marine sands on the Hohes Venn (690 m) and on several etchplain levels require a subdued relief of a levelled landscape still at the time of the high Oligocene sea level. A flooded relief with great differences in altitude ought to have been covered by thicker sediments. The decomposition residues on the etchplain levels, however, require a pre-Oligocene dating. The constant level of the individual etchplains demands for parts of them en-bloc uplifts after the Oligocene with a vertical displacement of 80-100 m in each case from the Mosel to the Hohes Venn (Fig. 1). The borders between the blocks are assumed in the area of Kyllburg and along the line Prüm-Jünkerath.

In consideration of all geomorphologic and geological facts, in which constant levels, the decomposition residues and traces of marine Oligocene have key positions, the steps can be recognized as having been caused by tectonics.

In the *southwestern and northern Eifel* and the Neuwied Basin and its surroundings, deposits of a river system are widespread (Vallendar beds). They have an age of Upper Eocene to Lower Oligocene (Löhnertz, 1978). Kurtz (1938) constructed a Vallendar River System crossing the western Eifel region. New results of mapping and sedimentary petrology (heavy mineral analysis, determination of smoky quartz and roundness) are not in agreement with this hypothesis. The main stream gravels of the northern Eifel (with andalusite, staurolite and kyanite, presence of smoky quartz) differ distinctly from those of the southern Eifel (absence of these components). Angular gravels of tributaries can be distinguished from the sediments of the main stream (rounded particles), which came from the Vosges area and ran parallel to the recent Mosel valley to the northeast. This means that the main divide of the Eifel already existed in Vallendar time.

In the Neuwied Basin, first subsidence took place at the beginning of the Upper Eocene. Sandy clays with lignite seams were deposited in small basins in the region of Neuwied and the southern Westerwald up to the Limburg basin. Thus an E-W to NE-SW trending depression as part of the Bitburg-Kassel depression (Pflug, 1959) can be constructed. At the end of the Eocene, subsidence increased and during the uppermost Eocene and Lower Oligocene, clays of nearly 100 m thickness, sands and conglomerates were deposited. The clays are sediments of tectonic basins, the conglomerates of the Vallendar river system.

North of the Neuwied basin clays and sands were deposited during Middle or Upper Oligocene time. They are overlain by trachyte tuffs, which erupted in the early Miocene (Todt and Lippolt, 1980).

In the Neuwied Basin, near some of the West Eifel maar volcanoes and in the southwestern Eifel, sediments or residues of sediments with brackish to marine Oligocene faunas have been found (Kadolsky, 1975; Kadolsky et al., unpublished; Weiler, unpublished). This demonstrates that at least large areas of the western Rhenish Massif were below sea level during the Rupelian or Chattian. Until the Chattian an etchplain landscape with only slight differences in height (up to 40 m) extended between the Siebengebirge and the northern part of the Westerwald. It was covered by in situ weathering products (grey loams). During the Rupelian lignites were deposited, and fluvial sediments during the Chattian. They formed an alluvial plain linked to the coastline in the Lower Rhine Embayment. This plain was only a few meters above sea level. In the northern part of the Westerwald no Upper Eocene and Lower Oligocene sediments have been found, although they are known from the Neuwied Basin and the Westerwald clay area near Montabaur. A transgression of the Oligocene sea can be excluded for the Westerwald (Fig. 2).

At the end of the Chattian intensive volcanic activity started in the Westerwald area. Basalts covered the older sediments and weathering products and protected them from erosion (Weitefeld; Berod near Montabaur). The volcanism is connected with tectonic activity which broke the Westerwald into a number of fault blocks, with only small vertical displacement.

Uplift During Neogene Times

During the Miocene, clays with lignite seams were deposited in small grabenlike depressions in the northern part of the Rhenish Massif (Antweiler, Bonn, Adendorf, Ringen), while systems of plains developed in the higher parts (Quitzow, 1978, 1982). Today the western Westerwald reaches altitudes of 550 m, the High Westerwald altitudes above 600 m. Between them a basin area extends from Limburg to Siegburg with an appendix towards Hachenburg. The basin area is further subdivided by NNW-SSE and WSW-ENE fault lines, which were active mainly from the Upper Miocene to the Pleistocene. The amount of displacement reaches more than 100 m. About half the movement took place in the early Pleistocene.

A brackish or marine connection between the Lower Rhine Basin and the Rhine Graben seems to have been open for a short time during the Lower Miocene (Hemmoorian) (Fig. 11 in Murawski et al., 1983).

In broad zones on both sides of the valleys of Mosel and Rhine, gravels rich in quartz pebbles are found. They contain staurolite as the typical heavy mineral and are called Kieseloolith beds, because of their content of silified oolithic Jurassic limestones (Schnütgen and Spaeth, 1978). Their exact age is unknown (Miocene and Pliocene).

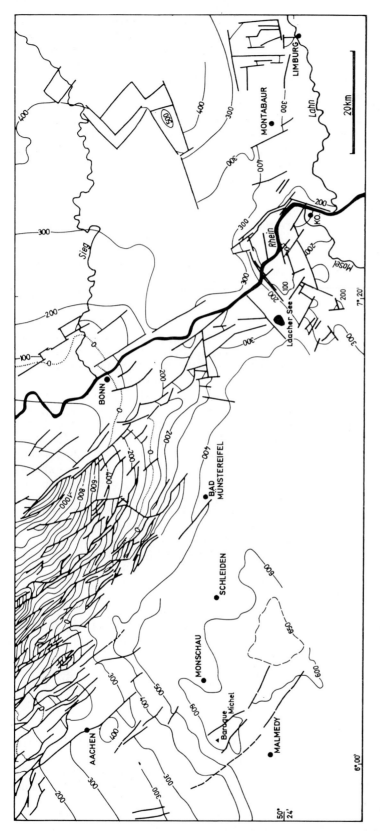

Fig. 2. Contours of the base of the Tertiary cover (Eocene or Oligocene) in the northwestern and central Rhenish Massif (Albers, unpublished; Hager in Knapp, 1978; Glatthaar, unpublished; Meyer, unpublished)

Development of the River Systems

In the central part of the Rhenish Massif, the development of the rivers is dominantly influenced by the Oligocene transgression and by tectonic movements during the taphrogenesis of the Lower Rhine Basin. For the Eocene and Lower Oligocene, the Vallendar river is known as a river which flowed from SW to the Neuwied Basin parallel to the present Mosel. It is not precisely known where its course ended. Part of the Vallendar drainage basin, which today lies west of the Rhine, was covered by the Oligocene transgression. East of the Rhine an alluvial coastal plain was formed. On this sedimentary surface a new river system developed: this was the origin of the Rhine valley (Quitzow, 1974; Boenigk, 1981). Towards the end of the Miocene the strong uplift of the Rhenish Massif began and consequently the river systems eroded and were fixed in their positions. The Tertiary Mosel and Rhine systems are characterized by the Mio-Pliocene Kieseloolith beds.

Pre-Quaternary Tectonic Movements of the Rhenish Massif in Relation to Eustatic Changes of Sea Level

During the Maastrichtian (Upper Cretaceous), the sea level was more than 300 m higher than the recent one (Pitman in Degens and Kempe, 1979). The western part of the Massif (Aachen, Hohes Venn) was covered by the sea, the eastern part was a land massif covered by etchplains with elevations only some meters above sea level. During the Upper Maastrichtian, the northwestern part was involved in active uplifting of more than 160 m, which did not affect most other regions. This paleographic configuration continued. During the Upper Eocene, the sea level descended to about +120 m (Vail and Hardenbol, 1979), and the northwestern part of the massif was about 200-300 m above the sea. Here karstification has been found to extend 120 m below the top of Maastrichtian and Paleocene carbonates. In the southwestern Eifel, the Vallendar river system was cut 80 m deep into the landscape.

The NE-SW trending Bitburg-Kassel depression (Pflug, 1959) subsided in the Upper Eocene; sediments are preserved in the southwestern Eifel, Neuwied Basin, southern Westerwald and Limburg Basin.

During the Rupelian (Middle Oligocene), the sea level rose to about +250 m and probably the whole Eifel region was flooded. The eastern part of the Rhenish Massif was a coastal alluvial plain not much above sea level. No facts are known which show an active uplift of the massif. Since the Rupelian, the subsidence of the Lower Rhine Basin took place. No morphologic evidence is known for the quick and extreme drop of the sea level to -120 m at the Chattian/Aquitanian (Oligocene/Miocene) boundary except for the hiatus in the sedimentary sequence (Hager, 1981).

After a new maximum during the Reinbeck (Middle Miocene) (+120 m), the sea level dropped again discontinuously down to its present level. Erosion with many phases took place developing a sequence of morphologic elements (intramontane plains, fluviatile terraces etc.). The present altitude of these geomorphologic elements, however, clearly exceeds the amount of eustatic processes, demonstrating active uplift of the Rhenish Massif up to the Recent. The post-Oligocene uplift of the Hohes Venn is at most 460 m, in the Siebengebirge area about 0 m, in the northeastern Westerwald 200 m. The maximum subsidence of the central part of the Neuwied Basin during the Tertiary and Quaternary is about 350 m.

Acknowledgements. We are indebted to Dr. Busche, Würzburg, for checking some English expressions.

References

Albers, H.J. and Felder, W.M., 1979. Litho-, Biostratigraphie and Palökologie der Oberkreide und des Alttertiärs (Präobersanton-Dan/Paläozän) von Aachen-Südlimburg (Niederlande, Deutschland, Belgien). Aspekte der Kreide Europas. IUGS, A 46:47-83

Albers, H.J. and Felder, W.M., 1981. Feuersteingerölle im Oligomiozän der

Niederrheinischen Bucht als Ergebnis mariner Abrasion und Carbonatlösungsphasen auf der Kreide-Tafel von Aachen-Südlimburg. Fortschr. Geol. Rheinland Westfalen, 29:469-482.

Altmeyer, H., 1982. Feuersteinfunde in der südlichen und östlichen Eifel. Aufschluss, 33:241-244.

Boenigk, W., 1981. Die Gliederung der tertiären Braunkohlendeckschichten in der Ville (Niederrheinische Bucht). Fortschr. Geol. Rheinland Westfalen, 29:193-263.

Bremer, H. and Pfeffer, K.-H., 1978. Zur Landschaftsentwicklung der Eifel. Köln. Geogr. Arb., 36:225.

Büdel, J., 1982. Climatic Geomorphology. Princeton University Press, Princeton N.J., 443 pp.

Degens, E.T. and Kempe, ST., 1979. Heizen wir unsere Erde auf? - Bild Wiss., 16(8):38-59.

Hager, H., 1981. Der Tertiär des Rheinischen Braunkohlereviers, Ergebnisse und Probleme. Fortschr. Geol. Rheinland Westfalen, 29:529-563.

Kadolsky, D., 1975. Zur Paläontologie und Biostratigraphie des Tertiärs im Neuwieder Becken. Decheniana, 128:113-137.

Knapp, G., 1978. Erläuterungen zur Geologischen Karte der nördlichen Eifel 1:100000. 2. Aufl., Krefeld, 152 pp.

Kurtz, E., 1938. Herkunft und Alter der Höhenkiese der Eifel. Z. Dtsch. Geol. Ges., 90:133-144.

Löhnertz, W., 1978. Zur Altersstellung der tiefliegenden fluviatilen Tertiärablagerungen der SE-Eifel (Rheinisches Schiefergebirge). Neues Jahrb. Geol. Paläontol. Abh., 156:179-206.

Lucius, M., 1948. Das Gutland. Erl. Geol. Spezialkarte Luxemburgs, V, 405 pp.

Meyer, W., 1979. Influence of the Hercynian structures on Cainozoic movements in the Rhenish Massif. Allg. Vermessungsnachr., 86:375-377.

Mückenhausen, E., 1979. Die Paläoböden der Eifel in Abhängigkeit von der Geomorphologie. Z. Geomorphol. N.F., Suppl., 33:16-24.

Mückenhausen, E. and Schalich, J., 1982. Paläoböden der Eifel. In: Inventur der Paläoböden der Bundesrepublik Deutschland. Geol. Jahrb., F 11. (in press).

Murawski et al., 1983. Regional tectonic setting and geological structure of the Rhenish Massif (this vol.).

Pflug, H.D., 1959. Die Deformationsbilder im Tertiär des rheinisch-saxonischen Feldes. Freiberg. Forschungsh., C71:110.

Quitzow, H.W., 1974. Das Rheintal und seine Entstehung. Bestandsaufnahme und Versuch einer Synthese. Cent. Soc. Géol. Belg., 53-104.

Quitzow, H.W., 1978. Der Abfall der Eifel zur Niederrheinischen Bucht im Gebiet der unteren Ahr. Fortschr. Geol. Rheinland Westfalen, 28:9-50.

Quitzow, H.W., 1982. Die Hochflächenlandschaft der zentralen Eifel und der angrenzenden Teile des Neuwieder Beckens. Mainz. Geowiss. Mitt. (in press).

Schnütgen, A. and Späth, H., 1978. Der Vergleich von Altschottern der südlichen Eifel mit tropischen Schottern aus dem nordwestlichen Sri Lanka. Köln. Geogr. Arb., 36:155-186.

Todt, W. and Lippolt, H.J., 1980. K-Ar Age Determinations on Tertiary Volcanic Rocks: V. Siebengebirge, Siebengebirge-Graben. J. Geophys., 48:18-27.

Vail, P.R. and Hardenbol, J., 1979. Sea-level changes during the Tertiary. Oceanus 22:71-79.

Voisin, L., 1981. Analyse geomorphologique d'une région type: L'Ardenne occidentale. Serv. Repr. Theses Univ. Lille, 3, T II:499-884.

4 Plateau Uplift During Pleistocene Time

4.1 Plateau Uplift During Pleistocene Time – Preface

A. Semmel[1]

The reconstruction of the tectonic behaviour of a Central European upland region seems to be fairly easy, provided there are valleys traversing the area concerned. The development of these valleys in Central Europe was determined to a great extent by the Pleistocene periglacial climate. The variable morphological intensity of this climate led to the alternation of periods with prevailing erosion, and of periods with prevailing accumulation. The terraced valley slopes, which document the course of former valley floors, are the result of these alternations. By the correlation of terrace remnants, which belong to the same age, a longitudinal profile of a former valley floor can be constructed, which then allows statements on subsequent tectonic movements. In the Rhenish Massif, this method has long been applied. Up to the most recent past it had been generally accepted, with one exception (see Chap. 4.2), that in the course of the Quaternary the Rhenish Massif was not only uplifted relative to its surroundings, but also that distinct tectonic upwarping could be proven along individual valleys, for example, in the Rhine valley. These assumptions, however, were recently placed in doubt, as a result of which the valleys of the rivers Rhine, Lahn and Mosel were re-examined. Locally, a great number of drillings were involved. This occurred within the framework of the research program described in the following.

One result of this re-examination is that the major river terraces within the Rhenish Massif are not affected by an appreciable tectonic. The boundary zone between the Lower- and Middle Mosel valley is an exception. In contrast to Bibus (1983), Brunnacker and Boenigk (1983) came to the conclusion that a strong tectonic upwarping of the Rhine terraces can also be demonstrated north of the Neuwied basin. The analysis of upland areas located far from the major valleys provided indications of considerable Quaternary tectonic movements, as, for example, in the northern Eifel Mountains and in adjacent areas. As far as distinct dislocations of terraces relative to their southern and eastern forelands are recognizable, these are also affected by subsidence outside the Rhenish Massif, for example, in the Upper Rhine Graben. Along the south-western boundary of the Rhenish Massif there are indications of minor tectonics, and along its north-western boundary indications of very strong tectonic movements in the course of the Late Tertiary and the Quaternary.

When interpreting these results, it should be kept in mind that, within the framework of this research program, not all areas adjacent to the Rhenish Massif were re-examined.

References

Bibus, E., 1983. Distribution and dimension of young tectonics in the Neuwied Basin and the Lower Middle Rhine. This vol.
Brunnacker, K. and Boenigk, W., 1983. The Rhine valley between the Neuwied Basin and the Lower Rhenish Embayment. This vol.

---

[1] Institut für Physische Geographie, Universität Frankfurt, Senckenberganlage 36, D-6000 Frankfurt 11, Fed. Rep. of Germany

4.2 The Early Pleistocene Terraces of the Upper Middle Rhine and Its Southern Foreland – Questions Concerning Their Tectonic Interpretation

A. Semmel[1]

Abstract

By analysing the main terrace sequence (Hauptterrassen) of the upper Middle Rhine valley, the author tries to prove that, in the course of the Quaternary, no important vertical tectonic dislocations have affected this area. On the other hand, a vertical dislocation of approximately 70 m has taken place since the Pliocene in the area between the Rhenish Massif near Bingen and the eastern margin of the Rhine Hessian Plateau near Mainz. Even stronger Quaternary dislocations can be evidenced for the area between the Taunus mountains north of Wiesbaden and the mouth of the river Main, and, above all, for the area adjacent to the east, which lies on the western boundary of the Upper Rhine Graben.

For a long time the terraces of the Middle Rhine valley have been regarded as evidence of tectonic behaviour of the Rhenish Massif during the Quaternary. An especially striking example, frequently cited in subsequent publications, is the Rhine gravels southwest of Trechtingshausen (Fig. 1), which are situated at elevations of 280-290 m above s.l., and which were first described by Leppla (1904). Within these sediments, Steuer (1906) found alpine radiolarite cherts. Wagner (1930a) supposed that these gravels were sediments of the oldest main terrace of the Rhine, and assumed, because of their actual position, a tectonic uplift of 35 m relative to the southern foreland of the Rhenish Massif. Already before him, Oestreich (1909) had interpreted these Trechtingshausen gravels as strongly uplifted Rhine terrace sediments, an interpretation which was accepted by many subsequent scientists. Gallade (1926) was an exception, as he denied an upwarping of these terraces, because of the undisturbed, level course of some older erosional surfaces in the same area. Birkenhauer (1971) also rejected the hypothesis of upwarped terraces, and furthermore concluded that the present remnants of the main terraces of the river Rhine do not possess any longitudinal gradient (1971).

In order to resolve these contradictory statements, the Rhine terraces were mapped again. To make statements on the problem of tectonic dislocations, *definite correlations* of terrace remnants, which constitute a former valley floor, are absolutely necessary. In my opinion, such a procedure is only applicable for the group of the main terraces of the Rhine, because only with these does the number and the state of preservation of these forms allow an almost certain reconstruction of former valley floors.

Recent mapping showed that in many places along the upper Middle Rhine five to six individual terraces are situated above the narrow Rhine gorge (t_{R1}-t_{R6}). This group of main terraces sets in at 190 or 200 m above s.l., and ends below an ero-

---

[1] Institut für Physische Geographie, Universität Frankfurt, Senckenberganlage 36, D-6000 Frankfurt 11, Fed. Rep. of Germany

---

Plateau Uplift, ed. by K. Fuchs et al.
© Springer-Verlag Berlin Heidelberg 1983

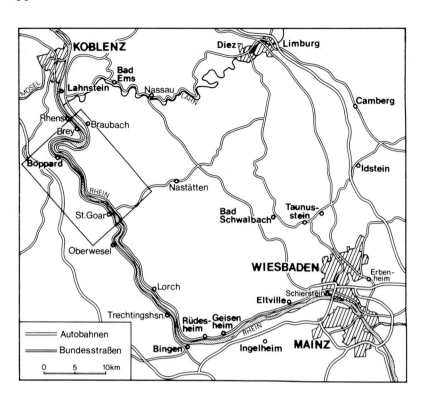

Abb. 1. Geographic sketch of the area

sional surface, which is situated at about 300 m above s.l. (cf. Fig. 2). The gravels on the latter surface no longer show typical Pleistocene properties: their petrographic content is characterized by the lack of slate, as well as by the absence of large sandstone blocks, which are explained as ice-drifted components. Their heavy mineral association is characterized by the prevalence of stable components (zircon, tourmaline), whereas the less stable components (epidote-zoisite, garnet, green hornblende) are present in quantities of less than 10%. On the other hand, the gravels below 300 m above s.l. contain, with decreasing age, increasing amounts of slate and more easily weathering heavy minerals. Palynologically dated findings from the Upper Rhine Graben repeatedly demonstrated that these differences occur around the Pliocene/Pleistocene boundary, e.g., Bartz (1976); Semmel (1972).

Some separate gravel accumulations well below 300 m above s.l., which because of their sediment properties, are distinctly of Tertiary age, can be found on the Kieselberg west of Rhens (TM 25, Sheet 5711 Boppard), and near Aulhausen (TM 25, Sheet 6013 Bingen). Their heavy mineral associations (prevalence of zircon over tourmaline, absence of staurolite) in my opinion indicate rather an Oligocene age of these sediments, which were most likely downfaulted before the Pleistocene, at least at the first of the above-mentioned sites. From the area of the upper Middle Rhine valley, distinctly Pliocene fluvial sediments 300 m below s.l. are not known to me. Hüser (1972) has a different opinion concerning the Aulhausen-gravels.

Bartz (1936) described gravels near Reitzenhain (TM 25, Sheet 5812 St. Goarshausen), which are situated higher than 300 m above s.l., but which look nevertheless more like Pleistocene terrace accumulations. A similar opinion was expressed by Birkenhauer (1971), who accepted these gravels between 325 and 330 m above s.l. as evidence of an Early Pleistocene valley aggradation. According to him, the accumulation of these sediments reached topographically higher than the Late Pliocene valley floor. Hüser (1972) avoided a definite, stratigraphical classification of these gravels. As far as I can see (Semmel, 1977), the Pleistocene age of these sediments cannot be proven with absolute certainty.

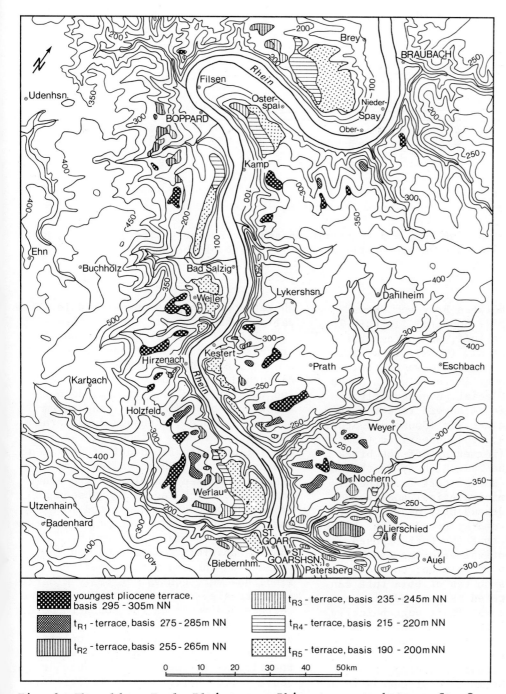

Fig. 2. The oldest Early Pleistocene Rhine terraces between St. Goar and Boppard

In the Upper Middle Rhine valley, there exist only few exposures, where main terrace sediments are dislocated by minor faults. As far as the age of these faults can be determined, the vertical dislocations occurred already more than 250,000 years ago. Horizontal movements still occurred in the course of the later Pleistocene.

In general, our present studies demonstrate that *there is no certain evidence for* considerable tectonic *dislocations* of the main terraces *of the upper Middle Rhine valley* (Fig. 3).

Within this context, two exceptions are especially worth mentioning: firstly, the situation of the so-called Upper

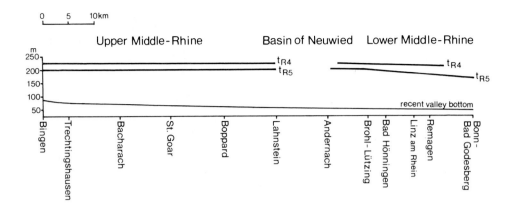

Fig. 3. Altitude of the t_{R4}- and t_{R5}-terraces and the Recent valley bottom in the Middle Rhine Valley (schematic profile). Section north of the Neuwied Basin after Bibus, 1983

Lower Terrace between Brey and Rhens (TM 25, Sheet 5711 Boppard). In this locality, a pumice tuff, which, according to Brunnacker et al. (1979), is at least of Rissian age, plunges down to the level of the Upper Lower Terrace, which dates from the last Glacial. Consequently, this would indicate a more recent subsidence. Other scientists (Sonne and Stöhr, 1959; Schönhals, 1961) described from this locality (former pit of the brickyard west of the Fed. Highway 9) the so-called Laacher pumice tuff from the Alleröd. Recent mineralogical analyses, carried out independently by Prof. Dr. Frechen, Bonn, and by Dr. Juvigné, Liège, both confirmed that the tuff exposed at present is identical with the Laacher pumice tuff. In my opinion, the covering sediment layers on top of this tuff can also be interpreted as Late Würmian or Holocene sediments. The topographically low position of this tuff could thus be explained without tectonic subsidence.

Secondly, the situation southwest of Trechtingshausen must be discussed in detail. There, two gravel accumulations with bases at about 260 m and 280 m above s.l., which have until now been interpreted as strongly uplifted terrace remnants, can be distinguished from younger main terraces, especially by reason of their heavy mineral content (larger quantities of stable minerals, and of epidote zoisite). Unfortunately, and to all appearances, similarly thick sediments of equivalent age, and in a comparable elevated position are nowhere known along the upper Middle Rhine valley, which favours the hypothesis of an uplift of these Trechtingshausen gravels. Their sediment properties, however, and the already-mentioned, undisturbed level course of both older and younger fluvial terraces, do not allow such a conclusion. These accompanying terraces, although in general mere rock terraces without gravel accumulations, are frequently found along the upper Middle Rhine valley. The Trechtingshausen gravels, on the other hand, were, with regard to subsequent erosional processes, deposited in a sheltered position on top of a quartzite ridge. This special "petrographic" situation possibly also exerted an influence on the sediment properties. We find, however, equivalent heavy mineral contents also in some Early Pleistocene terrace sediments on the southern border of the Rhenish Massif. These findings also refute the assumption of uplifted terraces. Otherwise, the preservation of the fluvial terraces is best outside the above-named, harder quartzite series, because of the wider valley development on softer rocks.

The transition of the main terraces from the southern Rhenish Massif foreland into the upper Middle Rhine valley cannot be reconstructed with the same certainty, as the terrace sequence of the Rhenish Massif proper. The comparison of the 220-m terrace of the Middle Rhine (t_{R4}) with the Mosbach sands near Wiesbaden-Biebrich, discussed by Semmel (1977), remains doubtful, although palaeomagnetic and palaeontological findings (Bibus, 1983) seem to favour this assumption. It seems more probable, however, to regard the t_{R4}-sediments as equivalents of the Younger Weisenau sands (Semmel, 1983, in press), whose basis lies at about 200 m above s.l. west of Mainz. This would indicate a post-sedimentary dislocation of approximately 20 m between the Rhenish Massif and the surroundings of Mainz (Eastern Rhine Hessian Plateau). A comparison of the elevations of the latest Tertiary surfaces in both areas provides a distance of about 70 m, available for the relative uplift of the Rhenish Massif during the Quaternary. This uplift is achieved in several steps downstream from Mainz.

According to Sonne (1978) the basis of the youngest, last glacial lower terrace of the Rhine shows no gradient between Mainz and Bingen-Kempten, which is interpreted to be the result of recent tectonics. Kandler (1970) also assumed a post-sedimentary dislocation of this lower terrace, caused by a stronger uplift of the western section of this area. But according to Kuemmerle (1982) the recent measurements provide no evidence for this assumption. Indications of some very recent block faulting are indeed frequently found here, as, for example, in the kaolin pit on the Roter Berg in Geisenheim, where gray, carbonate-containing sands are dislocated by faults, running parallel to the Rhine. The dislocations amount to some decimetres. These sands are most likely equivalents of the Mosbach sands, which also show such dislocations in the vicinity of Wiesbaden-Schierstein, Wiesbaden-Biebrich, and Wiesbaden-Erbenheim. In this area, they were downfaulted in a graben running west to east, the Mainz-Binger-Graben in the sense of Wagner (1930). On the opposite side, that is south of the Rhine, the above-mentioned Weisenau-sands were downfaulted synsedimentarily. They were, moreover, later affected by faults running east to west and north to south, which led to an increase of downfaulting from the south to the north, or from the west to the east. Contrary to this downfaulting, there exist areas of uplifted horsts, where Pleistocene terraces have been affected, for example, on the margins of the Eppstein-horst, which lies on the western flank of the Upper Rhine Graben. Here, Middle Pleistocene terraces have been uplifted for 20 m, respectively for 40 m (Semmel, 1978). These rather local tectonic movements in many places prevent statements on the general amount of uplift of the Rhenish Massif relative to its southern foreland. Nevertheless, the Pliocene and Early Pleistocene terraces west of Mainz - with the exception of the above-described areas of subsidence - provide reliable data, indicating a relative subsidence of 70 or 20 m. The subsidence within the Upper Rhine Graben amounted to more than 100 m during the Quaternary. In general, the influence of local block faulting has to be taken into account, which provides, on the one hand, indications of Holocene faulting, on the other hand also evidence that no more vertical dislocations occurred within the last 20,000 years (Semmel, 1979).

References

Bartz, J., 1936. Das Unterpliocän in Rheinhessen. Jahresber. Mitt. Oberrh. Geol. Verein, 25:121-228.

Bartz, J., 1976. Quartär und Jungtertiär im Raum Rastatt. Jahrb. Geol. Landesamt Bad.-Württ. (Freiburg i. Br.), 18:121-178.

Bibus, E., 1983. Distribution and dimension of young tectonics in the Neuwied Basin and the Lower Middle Rhine. This vol.

Birkenhauer, J., 1971. Vergleichende Betrachtungen der Hauptterrassen in der rheinischen Hochscholle. Kölner Geogr. Arb. (Sonderbd.) Wiesbaden: 99-140.

Brunnacker, K., Bosinski, G., and Windheuser, H., 1979. Bimstuffe als Leithorizonte im Quartär am Mittelrhein. Mainer Naturwiss. Arch., 17:13-28.

Gallade, M., 1926. Die Oberflächenformen des Rhein-Taunus und seines Abfalles zum Main und Rhein. Jahrb. Nass. Verein Nat.-Kde., 78 Wiesbaden:1-100.

Hüser, K., 1972. Geomorphologische Untersuchungen im westlichen Hintertaunus. Tübinger Geogr. Stud., 50:184 pp.

Kandler, J., 1970. Untersuchungen zur quartären Entwicklung des Rheintales zwischen Mainz/Wiesbaden und Bingen/Rüdesheim. Mainzer Geogr. Stud., 3: 92 pp.

Kuemmerle, E., 1982. Beobachtungen zur Tektonik im Rheingau. Geol. Jahrb. Hessen, 110:101-115.

Leppla, A., 1904. Erl. Geol. Karte Preußen, Bl. Preßberg - Rüdesheim: Berlin, 67 pp.

Oestreich, K., 1909. Studien über die Oberflächengestaltung des Rheinischen Schiefergebirges. Petermanns Mitt., 54:73-78.

Schönhals, E., 1961. Spät- und nacheiszeitliche Entwicklungsstadien von Böden aus äolischen Sedimenten in Westdeutschland. Proc. 7. Int. Congr. Soil Sci., Madison USA:283-290.

Semmel, A., 1972. Fragen der Quartärstratigraphie im Mittel- und Oberrhein-Gebiet. Jahresber. Mitt. Oberrh. Geol. Verein, 54:61-71.

Semmel, A., 1977. Das obere Mittelrhein-Tal. In: Bibus, E., Semmel, A. (eds.) Über die Auswirkung quartärer Tektonik auf die alt-pleistozänen Mittelrhein-Terrassen. Catena, 4:385-408.

Semmel, A., 1978. Untersuchungen zur quartären Tektonik am Taunus-Südrand. Geol. Jahrb. Hessen, 106:291-302.

Semmel, A., 1979. Geomorphological criteria for recent tectonic - A discussion of examples from the northern Upper Rhine area. Allgem. Vermess. Nachr., 86:370-374.

Semmel, A., 1983. Die pliozänen und pleistozänen Deckschichten im Steinbruch Mainz-Weisenau. Geol. Jahrb. Hessen, 111 (in press).

Sonne, V., 1978. Tiefenlinien des Talbodens der Rhein-Niederterrasse zwischen Budenheim bei Mainz und Bingen-Kempten. Mainz. Naturwiss. Arch., 16:83-90.

Sonne, V. and Stöhr, W., 1959. Bimsvorkommen im Flugsandgebiet zwischen Mainz und Ingelheim. Jahresber. Mitt. Oberrh. Geol. Verein, 14:103-116.

Steuer, A., 1906. Über das Vorkommen von Radiolarien-Hornsteinen in den Diluvialterrassen des Rheintals. Notizbl. Verein Erdkde., Darmstadt, 4:27-30.

Wagner, W., 1930a. Diluvium. In: Erl. Geol. Karte Hessen 1:25000, Bl. Bingen-Rüdesheim, Darmstadt, 64-86.

Wagner, W., 1930b. Die ältesten linksrheinischen Diluvialterrassen zwischen Oppenheim, Mainz und Bingen. Notizbl. Verein Erdkde. Darmstadt, 5:177-187.

4.3 Distribution and Dimension of Young Tectonics in the Neuwied Basin and the Lower Middle Rhine

E. Bibus[1]

Abstract

In the Neuwied Basin the displacement of the oldest Tertiary sediments is widespread, whereas in the Lower Middle Rhine area this occurred locally along a structurally weak line (Ahr Valley-Laacher Lake). This displacement continued during the Pliocene period in the latter area with the result that the Pliocene siliceous oolite gravels overlay the Oligocene sediments or they are to be found close together. On the edge of and outside the Neuwied Basin the displacement of the Tertiary sediments lies between 20 and 130 m. In the case of the old Pleistocene terraces in the Neuwied Basin, the displacement values are considerable, whereas one is confronted with an unimportant single-block tectonic process in the Lower Middle Rhine area. In contrast to the older main terraces (t_{R1}-t_{R4}) with their constant altitude, the younger main terrace (t_{R5}) in the Lower Middle Rhine area is characterised by a very large falling gradient which can be traced back to tectonic processes. Proof of displacements belonging to the Middle or Young Pleistocene period can only be found on the northern edge of the Rhenish Massif and in parts of the Neuwied Basin.

Introduction and Methods

Young tectonics in the Middle Rhine Valley can be defined by direct proof of shifts in outcrops and the disturbed patterns of individual terraces and their sediments. In order to avoid mistakes as far as possible in the terrace-morphological methods, the terraces have to be well-structured, stratigraphically exactly arranged and characterized by typical sedimentological features. Based on older contributions (geological charting, Jungbluth, 1918; Kaiser, 1961; Quitzow, 1974) recent investigation in the Middle Rhine valley revealed at least 12 Quaternary (t_{R1}-t_{R12}) and 2 to 3 Pliocene terraces (P_1 to P_3) (cf. Figs. 1, 2; Bibus and Semmel, 1977; Bibus, 1980 including references to older literature) of which only a few terraces allow conclusions on young vertical movements on the basis of the following typical characteristics.

The Pliocene siliceon oolite gravels differ from Quaternary gravels in their residual gravel and heavy mineral spectra (Epidote, garnet, green hornblende). Within the structure of the main terraces (t_{R1}-t_{R6}), the oldest terraces are characterized by an Epidote Maximum. The t_{R4}-terrace (oldest main terrace) deserves special attention because here calcareous, fossiliferous fine sands of the type of Cromerian-age Mosbach sands can be found. Palaeomagnetic measurements (Fromm, pers. commun.) reveal that in the Middle Rhine the Matuyame-Brunhes boundary lies in this terrace. The Jaramillo Event is to be found in the adjoining high-flood loam on the next

---

[1] Geographisches Institut, Universität Tübingen, Hölderlinstraße 12, D-7400 Tübingen, Fed. Rep. of Germany.

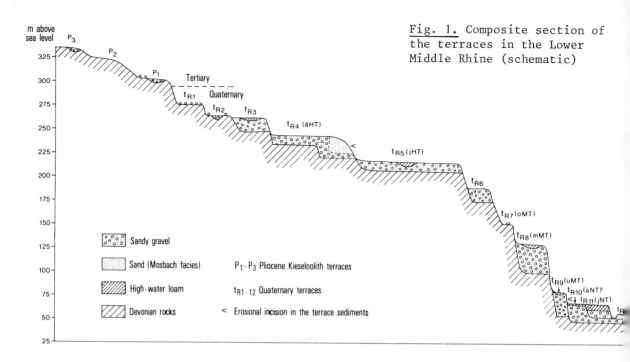

Fig. 1. Composite section of the terraces in the Lower Middle Rhine (schematic)

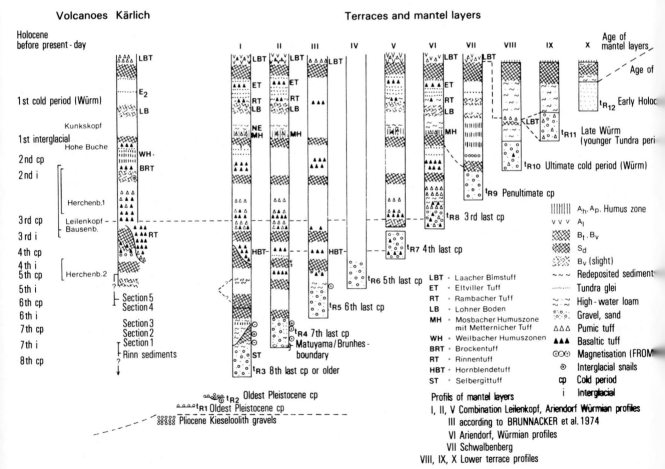

Fig. 2. Correlation diagram and dating of terrace, loess and tuff in the Middle Rhine

oldest t_{R3}-terrace. The t_{R4} is of a Cromerian age which can be proved through micro-mammal and mollusc findings (Geissert, Remy, Storch, pers. commun.). Tectonically, the t_{R7} terrace is important because here the brown hornblendes produced by Eifel volcanism increase suddenly and extensively (Frechen, 1975).

Dislocation of the Siliceous Oolite Gravels and the Older Tertiary Sediments

The Kieseloolith terraces and their sediments can be traced morphologically around the whole Neuwied Basin at an average altitude of 300 to 330 m. Various places reveal a significant division into a terrace at 318 ± 4 m (P_2) and 330 ± 3 m (P_3). There is an occasional appearance of a lower level at 300 ± 5 m (P_1). In the eastern surroundings of the basin the three terraces and their sediments are widely spread at a constant altitude which proves a far-reaching tectonic stability of the eastern periphery. Some Pliocene pebbles in a lower position in a narrow transitional area to the southeast of the basin indicate tectonic downwarps and very rarely secondary redeposition. Small, encheloned blocks exist east of Vallendar and southeast of Höhr-Grenzhausen (Fig. 3) with lowering amounts of at least 20, 50, 70 and 100 m. (Striking of exposed faults: 50°, 270°. All figures are based on correlation with the lowest terrace. Through the ordering of the sediments to the highest Pliocene terrace the figures increase by 30 m).

In the northeast surroundings of the basin the Kieseloolith terraces border immediately on the morphologically significant Sayn fault (Ahrens, 1953) from which no downwarped Kieseloolith gravels can be found in the direction of the basin.

On the west border of the Neuwied Basin Kieseloolith gravels lie undisturbed between the Pellenz depression and the Mosel valley. Towards the Mosel valley, however, there are blocks with displacements of 20 to 50 m. The low number of sediment relicts and denudation plains on the northern border of the basin allows the conclusion that the area north of the Andernach fault is not affected by young tectonics.

The distribution of the siliceous oolite gravels shows a wide extension of the Neuwied Basin in the Pliocene as well as a post-Pliocene stability in large areas of the present-day basin. An instable zone can only be found at the edge of the basin opposite the mouth of the River Mosel and in the lower part of the Mosel valley. These areas are characterized by single block tectonics with occasional huge fault throws. So far no primary deposition of Kieseloolith gravels has been found in the basin itself. This fact explains why the Quaternary erosion must have occurred faster than the assumed lowering of the Pliocene gravels.

North of the Neuwied Basin there are downwarped Kieseloolith gravels in small amounts and assembled in a line which runs north from the Laacher See parallel to the River Rhine (Ahrens, 1953). It is remarkable that the pebbles are superimposed on local Oligocene sediments or lie in the vicinity of obvious dislocated Oligocene (Köhlerhof-Ahrtal). In a small trench below the Middle Pleistocene Herchenberg Volcano the Pliocene was lowered by 70 m and the Oligocene by 100 m at a distance of only 3 km. In a small tilted block near Koisdorf, lowering figures for the Pliocene and the Oligocene sediments of about 130 m can be traced. North of the river the Pliocene deposits lie only 60 m below the lowest undisturbed terraces of the right bank of the Rhine valley. In the immediate neighbourhood Oligocene sands and gravels appear, which in outcrops are split into encheloned trenches and horsts. These sediments which have obviously been disturbed can therefore not be identified as remains of an Oligocene filling-up of the valley as Birkenhauer (1973a) suggested.

Looking at the lowered Tertiary sediments in a wider context, it becomes apparent that the locations at the edge of the Ringen depression and around the Köhlerhof are situated in a direct elongation of the Swist fault system which separates

Fig. 3. General map and sites of tectonically displaced Pliocene and Quaternary sediments

the southern part of the Ville from the Erft-Swist block. This zone of weakness seems to turn in the Eggian direction north of the River Ahr and then runs towards the Laacher See (Fig. 3). Since the Pliocene gravels are mostly underlain by Oligocene but are not overlain by huge Quaternary Rhine sediments, although situated in the main-terrace area of the River Rhine, the following can be stated about the tectonic-morphological evolution.

During the Oligocene local lowerings must have already started and continued until the Pliocene. These must have caused palaeosoils and Oligocene sediments to be positioned at varying depths. Afterwards various phases of erosion followed, particularly in the Pliocene, through which the landscape in the original Rhine valley area was lowered at a great width and the siliceous oolite gravels have either incised in Devonian rocks or downwarped Oligocene sediments and palaeosoils. After sedimentation of the Pliocene pebbles it came to further, limited downwarps of local blocks (60 - 130 m) at a simultaneous deep erosion of the River Rhine. In this case the erosion of the River Rhine must have preceded the lowering, because the River Rhine was not drawn into this zone of weakness and there are no indications of huge Quaternary river sediments.

The Tectonic Dislocation of Quaternary Terraces

To date it has not been possible to prove the existence of the rare relics of the oldest Pleistocene terraces (t_{R1} to t_{R3} with Epidote-Maximum) in a tectonically downwarped position in the Middle-Rhine Valley. Many of these localities were formerly dated by Ahrens (1939) as belonging to the Pliocene (e.g., Brohltal) and have been wrongly used as proof of differentiated tectonics (Birkenhauer, 1973b). It is surprising that so far no downwarped sediments of the t_{R1} to t_{R3} could be traced in the Neuwied Basin, in which old Pleistocene tectonics caused a storing of terraces (claypit Kärlich, e.g., Brunnacker et al., 1969). If it is not possible to conclude that this phenomenon results from young sedimentary lapping and thus less weathering, then it follows that all river sediments examined to this day in the central main terrace area of the Neuwied Basin must belong to younger main terraces.

The t_{R4} terrace (older main terrace), however, allows important tectonic conclusions at such places, at which it appears in calcareous Mosbach facies. This is the case at the upper Middle Rhine, at the eastern Neuwied Basin and above all at the eastern lower Middle Rhine Valley up to the Siebengebirge. The basis of this terrace, which is up to 30 m thick in places, lies approximately 220 m above sea level outside the Neuwied Basin and thus reveals a nearly constant altitude which was first described by Birkenhauer (1971). The relics of the terraces in the Andernach Pforte also lie at a corresponding alitutde (e.g., Krahnenberg). Therefore there is no proof of any stronger young arching or complex folding in this area (in contrast Neumann, 1935; Quitzow, 1974). This fact provides an essential reference to the young lowering of the basin. The basis of the typical t_{R4} sediments lies at about 185 m above sea level in the eastern part of the basin. This fact explains why this region was tectonically lowered by about 35 ± 5 m in contrast to the adjacent high altitude areas within a period of less than 700,000 years. Also at the lower Middle Rhine opposite the mouth of the River Ahr a local dislocation of the t_{R4} of about 20 m can be traced. It is much more difficult to obtain information on the western part of the research area, because there the t_{R4} does not appear in calcareous facies. Based on palaeomagnetic and palaeopedological results a lowering of about 60 m may be expected in the Kärlich-Saffig area in a time period of less than 700,000 years. In this area outcrops reveal echeloned dislocation towards the Neuwied Basin. Recently in the Kärlich claypit, a fault along the strike of 120° was opened, at which the Tertiary sediments were lowered by about 6 m. In the overlying gravels of the main terrace and the loess layers the fractured zone continued with many striking faults of 320° to 350° within a decreasing tendency towards displacements with regard to the younger strata. Brunnacker (1980) connects the above-mentioned faults with inbreak craters developed as a result of the eruption of the Brockentuff. A widespread distribution of the faults at the western border of the basin as well as the decreasing fault throws in the younger layers speaks against this volcano-tectonic explanation.

The interpretation of the gravels lying in the altitude of the t_{R4} in the western lower Middle Rhine Valley is also problematic because there the gravels differ from the typical t_{R4}-sediments by more resistent components. This local facies has already been explained as being a reworking of Tertiary gravels and palaeosoils without any sedimentary influence of the River Rhine (Bibus, 1980). A connection with the Kieseloolith gravels must be excluded because of differing composition (cf. in contrast Frechen and van den Boom, 1959; Boenigk, 1979). In the Waldorf-Dietenkopf area the more resistant gravels are downwarped by about 15 m in the zone of weakness between the Laacher See and the Ahr Valley.

The t_{R5} terrace (younger main terrace) is widespread in the lower Middle Rhine Valley and forms the edge of the narrow valley at about 200 m above sea level. North of the Neuwied Basin its terrace lies slightly higher than at the Upper Middle Rhine. In the area of the Brohl

Valley it reaches the terrace basis again at about 200 m above sea level so that in the t_{R5} a consistent altitude extending to the start of the lower Middle Rhine is indicated. A longitudinal gradient appears at the surface of the terrace for the first time between the Brohl- and Ahr Valley, which significantly increases at the northern edge of the Rhenish Massif extending to the Ville. The surface of the t_{R5} (= HT_3 at the lower Rhine according to Schnütgen, 1974) continues without any apparent fault scarps to the Ville, excepting some small tilted blocks south-west of Bonn. According to Schünemann (1958) remarkable dislocations are said to exist. This permits the conclusion that, during the sedimentation of the t_{R5}, very strong synsedimentary tectonics existed at the northern edge of the Rhenish Massif. After the sedimentation of the terrace the relative uplift of the northern Rhenish Massif, including parts of the foreland (Ville), took place. Significant postsedimentary downwarps of the t_{R5} with underlying siliceous oolite gravels occurred only where depression areas (the Duisdorf trench near Bonn, e.g., according to Fliegel, 1922, Meckenheim Bucht) reach the Rhenish Massif or intrude into it. In some trenches the tectonic activity can continue up to the Younger Würm and the Holocene, as is proved by the dislocation of about 3 m of the Lohner-Soil (about 30,000 years) at the Hardtberg near Bonn. The strong longitudinal gradient at the lowest section of the Middle Rhine cannot be attributed to a frequently assumed northward tilt of the Rhenish Massif or a slanting of the West German Block (Quiring, 1926) northwestwards because the older t_{R4} terrace and the rock basis of the siliceous oolite terraces can be traced at the same altitude to the Siebengebirge. Instead, an extensive uplift after the sedimentation of the t_{R4} must have taken place by which the longitudinal gradient of the older terraces was levelled off. The longitudinal gradient of the t_{R5} at the lower Middle Rhine can only be traced back to strong tectonic downwarps in the Lower Rhenish Embayment shortly before or during the sedimentation of the terrace (cf. Ahorner, 1962). Clear indications of a deposition of the t_{R5} sediments in a wedge-shaped trench, intruding southward into the Rhenish Massif between Bonn and Andernach with a crossing of individual main terraces, as assumed by Philippson (1903) and Birkenhauer (1971), could not be verified, although few local faults cannot be excluded in this area. So far faults could only be detected in the Neuwied Basin within the region of the middle and lower terraces. Noll and Ahorner (1978, pers. commun.) mentioned an upper middle terrace west of Gladenbach which has been disturbed by a downthrown fault (150° to 170°). Young tectonics of this kind are not surprising in the Neuwied Basin since in the claypit of Kärlich pumice tuff of the Alleröd and its redeposited sediments of the younger Tundrenzeit were disturbed (I am most grateful to K. Würges, Mülheim-Kärlich, for this latter information).

References

Ahorner, L., 1962. Untersuchungen zur quartären Bruchtektonik der Niederrheinischen Bucht. Diss. Univ. Köln 1961. Eiszeitalter Gegw., 13:24-105.

Ahrens, W., 1939. Erläuter. Geol. Karte Preußen 1:25000. Bl. Linz, 3157:47.

Ahrens, W., 1953. Bau und Entstehung des Neuwieder Beckens. Z. Dtsch. Geol. Ges. 104:152-153.

Bibus, E., 1980. Zur Relief-, Boden- und Sedimententwicklung am unteren Mittelrhein. Frankfurter Geowiss. Arb., Ser. D, 1:296.

Bibus, E. and Semmel, A., 1977. Über die Auswirkung quartärer Tektonik auf die altpleistozänen Mittelrhein-Terrassen. Catena, 4 (4):385-408.

Birkenhauer, J., 1971. Vergleichende Betrachtung der Hauptterrassen in der rheinischen Hochscholle. Kölner Geogr. Arb. (Sonderbd. Festschrift K. Kayser) 99-104.

Birkenhauer, J., 1973a. Die Entwicklung des Talsystems und des Stockwerkbaus im zentralen Rheinischen Schiefergebirge zwischen dem Mitteltertiär und dem Altpleistozän. Arb. Rhein. Landeskde., Bonn, 34:217.

Birkenhauer, J., 1973b. Zur Chronologie, Genese und Tektonik der plio-pleistozänen Terrassen am Mittelrhein und seinen Nebenflüssen. Z. Geomorphol., 17:489-496.

Boenigk, W., 1978. Gliederung der altquartären Ablagerungen in der Niederrheinischen Bucht. Fortschr. Geol. Rheinland Westfalen, 28:135-212.

Boenigk, W., 1979. Protokoll über das 4. Kolloquium, im Schwerpunkt Vertikalbewegungen und ihre Ursachen am Beispiel des Rheinischen Schildes. Neustadt a.d. Weinstraße, 16.-17. Nov. 1979, pp. 48-49.

Brunnacker, K., 1980. Forschungsentwicklung in Kärlich. In: Bosinski, G., Brunnacker, K., Lanser, K.P., Stephan, S., Urban, B., Würges, K. (eds.) Altpaläolithische Funde von Kärlich, Kreis Mayen-Koblenz (Neuwieder Becken). Archäol. Korrespondenzbl. 10, 4:295-314.

Brunnacker, K., Streit, R., and Schirmer, W., 1969. Der Aufbau des Quartärprofils von Kärlich/Neuwieder Becken (Mittelrhein). Mainz Naturwiss. Arch., 8:102-133.

Fliegel, G., 1912. Zum Gebirgsbau der Eifel. Verh. Naturhist. Verein Rheinland Westfalen, 68:489-504.

Fliegel, G., 1922. Der Untergrund der Niederrheinischen Bucht. Abhandlg. Preuss. Geol. Landesanstalt, Berlin, 92:1-155.

Frechen, J., 1975. Tephrostratigraphische Abgrenzung des Würmlösses und der älteren Lösse im Quartärprofil der Tongrube Kärlich, Neuwieder Becken. Decheniana, 127:157-194.

Frechen, J. and Boom, G. van den, 1959. Die sedimentpetrographische Horizontierung der pleistozänen Terrassenschotter im Mittelrheingebiet. Fortschr. Geol. Rheinland Westfalen, 4:89-125.

Jungbluth, F.A., 1918. Die Terrassen des Rheins von Andernach bis Bonn. Verh. Naturhist. Verein Rheinland Westfalen, 73:1-103.

Kaiser, K., 1961. Gliederung und Formenschatz des Pliozäns und Quartärs am Mittel- und Niederrhein sowie in den angrenzenden Niederlanden unter besonderer Berücksichtigung der Rheinterrassen. Festschrift 33 Dtsch. Geogr. Tagung, Köln:236-278.

Neumann, K.L., 1935. Fragen zum Problem der Großfaltung im Rheinischen Schiefergebirge. Z. Ges. Erdkde.:321-352.

Philippson, A., 1903. Zur Morphologie des Rheinischen Schiefergebirges. Verh. XIV Dtsch. Geogr. Tagung Köln, Berlin, pp. 193-205.

Quiring, H., 1926. Die Schrägstellung der westdeutschen Großscholle im Känozoikum in ihren tektonischen und vulkanischen Auswirkungen. Mit dem Versuch einer Terrassenchronologie des Rheines. Jahrb. Preuss. Geol. Landesanst., Berlin, 47:456-558.

Quitzow, H.W., 1974. Das Rheintal und seine Entstehung. Bestandsaufnahme und Versuch einer Synthese. Cet. Soc. Géol.:53-104.

Schnütgen, A., 1974. Die Hauptterrassenabfolge am linken Niederrhein aufgrund der Schotterpetrographie. Forsch.-Ber. Nordrhein-Westfalen, 2399:150.

Schünemann, W., 1958. Zur Stratigraphie und Tektonik des Tertiärs und Altpleistozäns am Südrand der Niederrheinischen Bucht. Fortschr. Geol. Rheinland Westfalen, 2:457-472.

4.4 The Rhine Valley Between the Neuwied Basin and the Lower Rhenish Embayment

K. Brunnacker and W. Boenigk[1]

Abstract

The older Rhine terraces rise with increasing steepness from N to S caused by tectonic tilting.

The elevation of the terraces is modified by the subsiding Neuwied Basin, by the saddle form upward doming area north of it and by some flexure zones; one zone is located near the southern end of the Lower Rhenish Embayment and the second is where the Palaeozoic basement submerges under the Cretaceous and Cenozoic coverbeds.

This tectonic pattern is again modified by local block faulting along the Middle Rhine valley.

Introduction

From the Neuwied Basin, lower Middle Rhine, and Lower Rhenish Embayment (Fig. 1), many investigations regarding Quaternary stratigraphy based on several different methods have been undertaken. As a supplement, recent studies concerning tectogenesis and volcanism, especially in connection with DFG-research project must be added *Plateau Uplift and its Causes in the Rhenish Massif*.

---

[1] Geologisches Institut der Universität zu Köln, Zülpicher Str. 49, D-5000 Köln 1, Fed. Rep. of Germany.

Along the Rhine the depression of the Neuwied Basin is followed by the lower Middle Rhine with its uplift character. Then follows the Lower Rhenish Embayment with its wide depression, which west of the Erft with its accompaning faults was effective as a sediment trap during the upper Tertiary and lower Quaternary (Fig. 1). In contrast to this we find the lower Quaternary terrace flight from the mouth of the Sieg via the Bergische Randhöhen up to Bocholt.

Stratigraphy

The fundamental facts as well as the details regarding the stratigraphy of the Pliocene and Quaternary were mainly based on data from the Lower Rhenish Embayment.

Lower Rhenish Embayment

Of great importance were the large lignite pits along the west border of the Ville (Fig. 2), especially the lignite pit Frechen. The complete upper Tertiary and the lower Quaternary could be found here overlaying each other. To the east follows the terrace flight of the upper Quaternary. It is true that towards the northwest from about Krefeld young tectonic movements have led to the fact that the Lower Terrace (NT), belonging to the last ice age, is covered increasingly by holocene deposits of the Rhine (Brunnacker, 1978). The Pliocene which is made up of gravel and clay ends with the Reuverium C. Like the underlying

Plateau Uplift, ed. by K. Fuchs et al.
© Springer-Verlag Berlin Heidelberg 1983

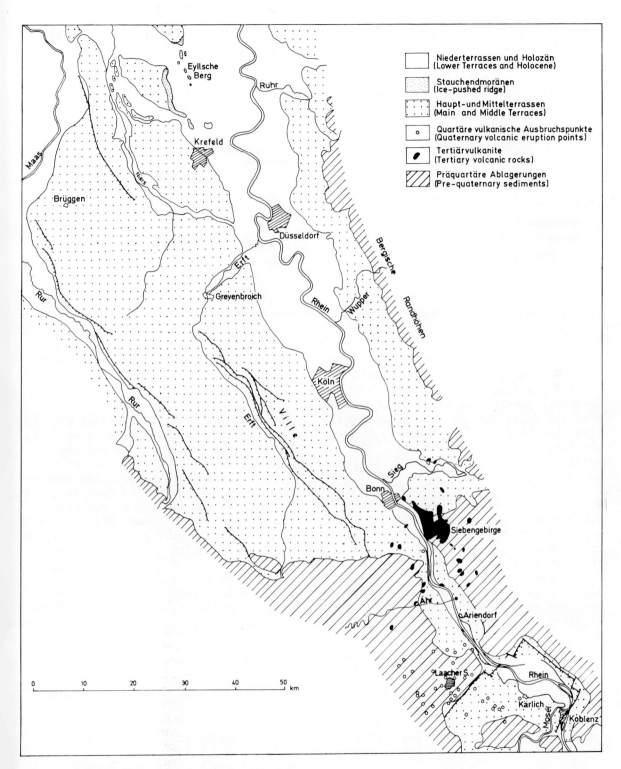

Fig. 1. Geographic sketch map of the area

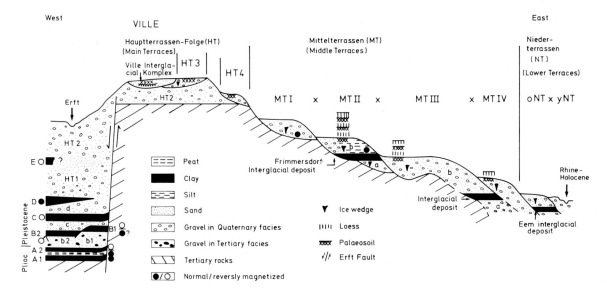

Fig. 2. Schematic crossection through the Ville in the southern Lower Rhine Basin

Reuverium B, it contains a typical pliocene mollusc fauna. The Reuverium B and C are magnetized normally. Only in the uppermost part of the Reuverium C we find the Gauss/Matuyama change in magnetization. Therefore the boundary between the Tertiary and Quaternary must lie somewhat higher in the profile. It is estimated about 2.3 m.y.B.P.

The lowest Pleistocene reaches a thickness of 40 m with gravel and the interbedded clays of warmer periods. The underlaying gravel is still a siliceous oolite (Kieseloolith) type gravel. This is followed by mixed gravel. As far as palaeomagnetic data is available, the clay horizons as a rule show a reverse magnetization. Only in the clay horizon B 1 does a partially normal magnetization give an indication of the Reunion Event. In the clay horizon D, there is another change in magnetization pointing to the Olduvai Event, or according to Boenigk (1978) to the Jaramillo Event. This lowest Pleistocene corresponds to the Pretiglium and to the Tiglium Complex of the Netherlands.

The overlaying main terraces 1 and 2, about 60 m thick, are subdivided by a horizon of erosion. Here garnet is strongly reduced. The heavy mineral spectrum changes again only from the middle terraces with an increase in the amount of garnet and the introduction of volcanic heavy minerals.

Up to now the erosional horizon was correlated with the Regensburg Interglacial Complex (Brunnacker, 1978) and thus with the Jaramillo Event; further with the Waalium warm periods of the Netherlands. But as Zagwijn and Doppert (1978) could show, the Waalium is yet older than the Jaramillo Event. Boenigk (1978) correlated this horizon with a stage of the Cromer Complex in the Netherlands.

The quality of the cold periods hidden in the main terraces 1 and 2 is not known according to the findings in the Lower Rhenish Embayment. Only driftblocks are present as cold climate indicators in the gravel d. The first synsedimentary ice wedges follow in the wide trough overlaying the main terrace 3. Here for the first time we also have volcanic heavy minerals in a larger quantity, which have been transported from the East Eifel volcanic area.

The next sequence is only documented in local gravels from the Erft with interbedded valley soils. They are combined into what is known as the Ville Interglacial Complex.

Then follows along the present Rhine valley rests of the main terrace 4, which can be found in local not too wide deposits. Then comes the middle terrace flight I to IV and in the present valley floor the Lower Terrace covered by holocene deposits.

The nordic inland ice has twice reached the Lower Rhine. The first advance of the inland ice is documented through boulder detritus in the Middle Terrace IIb and the second in ice-pushed ridges, which correspond to the Middle Terrace IV. The first advance is correlated with the Elster glacial and the second with the Saale glacial. Therefore, we must have another glacial between the mentioned advances of the inland ice but without any glacial evidence.

All the quaternary warm periods were, according to palaeobotanical and palaeopedological findings, of about the same length and intensity. The cold periods, on the other hand, hint at an escalation in so far as the first two glacials already eliminated the mollusc fauna of the Pliocene, but only from the middle terrace I onwards are ice wedges generally present and only in the middle terraces II and IV are large ice advances documented. Generally the number of quartz pebbles diminishes from the older to the younger terraces. But it is remarkable that in the gravel b2 the main terrace 3 and the middle terrace I, i.e., at every period of extreme cold, the percentage of quartz pebbles is reduced disproportionally and is replaced by a mixed spectrum. This is due to the fact that the erosion was stronger in the area of the Rhenish Massif.

The stratigraphy can be well fitted into an exact time table with the help of numerous palaeomagnetic data (Brunnacker and Boenigk, 1976; Boenigk et al., 1979).

By such a classification one can record about 15 cold-warm cycles during the Quaternary. Each of them had a duration of about 100,000 years during the youngest Pleistocene and in the Brunhes epoch. From this span only about 10,000 to 15,000 years were necessary for each warm period (this refers to the terrestrial development of the Quaternary).

If we start from the fact that in the main terrace I, which is equivalent to the Waalium, as well as in the main terrace II there are several such cycles hidden, then one can clearly assume, according to Zagwijn and Doppert (1978), a large number of such cycles for the complete Quaternary.

Bergische Randhöhen

The terrace flight of the Bergische Randhöhen (mountainous rim; cf. Fig. 3) can be integrated without any problems into the scheme mentioned above (Brunnacker et al., 1982). Four upper terraces represent the Lowest Pleistocene. The main terraces I and II are represented through one well marked terrace. The gravel spectrum of these two main terraces is characterized by a somewhat lower percentage of quartz than the spectrum in the basin. This is due to the fact that the relatively soft material from the tributaries could not be fully incorporated into the Rhine spectrum.

Neuwied Basin

The oldest member in and around the area of depression near Neuwied is made up of Pliocene siliceous oolite gravel. To this come older tertiary gravels, which to a large extend have been influenced tectonically.

The most important quaternary profile is in Kärlich. This profile stands in close relation to the classification developed for the Lower Rhenish Embayment. In Kärlich, as mostly in the Lower Rhenish Embayment, the underlying sequence of gravels is stacked up due to tectonic depression. The gravel forming the lowest part of the sequence A is correlated with the lower Pleistocene High Terrace (Boenigk, 1978). It appears that the soil which completes this section belongs to an older event of the Matuyama epoch, and that the overlaying gravel Ba can be subdivided further, according to its heavy mineral spectrum, so that it comprises the period of the main terraces 1 and 2 and on top is completed with gravels of the main terrace 3, with ice wedge pseudomorphoses and pyroxenes.

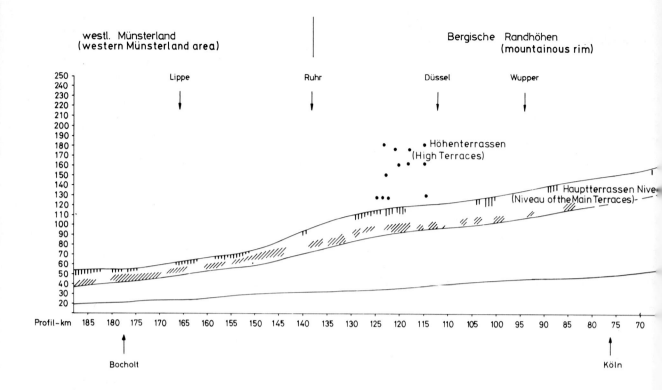

Thus a sequence which represents a considerable period is documented by a 10-m-thick layer of gravel. In Kärlich this development lasts until the Brunhes epoch, then a Mosel gravel (Bb) follows, which strongly reduces the influence of the Rhine. In this portion lies the Matuyama/Brunhes boundary. The Mosel gravel corresponds to the lower section of the Ville Interglacial Complex of the Lower Rhine. Superimposed we find a number of loess palaeosoils, which can be loosely correlated with the sequence in the Lower Rhine area.

Lower Middle Rhine

To clarify the stratigraphy and thereby also the tectogenesis in the uplift area between the Neuwied Basin and the Lower Rhenish Embayment the following procedure was valid: Along the present Rhine extensive gravel analysis were made (Spoerer, 1982) to see how the local material from the tributaries is incorporated into the Rhine spectrum. The percentage of quartz, again a simple indicator, is generally less than 30%. The local influence is wiped out as soon as it reaches the section dominated by the Rhine.

The other extreme is the Tertiary gravel with more than 75% quartz (Langer and Brunnacker, 1983). The Pleistocene terrace flight which lies between the two extremes is characterized by a gradual reduction in percentage of quartz pebble from terrace to terrace. With the help of a great number of gravel analyses, the full homogenizing effect of the Rhine can also be shown here. If the localities are ordered according to the altitude and percentrage of quartz gravel, then the high terraces can be clearly differentiated as a group. This is also the case for the two main terraces (upper terrace and main terrace in its true sense) with a vertical band width of 30 m (Fig. 3). As the thickness of each gravel bed lies by or under 10 m, also here a subdivision is necessary, as is the case in the Bergische Randhöhen (upper terrace and main terrace in the true sense). However, the group cannot be subdivided on the basis of quartz percentages.

Of course, the percentage of quartz is still more reduced in the younger ter-

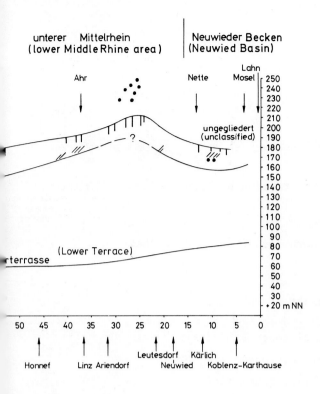

Fig. 3. Morphologic position of the lower Quaternary gravel terraces between Neuwied Basin and Bocholt (between river Sieg and Bocholt; right valley side only)

races. Here a further classification is possible with the help of the superposed strata and tephrostratigraphy.

The disadvantage of this method in the Middle Rhine area is that local tectonics can only be grasped in a limited form or not at all, due to lack of outcrops. Rock terraces, petrovaricance of underlying layers and areas without an outcrop cannot be classified.

Volcanism

The tuffs of the East Eifel volcanism are important stratigraphic indicators. It is, moreover, only natural to consider a connection between the volcanism and the depression of the Neuwied Basin, but this cannot be confirmed (Windheuser et al., 1982). The depression of the basin is clearly older than the main phase of East Eifel volcanism. It is remarkable, however, that the main phase corresponds to periods during the Quaternary with extreme permafrost. This is the case with the main terrace 3 and from the middle terrace onwards. It seems further that the main volcanic activity appears to fall together with the warm periods that subdivide the period with permafrost.

Results

By correlation within the complete area from the Neuwied Basin northwards, the number of methods is reduced considerably in certain localities. Here the gravel analysis, especially the percentage of quartz pebbles, and the analysis of heavy minerals are of further help.

The percentage of quartz pebbles becomes smaller from the older to the younger terraces (Fig. 4). However, only the lowest percentage rates within the respective band width are of importance, as they only reflect freshly delivered material. Along the Middle Rhine and Bergischen Randhöhen (mountainous rim) the percentage rates are relatively lower, as compared to those of the Lower Rhenish Embayment. The trend, however, remains the same. The difference is due to the influence of a local delivery from the Rhenish Massif. Only after the Rhine enters the Lower Rhenish Embayment does this local influence decline. The non-resistant components such as slate, silt

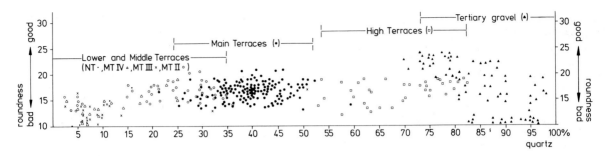

Fig. 4. Percentage of quartz pebbles in the fraction between 20-50 mm ⌀ in the lower Middle Rhine area

Middle and lower terraces:	Pyroxenes, hornblende, epidote, garnet
Main terraces 2 and 3:	Epidote
Main terrace 1, Lowest Pleistocene and uppermost Reuverium:	Epidote, garnet, hornblende
Pliocene:	Tourmaline, zircone, staurolite
Miocene:	Zircone, rutile, tourmaline, Sphen
Oligocene:	Zircone, rutile, tourmaline

and sandstones are eliminated by further transport, resulting in a relative increase of quartz.

The present data available on heavy mineral spectra from Tertiary and Quaternary sediments in the Lower Rhenish Embayment and in the Rhenish Massif clearly show the dependence on age (Boenigk, 1978) due to weathering and provenance (see above).

Only stable heavy minerals exist during the Pliocene; this points to a limited source area for the Rhine. The well-rounded siliceous oolite gravel derives mainly from the Mosel. This situation changes rapidly during the Reuverium B to an instable Rhine spectrum indicating a source of delivery south of the Upper Rhine Graben (Boenigk, 1978).

A stronger Mosel influence is again present at the lower part of the Ville Interglacial Complex (Kärlich, Sect. Bb), then again the Rhine dominates. Of importance is the fact that from the erosional horizon onwards, which divides the main terraces 1 and 2, garnet is reduced in the heavy mineral spectrum and, therefore, the spectrum of the upper main terrace is also altered.

For a correlation of the Rhine terraces over a large area and their partial tectonic displacements, the levels of the high terrace and the main terraces are of importance, especially the main terraces 1 and 2 in the Lower Rhenish Embayment and their corresponding upper terrace and the main terrace - in the true sense - in the Middle Rhine area and along the Bergische Randhöhen (mountainous rim). The two terraces lie on the one hand morphologically very near together, on the other hand quite a time-lap is represented by them. This morphological situation cannot only be explained by a phase of tectonic inactivity, but also by the still little understood intensity of the enclosed cold periods.

On the basis of their typical quartz rate supported by heavy mineral analysis for the upper and main terrace an inclination to the north results, which is modified through flexures along the lower Middle Rhine and near the Ruhr (Fig. 3).

Contrary to this is the interpretation of Bibus (1980) regarding the lower Middle Rhine. He does not find an inclination to the north, but a horizontal arrangement which is believed to be due to an uplift in the north. Misinterpretations of the field findings are the cause. His quartz rates as well as his heavy mineral analysis are too heterogenous within one terrace. His sites belong to different terraces according to our scheme (Fig. 4).

Tectonics

The main element of the investigated area are the Rhine valley starting at the Neuwied Basin and the Lower Rhenish Embayment. As a time marker for the tectonic movement the Pliocene sediments and the main terrace sequence present in most areas can be used.

The presentation of minor tectonics is disregarded. Besides volcanotectonic effects (e.g., in Kärlich), some small faults with a displacement of less than 1 m can be observed. This possibly only indicates, however, the effects of gravity sliding (e.g., near Miesenheim) or of "Loess tectonics" (e.g., in Gönnersdorf).

Lower Rhenish Embayment

The Lower Rhenish Embayment is the south end of a long trough which stretches from the North Sea up to here. During parts of the Tertiary it extended to the Neuwied Basin.

Within this basin at the contact between the Erft block and the Cologne (Ville) block a system of faults is well exposed (Figs. 2 and 5), and movements can be measured (Boenigk et al., 1979).

The attempt to measure the rate of movement along single faults (Erft and Horrem fault) resulted in a more or less constant rate of about 0.07 mm per annum during the Pliocene and Quaternary. The situation changes, however, when the time intervals are reduced. Periods of relative inactivity (Lower Pliocene) can then be differentiated from those periods with stronger movement (Quaternary).

The attempt was therefore made to compare the rate of movements of different faults with each other. The result was that active and inactive periods along different faults are not equal to each other in timing. This means that movements along a single fault are not representative for the uplift or subsidence of the whole Lower Rhenish Embayment.

It can be established that there are only a few larger faults which were already strongly active during the Oligocene and Miocene, but quite a number of smaller and larger faults, which were active almost solely during the Pliocene and Quaternary.

The facies and thickness of a sediment layer allow conclusions regarding the tectonic movement in the whole Lower Rhenish Embayment. Hence the following movements, with respect to the sea-level, can be reconstructed:

Pliocene to Quaternary:

> Uplift in the Basin in connection with stretching and block faulting. Tectonic elements are tilted, so that local depressions are formed, e.g., parts of the Erft and Rur blocks

Miocene:

> As in the Oligocene, but weaker

Oligocene:

> Strong depression of the Basin without many faults.

The uplift tendency of the Lower Rhenish Embayment is illustrated in Fig. 6. Here the areas are shown where the basis of the main terrace lies above or below the present erosional horizon. The picture would not change too much, if the Quaternary basis were taken into account.

A comparison between the Basin and the Rhenish Massif is difficult, mainly because not enough Oligocene and Miocene sediments can be found in the Rhenish Massif. The results allow, however, the conclusion that the movement along the lower Middle Rhine correspond to the movements in the Lower Rhenish Embayment. Only the uplift was stronger and the subsidence weaker than in the Basin. The block tectonics during the Pliocene and

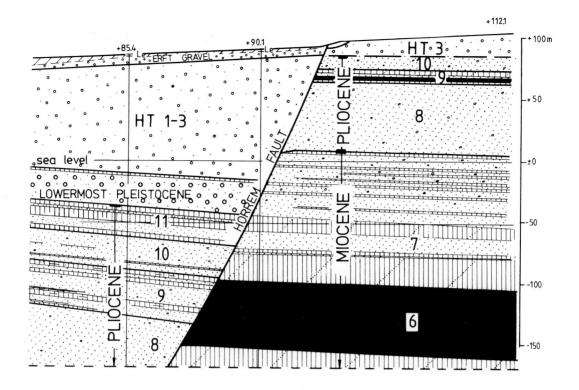

Fig. 5 (Caption see opposite page)

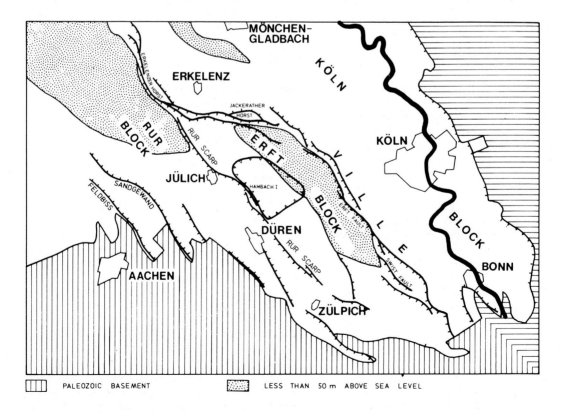

Fig. 6 (Caption see opposite page)

Fig. 5. Stratigraphic sequence and tectonic pattern at the Horrem Fault (southern Lower Rhenish Embayment). 6 main brown coal seam; 7 Inden beds; 8-10 siliceous oolite (Kieseloolith) formation; 8 Hauptkies series; 9 Rotton series; 10-11 Reuver series; HT 1-3 main terrace 1-3

Fig. 6. Morphologic position of the main terrace basis compared with the present erosion level (+50 m above sea level) in the Lower Rhenish Embayment

Quaternary is of about the same magnitude in both areas.

The regions of the lower Middle Rhine valley can be regarded as the continuation of the Lower Rhenish Embayment.

Rhine Valley

The position of the main terrace from the Middle Rhine northwards is very suitable for understanding tectonic movements up to the present, if we use the same gradient as the Rhine has today. However, in such a proceeding one must accept the not yet proven fact that the river has had an idela gradient (Fig. 3).

The vertical distance between the two main terraces is very small, even though they represent a considerable interval of time. The similarity in petrography (quartz rate) of the two lower levels points to a lack of incision into the Rhenish Massif. Otherwise the percentages of new and not too resistant pebbles must be clearly higher in the younger main terrace.

For the Quaternary the following results can be presented:

1. The Rhenish Massif was integrated into the general Quaternary uplift.
2. The uplift was increasing from north to south in the investiated area of the Rhenish Massif.
3. Modifications are found in the north where the Rhenish Massif is covered by Cretaceous and Cenozoic beds because here the main terrace level is bent like a flexure. Another flexure is present near the mouth of the Sieg, which belongs to the large flexure at the south end of the Lower Rhenish Embayment. Further local movements can be found in the lower Ahr valley, near Linz and in the vicinity of Waldorf/Herchenberg.
4. Just north of the Neuwied Basin we have a local anticlinorium, which tilts the main terraces. Here the uplift tendency already appears to be effective during the lower Tertiary. A connection with the East Eifel volcanism must be excluded for the time being. Obviously there is a regional connection with the horst, which Meyer (pers. commun.) has discovered in the area around the Laacher Sea. The diminution of the uplift north of the Neuwied Basin immediately after the main terrace period (the middle terraces do not clearly show flexural tendency) is in agreement with the weakening tendency of depression in the Neuwied Basin.

Trying to quantify the uplift of the Rhenish Massif from north to south in time and space, the following average values for the upper boundary of the main terrace (HT 2 and 3) can be expected (Brunnacker et al., 1982; Langer and Brunnacker, 1983; Windheuser et al., 1982):

South of the river Ruhr: about 0.035 mm a^{-1}

South of the river Wupper: about 0.05 mm a^{-1}

South of the river Sieg: about 0.082 mm a^{-1}

Gorch of Andernach: about 0.15 mm a^{-1}

Neuwied Basin

The Neuwied Basin, which started subsiding already in the lower Tertiary, experienced during the lower and lowest Pleistocene relatively stronger rates of depression than during the Pliocene, or a relative slower uplift. Consequently the complete older Quaternary in Kärlich is piled up in only a 10-m-thick gravel layer, or is documented in the same geomorphological position (Boenigk, 1978).

Here also it is only possible to give general quantitative data (HT 2 and HT 3):

Uplift of Neuwied Basin: 0.1 mm a^{-1}

Relative depression in relation to the Lower Middle Rhine: 0.04 mm a^{-1}

Results

In comparison to the Upper Tertiary, uplifts and local depressions developed more quickly during the Quaternary. Exceptions are some faults in the Lower Rhenish Embayment with more or less continuous rates of movement. The general relative uplift during the Quaternary was modified along the Rhine Valley by tilting upwards from N to S and additionally by flexures in the vicinity of the rivers Ruhr and Sieg. To this must be added an uplift of local character in the lower Middle Rhine area in the vicinity of the Neuwied Basin, which can also be seen in the Laacher See area. The cause remains unclarified.

The Neuwied Basin shows a depressional tendency in relation to the surrounding area, mainly in the lower Quaternary.

The period of the main terrace sequence is characterized by its very long duration and the slowing-down of tectonic activities. As the character of the enclosed cold periods is poorly known (drift blocks are present) it can be stated only with reservation that beside the slowing-down of the tectogenesis also climate-morphological aspects may have played a role.

During the Brunhes epoch there is a certain connection between more active crustal movements and climate morphology with strong incision into the Rhine Valley and with local flexures (Sieg and Ruhr flexures), as well as an anticlinorium south of the Neuwied Basin.

References

Bibus, E., 1980. Zur Relief-, Boden- und Sedimententwicklung am unteren Mittelrhein. Frankfurter Geowiss. Arb. D 1: 296.

Boenigk, W., 1978. Gliederung der altquartären Ablagerungen in der Niederrheinischen Bucht. Fortschr. Geol. Rheinland Westfalen, Krefeld, 28:135-212.

Boenigk, W., Koci, A., and Brunnacker, K., 1979. Magnetostratigraphie im Pliozän der Niederrheinischen Bucht. Neues Jahrb. Geol. Paläontol. Mh., 1979:513-528.

Brunnacker, K., 1978. Gliederung und Stratigraphie der Quartärterrassen am Niederrhein. Kölner Geogr. Arb., 36: 37-58.

Brunnacker, K. and Boenigk, W., 1976. Über den Stand der paläomagnetischen Untersuchungen im Pliozän und Pleistozän der Bundesrepublik Deutschland. Eiszeitalter und Gegenwart, 27:1-17.

Brunnacker, K., Farrokh, F., and Sidiropoulos, D., 1982. Die altquartären Terrassen rechts des Niederrheins. Z. Geomorphol. (Suppl.) Berlin, 42:215-226.

Langer, Cl. and Brunnacker, K., 1983. Schotterpetrographie des Tertiärs und Quartärs im Neuwieder Becken und am unteren Mittelrhein. Decheniana, Bonn, 136 (in press).

Spoerer, H.E., 1982. Petrographische Untersuchungen entlang dem Rhein-Lauf zwischen Main-Mündung und Bonn. Mainzer Naturwiss. Arch. Main (in press).

Windheuser, H., Meyer, W., and Brunnacker, K., 1982. Verbreitung, Zeitstellung und Ursachen des quartären Vulkanismus in der Osteifel. Z. Geomorphol. (Suppl.) Berlin, Stuttgart, 42:177-194.

Zagwijn, W.H., and Doppert, W.Chr., 1978. Upper Cenozoic of the southern North Sea Basin: palaeoclimatic and palaeogeographic evolution. Geol. Mijnbouw 's Gravenhage, 57:577-588.

4.5 The Tectonic Position of the Lower Mosel Block in Relation to the Tertiary and Old Pleistocene Sediments

E. Bibus[1]

Abstract

The results stated here the following spatial and temporal pattern of movement within the Lower Mosel Valley to be outlined: the various old Tertiary deposits in the study area have undergone severe tectonic disturbances. The most extreme downwarping values are to be found by the Vallendar gravels, whereas the calcareous sandstone of the Münstermaifeld area have been subjected to relatively small disturbances.

The siliceous oolite gravels are to be found at a constant elevation, excepting a few local gravel deposits, thus suggesting that there was tectonic stability in the Lower Mosel Valley during the post-Pliocene period. This is underlined by the course of the main terraces, as long as one accepts that there are six terraces. The tectonic separation of the Middle Mosel Block from the Lower Mosel Block occurred withhin a short distance in the Trais-Karden-Cochem valley section and took the form of a stair-like uplifting which resulted in post-Pliocene dislocation values of 50-65 m and Quaternary values of ca. 40 m. The main tectonic activity can be limited to the oldest Pleistocene. The entire terrace area forms an elevated surface near Cochem. It does not, however, form an arched structure (as suggested by Negendank, 1977, 1978) which slopes continually or stepwise in the direction of the Neuwied Basin.

---

[1] Geographisches Institut, Universität Tübingen, Hölderlinstraße 12, D-7400 Tübingen, Fed. Rep. of Germany.

Introduction

In the Middle Mosel Valley the border between the Tertiary and Quaternary sediments has been laid higher than in the other main valleys in the Rhenish Massif (Kremer, 1954; Löhnertz, 1982). It is, therefore, to be expected that there are tectonic lines in the Lower Mosel Valley along which central areas of the western Rhenish Massif have been relatively raised. Up to now no satisfactory connection has been found, from the tectonic point of view, between the Middle Mosel terraces above the Lower Mosel and the Rhine system. Thus the question remains open as to where and in which form the tectonic separation of the Middle Mosel Block took place (Borgstätte, 1910; Osmani, 1976). The answer is to be found in the position of the Tertiary-Quaternary border, the course of the Plio-Pleistocene terrace relics, as well as the distribution of the Lower Oligocene Münstermaifeld layers, as described by Kadolsky (1975) and of the probably older Vallendar layers.

The Pliocene and Old Pleistocene Terraces and Their Displacement

In contrast to earlier studies, this analysis has proved that there are at least 6 main terraces ($t_{M1}-t_{M6}$) and not only 2-4 in the Lower Mosel Valley. In the debochure area of the Mosel there are three terraces (170-180 m = t_{M6}, 197-203 m = t_{M5}, 220-225 m = t_{M4}, cf.

Plateau Uplift, ed. by K. Fuchs et al.
© Springer-Verlag Berlin Heidelberg 1983

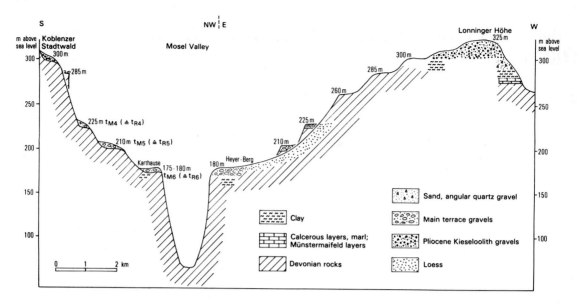

Fig. 1. Terrace relics in the Lowest Mosel Valley

Fig. 1) which can be compared with the lower main terraces in the Rhine Valley (t_{R4}-t_{R6}) (Bibus, 1980). The youngest t_{M6} terrace passes partially through slate, weathered slate, clay and older gravels near Koblenz and thus terrace stacking is present near Koblenz and thus terrace stacking is present near Metternich (Heyerberg), as can be recognised through three interpolated high-water loams. In addition, there is also a final high-water loam which has a complex soil system to be taken into consideration (*Mammontheus trogontherii* molar find verified by Lanser, pers. commun.). The identification of Cromerial micromammals (verified by Storch, pers. commun.) in the second interpolated highwater loam from the top along with the positive magnetisation in this layer and the final high-water loam prove that there must have been Cromerian deposition of the upper part of the terrace system during the Brunhes epoch. As far as the two older terraces are concerned, to date if can only be stated that the palaeomagnetic analyses carried out so far (Fromm, pers. commun.) on the high-water loams and fine-corn inclusions have shown positive magnetisation.

Above the lower main terrace group one also finds local relics from older main terraces in the Lower Mosel Valley which have developed into rock surfaces or which are covered by loess and pumice tuff. The development of Quaternary terraces up to 285 m above sea-level is proved by the existence of a rock terrace covered by Quaternary gravel at this altitude to the South of Lay (Fig. 1).

The situation becomes much clearer when one includes the Münstermaifeld in the discussion because the older main terrace group is much more widely developed here and the younger main terraces intrude into the unclassified section of the narrow valley (Fig. 2). The continuity of the t_{M3} terrace at the junction point near Lehmen proves that these high terraces in the Münstermaifeld are indeed older main terraces and not part of the raised younger main terrace sequence. The upper main terrace group consists of one terrace body at 240-250 m (= t_{M3}), 260-274 m (= t_{M2}) and at ca. 280 m (= t_{M1}). The t_{M1} in the Münstermaifeld is not entirely certain, as it is partly deeply covered by loess or there are only few scattered gravel relics to be found. On the opposite side of the valley however, to the West of the Beierberg

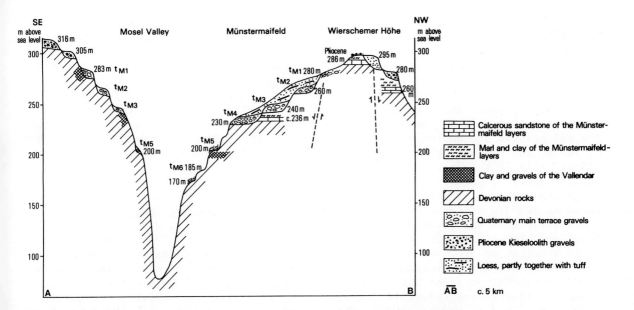

Fig. 2. Position of the main terraces, the Münstermaifeld layers and the Vallendar in the lower Mosel (Münstermaifeld-Löf-Burgen area; roughly schematic)

near Burgen, a little way off from the present-day Mosel Valley, there is a flute with Quaternary Mosel gravel rising to a height of 283 m and lying directly below the Pliocene valley relics, thus offering clear proof of the existence of this oldest Quaternary terrace through a gravel body. The conditions are, therefore, identical with those in the Middle Rhine Valley, excepting the Neuwied Basin. The similarity is such that red-tinged palaeosoils over the terrace sediments can only be followed as far as the t_{M3} terrace.

An interesting phenomenon can be observed in the formation of the main terraces. The respectively lower-lying main terraces are especially well-developed in the direction of the Neuwied Basin and are partly missing in the narrow valley sections of the Mosel upstream, whereas the older main terraces upstream are wider but are missing or only just existent downstream. This is due to the broad expansion of the lower terraces. Borgstätte (1910) thus mentions only a few local terrace flights through whose direct comparison an echeloned tectonic process in the direction of the Neuwied Basin would result. On account of this study, however, such a hypothesis can no longer be accepted.

Relics of the varied Pliocene Kieseloolith gravel are to be found above the Quaternary terraces at a height of between 295 and 325 m. At the same time there is a petrographical change to virtually entirely stable components. Above all, the Kieseloolith gravel is characterized by extremely well-rounded quartz and subordinate quartzite (Kurtz, 1926).

Since the relics of the Kieseloolith terraces can be followed at a constant elevation between 300 and 325 m from the Neuwied Basin to the Elz Valley, there could not have been any large-scale cross-faulting in the form of block-flighting in the Lower Mosel Valley. This does not, however, exclude the possibility of small local blocks (e.g., Lonning-Wolken and Wierschem areas) in which Pliocene graveal was downwarped. Moreover, Kieseloolith gravels were found on the western edge of the elevated surface between the Mosel and the Rhine valleys up to a height of 345 m. One must, therefore, assume that the northern tip of the Hunsrück was raised by about 20 m during the Quaternary period.

To the West of the Elz-Bach line there is the Trais-Karden-Cochem valley section.

On both sides of the Elz debouchure the t_{M1} and t_{M2} terraces are at the same altitude as the Pliocene degradation plain in the lower valley section. Quaternary faulting along the Elz-Bach line can, therefore, no longer be accepted. The course of the individual terraces changes abruptly to the West of the Brohl Valley. The t_{M1} gravels are no longer to be found at 285 m, but have risen to 310, 315 and 325 m, while at the same time local Quaternary gravels, occassionally with very gross iron-crusted blocks, are to be found at 295-310 m at the NW edge of the Mosel Valley. To the South of the River Mosel one also finds local Mosel gravels up to a height of 310 m above sea level, which, due to their petrographical formation, are sometimes found between Pliocene and Pleistocene sediments (e.g., in Macken, cited in Osmani, 1976). The great height at which these gravels are to be found can, however, be connected with the Quaternary uplifting of the Hunsrück mentioned above or, in the case of the local gravels, with the distance from the main river.

The siliceous oolite gravel relics show an even larger displacement tendency. From their undisturbed position at 300-315 m they rise to the West in the direction of the Elz-Bach line to 325 m, 335 m, 355 m and finally to 365-380 m (Ediger Forest) within a distance of 10 km. Post-Pliocene dislocation of a minimum of at least 50 and a maximum of 65 m can, therefore, be observed (Bibus, 1983, Fig. 3). On the western edge of the uplift area, the Kieseloolith gravels reach a height of 380 m and the oldest Pleistocene terrace a height of 325 m. These altitudes are similar to those expressed by Löhnertz (1982) and Kremer (1954) for the Middle Mosel Block. The tectonic separation of the Middle Mosel Block from the Lower Mosel Block took place through a few echeloned faults in the short Trais-Karden-Cochem valley section. Moreover, there is proof that the t_{M2} most definitely, and the t_{M1} most probably, cross into the uplift area without important dislocation values being registered. The tectonic uplifting can, therefore, be placed between the sedimentation of the siliceous oolite gravels and the old Pleistocene terraces. This would, however, mean that the development of the oldest Pleistocene terraces in both blocks did not occur in the same manner, a fact which has to be taken into consideration when large-scale comparisons are being undertaken. The altitude segment between 280 and 300 m to the East of the displacement zone could be the temporal equivalent of the altitude distance between 280 and 325 m to the West of the fault zone.

The Older Tertiary Sediments

The calcareous sandstone deposits of the Lower Oligocene Münstermaifeld layers (Kadolsky, 1975) were originally deposited at the same height, but now lie at 286 m near Wierschem (Neuhof) and at 276 m above sea level on the western edge of the Neuwied Basin (Waldorfer Höfe) overlying unweathered Devonian rock. Important displacements between the widely separated deposits can, therefore, not be documented. Within the Neuwied Basin a boring was carried out between 200 and 90 m above sea level near Kärlich which produced no calcareous sandstone, and thus no certain statement can be made concerning the amount of dislocation towards the basin.

On the other hand, there are considerable height differences within the Münstermaifeld itself (Fig. 2). The calcareous sandstone, which overlies Devonian rock in both cases, is found at 236 m above sea level 4 km North of Neuhof in its locus typicus in the Schrumpfbach valley, whereas at the periphery of Wierschem it lies at 260 m above sea level. The difference in altitude is, therefore, 50 m or rather 36 m. It is remarkable that the siliceous oolite gravels at the periphery of Wierschem have only sunk by 20 m. This means that the downwarping must have started during the Tertiary period before the deposition of the siliceous oolite gravels. To the West of Wierschem one also finds downwarped calcareous sandstone and siliceous oolite gravel relics. This suggests a NW longitudinal faulting of the Münstermaifeld as compared with the relatively wide erosion valley of the Noth-Bach. This wide erosion zone is, without doubt, connected with the Vallendar gravels on the western edge of the valley which are very thick and well preserved. The difficult and disputed question must now be raised as to whether the deep-lying Vallendar relics indicate a valley-burying process or local downwarping processes (e.g., Louis, 1953; Birkenhauer, 1973; Discussion by Semmel, 1972). There is an altitude difference of 245 m between the highest deposits on the western edge of the Mosel trough at 395 m above sea-level near Gondershausen (I am very grateful to Herr Löhnertz for this information) and the lowest Vallendar gravel deposits known to me near Lehmen at ca. 140 m above sea level. The linear connection between individual deposits, as is, for example, the case between Brieden, Pilligerheck and Lehmen and the small zone consisting of mottled kaoline clays with direct contact with Devonian rocks could offer proof of a back-filled valley. The many isolated deposits and the abrupt altitude differences, however, tend, in my opinion, to suggest that downwarping has taken place. I, therefore, believe that the Vallendar sediments have undergone large-scale downwarping with respect to their original place of deposition and have also been affected by an individual block tectonic process.

References

Bibus, E., 1983. Distribution and Dimension of young Tectonics in the Neuwied Basin and the Lower Middle Rhine. This vol.

Bibus, E., 1980. Zur Relief-, Boden- und Sedimententwicklung am unteren Mittelrhein. Frankf. Geowiss. Arb., Ser. D, Frankfurt a.M., 1:296.

Birkenhauer, J., 1973. Die Entwicklung des Talsystems und des Stockwerkbaus im zentralen Rheinischen Schiefergebirge zwischen dem Mitteltertiär und dem Altpleistozän. Arb. Rhein. Landeskde., Bonn, 34:217.

Borgstätte, O., 1910. Die Kieseloolithschotter- und Diluvialterrassen des unteren Moseltales. Thesis, Bonn, p. 55.

Kadolsky, D., 1975. Zur Paläontologie und Biostratigraphie des Tertiärs im Neuwieder Becken. Decheniana, Bonn, 128:113-137.

Kremer, E., 1954. Die Terrassenlandschaft der mittleren Mosel - als Beitrag zur Quartärgeschichte. Arb. Rhein. Landeskde., Bonn, 6:109.

Kurtz, E., 1926. Die Leitgesteine der vorpliozänen und pliozänen Flußablagerungen an der Mosel und am Südende der Kölner Bucht. Verh. Naturhist. Verein Preuß. Rheinland Westfalen, 33:97-159.

Löhnertz, L.W., 1982. Die altpleistozänen Terrassen der Mittelmosel - Überlegungen zur "Horizontalkonstanz" der Terrassen der "Rheinischen Hochscholle". Catena, Braunschweig, 9, 1/2:63-75.

Louis, H., 1953. Über die ältere Formenentwicklung im Rheinischen Schiefergebirge, insbesondere im Moselgebiet. Münchner Geogr. Hefte, Regensburg, 2:97 p.

Negendank, J.F.W., 1977. Argumente zur känozoischen Geschichte von Eifel und Hunsrück. Neues Jahrb. Geol. Paläontol., Mh. Stuttgart, 1977:532-548.

Negendank, J.F.W., 1978. Zur känozioschen Geschichte von Eifel und Hunsrück - Sedimentpetrographische Untersuchungen im Moselbereich. Forsch. Dtsch. Landeskde., Trier, 211:90.

Osmani, G.N., 1976. Die Terassenlandschaft an der unteren Mosel - eine geologische Untersuchung. 2. Bd.: Textbd. Thesis, Bonn, p. 125

Semmel, A., 1972. Geomorphologie der Bundesrepublik Deutschland. Geogr. Z., Wiesbaden, 30:149.

4.6 Cenozoic Deposits of the Eifel-Hunsrück Area Along the Mosel River and Their Tectonic Implications

J. Negendank[1]

Abstract

Investigations on stratigraphy and sedimentology of Tertiary (marine, lacustrine and fluvial deposits) and Quaternary (Moselriver terrace and Maar lake deposits) deposits of the Hunsrück-Eifel area revealed new stratigraphic results.

Its interpretation allows inferences on different vertical movements during the Cenozoic.

Introduction

The Cenozoic remnants of the Eifel-Hunsrück-area are deposits of different origin. It has therefore been difficult until now to explain the real stratigraphic position of these sediments, to relate single occurrence to each other, and to pinpoint their formation.

Detailed mapping, as well as sedimentological and palaeontological studies, revealed new stratigraphic and palaeogeographic data, but gave only few results in fixing vertical tectonic movements along definite faults.

---

[1] Universität Trier, Abteilung Geologie, Postfach 3825, D-5500 Trier, Fed. Rep. of Germany.

Geologic Setting

Figure 1 (see pocket inside back cover) shows the geographic distribution of the investigated deposits and their stratigraphic position in relationship to the new stratigraphic nomenclature. Roughly speaking, one can distinguish between the following types of sediments.

1. White clays and silts from the localities Speicher, Binsfeld, Herforst and Eckfeld and of small occurrences north of Trier (lacustrine, in some cases with a tropical soil at the base).

2. Quartz gravel and sands with small interbedded coal seams and clays with Vallendar Flora (located in Gut Heeg) northeast of Trier around Manderscheid Arenrath-Bergweiler and on the Mosel River south of Neuwied Basin (fluvial) (Löhnertz, 1978).

3. Cherts and quartzites near Idenheim north of Trier with a special fauna and flora (Kadolsky et al., 1979) which are correlated by the same authors with clay and marl occurrences of the Schrumpfbachtal at the lower Mosel River (lacustrine-brackish?).

4. Old Maar lake deposits from Eckfeld (near Manderscheid) (Pyroclastic deposits, diatomites and oilshales) and young Maar lake deposits of Meerfeld with a detrital marine Tertiary fauna and flora (Sonne and Weiler in Negendank et al., 1983).

5. Miocene-Pliocene-Quaternary river terrace deposits of the Mosel River and some tributaries.

Plateau Uplift, ed. by K. Fuchs et al.
© Springer-Verlag Berlin Heidelberg 1983

Stratigraphy of the Marine, Lacustrine and Fluvial Deposits

Figures 2 and 3 summarize this deposit stratigraphy and its encompassing area. However, it is useful to mention several sequential zones and significant palaeogeographic discoveries.

a) Tertiary Deposits

1. White clays of Speicher-Binsfeld-Eckfeld (Middle Eocene) (De Brelie et al., 1969) explained these occurrences palynologically as deposits of the Middle Eocene, including the clays which are underlain by the Eocene Maar lake deposits from Eckfeld.

2. Quartz gravel and sands (Upper Eocene-Lower Oligocene, Löhnertz, 1978). Under this heading are included gravel, sand and clay/silt deposits, which occur in different altitudes from ±280 m to ±400 m above M.S.L. They seem to belong to an Eocene river system. In different localities but especially in the territory Gut Heeg, the Vallendar Flora was found, which has been classified as the Middle Oligocene Vallendar territory near Koblenz. This deposit of white quartz gravel in the lower and quartz sand in the upper parts contains from top to bottom clay/silt lenses in addition to flora (Löhnertz, 1978). However, the palyonological studies of De Brelie, published in the same paper, revealed the same pollen spectrum as was determined in the clay occurrences. Therefore, these two dates, which evolved from a study concerning different fossils (plants and pollen), contradict each other. Consequently one must keep in mind that this stratigraphic classification as Upper Eocene to Lower Oligocene is somewhat doubtful; but aside from this, the geological interpretation appears to be valid.

3. Cherts and Quartzites of Idenheim (Miocene Baeckeroot, 1929; Middle Oligocene Kadolsky et al., 1979). Kadolsky et al. (1979) revised the stratigraphic classification of these cherts and quartzites near Idenheim. These deposits lie almost on the mountain top (±380 m). The incorporated fauna and flora seem to indicate a lacustrine to brackish environment and can be correlated with the occurrence of the Schrumpfbachtal (±235 m) at the lower Mosel River covered by a river-terrace-deposit of the Moselle. The correlation of these two horizons as Middle Oligocene allows one to draw a line determining different vertical movements if we neglect the unknown original thicknesses of these deposits.

4. Upper Middle Oligocene-Upper Oligocene marine (brackish) detrital fauna and flora in Pleistocene-Holocene sediments of Meerfeld Maar lake (Sonne and Weiler in Irion and Negendank, 1983).

These micro- and several macrofossils are well known as sediments of the Mainz-Basin because they have an Upper Middle - Upper Oligocene age. Therefore a transgression of the sea which could have originated in the Mainz-Basin may be inferred. This does not exclude possible connecting passageways between Meerfeld and the Paris Basin or the Lower Rhine Basin, which have still to be studied. The Middle Oligocene sea prograded passing the Neuwied Basin, although additional microfossils (Hystrichosphaeroidea) were found in old Maar deposits from Dehner near Blankenheim (Northern-Eifel) (Weiler, pers. commun.).

5. Miocene? Pliocene "Kieseloolith-formation. These deposits are found in different positions along the Mosel River and were described in detail in earlier publications (Kremer, 1954; Kurtz, 1929, 1932, 1938; Louis, 1953; Quitzow, 1969; Negendank, 1977, 1978).

Their occurrence is given in Figs. 1, 2 and 3 together with the Mosel River terrace deposits, indicating the beginning of the Pliocene-Pleistocene uplift.

But the base of these formation is deposited in different elevations along the Mosel River. This also caused the subdivision in the Middle Mosel block, which had a ±340 m base of this formation, and the Lower Mosel block measured at the base 290-300 m a.M.S.L. The terrace sequence of the Lower Mosel block corresponds the terrace-sequence of the Middle Rhine area (Bibus and Semmel, 1977).

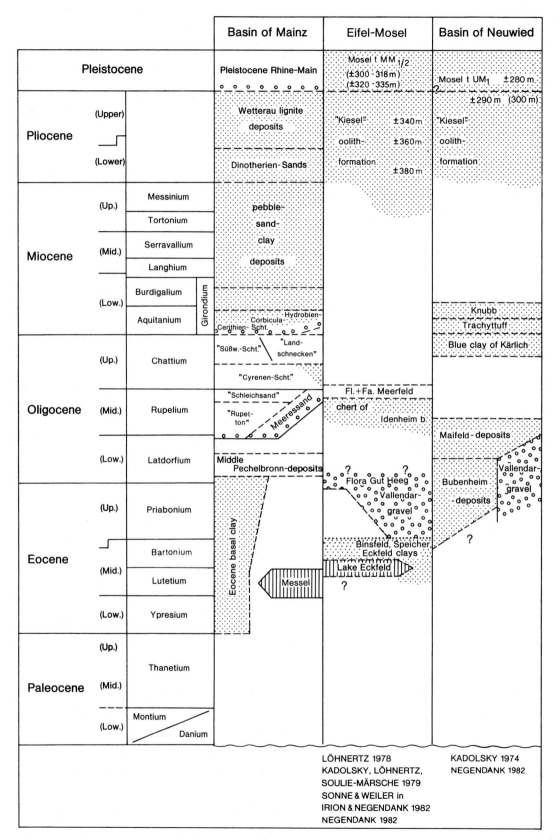

Fig. 2. Timetable of Tertiary deposits

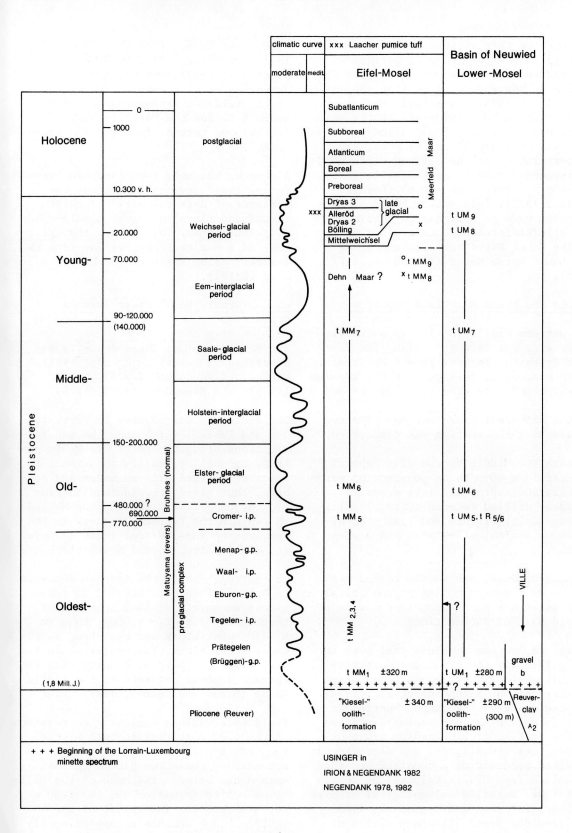

Fig. 3. Timetable of Quaternary deposits

The base of this siliceous oolite (Kieseloolite) formation is defined as the Pliocene/Pleistocene boundary by the change in heavy mineral content (Negendank, 1977, 1978). Especially the contained opaque heavy minerals (Lorraine-Luxemburg Oolite-spectrum) allow one to fix this boundary along the Mosel River in correlation to the defined Pliocene/Pleistocene boundary of the Ville (clay horizon A_2/gravel deposit b) (Boenigk et al., 1972). This opaque heavy mineral spectrum made it possible to determine the Plio/Pleistocene boundary not only at the Mosel River but also at the Maas and Lower Rhine River (Negendank, 1977, 1978).

b) Old and Young Maar Lake Deposits

The western Eifel district is a classic study area for Maar volcanism. The Maar lake deposits preserve proofs of Cenozoic Earth history of this part of the Rhenish Massif which have otherwise been eroded.

In two different Maar lake deposits new interesting information was discovered.

Beneath the Middle Eocene clay deposit in Eckfeld a 60-m thick sequence of pyroclastic and sapropelic lake deposits could be drilled (graded tuffs, lapillituffs, lapillibreccia and laminated bituminous shales, oil-shales and diatomites).

This occurrence, previously termed Kesseltal with regard to its morphological depression has been recognized now to be an old Maar of Eocene times.

Therefore, we can conclude that even in early Tertiary times volcanic activity of this explosive style had taken place in this region (Negendank et al., 1982). The sediments of the famous Meerfelder Maar lake (W of Manderscheid) disclosed a 20-m profile of fine-medium grained, graded beds (turbidites) and diatomaceous varvites documenting a complete sequence deposited from Middle Weichselian to recent time. Detailed palynological studies (Usinger in Irion and Negendank, 1983) verified the Upper Alleröd-position of the Laacher pumice-tuff, an important stratigraphic marker in young Pleistocene sediments of Middle and now also in parts of Western Europe (Erlenkeuser et al., 1972; Negendank, 1977, 1978; Irion and Negendank, 1983; Juvigné, 1977, 1980a,b; Jungerius et al., 1968). Various age tests yielded a minimum age of 29,000 years B.P. for the Meerfeld-Maar (Büchel and Lorenz, in Irion and Negendank, 1983).

A chance, but very important discovery during these investigations was the presence of detrital marine fossils (fauna and flora) of Upper Middle-Upper Oligocene times. This proves the existence of a marine transgression from the Tertiary Mainz Basin to the Meerfeld area (Eifel).

c) Quaternary Mosel River Terrace Deposits

The development of these deposits was described in detail by Kremer (1954), Müller and Negendank (1974), Müller (1976), and Negendank (1977, 1978).

The evolution can be seen in Figs. 1, 3-5 and Table 1 which show the two subdivisions along the Mosel River. In contrast to earlier publications the terraces were fixed with an index and numbers, thus giving the succession from old to young (1-9). The indices MM and UM explain the terrace deposits in the subdivision (MM=) Mittel (Middle) Mosel and (UM=) Unter (Lower) Mosel block.

The correlation of the terrace deposits is presented in both blocks. In these blocks we have a similar sequence of terrace deposits t_1-t_9, t_{MM1}-t_{MM9} in the Middle Mosel block and t_{UM1}-t_{UM9} in the Lower Mosel block. The evolution in the last block corresponds exactly with the terrace deposit sequence of the Middle Rhine (Bibus and Semmel, 1977).

Field observations suggest a correlation of the ±280 m terrace in both blocks, t_{MM3} and t_{UM1}, involving a vertical movement of these two blocks against each other between the base of the siliceous oolite formation and the ±280 m terrace deposit. The uplift of the western block amounts to approximately 50 m. This hypothesis indicates that the two oldest Pleistocene terrace deposits in the Middle Mosel block did not develo

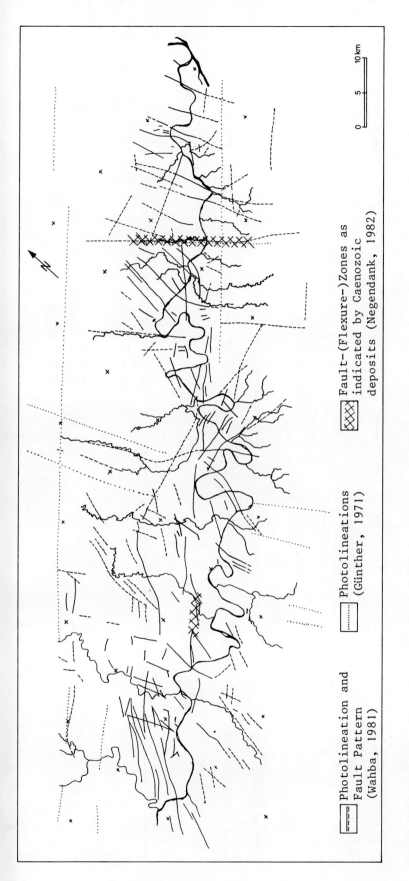

Fig. 4. Tectonic pattern of the Mosel area

Table 1. Stratigraphy of the terrace deposits along the Mosel River and comparison with the terrace of the Middle Rhine

Upper Mosel block (200–34 km)	Lower Mosel block (34–0 km) Münstermaifeld	Kobern-Dieblich-Koblenz	Middle Rhine (Bibus and Semmel, 1977)	
±360 m KOT Pliozän				
±340 m KOT Pliozän				
±300 m HÖT t_{MM} 1/2 (±320–335 m) (±300–318 m)	±290 m KOT Pliozän	±275–280–290–298 m (KOT Pliozän)	Pliozän	±300 m
±280 m OHT t_{MM} 3	±280 m t_{UM} 1	?	t_R 1	±280 m
±260 m MHT t_{MM} 4	±260 m t_{UM} 2	±260 m	t_R 2	±260 m
±240 m UHT t_{MM} 5	±240 m t_{UM} 3	?	t_R 3	±240 m
	±230 m t_{UM} 4	±230(−25) m ±215 m	t_R 4	±220 m
	±190–200 m t_{UM} 5	±190(185)–200 m	t_R 5 t_R 6	±200 m ±190 m
±180 m NN±160 m OMT t_{MM} 6	±150 m t_{UM} 6	±160 m		
±160 m NN±120 m UMT t_{MM} 7	±96–100 m t_{UM} 7	±96–100 m		
±115 m NN±100 m ONT t_{MM} 8	±80 m t_{UM} 8	±76 m		
±120 m NN±80 m NT t_{MM} 9	±60 m t_{UM} 9	∼60 m		

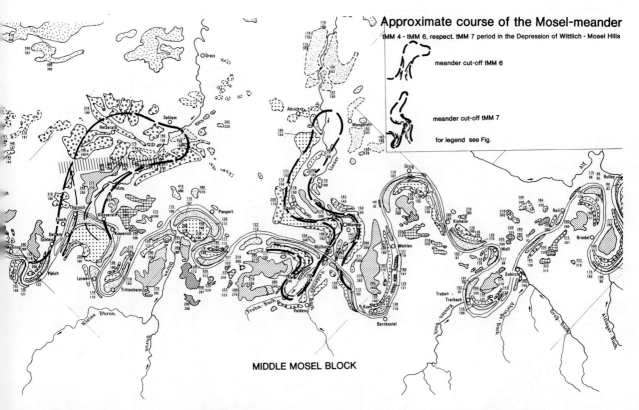

Fig. 5. Meandering of the Mosel River in the area of the Wittlicher Graben during t_{MM4}-t_{MM6} and t_{MM7} times (Quaternary)

in the Lower Mosel and the Rhine sequence. Consequently the Pliocene/Pleistocene boundary does not have the same meaning in both blocks, at the Lower Moselle and at the Rhine there must be a stratigraphic gap which seems unlikely.

Two further explanations are possible. A comparison of the terrace sequence t_{MM1}-t_{MM5} and t_{UM1}-t_{UM5} in both blocks seems to suggest a vertical movement post t_{MM5} (t_{UM5}) and pre t_{MM6} (t_{UM6}). The relative uplift of the Middle Moselle block in a tilt-like movement could possibly have compensated for the fall in the original river line (±55 m on a distance of ±160 km). This may explain the fact that all old terraces along the Mosel River are now deposited horizontally. Fortunately this interpretation includes the same Plio/Pleistocene boundary in both blocks.

Louis (1953) favored a third hypothesis supposing a pre-siliceous oolite formation subsidence of the Lower Mosel Rhine block causing intensive sedimentation in this area in comparison to the Middle Moselle block. This explanation demands that Quaternary terrace deposits are covering siliceous oolite formation deposits which thus cannot be observed.

However, the tectonic argument can be disregarded. The assumption concerning an erosive, possibly marine depression, in the Lower Mosel and Rhine block which afterwards was used by the Mio Pliocene River depositing the siliceous oolite formation is probably false.

d) Palaeogeography

During the early Tertiary (Middle Eocene) white clays were deposited in widespread lakes accompanied by the Maar volcanism, while the following Upper Eocene Lower Oligocene fluvial sediments of the Eifel-Hunsrück region with Vallendar Flora (Gut Heeg) seem to indicate a first erosive phase of the Cenozoic as a possible result of a first uplift of this area.

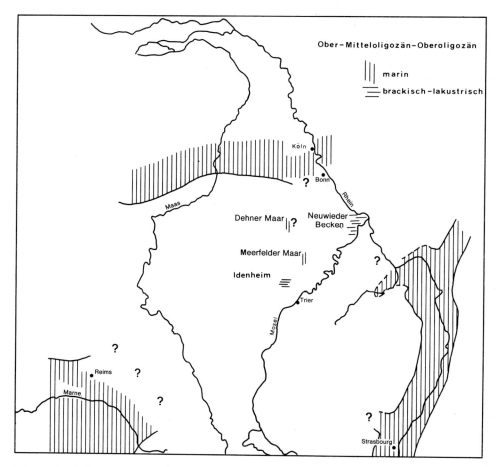

Fig. 6. Palaeogeography of Upper-Middle Oligocene Upper Oligocene times

These deposits are filling up the "river" depressions up to ±400 m of their present altitude. This could be caused by the (in this area) prograding sea penetrating the river valleys and drowning them.

This seems to have been the hypothetical niveau of the following Upper-Middle-Oligocene to Upper Oligocene marine transgression coming from the southern Mainz Basin (Fig. 6).

A preceding or the same transgression may have deposited the cherts and quartzites of Idenheim as marginal facies. The same stratigraphic position also appears to be valid for the Tertiary sediments of the Schrumpfbachtal in the southern neighbourhood of the Neuwied Basin tracing the possible pattern of the prograding sea.

A second uplift starts in Miocene times which is verified by Miocene-Pliocene fluvial terrace deposits belonging to the siliceous oolite formation. These sediments are deposited along the Mosel River in an altitude of 380, 360 and 340 m, at the Lower Mosel River down to an altitude of 290-300 m.

This uplift continues during the Pleistocene and Holocene and is documented by the river-terrace deposits of the Moselle revealing the Quaternary history of the Mosel River.

Tectonic Implications

Under the condition of one identical Pliocene/Pleistocene boundary in both blocks (Middle and Lower Mosel block) we can infer vertical tectonic movements only in one 140°-direction, the so called Elzbachtal line (Figs. 1, 4). The field observations support a relative uplift of the Middle Mosel block of about 50 m during the Quaternary, independent of each of three, possible hypothesis determining the time interval of this event.

Field observations do not support the assumption of a fault which demands the sedimentary sequence, but Erts-Landsat and photogeological interpretations seem to trace the same line (Wahba, 1979). Therefore, it must be taken into consideration that this Elzbachtal line may be an inflexion or an old escarpment.

Further lineations and faults can be seen in Figs. 4 and 5. During the Quaternary the t_{MM6} terrace in the area of the Wittlicher Senke subsided around 5-15 m in comparison with the occurrence in the southern Mosel Mountains. This can be determined by the cutoff of the meander river Bekond and Mosel at t_{MM6} times.

The position of the fault line can only be estimated (Figs. 1, 5). No further tectonic movements can be deduced with the exception of the different elevation of the correlated Middle Oligocene. This Tertiary movement has uplifted the Middle Mosel block around 30 m in comparison with the occurrence in the neighbourhood of town of Wierschem which lies in the Lower Mosel block. In this block an undetermined system of faults has caused the subsidence of the Middle Oligocene of the Schrumpfbachtal at around 50 m.

Summarizing the results, we can infer the following vertical movements.

1. First Cenozoic uplift and undercutting of the Eifel-Hunsrück-area during Eocene-Oligocene.

2. Relative uplift of the Middle Mosel block in contrast to the Lower Mosel block around 30 m, only with respect to the Middle Oligocene deposits in both blocks, Post Middle Oligocene along the Elzbachtal line. Special movements inside the Lower Mosel block at around 55 m (Wierschem-Schrumpfbachtal).

3. Uplift of the Mosel River area and undercutting during Miocene-Pliocene times at around 40 m (380-340 m), only to be determined in the Middle Mosel block.

4. Uplift and undercutting during the Quaternary approximately 200-250 m in both blocks (along the River Mosel from Trier to Koblenz).

5. Relative uplift of the Middle Mosel block in comparison with the Lower Mosel block about 50 m along the Elzbachtal line (post-KOT - pre t_{MM3} = t_{UM1} or post t_{MM5} - pre t_{MM6}).

References

Baeckeroot, G., 1929. Sur la présence des fossiles d'âge aquitanien dans de grès quartzites épars à la surface du Plateau Mosellan. C.R. Nebd. Sé. Acad. Sci., 189:804-805.

Bibus, E. and Semmel, A., 1977. Über die Auswirkungen quartärer Tektonik auf die Altpleistozänen Mittelrhein-Terrassen. Catena, 4:385-408.

Boenigk, W., Kowalcyk, G., and Brunnacker, K., 1972. Zur Geologie des Ältestpleistozäns der Niederrheinischen Bucht. Z. Dtsch. Geol. Ges., 123: 119-161.

Brelie, G.v.de, Quitzow, H.W., and Stadler, G., 1969. Neue Untersuchungen im Alttertiär von Eckfeld bei Manderscheid (Eifel). Fortschr. Geol. Rheinland Westfalen, 17:27-40.

Erlenkeuser, H., Frechen, I., Straka, H., and Willkomm, H., 1972. Das Alter einiger Eifelmaare nach neuen petrologischen, pollenanalytischen und Radiokarbon-Untersuchungen. Decheniana, 125:113-129.

Irion, G. and Negendank, J.F.W., 1982. Das Meerfelder Maar - Untersuchungen zur Entwicklungsgeschichte eines Eifelmaares mit Originalbeiträgen von Büchel, Hansen, Haverkamp, Hofmann, Irion, Lorenz, Negendank, Scharf, Sonne, Usinger, Weiler. Senckenberg, Frankfurt, submitted.

Jungerius, P.D., Riezebos, P.A., and Slotboom, R.T., 1968. The age of Eifel Maars as shown by the presence of Laacher See ash of Alleröd age. Geol. Mijnbouw, 47:199-205.

Juvigné, E., 1977. La zone de dispersion des poussières émises par une des dernieres éruptions du volcan du Laachersee (Eifel). Z. Geomorphol., 21:323-342.

Juvigné, E., 1980a. Révision de l'âge de volcans de l'Eifel occidental. Z. Geomorphol., 24:345-355.

Juvigné, E., 1980b. Vulkanische Schwerminerale in rezenten Böden Mitteleuropas. Geol. Rundsch., 69:982-996.

Kadolsky, D., Löhnertz, W., and Soulie-Märsche, J., 1979. Neue Erkenntnisse zur tertiären Entwicklung von Eifel-Mosel-Hunsrück (Rhein. Schiefergebirge) aufgrund faunenführender mitteloligozäner Hornsteine. (unpublished manuscript).

Kremer, E., 1954. Die Terrassenlandschaft der mittleren Mosel als Beitrag zur Quartärgeschichte. Arb. Rhein. Landeskde., 6:100.

Kurtz, E., 1929. Die Leitgesteine der vorpliozänen und pliozänen Flußablagerungen an der Mosel und am Südrande der Kölner Bucht. Verh. Naturhist. Verein Rheinland Westfalen, 83:97-159.

Kurtz, E., 1932. Die Spuren einer oberoligozänen Mosel von Trier bis zur Kölner Bucht. Z. Dtsch. Geol. Ges., 83:39-58.

Kurtz, E., 1938. Herkunft und Alter der Höhenkiese der Eifel. Z. Dtsch. Geol. Ges., 90:133-134.

Löhnertz, W., 1978. Zur Altersstellung der tiefliegenden fluviatilen Tertiärablagerungen der SE-Eifel (Rhein. Schiefergebirge). Neues Jahrb. Geol. Paläontol., Abh., 156,2:179-206.

Louis, H., 1953. Über die ältere Formenentwicklung im Rheinischen Schiefergebirge, insbesondere im Moselgebiet. Münchner Geogr. Hefte Regensburg, 2:97.

Müller, M.J., 1976. Untersuchungen zur pleistozänen Entwicklungsgeschichte des Trierer Moseltals und der "Wittlicher Senke". Forsch. Dtsch. Landeskde. 207:185.

Müller, M.J. and Negendank, J.F.W., 1974. Untersuchung von Schwerminalen in Moselsedimenten. Geol. Rundsch., 63: 998-1035.

Negendank, J.F.W., 1977. Argumente zur känozoischen Geschichte von Eifel und Hunsrück. Neues Jahrb. Geol. Paläontol. Mh., 1977:532-548.

Negendank, J.F.W., 1978. Zur känozoischen Geschichte von Eifel und Hunsrück. Sedimentpetrographische Untersuchungen im Moselbereich. Forsch. Dtsch. Landeskde., 211:90.

Negendank, J.F.W., Irion, G., and Linden, J., 1982. Ein eozänes Maar bei Eckfeld nordöstlich Manderscheid (SW-Eifel). Mainzer Geowiss. Mitt., 11:157-172.

Quitzow, H.W., 1969. Die Hochflächenlandschaft beiderseits der Mosel zwischen Schweich und Cochem. Beih. Geol. Jahrb., 82:79 pp.

Wahba, Y., 1979. Unveröffentlichter, interner Arbeitsbericht.

4.7 Neotectonic Movements at the Southern and Western Boundary of the Hunsrück Mountains (Southwestern Part of the Rhenish Massif)

L. Zöller[1]

Abstract

Since the Oligocene, the Hunsrück was uplifted about 200-250 m, in relation to the Rheinhessen Plateau (Mayence Basin), and tilted southwards with amounts of up to 50 m. The plateau uplift includes parts of the Saar-Nahe Basin. The quaternary partial uplift does not exceed 40-50 m.

Geomorphology of Particular Regions

Southwestern and Western Boundary of the Hunsrück

According to the petrography of gravels and the geomorphological face, it has been possible to reconstruct a tertiary valley from the northeastern Graben of Merzig into the western Hunsrück Mountains continuing to the Mosel region. This river course was abandoned in the younger Tertiary. Since that time the valley bottom was uplifted about 50 m along a SW-NE-directed flexure at the southern border of the Devonian outcrops. Towards the river Saar, further west, the uplift decreases rapidly. The flexure zone was broken into several blocks in a NW-SE direction, which were affected to various degrees by tipping movements. These movements still displaced middle Pleistocene river terraces up to 10-15 m (Müller et al., 1981).

From the longitudinal sections of the terraces of the rivers Saar and Ruwer, a middle and younger Pleistocene (perhaps even Holocene) tilt of 20-30 m of the western Hunsrück Mountains towards the south can be clearly recognized. The maximum uplift did not occur along the Mettlach-Sierck Anticline in the southern part of the Hunsrück Mountains, but rather in the northern part. Close to the mouth of the Saar to the Mosel, it is obvious that there has even been a Holocene uplift of 5 m of the Hunsrück relative to the Trier-Luxembourg Embayment and the Wittlich Graben. The sum of the younger Tertiary and Quaternary maximal uplift, in relation to the Merzig Syncline, amounts to 70-80 m, and in relation to the central part of the Luxembourg Embayment to about 100 m (Zöller, 1983a, in press).

The Eastern Hunsrück Area and its Southern Border (Between Bingen and Kirn)

Due to the altitude and the thickness of middle Oligocene to lower Miocene ("Aquitanian") layers, it can be concluded that synsedimentary subsidence of the Hunsrück was followed by a strong uplift (up to 250 m) in relation to the Rheinhessen Plateau (Mayence Basin; Fig. 1). The pleistocene terraces of the river Nahe, however, maintain the same altitude from its mouth at the Rhine far into the Saar-Nahe Basin without any significant displacement (Zöller, 1983a, in press). Therefore, only small amounts of vertical block movements must have occurred since the beginning of the middle Pleistocene, if they are related to the middle Rhine valley.

---

[1] Kirchstraße 16, D-5501 Thomm, Fed. Rep. of Germany.

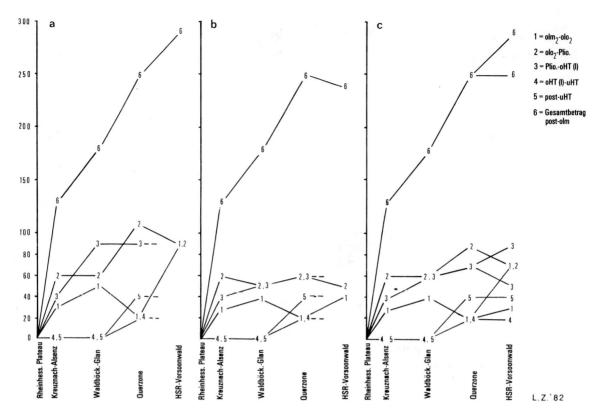

Fig. 1. Estimated amounts of the uplift of the Hunsrück and the Saar-Nahe-Mountains relative to the Rheinhessen Plateau. a) maximal amounts, b) minimal amounts, c) amounts estimated to be realistic. 1 = middle Oligocene to upper Oligocene; 2 = upper Oligocene to Pliocene; 3 = Pliocene to older Pleistocene; 4 = older Pleistocene to middle Pleistocene; 5 = since middle Pleistocene; 6 = total amount. Rheinhess. Plateau Plateau of Rheinhessen; Waldböck. horst of Waldböckelheim; Querzone Transverse Zone; HSR southern boundary of the Hunsrück

(It must be considered, however, that this occurred within 1 m.a. only.)

Recent studies prove that the middle Oligocene marine transgression intruded from the Mayence Basin far into the eastern Hunsrück along pre-existing valleys. Therefore, later vertical block movements can be related to the maximum altitude of the Oligocene sea level. Two facts seem to be evident:

1. the eastern Hunsrück - except for the Mosel Synclinorium - was tilted so that close to the watershed between the rivers Mosel and Nahe the shoreline is about 50 m higher than in the southern parts; generally it is slightly lower in the east (towards the Rhine);

2. the central Hunsrück was already lifted higher above sea level, in pre-middle Oligocene times than the eastern Hunsrück (Zöller, 1983b, in press).

The Southern Border of the Central Hunsrück Area (Between Kirn and the River Prims)

The Pleistocene terraces and the Tertiary reference levels along the river Nahe are significantly uplifted in a Transverse Zone which is part of the photogeological Zone Nordbrabant - Swabian Alb (Günther, 1977). Since the upper Oligocene the uplift has amounted to a maximum of 130 m, relative to the area between Bad Kreuznach and the river Glan (Saar-Nahe-

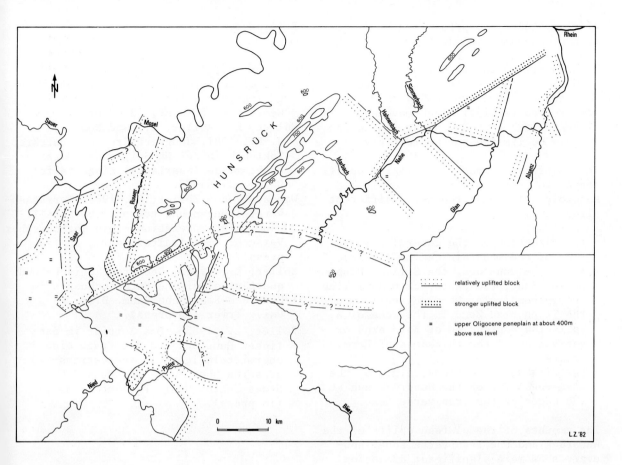

Fig. 2. Younger Tertiary and Quaternary relative block movements in the Hunsrück-Saar-Nahe Region. (Due to different reference levels, no absolute amounts of block movements can be illustrated in this figure)

Mountains), and to about 40 m since the beginning of the middle Pleistocene. If related to the Plateau of Rheinhessen, the uplift may have reached even 250 m since the Oligocene. Thus, the Transverse Zone substitutes the entire amount of the uplift at the southern boundary of the Hunsrück between Bingen and Kirn.

Between Kirn and the river Prims neither Pleistocene terraces nor tertiary peneplains show any significant displacement at the boundary between the Hunsrück and the Saar-Nahe Basin (see Schwab, this vol.).

Tectonic Inferences of the Geomorphological Results

Figure 2 tries to illustrate the relative vertical block movements in the entire region which has been considered. In the area of the Transverse Zone, the southern edge of the plateau uplift does not coincide with the fault system at the southern boundary of the Hunsrück but probably widens to the south.

The neotectonic inclination from the Transverse Zone towards the Graben of Merzig and the Lothringian Depression probably occurs along diagonal WNW-ESE faults or flexures which up to now can only be presumed by means of photogeological methods. The perpendicular direction of SSW-NNE running faults is to be seen at the western boundary of the Rhenish Massif to the Lothringian Depression as well as between the Transverse Zone and the Upper Rhine Graben. In the latter area vertical block movements along SSW-NNE to S-N running faults can largely be limited to the Tertiary.

The illustrated zones of relative uplift appear to coincide with Palaeozoic and Mesozoic sills and troughs of different directions:

- the direction of the general orientation of the mountain chains (SW-NE), e.g., the Hunsrück, the Sill of Düppenweiler/Saar, the Saarbrücken Anticline
- the perpendicular NW-SE direction, e.g., the Transverse Zone in the zone of a Variscian Depression of fold axes or variations of the thickness of Permian layers
- the diagonal directions, e.g., at the western border of the Hunsrück and at the edges of the Transverse Zone.

The amounts of the plateau uplift of the Hunsrück and the Transverse Zone, however, show more significant dimension than those of the vertical movements of the other blocks in the foreland.

References

Günther, R., 1977. Großfotolineationen des mitteleuropäischen Raumes und ihre geologisch-tektonische Bedeutung. Geotekton. Forsch., 53:42-67.

Müller, E., Zöller, L., and Konzan, H.-P., 1981. Jungtertiäre und quartäre Tektonik in der NE-Spitze der Merziger Grabenmulde (Saarland). Eiszeitalter Gegw., 31:65-78.

Wahba, Y. and Zöller, L., 1983. Terrassenverstellungen und tektonische Satellitenbildinterpretation - ein methodischer Versuch. Eiszeitalter Gegw., 33 (in press).

Zöller, L., 1983a. Geomorphologische und quartärgeologische Untersuchungen im Hunsrück-Saar-Nahe-Raum. Diss. FB 3 Univ. Trier (in press).

Zöller, L., 1983b. Das Tertiär im östlichen Hunsrück und die Frage einer obermitteloligozänen Meerestransgression über Teile des Hunsrücks. Neues Jahrb. Geol. Paläontol. Mh., (in press).

4.8 The Lower Pleistocene Terraces of the Lahn River Between Dietz (Limburg Basin) and Laurenburg (Lower Lahn)

W. Andres and H. Sewering[1]

Abstract

Along the Lower Lahn valley between Diez (Limburg Basin) and Laurenburg, the base levels of the lower Quaternary main terraces which are mostly covered by fluvial sediments and loess sheets, were mapped out exactly by the hammer refraction seismic method. It can be shown that there is a fault of 20 m at the western margin of the Limburg Basin while the younger terraces pass this area with a continuous gradient. Moreover, it can be deduced that along the Lower Lahn a vertical erosion of 120 m to 150 m took place in a period of about 650.000 years.

The terraces of the Lower Lahn valley between Diez and the mouth to the Rhine have been subject to previous investigations (Lauterbach, 1914; Ahlburg, 1915). Their course and typical formation in comparison with the remains of valley floors in the Limburg Basin caused Ahlburg (1915) to assume tectonic influences in the form of warping at the transition from Limburg basin to the Lower Lahn valley. More recent publications (Andres, 1967; Müller, 1974, 1975) also suppose dislocations of the terrace basis in this area.

Müller (1974) mapped the Lower Lahn terraces and presented a detailed differentiation of the Quaternary valley floor levels which made possible a parallelization with the Rhine terraces. Altogether nine different quaternary valley-floor levels could be identified. The complex of the main terrace was subdivided into three levels (T_7 - T_9). [Müller's terrace succession begins from the valley floor up to the older levels (T_1 to T_{10})]. An older level (T_{10}) could not be characterized with certainty as Quaternary. It can be presumed that the basis of Müller's T_7 terrace corresponds with that of the younger Rhine main terrace (in the sense of Kaiser, 1961).

In addition, in the Limburg Basin all lower Quaternary valley floors should have subsided in comparison with the Lower Lahn valley by an amount of 9-12 m (Müller, 1975)

These preliminary examinations, the possibility of a connection with the Rhine terraces, and in this context the comparison with the results of Bibus and Semmel in the Upper and the Lower Middle Rhine valley (Bibus and Semmel, 1977) caused a renewed detailed mapping of the main terrace levels between Diez and Laurenburg. This was done especially to find a possible tectonic dislocation in the morphographically obvious bend area near Altendiez/Fachingen between the Limburg Basin and the deeply incised Lower Lahn valley. The aim was to map exactly the erosion surface in the rock in situ as the only reliable reference horizons, which are mostly hidden by a cover of fluvial sediments and loess sheets with a thickness of several meters. This was the only way to attain a differentiation and parallelization which exceeded previous knowledge. This aim was reached by using the hammer refraction seismic method. Twelve profile

---

[1] Geographisches Institut, Universität Marburg, Deutschhausstr. 10, D-3550 Marburg, Fed. Rep. of Germany.

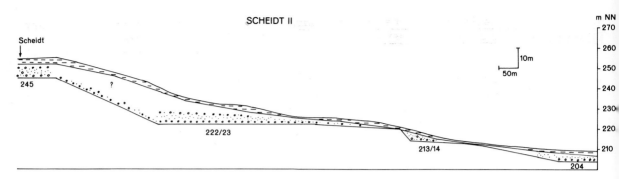

Fig. 1. Cross-section of the main terraces based upon refraction-seismic measuring (legend see Fig. 2)

series with a total length of 8300 m were measured in the river segments mentioned above between Diez and Laurenburg in the main terrace level. Most of the different substratums could be identified beyond doubt. The sound travel times in the loess sheets (between 250 and 300 m s^{-1}) and the fluvial sands and gravels (usually between 350 and 700 m s^{-1}) could be distinguished from the rock in situ (more than 2000 ms^{-1}). Problems of interpretation appeared only at transitions with deeply decomposed schists. To solve this problem, borings up to a depth of 12 m were carried out as a control.

Those results show that a number of nearly horizontal or only slightly inclined erosion surfaces in the bedrock are hidden by gently sloped profiles in the main terrace level between 190 and 260 m above M.S.L. They are covered mostly by sands and gravels, and should be former Lahn valley floors. For example, a 400 m wide erosion surface can be found in the rock in situ at about 222 m below the gently ascending plain (from 224 to 239 m) SE of Scheidt. The sediment cover increases constantly from SE to NW (Fig. 1).

The different levels are separated from each other by erosion surfaces in the rock in situ which ascend to the upper terraces, which means distinct steps caused by erosion.

A comparison of the various profiles showed that all levels appear several times in the same succession, one upon another in different series. This fact supports the assumption that they represent independent valley floors, except for the three lowest levels (basis at 191, 195 and 198 m), which are existent only at the Cramberg spur (Fig. 2). The vertical distances, the width of the valley floor, and the succession which is demonstrated in Fig. 2 makes sure that this terraces represent the main terrace complex.

The sequence of this complex, which was published by Müller (1974), could be verified in its essential traces. Also the subdivision of his T_7 terrace (t_{L5}) into three levels at the Cramberg spur could be confirmed (Müller, 1974). Decisive information was the discovery that the t_{L6} terrace between 190 and 200 m in the Cramberg area may also be divided into three levels. This sequence is to be put between Lahn terraces T_6 and T_7 (Müller, 1974) and represents the transition between the wide high valley and the narrow valley. Also the T_8 (t_{L4}) and the T_{10} (t_{L1} and t_{L2}) can be divided into two levels each. But it is still impossible to determine their age. Consequently the main terrace complex consists of six independent terrace levels. With regard to the whole succession, as well as to the absolute altitude and the typical dominance and form of certain levels, the correspondence with the main terrace complex of the Rhine is striking. Only the vertical distance of the older terraces is smaller at the Lahn than at the Rhine. Therefore, it seems to be justified to parallelize the terrace suite t_{L1}-t_{L6} with the corresponding Rhine terraces t_{R1}-t_{R6} (Bibus and Semmel, 1977; Semmel,

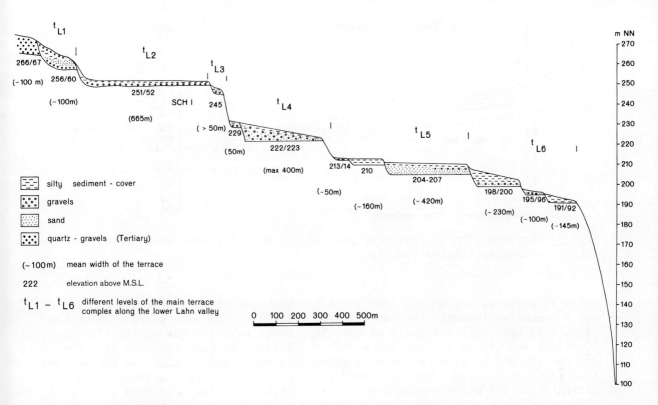

Fig. 2. The complex of the main terraces between Balduinstein and Laurenburg (Lower Lahn area)

1979; Bibus, 1980) considering the fact that Müller (1974, 1975) could correlate his T7 terrace (t_{L5}) without any doubt with the younger main terrace of the Rhine (t_{R5}) (Bibus, 1980).

Sofar these results refer to findings in the profile series in the main terrace level near Scheidt, Holzappel, Cramberg, Langenscheid and west of Altendiez from the Kehrberg to the Unterlohskopf. Another profile, which was surveyed between Altendiez and Heistenbach in an area which morphologically belongs to the western margin of the Limburg Basin, requires a special interpretation (Fig. 3). Beneath a gently ascending surface (from 177 to 210 m) is hidden a succession of three almost horizontal platforms eroded into the rock in situ. Each of them is covered by terrace gravels and altogether they are overlain by loess. There base levels are found at 171, 178 and 185 m. At this locality exists also a lower level with a basis at about 160 m. It can be identified beyond doubt as a terrace which upstream gains exceptional importance and which Ahlburg (1915) called the Limburger main terrace. Later it was included as T_5 into the system of Pleistocene Lahn terraces (Andres, 1967; Müller, 1975).

The fact that above these T_5-valley floors the main terrace level sensu strictu starts with a distinctly differentiated succession so far had been hidden by the sediment cover and the gentle slopes. Instead a uniform main terrace level at about 180-185 m had been proposed (Andres, 1967; Müller, 1975), which should possess a sublevel at some places (Ahlburg, 1915; Müller, 1974). The newly achieved knowledge of the existence of a repeatedly stepped lower main terrace level east of the line Fachingen/Altendiez, which may be found above the T_5-terrace (in the sense of Andres, 1967, and Müller, 1974), makes a parallelization of the main terrace complex possible with regard to the whole succession of terraces. It is striking that the lowest platforms at Cramberg and Heistenbach correspond even in the

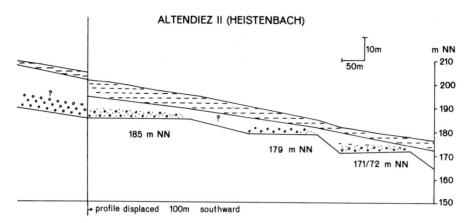

Fig. 3. Cross-section of the lower part of the main terraces based upon refraction-seismic measuring (legend see Fig. 2)

width of their valley floors (145-150 m). But a comparison between the base levels shows that here exists a fault of exactly 20 m (171 and 191 m, 178 and 198 m and 185 and 204-207 m) between the main terrace complex upstream from Fachingen/Altendiez at the western margin of the Limburg Basin and the Lower Lahn between Balduinstein and Laurenburg. This fault of almost exactly 20 m corresponds with the theory of Lauterbach (1914) and considerably exceeds the values preposed by Ahlburg (1915) and Müller (1974, 1975). This takes place in a rather narrow area between Lahnkilometer 86 and 89, where the "normal" altitude of the main terrace at the Lower Lahn can already be found again.

A direct consequence of this relative subsidence was the fact that the formation of wide valley floors still continued in the Limburg Basin, while in the Lower Lahn valley area the dissection of the t_{L6} began to take place marked by a strongly vertical erosion. As the T_4 terrace which corresponds with the lower middle terrace of the Rhine (Müller, 1974) already passes the bend area with a continuous gradient, it must be presumed that in the upper Quaternary no major faultings occurred. In relation to the lower edge of the t_{L6} the following vertical erosion at the Lower Lahn near Cramberg reached a value of about 95 m, but near Diez (Limburg Basin) only about 70 m.

With regard to the almost continuous level of the main terrace along the Lower Lahn and the possibility to parallelize the Lahn terraces with the Rhine terraces and considering the age determination of the t_{R4} Rhine terrace, which has been performed by Bibus (1980), at the Lower Lahn may have taken place a vertical erosion of about 150 m at Lahnkilometer 134 (Wolfsmühle) until about 120 m at Lahnkilometer 89 (Kehrberg) after the dissection of the t_{L4} terrace (222 m terrace). For this a period of about 650,000 years must be taken into account (see also Bibus, 1980).

References

Ahlburg, J., 1915. Über das Tertiär und das Diluvium im Flußgebiet der Lahn. Jahrb. Königl. Preuß. Geol. Landesanst. 36:269-373.
Andres, W., 1967. Morphologische Untersuchungen im Limburger Becken und in der Idsteiner Senke. Rhein-Main. Forsch., 61.
Bibus, E., 1980. Zur Relief-, Boden- und Sedimententwicklung am unteren Mittelrhein. Frankfurter Geowiss. Arb., Ser. D, 1.
Bibus, E. and Semmel, A., 1977. Über die Auswirkung quartärer Tektonik auf die altpleistozänen Mittelrhein-Terrassen. Catena, 4:385-408.

Kaiser, K., 1961. Gliederung und Formenschatz des Pliozäns und Quartärs am Mittel- und Niederrhein, sowie in den angrenzende Niederlanden unter besonderer Berücksichtigung der Rheinterrassen. Festschrift 33. Dtsch. Geogr.-Tag, Köln, pp. 236-278.

Lauterbach, W., 1914. Das Diluvium zwischen Limburg und Koblenz. Thesis Gießen, 53 pp.

Müller, K.-H., 1974. Zur Morphologie der plio-pleistozänen Terrassen im Rheinischen Schiefergebirge am Beispiel der Unterlahn. Ber. Dtsch. Landeskde., 48:61-80.

Müller, K.-H., 1975. Tektogenetische und klimagenetische Einflüsse auf die Talentwicklung an der Unteren Lahn. Z. Geomorphol., 23:75-81.

Semmel, A., 1979. The terraces of the Rhine river: Reference level for the shield uplift. In: Illies, H., Prodehl, C., Schmincke, H.-U., Semmel, A. (eds.) The quaternary uplift of the Rhenish Shield in Germany. Tectonophysics, 61: 197-225.

4.9 Quaternary Tectonics and River Terraces at the Eastern Margin of the Rhenish Massif

K. H. Müller and S. Lipps[1]

Abstract

The Zwester Ohm, a tributary river of the Lahn, connects the Hessen depression with the eastern margin of the Rhenish Massif. Eleven Pleistocene erosion surfaces can be distinguished. The younger terraces fade in valley-floor steps, upstream. The older levels subside in step faults down to the Hessen Depression. The fault pattern mainly displays three directions. Movements are still taking place at the NNE-SSW-directed dislocations.

The river Zwester Ohm, coming from the Vogelsberg Mountain, flows from the east to the Lahn south of Marburg (Fig. 1), and thus connects parts of the Hessen Depression with the eastern margin of the Rhenish Massif. The history of the Zwester Ohm Valley is thus to a great extent influenced by the tectonic development at the boundary between those natural areas.

The Zwester Ohm first flows from the Vogelsberg to the subsiding area of the Ebsdorfergrund, which belongs as a southern spur of the Amöneburger Basin to the European graben system. It is covered mainly by Tertiary and Quaternary sediments. To the west the Ebsdorfergrund is bounded by the pedestal of the Lahnberge consisting of bunter sandstone. Here the Zwester Ohm has incised very intensively and forms a nar-

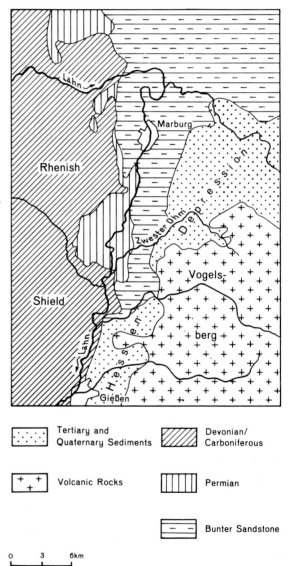

Fig. 1. Generalized geologic map of study area

[1] Geographisches Institut, Universität Marburg, Deutschhausstr. 10, D-3550 Marburg, Fed. Rep. of Germany.

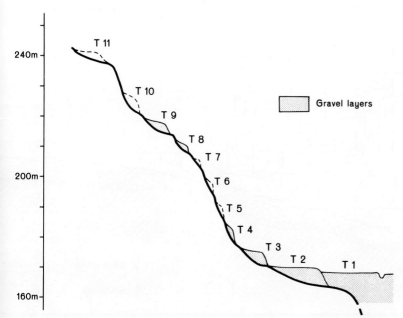

Fig. 2: Schematic succession of terraces

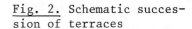

Fig. 2. Schematic succession of terraces

row valley. Near its mouth, the river Lahn cuts down through Permian layers. The Zwester Ohm does not reach the adjacent Rhenish Massif in the west.

The development of the Zwester Ohm valley has been influenced to a great extent by the quickly changing rock in place as well as by climatic and tectonic events. The morphogenesis in this area proved thus to have been extremely complex. Huckriede and Zachos (1969) could prove residuals of a Pliocene river in the Amöneburger Basin and Ebsdorfergrund. The Lower Tertiary sediment series in the central parts of these subsiding areas are situated in lower layers than the Pliocene sediments.

In the Pliocene the subsidence factor was first exceded by the vertical erosion of the waters. This may be seen in connection with the gradual changing of climate at the Tertiary/Quaternary boundary. The changing from dominating accumulation during the Tertiary to dominating fluvial erosion during the Quaternary is not documented in the Ebsdorfergrund. At that time neither thick sediment layers nor strongly marked fluvial terraces could develop.

From the upper Early Pleistocene (probably Tiglium C) terrace residuals are preserved which can be differentiated with certainty and followed over significant river sections. From now on a terrace sequence exists which may be parallelized with the sequences at the Lower Lahn (Müller, 1974) and at the Middle Rhine (Bibus, 1977, 1980; Semmel, 1977) and also with the Quaternary stratigraphy of Brunnacker et al. (1975, 1979, etc.) at the Lower Rhine. Differentiated were the main terrace complex (T_{10} to T_8) with wide residual surfaces and partly mightly sediment bodies, the narrow valley terraces (T_7 to T_3) with mainly small rock benches and the valley floor terraces (T_2, T_1) (Fig. 2).

The lower terraces up to the T_8 disappear under the valley floor level upstream. Each terrace level can be correlated with a step of the valley floor. These steps are also preserved in the native rock below the Quaternary fluvial sediments.

Müller (1974) described an increasing acclivity of the terraces with decreasing age in the Lower Lahn area. This is only partly the case for the Zwester Ohm valley terraces. For the younger terraces, which exist only at the lower course, a certain parallelism to the present Ohm course may be observed. The older terraces have almost no gradient in this area.

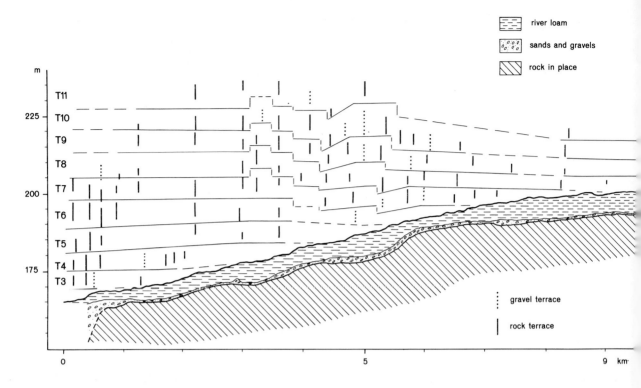

Fig. 3. Longitudinal profile up Zwester Ohm Valley

In contrast, the upstream section of the river resembles the western margin of the Limburg Basin (Müller, 1974, Fig. 1). Here the erosion surfaces show gradient opposite to that of the present Zwester Ohm Valley bottom, stepped down northeastward to the Ebsdorfergrund. The upper terraces are always stronger affected than the lower.

Here the subsidence of the Ebsdorfergrund is clearly documented as part of the Hessen Depression, which continued during the Quaternary and, from the Tiglium C till today attained the proved total amount of 10 m, in relation to the Lahnberge. The central depressed area of the Ebsdorfergrund probably has subsided even more strongly.

Strictly, the tectonic development has been much more complicated. At some fault lines terrace dislocations with an amount of more than 10 m could be identified. The general northeastward stepped incline is interrupted by tilted and uplifted blocks (Fig. 3).

The pattern of disturbances in the Zwester Ohm Valley area, which results from the analysis of the dislocations of the Quaternary terraces, shows three marked directions (Fig. 4). The rarely found, but conspicuous NE-SW direction corresponds to Hölting and Stengel-Rutkowski's (1964) Ebsdorfergrund line, which is connected with the antithetic tipping of the blocks between the Vogelsberg and the eastern margin of the Rhenish Massif.

A number of faults which follow the NNW-SSE direction, cut those blocks several times in the bunter sandstone area. A younger development can be proved because of the terrace displacements. The activity of both dislocation generations ends, however, already in the Middle Quaternary.

Another fault generation appears mainly at the boundary between bunter sandstone area and Ebsdorfergrund, and follows the NNE-SSW direction. These are the youngest faults which generally show the highest

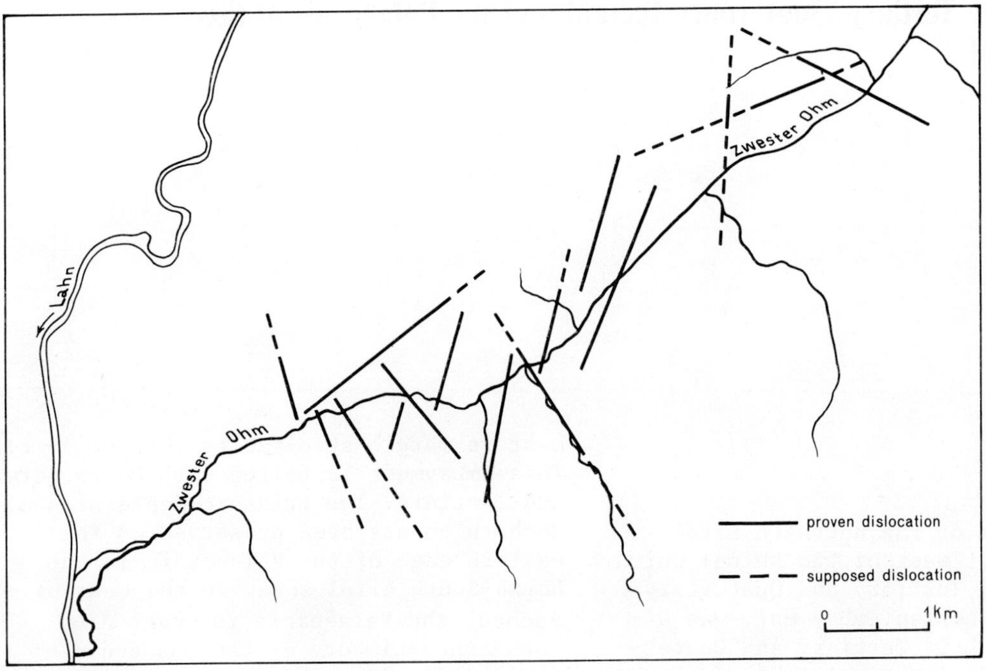

Fig. 4. Tectonics

— proven dislocation
--- supposed dislocation
0   1 Km

amounts of terrace displacement. From the magnitude of tectonic activity during the Quaternary it is possible to conclude a continuation of dislocations at the NNE-SSW directed faults until today.

The lacking gradient of the native rock below the valley bottom in the middle course area of the Zwester Ohm and the consequent development of bogs, which was similarly interpreted by Hölting and Stengel-Rutkowski (1964) in the Lahn Valley, hint at recent tectonic movements at the NNE-SSW directed faults.

Recent tectonic activity in an area which is covered by relatively thin sediment layers belonging to the marginal facies of the Germanic basin (Permian and Triassic) must be seen in direct connection with the tectonic events in the Rhenish Massif. A relative depression of its marginal areas has taken place until today.

References

Bibus, E., 1977. Das untere Mittelrhein-Tal. In: Bibus, E., Semmel, A. (Hrsg.) Über die Auswirkung quartärer Tektonik auf die altpleistozänen Mittelrhein-Terrassen. Catena, 4:385-408.

Bibus, E., 1980. Zur Relief-, Boden- und Sedimententwicklung am unteren Mittelrhein. Frankfurter Geowiss. Arb., Ser. D, 295 p.

Brunnacker, K., 1975. Der stratigraphische Hintergrund von Klimaentwicklung und Morphogenese ab dem höheren Pliozän im westlichen Mitteleuropa. Z. Geomorphol. 23 (Suppl.):82-106.

Brunnacker, K., Urban, B., and Zaiss, S., 1979. Dünnschicht-chromatographisches Verhalten quartärer Altwassersedimente am Niederrhein. Catena, 6:63-71.

Hölting, B. and Stengel-Rutkowski, W., 1964. Beiträge zur Tektonik des nordwestlichen Vorlandes des basaltischen Vogelsberges, insbesondere des Amöneburger Beckens. Abh. Hess. Landesanst. Bodenforsch., 47:1-37.

Huckriede, R. and Zachos, S., 1969. Die pliozänen Flußschotter auf den Lahnbergen bei Marburg - ein wichtiges Dokument zur hessischen Landschafts- und Flußgeschichte. Geol. Palaeontol., 3:195-206.

Müller, K., 1974. Zur Morphologie der plio-pleistozänen Terrassen im Rheinischen Schiefergebirge am Beispiel der Unterlahn. Ber. Dtsch. Landeskde., 48:61-80.

Semmel, A., 1977. Das obere Mittelrhein-Tal. In: Bibus, E., Semmel, A. (Hrsg.) Über die Auswirkung quartärer Tektonik auf die altpleistozänen Mittelrhein-Terrassen. Catena, 4:385-408.

4.10 The Late Tertiary-Quaternary Tectonics of the Palaeozoic of the Northern Eifel

R. Müller[1]

Abstract

The Palaeozoic of the northern Eifel (Rhenish Massif west of the Rhine) which borders on the Tertiary and Quaternary of the Lower Rhenish Embayment, was affected in the Late Tertiary and Quaternary by powerful vertical tectonic movements. Rotations of part areas took place. The uplift did not take place en bloc, but occurred rather in zones, which are bordered by tectonic structures (faults, flexures). The formation of the faults is Variscan and their movement has been revived.

In comparison with the Late Tertiary-Early Quaternary debris of the Lower Rhenish Embayment (+120 m above mean sea level) the Palaeozoic area shows a relative uplift of over 300 m in the Late Tertiary and Quaternary.

Earthquakes, important recent uplift processes and active faults all argue for the continuing tectonic activity in the Palaeozoic area of the northern Eifel.

1. Introduction

As a part of the Rhenish Massif west of the Rhine, the Palaeozoic in the northern Eifel borders, in the Aachen-Düren area, on the Lower Rhenish Embayment with its western margin step-faults (Knapp, 1978). This Embayment is filled with Quaternary and Tertiary. The triassic depression of Mechernich has been preserved on the eastern edge of the Palaeozoic in the North-South Eifel zone. To the West of Aachen, the Palaeozoic is covered by the Aachen-Limburg cretaceous sediments (Fig. 1).

The Palaeozoic, which includes sediments from Cambrian to Upper Carboniferous, is Caledonian and Variscan folded. The most important tectonic structure is the Stavelot-Venn anticline, which strikes NE-SW and plunges to the NE. Parallel folds (Inde Syncline, Aachen Anticline, Wurm Syncline and others) join the area to the Northwest, and are perturbed by upthrusts and overthrusts (Venn Overthrust, Aachen Overthrust, etc.) and perpendicularly running NW-SE faults. The Variscan folds plunge at the North-South Eifel zone, in the Southwest of the Stavelot-Venn Anticline. The Lower Rhenish Embayment, connected to the Palaeozoic Massif with its Mesozoic cover represents an important NW-SE-running dilatation structure in northern central Europe. Its block tectonic development belongs to the Tertiary to Quaternary periods. It reaches from the North Sea to the Palaeozoic basement of the Rhenish Massif.

At the edge of the northern Eifel, the Palaeozoic sinks under the Caenozoic of the Embayment, at the points of important NW-SE striking faults. Narrow trenches filled with Tertiary sediments reach into the Palaeozoic like fingers and end there. Tectonically, the Palaeozoic and its Mesozoic cover is thus in interrela-

---

[1] Geologisches Institut RWTH Aachen, Wüllnerstr. 2, D-5100 Aachen, Fed. Rep. of Germany.

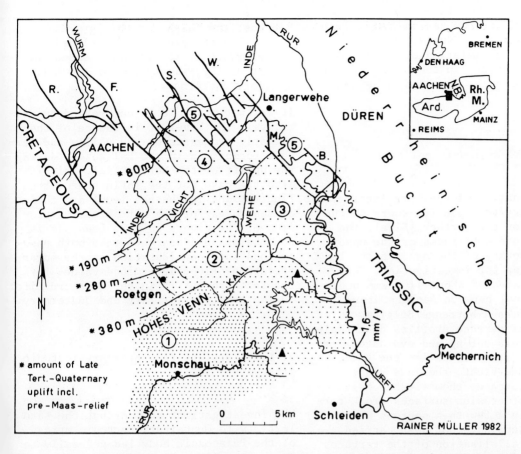

Fig. 1. The Late Tertiary-Quaternary uplift of the Palaeozoic of the northern Eifel. The relative uplift is presented by the zones bordered by the lines and the given amounts, taking into account the morphology of the "Pre-Maas Mountain Range". ■ area of investigation, ▲ carbonate spring - 1.6 mm/y - recent maximum of uplift. 1 Venn-Monschau zone; 2 Raffelsbrand-Rurberg zone; 3 Rott-Zweifall-Abenden zone; 4 Kornelimünster-Schevenhütte zone; 5 Eschweiler-Strass zone. L Laurensberg fault; R Richterich fault; F Feldbiss fault; S Sandgewand fault; W Weisweiler fault; M Merode fault; B Birgel fault; Niederrheinische Bucht = Lower Rhenish Embayment

tion with the western margin step-faults of the Lower Rhenish Embayment (Richtericher Fault, Feldbiss, Sandgewand and others), and to the East of the Bight with the Ruhr block, which is the southeastern continuation of the central Netherland Graben.

The uplift of the Palaeozoic of the northern Eifel had already begun in the Upper Oligocene/Lower Miocene, while the Lower Rhenish Embayment sank in relation to it (Richter, 1962; Knapp, 1978). Remains of the Oligocene/Miocene covering still exist today in various levels of the Eifel. In the Pliocene the uplift continued. Wide river systems developed on the Eifel Palaeozoic, and at some point the Maas and Rhine functioned as their receiving streams in the greatly subsiding Lower Rhenish Embayment. The characteristic block structure, like that of the Rur block, was formed there, especially on NW-SE striking downthrown faults.

The uplift of the Palaeozoic in principle continued in the Quaternary, along with a continued relatively sinking Lower Rhenish Embayment (Quiring, 1926). The climatic change during the transition to the Quaternary and the continuing and increasing uplift of the Palaeozoic Massif led to the Eifel rivers cutting themselves deeply into the landscape, forming deep valleys. Small and narrow fluvial terraces were preserved in the Palaeozoic erosion area (Kurtz, 1906; Quaas, 1917), while accumulation of the eroded material took place in the Lower Rhenish Embayment, over the Maas and Rhine terraces.

2. Method and Regional Application

For determination of the Late Tertiary to Quaternary tectonic framework in the Palaeozoic of the northern Eifel, the Plio-Pleistocene terraces of the upper Rur, a tributary of the Maas, presented the best area for investigation. This brought an area of about 1100 km^2 under investigation, between Aachen-Düren/Kreuzau-Heimbach-Botrange/Belgium. Longitudinal terrace profiles on a scale of 1:5000 were set up and evaluated for the Rur and its tributaries the Kall, the Wehe, the Vicht, the Inde and the Wurm. The terraces themselves consist primarily of erosion surfaces (rock terraces); only a few thin neptons on the lower terraces have not been carried away. A parallelisation of the terraces or the erosion surfaces took place by means of "Terrace Guide Groups" (Müller, 1978-1981).

Because a Late Tertiary-Quaternary basis in the Palaeozoic cannot be clearly defined by sediments, the clear morphological kink below the Tertiary plains (Richter, 1962) at the beginning of the steep or narrow valley was used to quantify the amount of uplift of the Palaeozoic. This kink is developed in all valleys. As it documents the beginning of the powerful uplift of the Rhenish Massif in the Late Tertiary-Early Quaternary, its Pliocene-Early Pleistocene age is assumed (Richter, 1962). In the Lower Rhenish Embayment the base of the Maas debris of similar age of the Rur block near Düren and north of Weisweiler, at plus 120 m above mean sea level is available to be used for the calculation. During sedimentation of the oldest terraces of the Maas in the Eifel foreland, an uplifted Palaeozoic mountain range, the morphology of which is not known in detail, had already existed since the Upper Oligocene/Lower Miocene. The height of this pre-Maas relief has therefore been included in calculations of the relative amount of uplift of the Palaeozoic for the Late Tertiary-Quaternary.

It was not possible to comprehend quantitatively the level of influence of the climatic processes on the valley and terrace formations in the Palaeozoic during the uplift of the North Eifel. This is also the case for the influence of the fluctuations in the sea level and their consequences for the retrogressive erosion of the rivers in this area. It is nonetheless necessary to take both phenomenona into consideration in the interpretation of the determined values for the tectonic uplift in the northern Eifel during the Late Tertiary and Quaternary.

3. Late Tertiary to Quaternary Tectonic Framework

The investigations showed that the Late Tertiary to Quaternary tectonic stress of the Palaeozoic body led generally to vertical uplift. In addition, rotation of parts of the area took place. Detailed dating of the epirogenic processes is not possible, because of the lack of datable sediments.

On the basis of the detailed field work on the terraces, however, it is possible to say that the vertical movements did not affect the whole Palaeozoic evenly. The upper Rur and Kall area was more extensively affected by uplifting than the area in the lower courses of the Vicht, Inde and Wehe. The uplifting can be described using two differently oriented axes [ca. NE-SW (60°) and NW-SE (130°)] which intersect amblygonaly. Thus an unhomogeneous tectonic picture in the Palaeozoic area developed in the Late Tertiary to Quaternary, with regard to the extent of vertical uplift its chronological effectiveness and its geographical arrangement. This picture can best be portrayed by use of zones, illustrating the different levels of uplift (Fig. 1). The zones are arranged in step faults like from the Hohe Venn Mountains down to the Lower Rhenish Embayment. Their boundary lines are structures of a tectonic nature (faults, flexures) with different directions and vertical throws. These structures can be trace back to Variscan tectonic stress and reappear again in the Late Tertiary/Quaternary. On the NW flank of the Stavelot-Venn Anticline, which is equivalent to the decline of the Hohe Venn Mountains, NE-SW striking structures dominate; these structures are dislocated by NW-SE striking faults in the Northwest. In contrast to these, several directions occur for the border structures on the SE flank of the Stavelot-Venn anticline: N-S, E-W and NW-SE. The border of these two differently structured areas runs NNE-SSW (Fig. 1). It defines the western edge of the N-S Eifel zone in the Late Tertiary to Quaternary for the Palaeozoic of the northern Eifel.

The area which includes the crest of the Hohe Venn Mountains and the towns of Monschau and Simmerath shows the greatest relative amount of uplift (over 300 m) in the Late Tertiary to Quaternary in the examined area. This Venn-Monschau zone is composed of Lower Palaeozoic of the core of the Stavelot-Venn Anticline and its SE flank, including the hanging Lower Devonian. A strong late Tertiary to Quaternary fractured or fractureless tectonic deformation has not developed in this zone (1). However, weak tectonic uplift processes which have taken place in the Subrecent have became indicated in a NNE-SSW striking zone between the Bois de Küchelscheid/Belgium and the Kall dam, which coincide with the SE edge of the Lower Palaeozoic in its transition to the Lower Devonian. NW-SE striking faults appear to have been revived, and to have influenced the convexly-formed river lines of the Rur and the Kall.

A zone with lower vertical uplift connects with the Venn-Monschau zone to the Northwest and Northeast. The Raffelsbrand-Rurberg zone coincides morphologically with the decline of the Hohe Venn Mountains to the Northwest and Southeast, and geologically with both flanks, including the hinge area, of the lower Palaeozoic Stavelot-Venn Anticline and with the Lower Devonian superposition up to the Mechernich Triassic. Neotectonically, the zone (2) is an intensively stressed area. Important uplift movements have been occurring there recently, shown by the 1.6-1.8 mm per year rise in the crust in the southeast section north of Gemünd in the Kermeter (Arbeitskreis DGK, 1979; Fig. 1). This represents the maximum uplift of the crust in the Federal Republic of Germany. Earthquakes, such as the one with its epicenter in Roetgen, also indicate the tectonic activity in this area (Ahorner, 1978). In addition, the carbonate springs of Dreiborn and Schmidt are present here. They lie along a NNE-SSW line. These, and the similar orientation of the isohypses of the recent uplift could indicate the latest movements on the many N-S striking faults in this zone.

In the direction of the Lower Rhenish Embayment and towards the area of its western margin step-faults to the East of Aachen, a zone (3) connects morphologically and neotectonically, in which are the villages of Rott, Zweifall, Grosshau and Abenden. Morphologically clearly separated from the rest of the mountains, the area can show an uplift up to 280 m. Its eastern part on the Rur between Hausen and Obermaubach is slightly rotated in an anticlockwise direction around a NW-SE striking axis.

In the northwest section of this zone (3), the lower Palaeozoic Venn-Stavelot anticline is Variscan up- and overthrust on the Lower Devonian of its NW flank, along several NE-SW striking faults. On the other hand, NW-SE faults of the same age characterise in the eastern part of the zone the tectonic framework of the Lower Devonian of the SE flank of the anticline.

Also along a NW-SE oriented fault zone, this zone (3) borders directly on the marginal unit of the Lower Rhenish Embayment the Rur block (zone 5). Since the block's western section is only lifted to a maximum of 80 m, there is an important tectonic throw (ca. 100 m) between Untermaubach and Ursprungsbach, which is made particularly apparent by the position of the Hochwald. The terraces of the Rur at Untermaubach/Schlagstein have also been shifted.

In the Northwest of the northern Eifel, in the area of the western margin step-faults of the Lower Rhenish Embayment a zone (4) with vertical uplift which can show up to 190 m is fitted in between the two above-mentioned units. Here, both the Palaeozoic of the SE flank of the Inde syncline between Kornelimünster, Stolberg and Gressenich, and the Lower Palaeozoic of the northeast-plunging Stavelot-Venn Anticline at Schevenhütte have been included in the uplift. The bordering of the Kornelimünster-Schevenhütte zone is defined by the NE-SW striking structural lines in the North, and by the NW-SE striking faults (downthrow faults) of the western margin step-faults such as Feldbiss, Sandgewand, and others. These faults, active in the Tertiary and Quaternary, continue to the north, deep into the Netherlands. To the south they lose their great importance for neotectonics at about the NE-SW to ENE-WSW running edge of this zone. In the East, the recently still active Merode fault (Knapp, 1978) marks the border.

On the edge of the Eifel in the transition to the Lower Rhenish Embayment, the Palaeozoic and parts of the northeastern Triassic triangle of Mechernich show their smallest uplift, with a maximum of circa 80 m. The Tertiary and Quaternary of the southeast edge of the Rur block up to the Stockheimer fault also belong to this zone (5).

The area of Stolberg, Eschweiler, Langerwehe and Strass is characterised by the NW-SE striking faults which can be traced deep into the Lower Rhenish Embayment and the Netherlands. These faults have shifted Lower Pleistocene Maas main terraces several meters and they have also affected the younger terraces of the Inde and the Rur. Upper Pleistocene loess near the open-cast lignite mine of Zukunft-West has been shifted for several centimeters along them. In general, the vertical throws of the faults increase to the northwest. The faults have also been active recently, as is made apparent by geodetically demonstrated movements on the Sandgewand (Paus, 1932) and parallel faults (Quitzow and Vahlensieck, 1955), and by earthquakes (Ahorner, 1962).

The intensive stressing in the Late Tertiary-Quaternary led, at the north-western Eifel margin, to the further development of the narrow northwest-southeast lying horst and graben structure of the Palaeozoic area. At the north-eastern margin, in contrast to the above, a step-like fault has developed. Thus it becomes clear that the marginal Palaeozoic of the northern Eifel is neotectonically a part of the Rur block, and that it is thus also a part of the Lower Rhenish Embayment.

In the central Palaeozoic area the strong neotectonic uplift (Fig. 1) which has taken place can be connected to the velocity anomaly in the mantle (Raikes and Bonjer, 1983), striking NW-SE. The contours of the uplift area and the great "uplift gradient" towards the Lower Rhine Embayment coincide with the northern limit of the mantle anomaly, with which, according to Fuchs (1981), uplift processes in the Rhenish Massif might be explain.

Acknowledgement. I am indebted to the Deutsche Forschungsgemeinschaft for financial support within the priority research program: The Rhenish Massif: mode and mechanism of plateau uplift.

References

Ahorner, L., 1962. Untersuchungen zur quartären Bruchtektonik der Niederrheinischen Bucht. Eiszeitalter Gegw., 13:24-105.
Ahorner, L., 1978. Untersuchung von Mikroerdbeben im Bereich des Rheinischen Schildes. Protokoll 3. Koll. Schwerpunktprogramm "Vertikalbewegungen und ihre Ursachen am Beispiel des Rheinischen Schildes" der DFG Bonn, pp. 40-42.
Arbeitskreis "Rezente Höhenänderungen" der Deutschen Geodätischen Kommission, 1979. Karte der Höhenänderungen in der Bundesrepublik Deutschland - Stand 1979 -. Allg. Verm. Nachr., 86:362-364.
Boenigk, W., 1978. Die flußgeschichtliche Entwicklung der Niederrheini-

schen Bucht im Jungtertiär und Altquartär. Eiszeitalter Gegw., 28:1-9.

Breddin, H., 1937. Lehrausflug in das Unterdevon zwischen dem Vennsattel und der Sötenicher Mulde am 28.8.1937. Z. Dtsch. Geol. Ges., 89:595-601.

Fuchs, K., 1981. Zur Diskussion zwischen Geophysik und Petrologie. Protokoll 6. Koll. Schwerpunktprogramm "Vertikalbewegungen und ihre Ursachen am Beispiel des Rheinischen Schildes" der DFG Bonn, pp. 186-187.

Kirchberger, M., 1917. Der Nordwestabfall des Rheinischen Schiefergebirges zwischen der Reichsgrenze und dem Rurtalgraben. Verh. Naturhist. Verein Preuss. Rheinland Westfalen, 74:102 pp.

Knapp, G., 1978. Erläuterungen zur Geologischen Karte der nördlichen Eifel (1:100 000), Geol. L.-Amt Nordrhein-Westfalen, 152 pp.

Kurtz, E., 1906. Geologische Beobachtungen über die Bildung des Rurtales. Beil. Progr. Gym. Düren, 540:15 pp.

Kurtz, E., 1909. Beziehungen zwischen Rur, Maas und Rhein zur Diluvialzeit. Beil. Progr. Gym. Düren, 588:23 pp.

Kurtz, E., 1914. Die diluvialen Flußterrassen am Nordrand von Eifel und Venn. Verh. Naturhist. Verein Preuss. Rheinland Westfalen, 70(1913):55-85.

Kurtz, E., 1914. Die Verbreitung der diluvialen Hauptterrassenschotter von Rhein und Maas in der Niederrheinischen Bucht. Verh. Naturhist. Verein Preuss. Rheinland Westfalen, 70 (1913):87-188.

Müller, R., 1978-1981. Jungkretazisch-tertiäre Schollenbewegungen und ihre mögliche quartäre Neubelebung im Stavelot-Venn-Massiv und angrenzenden Gebieten. Protokolle 2.-6. Koll. Schwerpunktprogramm "Vertikalbewegungen und ihre Ursachen am Beispiel des Rheinischen Schildes" der DFG Bonn.

Musa, I., 1973. Rhein- und Eifelschüttungen im Süden der Niederrheinischen Bucht. Sonderveröff. Geol. Inst. Univ. Köln, 23:151 pp.

Parting, H.-M., 1980. Geomorphologische Untersuchungen in der Rur-Eifel. Thesis RWTH Aachen, 174 pp.

Paus, H., 1932. Messungen an der Aachener Sandgewand. Thesis TH Aachen, 48 pp.

Quaas, A., 1917. Das Rurtal. Ein Beitrag zur Geomorphologie der Nordeife. Verh. Naturhist. Verein Preuss. Rheinland Westfalen, 72 (1915):180-309.

Quiring, H., 1926. Die Schrägstellung der Westdeutschen Großscholle im Kaenozoicum in ihren tektonischen und vulkanischen Auswirkungen. Jahrb. Preuss. Geol. L.A., 47:486-558.

Quiring, H., 1926. Quartäre Bodenhebung und -senkung am Nieder- und Mittelrhein. Z. Berg-, Hütten- und Salinenwesen, 74:B59-B75.

Quitzow, H.W. and Vahlensieck, O., 1955. Über pleistozäne Gebirgsbildung und rezente Krustenbewegungen in der Niederrheinischen Bucht. Geol. Rundsch., 43:56-67.

Raikes, S. and Bonjer, K.-P., 1983. Large scale mantle heterogeneity beneath the Rhenish Massif and its vicinity from teleseismic p-residual measurements. (This vol.).

Richter, D., 1962. Die Hochflächen-Treppe der Nordeifel und ihre Beziehungen zum Tertiär und Quartär der Niederrheinischen Bucht. Geol. Rundsch. 52:376-404.

Schnütgen, A., 1974. Die Hauptterrassenabfolge am linken Niederrhein aufgrund der Schotterpetrographie. Forsch. Ber. Land Nordrhein-Westfalen, 2399:150 pp.

Stickel, R., 1922. Der Abfall der Eifel zur Niederrheinischen Bucht. Beitr. Landeskde. Rheinland Westfalen, 3:96 pp.

Stickel, R., 1927. Zur Morphologie der Hochflächen des linksrheinischen Schiefergebirges und angrenzender Gebiete. Beitr. Landeskd. Rheinland Westfalen, 5:104 pp.

Zeese, R., 1978. Der präpleistozäne Formenschatz in der Rureifel und seine Beziehung zur Tektonik. Kölner Geogr. Arb., 36:121-128.

4.11 The Quaternary Destruction of an Older, Tertiary Topography Between the Sambre and Ourthe Rivers (Southern Ardennes)

H. P. Berners[1]

Abstract

In the southern and western Ardennes the rheno-ardennic shield is not uplifted en bloc. Fluvial morphologic investigations of the destruction of Tertiary plains and the terraces of Quaternary time show very different vertical tectonic movements for different times and regions. In the region of the Rocroi massif (west of the Meuse river), the massifs of Croix Scaille and St. Hubert (east of the Meuse river) and in the southern parts of the Dinant downfold these tectonic activities of vertical displacements had been studied by different authors. Recent tectonic movements show regions of higher uplift around Givet, in the area of the upper course of the Ourthe river and in the region between St. Vith - Bastogne - Libramont - St. Hubert-Croix Scaille. The amount of uplift decreases from the eastern to the western Ardennes, and at the southern margin of the Rocroi massif uplifting has ceased and subsidence takes place.

The uplift of the southern and western Ardennes has been caused by tectonic movements of vertical displacements since the late Pliocene. Therefore, a very different destruction of the pre-Quaternary plains is found for different times and regions. Decisive indications of the development of the uplift of the Ardennes are given by the pre-Quaternary plains as well as by the terraces of the rivers Meuse and Semois.

The highest terraces of the rivers Meuse and Semois were accumulated in a wide area of the Ardennes during the late Pliocene. These sediments were often cemented by iron hydroxide and contain a typical amount of silicious Oxfordian limestone and of rocks originating in the Vosges.

Below this oldest terrace another one, accumulated during the Mindel glacial time, is well developed. Its sediments contain often cherts and glauconite of the Cretaceous and show fossil soils. In these plains the rivers have incised in post Mindel time. Because of a strong uplift, terraces of the post Mindel time could not be formed or preserved.

In recent times the point bars of the meander loops were superimposed by earth slides caused by cryoturbation and solifluction. For this reason the point bars often show not a convex, but a concave surface (Voisin, 1981).

On the west side of the Meuse river the surface of the Rocroi massif was uplifted only a little. In this area a different, more or less intense uplifting is recognized for different Tertiary and Quaternary times by remains of continental Tertiary sediments and fluvial terraces. In connection with these movements, the W-E orientated axis of maximum uplift was displaced since the late Pliocene from the southern border of the Ardennes to the northern parts of the Rocroi massif (Voisin, 1981). For the Pleistocene time it is located near Deville, for the Holocene near Revin and nowadays north of Rocroi. These movements caused different erosional phases of the Cretaceo

---

[1] Lehr- und Forschungsgebiet allg. und hist. Geologie, RWTH Aachen, Wüllnerstraße 2, D-5100 Aachen, Fed. Rep. of Germany.

and Tertiary sediments at the Rocroi massif and can be recognized in the composition of the terrace sand and gravel. Numerous erosional remnants of Cretaceous and Tertiary age can be found at the exposed surface of the Meso-Caenozoic transgressions directly in the west of Rocroi. This surface itself has been eroded since the uplift of the Rocroi massif during Holocene time. A region with more intense uplifting is located on the east side of Rocroi. In this area remnants of Tertiary sediments have been preserved only on downthrown fault blocs (Voisin, 1981).

On the east side of the Meuse river, in the area of the Croix Scaille and St. Hubert massifs, the Quaternary uplift reaches its maximum height amounting to more than 500 m above present sea level. On these highlands, peat has formed from Würm to recent times. The pre-Pliocene relief was destroyed by solifluction, cryoturbation and fluvial erosion, because strong erosional action, caused by the high velocity of uplift, prevents a well-developed soil formation. Great blocs of non-transported, residual boulders of sandstones and quarzites are dispersed on the exposed surface of Tertiary age. These blocs can form boulder streams. The partly preserved periglacial elements are dislocated by recent tectonic movements (Voisin, 1981).

In the Croix Scaille and St. Hubert massifs the uplift has continued during the Plio-Pleistocene to recent times. Deeply incised valleys with terraces from late glacial time reach up to the upper course of the rivers. Post-glacial, alluvial deposits were found only in the lower course of the rivers, incised in the terraces of Pleistocene age (Voisin, 1981). The St. Hubert massif shows heights of more than 550 m above present sea level. The amount of erosion is still higher than in the Croix Scaille massif and all pre-Holocene plains and terraces have been destroyed.

At the southern border of the Dinant downfold a depression is located, which reaches from the Sambre river in the west to Givet at the Meuse river, and its eastern part extends to the upper course of the Ourthe river.

In the western part of this depression, including a region belonging to the drainage area of the Sambre river, a strong recent uplift caused fast retrogressive erosion to the east. Nowadays the eastern border of this retrogressive erosion is near Trélon. Erosion attacks the superposed Cretaceous sediments at the western margin of the depression.

The peneplain of Old Tertiary age is to be recognized at 280 m above present sea level between Trélon and Chimay. Remnants of marine sediments of Cuisium age are preserved in contact with the Variscan basement.

East of Chimay in the direction of Givet, the plains of Old Tertiary age (280 m level) were destroyed during the Younger Tertiary, when an uplift caused a strong fluvial erosion. Today only some ranges of Middle Devonian limestones reach the 280 m level of the Old Tertiary plain, forming a morphology called relief appalachien (Aubouin et al., 1968).

In Pleistocene time these rivers, belonging to the drainage area of the Meuse river, accumulated fluvial sediments and formed a plain between the ranges of limestones up to a present level of 200 m. The terraces of the Meuse river also substantiate a relative subsiding area near Givet during the Pleistocene. Leaving the area of the Rocroi massif, the Meuse river starts to meander in a wide valley while the terraces dip under the surface of the valley fill. Then, in Holocene time, a more intense uplift took place, especially near Givet. For this reason retrogressive erosion destroyed the 200 m plain of Pleistocene age. Now a new plain has formed at a level of 180 m (Voisin, 1981).

To the east of Givet, between the drainage area of the Meuse river and the Ourthe river, the peneplain of the Old Tertiary is preserved at 290 m above present sea level. But here only continental, fluvial Tertiary sediments were accumulated in contact with the Variscan basement. The marine transgressions during the Tertiary and probably those of the Upper Cretaceous did not reach the area on the east side of the Meuse river (Voisin, 1981).

In the region of the upper course of the Ourthe river the tectonic-morphologic development of destructional landforms is the same as in the region of Givet. In the southeastern part of the Dinant downfold, the terraces of the rivers Ourthe, Lesse and Amblève show a downstream intensified uplift, indicating a stronger uplift during the Younger Pleistocene and the Holocene for the northern part of the Dinant downfold.

These region descriptions, including fluvial-geomorphologic information and aspects of a Caenozoic plain development, show very different tectonic movements for different times and regions.

Especially recent tectonic movements of vertical displacements have been made visible by precise levelling and gravimetry. The results of these investigations confirm the previous statements to recent tectonic movements.

In the south of the Rocroi massif at Rimogne, gravimetric measurements have shown a maximum of gravity (+5 mgal). Positive gravity gradients are restricted to the parts of the Rocroi massif on the west side of the Meuse river (subsiding area).

On the east side of the Meuse river, a centrum of negative gravity gradients is situated in the triangle Rochefort-St. Hubert-Bouillon (-22 mgal) (uplifted area).

Another region of recent negative gravity gradients is located at the southern margin of the Dinant downfold with a minimum of gravity at Givet (-17 mgal) and at the upper course of the Ourthe river (-14 mgal) (uplifted areas) (Voisin, 1981).

Precise levelling substantiates an axis of recent uplift in the region Libramont-Bastogne. A comparison between the measurements of precise levelling from 1892 and 1948 (Jones, 1951) shows an uplift of more than 100 mm near Stavelot-St. Vith.

These amounts of uplift decrease westward to the region of the St. Hubert and Croix Scaille Massif (89-62 mm uplift). On the west side of the Meuse river a subsidence has been recognized in the area of the Rocroi massif. Here at the frontier from Belgium to France, the highest recent subsidence has been measured with the absolute amount of 656 mm (Voisin, 1981).

References

Aubouin, J., Brousse, R., and Lehmann, J.-P., 1968. Précis de géologie. Tome III tectonique, morphologie le globe terrestre. Dunod, Paris, 549 p.
Jones, L., 1951. Les premiers résultats de la comparaison du deuxième nivellement général (1948), du royaume avec les nivellements anciens. Bull. Soc. Belge de Géol. Palaeontol. Hydr. T LIX: 156-162.
Voisin, L., 1981. Analyse géomorphologique d'une région type: l'Ardenne occidentale. Serv. Repr. Theses Univ. Lille 3, T II:499-883.

5 Volcanic Activity

5.1 Distribution of Volcanic Activity in Space and Time

H. J. Lippolt[1]

Abstract

The Rhenish Massif is studied with an enormous number of occurrences of volcanic rocks, with ages reaching from the Middle Cretaceous to the youngest Pleistocene. K-Ar and ^{14}C dating are the main tools for establishing their distribution in time. The oldest post-Permian volcanic rock occurs in the Wittlich basin and has an age of more than 100 m.y. The main volcanicity of the Rhenish Massif starts in the Hocheifel area during the Upper Eocene, where it continues until the Middle Miocene. Volcanic activity in the Siebengebirge, Westerwald and Rhön unfolded during the Upper Oligocene and Lower Miocene. During the Middle Miocene, the Vogelsberg and the Northern Hessian volcanic province were active.

Quaternary volcanic activity is essentially confined to the West and East Eifel. K-Ar dates and paleomagnetic polarity narrow its time of activity down to the interval between 0.7 m.y. and about 10,000 a. The well-established sequence of the Rhine terraces can be dated in the younger section by its relation to the volcanic rocks, while for the older part one has to rely exclusively on paleomagnetic observations.

By now the distribution of the Tertiary volcanism in time and space is fairly well established, but the information on the Quaternary volcanism is still so scarce that further work is needed.

Introduction

The area of the Rhenish Massif and its surroundings is covered in the E and SE by an enormous number of volcanic eruption centers (Fig. 1). Wimmenauer (1974) discussed the petrological and petrographic peculiarities of the characteristic rock types as part of the "alkaline volcanic province of Central Europe and France". The post-Permian volcanism in this region started in Middle Cretaceous and culminated in Eocene, Miocene and Pleistocene times (Lippolt, 1982). Time relationships within and between the different volcanic areas are essential for the understanding of mode and dynamics of the magmatic processes. The sequence of eruptions can generally not be established by mere observations because of lacking stratigraphic relations and intercalations. Therefore a meaningful correlation in the relevant time span can only be attained by isotopic (K-Ar, etc.) and radiometric (^{14}C, etc.) dating. The potassium-argon decay is the main chronometer for the Cretaceous and Tertiary. Owing to the mineralogy and texture of the volcanic rocks, mainly whole-rock data have been reported. Since misinterpretation of the measurements because of argon excess or dif-

---

[1] Laboratorium für Geochronologie, Universität Heidelberg, Berliner Straße 17, D-6900 Heidelberg, Fed. Rep. of Germany

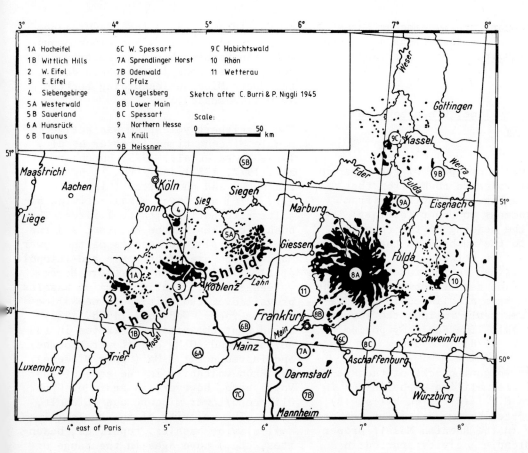

Fig. 1. Geographic distribution of volcanic centers in the Rhenish Massif and its surroundings. *Numbers* refer to the columns of Figs. 2 and 3

fusion deficiency cannot be excluded, the pertinent data summarized in Figs. 2 and 3 are considered as "apparent ages", with the exception of those for which detailed work was done. This holds true also for the ^{14}C ages (Fig. 3).

Distribution in Space

The volcanic areas of the Rhenish Massif proper exhibiting predominantly pre-Quaternary volcanic rocks are the Hocheifel (number 1A in Fig. 1), Siebengebirge (4), Westerwald (5A), and isolated volcanic dikes in Taunus and Hunsrück (6A, B). The western and eastern Eifel (2,3) are well known for their Pleistocene volcanic activity. There is also considerable volcanicity in the NE extension of the Rhenish Massif, where in Northern Hessen (9) various separate centers can be distinguished (Knüll, Meissner, Habichtswald: 9A, B, C). In the east of the Rhenish Massif the strato-volcano of the Vogelsberg (8A) is the dominant volcanic manifestation, followed some 50 km further to the east by the Rhön mountains (10). The Vogelsberg volcanism has some satellites in the south in the Spessart (8C) and the Lower Main region (8B). Volcanic dikes and pipes similar to the occurrences in the Taunus and Hunsrück are also found in the western crystalline part of the Spessart (6C). Volcanic rocks of post-Permian age also occur SE of the Massif, on the northeastern shoulder of the Upper Rhine Graben (Odenwald, 7B and Sprendlingen Horst, 7A) as well as in the northern parts of the Pfalz mountains (7C). Figure 1 does not display all known single occurrences and only serves as a guide to Figs. 2 and 3 and as an introduction to the locations of the following sections which deal with the individual regions. Some of the volcanic regions are rather large, for instance the northern most occurrences of the Westerwald associations are found 70 km to the north in the Sauerland (5B).

Distribution in Time: Cretaceous to Upper Tertiary

Figure 2 gives a synopsis of the apparent K-Ar ages of volcanic rocks from the area shown in Fig. 1. They range from Middle Cretaceous to Upper Tertiary. The numbers in the heading of the columns correspond to the numbers on the map. A rough distinction between rock types was made, as explained in the inset of Fig. 2. The number of samples comprised in the individual arrays are listed in the diagram. The time scale follows the suggestions of the 1976 Time Scale Symposium (Cohee et al., 1978) in Sydney.

The earliest volcanic rocks of the Rhenish Massif are melilite nephelinites in the southern Eifel (Wittlich hills, 1B), which were dated by two different methods at 108 m.y. In the Rhenish Massif proper the main volcanic activity started in Eocene times (about 45 Ma) in the Hocheifel and continued to the beginning of the Miocene, the youngest measured rock being 18 Ma old (Cantarel and Lippolt, 1977; Lippolt and Fuhrmann, 1980 and unpublished data).

Trachytes are among the first eruptions followed by basalts of variegated compositions. Rocks of Eocene age are also found in the mountains of the Upper Rhine Graben shoulders (6A-C, 7A-C). In the Siebengebirge, volcanic activity began with eruptions of trachytic tuff, followed by trachytes, latites and alkali basalts during a time span of 28 to 22 Ma (mainly Upper Oligocene). The extrusion of basaltic rocks continued for several Ma, the youngest date being around 15 Ma (Todt and Lippolt, 1980 and unpublished data). The Westerwald volcanism mainly produced basaltic rocks, with the exception of the southwest, where also trachytes occur. In this area two volcanic activity phases have been shown to exist during Tertiary times (Lippolt, 1976; Lippolt and Todt, 1978 and unpublished data); first, a dominant phase close to the Oligocene-Miocene boundary more or less contemporaneous with the Siebengebirge activity and second, a less conspicuous phase during the Upper Miocene (according to the 1976 time scale).

The Vogelsberg volcanism was active in Middle Miocene times (Kreuzer et al., 1974; Harre et al., 1975; Ehrenberg et al., 1977, 1981). At first it produced trachytes and phonolites (Lippolt et al., 1974) and later alkali-olivine basalts alternating with tholeiites. The basaltic dikes in the Spessart and the tholeiitic plateau basalts of the Lower Main region are of the same age (Lippolt et al., 1974). In the Northern Hesse volcanic area, alkaline-olivine basalts prevail over tholeiitic and nepheline-bearing rocks. Wedepohl (1978, 1982) found that the activity started in the north around 20 Ma ago (Lower Miocene) and lasted until 7 Ma (Upper Miocene). The Rhön volcanism in the east of the Vogelsberg, however, is again of Upper Oligocene/Lower Miocene age (Lippolt, 1978). The majority of the rocks (alkali-olivine basalts, tephrites, basanites, etc.) have ages in the range of 22 to 18 Ma. The Rockenstein near Oberweißenbrunn is so far the youngest rock of this suite (11 Ma). The ages of the post-Permian volcanic rocks of the mountain chains adjacent to the Upper Rhine Graben (6 and 7 on the map) are on the average much older than those of the main volcanic centers of the Central European volcanic arc. In the Sprendlingen Horst there exists a group of trachytes and alkali-olivine basalts with ages in the Uppermost Cretaceous, but the majority of the rocks are of Eocene age as in the Odenwald, the Palatinate mountains and occasionally also in the crystalline Spessart (Horn et al., 1972; Lippolt et al., 1974, 1975 1976). In the vicinity of the Otzberg fault zone of the Odenwald, ages were found which coincide with the Late Oligocene/Early Miocene main phase of the Rhenish Massif. Isolated occurrences of volcanic rocks in the Taunus and crystalline Spessart show ages up to 80 Ma.

To sum up briefly: in the area of interest volcanicity started during Middle

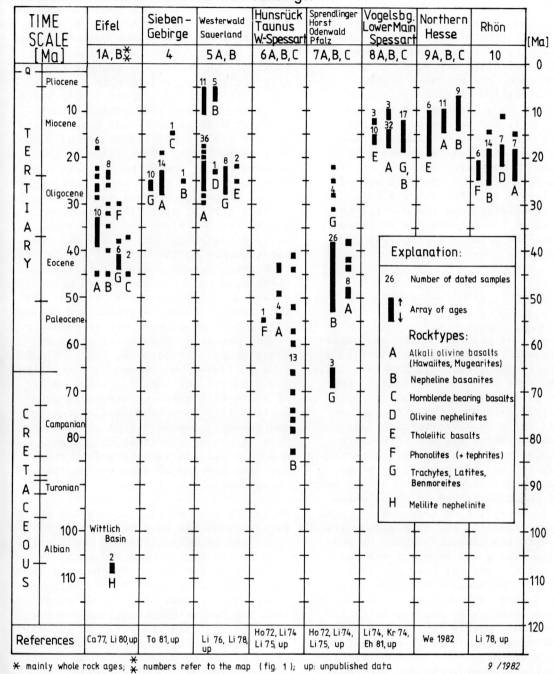

Fig. 2. Apparent K-Ar ages of Cretaceous and Tertiary volcanic rocks (mainly whole-rock ages) from the Rhenish Massif and its surroundings. *Numbers* in the table headings refer to Fig. 1

Cretaceous times in the Wittlich basin. At first its activity was very low and limited to the area SE of the Rhenish Massif proper until the beginning of the Eocene, then a more intense activity began in the northern Upper Rhine Graben. Subsequently, in Middle Eocene times the Hocheifel volcanics extruded. In the early Miocene, the volcanic activity culminated at various places (Siebengebirge, Westerwald, Rhön and also in regions more to the east). The Vogelsberg and part of the Northern Hessian volcanoes were active during the Middle Miocene. Some occurrences in Northern Hesse and in the western part of the Westerwald are of Upper Miocene age. For about 4 Ma no further activitiy took place until the beginning of the Pleistocene volcanic phases in the Eifel and Westerwald.

Distribution in Time: Quaternary

The main areas with volcanism of Quaternary age are the West and East Eifel (2 and 3 on the map, Fig. 1). However, there are also some minor occurrences in the western part of the Westerwald (5A). Further volcanic products are intercalated as tuffs in river terrace sediments of the Rhine (3) and in Pleistocene sedimentary rocks of the Wetterau (11), where they are exposed by brown coal and clay mining. The centers of eruption of the latter have not been established in detail. Some K-Ar and ^{14}C ages have been reported, but they are so few compared to the enormous number of outcrops that the question of proper age assignment has not yet been settled. Furthermore, methodical problems arise, as both dating methods have to be applied in ranges where their applicability is controversial, mainly because of chemical contamination (^{14}C) or the existence of radiogenic argon-40 from the pre-history of the present rocks and minerals. The limitation of the K-Ar method in this area because of excess argon has already been shown by Frechen and Lippolt (1965).

Figure 3 summarizes the majority of Quaternary isotopic data, comparing them to the scale of magnetic polarity (following Mankinen and Dalrymple, 1979), the Time Scale (Sibrava, 1976; in Cohee et al., 1978) and three of the suggested chronologies of the Rhine terrace system, as far as they are committed to absolute time (Brunnacker, 1978; Bibus, 1980; Frechen, 1980). The Quaternary age of the West Eifel volcanism can be established on geological grounds (amount of pyroclastics, relation to valleys, etc.). Frechen and Lippolt (1965) dated sanidine crystals from alkali basalt tuffs at 0.46 and 0.43 Ma. Cantarel and Lippolt (1977) reported an age of 0.96 Ma for an olivine nephelinite at the western rim of the Hocheifel (auf dem Beuel), which is considered to belong to the oldest Quaternary extrusions (column 2 right side, 1A). Schmincke and Mertes (1979) argued for a much earlier beginning (around 2.7 Ma) of the activity on the basis of nine whole-rock ages. However, the significance of these ages has been questioned meanwhile by the same authors (pers. commun. 1982). The observation by Haverkamp and Mertes (1981) that in general normal magnetization prevails, also casts doubt on these dates. Furthermore, Fuhrmann and Lippolt (1980) have shown by $^{40}Ar/^{39}Ar$ measurements that at least four rocks of the sample suite of Schmincke and Mertes (1979) are essentially younger and probably around 0.6 Ma old. Further work is needed to settle this question and to develop a safe method to overcome the problem of excess argon in these rocks.

As in the West Eifel, the general geological age of the East Eifel volcanic rocks can be established by their relation to the river terrace system, especially of the Rhine. Frechen and Lippolt (1965) considered 22 of their K-Ar dates on sanidine, biotite and whole-rock samples as chronologically relevant and confined the volcanic action which is datable by K-Ar dating to the range of 0.6 to 0.15 Ma (column 3, right side of Fig. 3). About one third of their samples clearly contained excess argon, the most dubious host mineral of this component being pyroxene. Schmincke and Mertes (1979) reported another 18 whole-rock data, which on average are older than the Frechen and Lippolt data by a factor of 1.5 to 2 (column 3, left side). Fuhrmann and Lippolt (1981) performed several $^{40}Ar/^{39}$

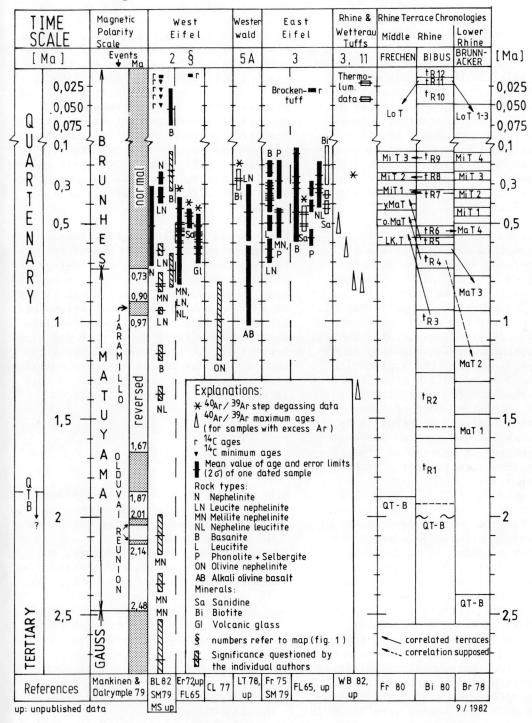

Fig. 3. Apparent K-Ar and ^{14}C ages for Quaternary volcanic rocks (*black bars*) and separated minerals (*open bars*) from the Rhenish Massif, compared with the paleomagnetic polarity scale and with the chronologies suggested by various authors for the Rhine terrace system

step-degassing datings on rocks and minerals from the East Eifel (unpublished in detail) and showed that this may be the way to reach unquestionable, geologically meaningful results. Their data fall in the same range as the Frechen and Lippolt data of 1965 (column 3, right side, open symbols or asterisks). Lippolt and Todt (1978) in their study on the Westerwald volcano-chronology also described two Pleistocene occurrences of this area and assigned ages to them of 0.45 and 0.8 Ma. They occur together geographically with Upper Miocene rocks in the Lower Westerwald. Fuhrmann and Lippolt (unpubl.) found another Pleistocene rock which yielded a biotite age of 0.28 Ma (column 5).

Sanidines from the intercalated tuffs are also tempting dating minerals for the K-Ar method. Frechen and Lippolt (1965) failed in their efforts because of excess argon (sample Leubsdorf, for instance). Fuhrmann and Lippolt (unpubl.) discussed the problem again and found that in six additional cases also excess argon was present, which resulted in increased ages. However, the application of the $^{40}Ar/^{39}Ar$ technique gives improved maximum values for the ages of host sediments of the tuffs. Most of these sediments are correlated by Bibus (pers. commun.) with terrace t_{R8} and the third-last glaciation. If this holds true, this glaciation should be younger than 0.46 Ma (columns 3,11).

The youngest glaciation is the field of the ^{14}C dating method. Organic material from the bottom of maar lake sediments yielded ages of about 10^4 a (Erlenkeuser) et al., 1972; Straka, 1975). This was taken as an argument for the recent geological formation of the maars. By means of an improved sampling technique, Büchel and Lorenz (1982) obtained considerably higher ages in several cases and therefore concluded that maar formation may have taken place throughout the Pleistocene volcanic period. In the East Eifel (Kärlich pit) Frechen (1975), by assigning an age of 3×10^4 a to a tree found in the Brocken Tuff, also reached an ^{14}C date which constrains stratigraphically. Thermoluminescence dating (Wintle and Brunnacker, 1982) of loesses appears to be a promising tool to check and support ^{14}C measurements.

The chronologies for the Rhine terraces are divergent (right side of Fig. 3), as they rely on different observations. Frechen (1980) combines his petrographic and field observations with the age data of Frechen and Lippolt (1965), thus confining three main terraces and the three subsequent middle terraces to the time span of 0.6 to 0.13 Ma. Bibus (1980) relies on the observation that parts of the sediments of the terrace called t_{R4} are inversely magnetized and that the first terrace was formed somewhere at the beginning of the Quaternary. Brunnacker (1978) bases his scale entirely on measurements of paleomagnetic directions observed in pertinent sedimentary rocks.

Summarizing Fig. 3: the Pleistocene volcanism in the Rhenish Massif is younger than 1 Ma and on the average probably younger than 0.6 Ma. It remains open, however, when it ended, if it did at all. As the Rhine terrace chronologies are more similar in the younger period when volcanicity is documented and divergent in the older period when it is lacking, we may relate the open question of the age of the old terraces to the lack of contemporaneous volcanism.

References

Bibus, E., 1980. Zur Relief-, Boden- und Sedimententwicklung am unteren Mittelrhein. Frankfurt. Geowiss. Arb., Ser. D, Bd. 1, 296 pp.

Brunnacker, K., 1978. Gliederung und Stratigraphie der Quartärterrassen am Niederrhein. Köln. Geogr. Arb., 36: 37-58.

Büchel, G. and Lorenz, V., 1982. Zum Alter des Maarvulkanismus der Westeifel. Neues Jahrb. Geol. Abh., 163, 1:1-22.

Burri, C. and Niggli, P., 1945. Die jungen Eruptivgesteine des mediterranen Orogens. Schweizer Spiegel Verlag, Zürich, 654 pp.

Cantarel, P. and Lippolt, H.J., 1977. Alter und Abfolge des Vulkanismus der Hocheifel. Neues Jahrb. Geol. Paläontol. Monatsh. 10:600-612.

Cohee, G.V., Glaessner, M.F., and Hedberg, H.D. (eds.), 1978. Contributions to the geologic time scale. Stud. Geol. 6. AAPG, Tulsa, Oklahoma, U.S.A., 388 pp.

Ehrenberg, K.H., Harre, W., and Kreuzer, H., 1977. Datierungen nach der K-Ar-Methode. In: Erläuterungen Geolog. Karte Hessen 1:25000, Bl. 5721 Gelnhausen:107-110.

Ehrenberg, K.H., Harre, W., and Kreuzer H., 1981. K-Ar-Datierungen an den Vulkaniten. In: Forschungsbohrungen im Hohen Vogelsberg (Hessen). Geol. Abh. Hessen 81:166 pp.

Erlenkeuser, H., Frechen, J., Straka, H., and Willkomm, H., 1972. Das Alter einiger Eifelmaare nach neuen petrologischen, pollenanalytischen und Radiokarbon-Untersuchungen. Decheniana, 125, 1/2:113-129.

Frechen, J., 1980. Stratigraphie und Chronologie des Pleistozäns am Vulkan Leilenkopf, Laacher-See-Gebiet. Neues Jahrb. Geol. Paläontol. Monatsh. 4: 193-214.

Frechen, J., 1975. Tephrostratigraphische Abgrenzung des Würmlösses und der älteren Lösse im Quartärprofil der Tongrube Kärlich, Neuwieder Becken. Decheniana, 127:157-194.

Frechen, J. and Lippolt, H.J., 1965. Kalium-Argon-Daten zum Alter des Laacher Vulkanismus, der Rheinterrassen und der Eiszeiten. Eiszeitalter Ggw. 16: 5-30

Fuhrmann, U. and Lippolt, H.J., 1980. Das Alter des jungen Vulkanismus der Westeifel aufgrund von $^{40}Ar/^{39}Ar$-Untersuchungen. Fortschr. Miner., 60, Beih. 1:80-82.

Harre, W., Kreuzer, H., Müller, P., Pucher, R., and Schricke, W., 1975. Datierungen nach der K/Ar-Methode und Paläomagnetik. Erläuterungen Geol. Karte Hessen 1:25000, Bl. 5319 Londorf, pp.67-73.

Haverkamp, U. and Mertes, H., 1981. Paläomagnetische Ergebnisse im Zusammenhang mit geologischen Untersuchungen in der quartären Westeifel. Vortr. 42. Jahrestag. Dtsch. Geophys. Ges.

Horn, P., Lippolt, H.J., and Todt, W., 1972. Kalium-Argon-Altersbestimmungen an tertiären Vulkaniten des Oberrheingrabens. I. Gesamtgesteinsalter. Ecol. Geol. Helv., 65/1:131-156.

Kreuzer, H., Kunz, K., Müller, P., and Schenk, E., 1974. Petrologie und K/Ar-Daten einiger Basalte aus der Bohrung 31, Rainrod I (Vogelsberg). Geol. Jahrb., D9:67-84.

Lippolt, H.J., 1976. Das pliozäne Alter der Bertenauer Basalte/Westerwald. Aufschluß, 27:205-208.

Lippolt, H.J., 1978. K-Ar-Untersuchungen zum Alter des Rhön-Vulkanismus. Fortschr. Miner. 56, Beih. 1:85.

Lippolt, H.J., 1982. K-Ar Age Determinations and the correlation of tertiary volcanic activity in Central Europe. Geol. Jahrb. Ser. D 52:119-141.

Lippolt, H.J. and Fuhrmann, U., 1980. Vulkanismus der Nordeifel: Datierung von Gang- und Schlotbasalten. Aufschluß 31:540-547.

Lippolt, H.J. and Todt, W., 1978. Isotopische Altersbestimmungen an Vulkaniten des Westerwaldes. Neues Jahrb. Geol. Paläontol. Monatsh. 6:332-353.

Lippolt, H.J., Todt, W., and Horn, P., 1974. Apparent potassium-argon ages of lower tertiary Rhine Graben volcanics. In: Illies, J.H. and Fuchs, K. (eds.) Approaches to taphrogenesis, Schweizerbart, Stuttgart, 460 pp.

Lippolt, H.J., Baranyi, I., and Todt, W., 1975. Die K/Ar-Alter der postpermischen Vulkanite des nord-östlichen Oberrheingrabens. Aufschluß (Sonderb.) 27:205-212.

Mankinen, E.A. and Dalrymple, G.B., 1979. Revised geomagnetic polarity time scale for the interval 0-5 m.y.B.P. J. Geophys. Res., 84, B2:615-626.

Schmincke, H.-U. and Mertes, 1979. Pliocene and quaternary volcanic phases in the Eifel volcanic fields. Naturwissenschaften 66:614-615.

Lippolt, H.J., Horn, P., and Todt, W., 1976. K-Ar-Altersbestimmungen an tertiären Vulkaniten des Oberrheingraben-Gebietes. Neues Jahrb. Miner. Abh. 127:242-260.

Straka, H., 1975. Die spätquartäre Vegetationsgeschichte der Vulkaneifel. Pollenanalytische Untersuchungen an vermoorten Maaren. Beitr. Landespfl. Rheinland-Pfalz, Beih. 3:1-163.

Todt, W. and Lippolt, H.J., 1975. K-Ar-Altersbestimmungen an Vulkaniten bekannter paläomagnetischer Feldrichtung. I. Oberpfalz und Oberfranken. J. Geophys. 41:43-61.

Todt, W. and Lippolt, H.J., 1980. K-Ar age determinations on tertiary volcanic rocks. V. Siebengebirge, Siebengebirge-Graben. J. Geophys. 48:18-27.

Wedepohl, K.H., 1978. Der tertiäre basaltische Vulkanismus der Hessischen Senke nördlich des Vogelsberges. Aufschluß (Sonderbd.) 28:156-167.

Wedepohl, K.H., 1982. K-Ar-Altersbestimungen an basaltischen Vulkaniten der nördlichen Hessischen Senke und ihr Beitrag zur Diskussion der Magmengenese. Neues Jahrb. Miner. Abh., 144, 2:172-196.

Wimmenauer, W., 1974. The Alkaline province of Central Europe and France. In: Sørensen, H. (ed.) The alkaline rocks. John Wiley & Sons, New York, pp.238-271.

Wintle, A.G. and Brunnacker, K., 1982. Ages of volcanic tuff in Rheinhessen obtained by thermoluminescence dating of loess. Naturwissenschaften 69:181.

5.2 Tertiary Volcanism of the Hocheifel Area

H.-G. Huckenholz[1]

Abstract

About 300 Tertiary eruptive centers are spread over a region of about 1000 km² in the Hocheifel Area. The volcanic rock suite reveals alkali basaltic features with strong sodium affinities and consists of primary alkali basalts (olivine basalts, alkali olivine basalts, nepheline basanites, olivine nephelinites) and fractionation products (hawaiites, mugearites, benmoreites, trachytes). Within the Tertiary volcanic field a systematic zoning of the volcanics due to variations of their specific petrology and geochemical parameters is obvious. These parameters project a N-S shaped body on to the surface from within the Upper Mantle which has released primary alkali basaltic melts 46 to 18 Ma ago. The depths of the alkali basalt generation is restricted below a cataclastic deformed but metasomatically altered spinel peridotite layer from which fragments record opx - cpx equilibration temperatures of 890° to 1040°C at about 50 km below the surface.

Introduction

The relics of about 300 Tertiary volcanoes are spread over a region of 1000 km² (40 km in N-S, and 25 km in

---

[1] Mineralogisch-Petrographisches Institut, TU München, Theresienstraße 41, D-8000 München 2, Fed. Rep. of Germany.

E-W direction) in the Hocheifel Area between Bad Münstereifel - Bad Neuenahr to the north and Daun-Ulmen to the south (Fig. 1). The eruptive centers are exposed as (deeply) eroded stocks and plugs penetrating folded Paleozoic (Devonian) sediments. They range in size from a few meters up to several hundred meters in diameter. The age of the volcanic activity is Eocene to Miocene (Cantarel and Lippolt, 1977; Lippolt and Fuhrmann, 1980). The older Tertiary volcanoes are intermingled with those of Pliocene, Pleistocene, and recent ages. The Tertiary Hocheifel Volcanic Province is connected across the Rhine River with the Tertiary Volcanic Province of the Siebengebirge to the NE, and with the Tertiary Volcanic Province of the Westerwald to the E, respectively (Lippolt, this vol.; Schmincke et al., this vol.).

The Tertiary volcanic rock suite of the Hocheifel Area reveals typical alkali basaltic features with strong sodium affinities and consists of primary alkali basalts, and derivative hawaiites, mugearites, benmoreites, and trachytes. The majority of the rocks are alkali basalts with ne in the norm, a few, however, bear hy. Thus, a distinction by norm (and by mode) in olivine basalts, alkali olivine basalts, nepheline basanites, and olivine nephelinites is made. The most essential features of the Hocheifel volcanics are briefly summarized in Table 1.

Potassium-argon dating on selected localities (31 samples) by Cantarel and Lippolt (1977), and Lippolt and Fuhrmann (1980) apparently restricts the

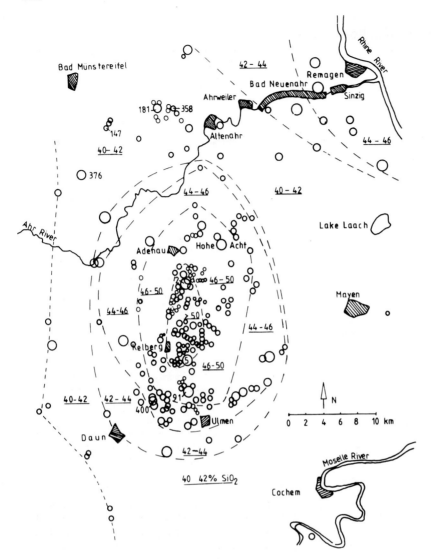

Fig. 1. Tertiary volcanics of the Hocheifel Area. Sizes of localities (*circles*) are exaggerated. Numbers *underlined* mark areas of equal SiO_2 contents (*long dashed lines*) of the volcanics as analyzed. Numbers *on or within circles* indicate peridotite xenolith-bearing nepheline basanites or olivine nephelinites, and are: (2) Nürburg; (5) Hochkelberg; (91) Kastelberg; (147) Michelsberg; (181) Hochthürmen; (358) Vellen, (376) Hümmel, and (400) Kapp (South).
For clarity, the Quaternary eruption centers of the East and West Eifel (Schmincke et al., this vol.; Seck, this vol.) which intermingle with the Tertiary Hocheifel volcanics are not shown. The Hocheifel Volcanic Area terminates to the West (*short dashed line*) but continues across the Rhine River toward the NE and E to the Siebengebirge and Westerwald, respectively)

trachytes (1 sample) and the benmoreites (2 samples) to ages of 41 to 44.5 Ma. The phonolitic benmoreite (= alkali trachyte) from the Selberg near Quiddelbach is 30 Ma old. Alkali basalts (23 samples), hawaiites (1 sample, which is a nepheline-and hauyne-bearing rock), and mugearites (4 samples) cover the total time span of 46 to 18 Ma with maxima at about 40 to 33 and about 27 to 25 Ma, respectively.

Within the Hocheifel Volcanic Area systematic zoning of volcanics which bear specific petrologic and geochemical characteristics is observed. Trachytes +

Table 1. Petrologic characteristics of the Hocheifel volcanics

Rock type	Normative composition	Modal composition	
		Phenocrysts	Groundmass
Primary Alkali basalts Mg·100/Mg+Fe^{2+}+Fe^{3+} = >60	ol > 10, di > 20		
Olivine basalt	hy	Olivine, clinopyroxene, plagioclase (An > 50); cpx "megacrysts"; peridotitic xenocrysts	Plagioclase (An > 50), alkali feldspar (Or 30-40), clinopyroxene, olivine, opaques, apatite, analcite, zeolites, (mesostasis, biotite, traces of amphibole, carbonate).
Alkali olivine basalt	ne < 5		
Nepheline basanite	ne > 5, ab > 2	Olivine, clinopyroxene, plagioclase (An > 50), (kaersutite, rhönite); periodotite inclusions; peridotitc xenocrysts	Plagioclase (An > 50), alkali feldspar (Or > 50), nepheline, clinopyroxene, olivine, opaques, apatite, analcite, zeolites (mesostasis, biotite, carbonate, hauyne).
Olivine nephelinite	ne > 5, ab < 2	Olivine, clinopyroxene (kaersutite, rhönnite); periodotite inclusions; periodotitic xenocrysts	Nepheline, alkali feldspar (Or 30-50), (traces of plagioclase), olivine, clinopyroxene, opaques, apatite, biotite, traces of amphibole, mesostasis, carbonate (hauyne), zeolites
Hawaiite Mg 58-45	ol < 10, di < 20 ne ≳ 5	Olivine, clinopyroxene, plagioclase (An ≤ 50) (kaersutite); (peridotitic xenocrysts)	Plagioclase (An < 50), alkali feldspar (Or 30-50), olivine, clinopyroxene, analcite, zeolites, (nepheline, mesostasis), opaques, apatite, carbonate, traces of amphibole.
Mugearite Mg 45-35	ol < 5, di < 10 ne ≳ 5	(Olivine) (clinopyroxene) plagioclase (An < 40)	Plagioclase (An < 35), alkali feldspar (Or ~ 30), olivine, clinopyroxene, analcite (nepheline), zeolites, opaques, apatite, traces of amphibole.

Table 1 (cont.)

Rock type	Normative composition	Modal composition	
		Phenocrysts	Groundmass
Benmoreite Mg 40-30	di < 5, (ol) ne or qz (hy) an < ab	Plagiclase (An < 40), clinopyroxene, amphibole (biotite)	Alkali feldspar (Or 30-40), clinopyroxene, opaques, apatite, (cristobalite, mesostasis, carbonate).
Phonolitic benmoreite (Alkalitrachyte) Mg 42-31	di < 5, (ol) ne < 10 an < ab	Kaersutite, clinopyroxene, plagioclase (An < 25), nosean, sphene, (olivine)	Alkali feldspar (Or ~ 40), clinopyroxene, opaques, apatite, mesostasis of analcite (nepheline) and clay minerals, carbonate
Trachyte Mg 30	qz, hy an < ab	Plagioclase (An < 20), alkalifeldspar (Or 40) (biotite)	Alkali feldspar (Or < 40), quarz, (cristobalite, sphene), opaques, apatite, mesostasis carbonate.

benmoreites + mugearites (all having SiO_2-values > 50%) occur in the central portion of the area between Adenau in the north and Kelberg in the south together with hawaiites and primary alkali basalts (SiO_2 < 50%). The N-S shaped central portion is surrounded by a 3 to 6 km wide zone containing mugearites + hawaiites (SiO_2 < 50% to 46%) intermingled with primary olivine basalts, alkali olivine basalts, nepheline basanites, and olivine nephelinites (SiO_2 < 46%). Zones of olivine basalts + alkali olivine basalts SiO_2 46% to 44%) together with nepheline basanits and olivine nephelinites (SiO_2 < 44%), nepheline basanites (SiO_2 44% to 42%) together with olivine nephelinites (SiO_2 < 42%), and olivine nephelinites (SiO_2 42% to 40%) alone follow toward outside.

Primary alkali basalts with Mg 100/Mg+ $Fe^{2+}+Fe^{3+}$ atomic ratios of > 70 (to 77) and intermingled with alkali basalts of < 70 are restricted to NE-SW zones in the central portion (Hohe Acht-Hochkelberg-Kapp) and to the NW (Hochthürmen-Aremberg) of the area. These zones are primarily composed of specific volcanics occupying N-S lines which are aligned in NE-SW echelon positions (e.g., Kapp S - Kapp N; Michelsberg S - Michelsberg N; Hochkelberg - Hünerbach). The volcanics are also characterized by abnormally high Cr, Ni, Co, and Sc contents. Alkali basalts with Mg-ratios < 70 can be mapped in zones of 70-68-65-58 Mg all spreading from the center to the NW and SE running in NE-SW directions.

Primary nepheline basanites and olivine nephelinites with P_2O_5 > 0.9 (weight per cent) depict a zone - about 15 km wide - passing through the center of the Hocheifel Area following a north-south direction. High K_2O values and high concentrations of REE (LREE ~ 200 times chondrites; HREE ~ 6 times chondrites; Huckenholz and Puchelt, in preparation) are consistently associated with this zone reflecting very low degrees of partial melting (≤ 7%) of an upper mantle source underneath the volcanic region. The concordance of P_2O_5 with K_2O and Ce, however, is low. The atomic ratio K·100/K + P is 73 ± 10, P_2O_5/Ce (wt.%) is 74 ± 10, and potassium (expressed as atomic ratio of K·100/K + Na) decreases from the center toward outer zones from > 28 to 10.

Alkali basalts highly ne-normative (ab < 5%) indicate pronounced solubility of Ca (probably) due to increasing activities of CO_2 (and probably CO_3^{2-}) which decreases the silica (and sodium) activities in primary alkali basaltic liquids (Eggler, 1978; Mysen, 1976; Frey, et al. 1978). Thus, Ca100/Ca+Na-ratios > 66 are (almost) identical with the local occupancy of olivine nephelinites (P_2O_5 > 0.9%, CaO/Al_2O_3 ⩽ 0.95) and reveal low degrees of partial melting (probably) due to the interaction of a H_2O-CO_2 vapor phase with the solids at the site of magma generation.

Inclusions of Ultramafic Rock Fragments

Inclusions of ultramafic rock fragments occur in alkali basalts of the Hocheifel area. They consist of peridotites (spinel lherzolites close to spinel harzburgites) and clinopyroxenites. There is also indication that kaersutitic amphibole may occur as assemblages of more than two grains and, therefore, as rock fragment.

Inclusions of peridotites are found in alkali basaltic hosts with Mg-ratios > 65, all belong to highly undersaturated (primary) rock types (e.g., olivine nephelinites or nepheline basanites, the latter with < 7 normative ab). Peridotite-bearing localities cluster along N-S-directed zones. One zone runs from the Hochthürmen-Vellen in the north via the Nürburg in the center to the Hochkelberg and the Kapp in the south (about 35 km). The other zone in the NW contains two peridotite-bearing localities (Michelsberg, Hümmel). Peridotite xenocrysts in the alkalibasalts of the Aremberg and Burgkopf, however, mark the continuation of this zone further to the South (about 15 km). The fragmental peridotite inclusions are of centimeter (Vellen, Nürburg, Michelsberg, Hümmel) and millimeter size (Hochthürmen, Hochkelberg, Kapp, Kastelberg). The inclusions in the cm size exhibit a four-phase mineralogy with olivine (80;70-90%), orthopyroxene (15;6-30), clinopyroxene (5;3-10), and spinel (1;0-3) but lack garnet. Physical and chemical breakdown of the peridotite fragments enrich primary alkali basalts with the crystal debris of upper mantle material generally resulting in Mg-ratios > 70.

The texture of the peridotite inclusions is porphyroclastic. Porphyroclasts of olivine and (unmixed) orthopyroxene are embedded in a neoblastic matrix of olivine + orthopyroxene + clinopyroxene (±spinel). Separated and analyzed orthopyroxene porphyroclasts (unmixed into orthopyroxene + clinopyroxene + spinel and with additional metasomatic (?) phlogopite) from xenoliths of the Nürburg olivine nephelinite yield a temperature prior to unmixing of about 1200°C (Lindsley and Dixon, 1975) and estimated pressures of 20 to 28 kbar. Orthopyroxene porphyroclasts of the Michelsberg olivine nephelinite apparently are not adjusted to the temperature of the neoblasts and reflect a temperature of about 1100°C. Orthopyroxene-clinopyroxene temperatures (Lindsley and Dixon, 1975; Wood and Banno, 1973) for the neoblastic matrix are 890±50°C for the Nürburg, 890±40°C for the Hochthürmen, and 995°-1040°C for the Michelsberg peridotite xenoliths. The corresponding pressures evaluated from the Ca contents of coexisting olivines (Finnerty and Boyd, 1978) are 12 to 14, 10 to 12, and 12 to 17 kbar, respectively. All peridotite inclusions analyzed reveal enrichment of incompatible elements (e.g., Ti, Al, Ca, Na, K, and P; Ba, Sr, Hf, Th, U), and have developed reaction rims to the host basalts. Three inclusions measured for REE show enrichment in LREE 16 to 44 times chondrites and HREE 1 to 2.5 times chondrites, respectively (Huckenholz and Puchelt, in preparation).

Fragments of clinopyroxenites are found in some alkali basalts (olivine basalts and nepheline basanites of Höchstberg, Barsberg, and Burgkopf). They vary in size from 1 cm up to 4 cm. They consist of grains of a green to dark green fassaitic clinopyroxene (Huckenholz, 1973). Cumulus textures are displayed and some fragments show sparse (basic) plagioclase as an intercumulus phase. Grains of magnetite and (dusty) apatite are abundant sometimes. All clinopyroxenite fragments are rimmed by different zones of clinopyroxenes having higher Mg-ratios than the enclosed fassaitic clinopyroxenes. The outer zone coalesces with the groundmass augite of the surrounding alkali basalt.

Xenocrysts, Megacrysts

Decreasing size of spinel peridotite fragments due to physical and chemical disintegration in a liquid environment (alkali basaltic) increases the Mg-ratios of the host basalts >70 because the spinel peridotite debris reacts with the (primary) alkali basaltic melt. A great number of alkali basalts and even (a few) hawaiites of the Hocheifel area contain the debris of spinel peridotite fragments as xenocrysts (i.e., crystals in the grain size of phenocrysts) and, thus, remnants of upper mantle material. These volcanics have been the subject of basalt + peridotite mixing which has obviously modified the primary condition of the alkali basalt. There is some indication that primary alkali basalts have incorporated upper mantle peridotite in the order of 10% and more (e.g., Hochkelberg).

Apart from visible criteria for xenocrystic olivine (anhedral and fractured crystals, transformation textures, undulatory extinction, picotite inclusions), orthopyroxene (reaction zones of olivine + clinopyroxene), spinel (magnetite reaction rims, corrosion), and clinopyroxene (cores of phenocrysts) there are physical and chemical constraints for the alien nature of these phases to the host basalt. Orthopyroxene and spinel occur as liquidus phases in alkali basalt only at high pressure, which in turn, would imply that they could also be megacrysts. The forsterite contents of xenocyrstic olivines are not in accordance with the Mg ratios of the host basalts if a K_D $(=Fe^{2+}/Mg)_{olivine}/(Fe^{2+}/Mg)_{liquid}$ of 0.3 to 0.35 (Roeder and Emslie, 1970) is applied. Narrow but iron-rich rims around olivine xenocrysts indicate (beginning) reequilibration of forsterite-rich (xenocrystic) olivine and Fe-rich liquid (basalt) by $Mg \rightleftarrows Fe^{2+}$ exchange (Vieten, 1980). The partition coefficients of Ni and Co between the olivine xenocrysts and the host basalt (referred to as liquid) do not imply equilibrium. The MgO contents of the host basalts are too low when plotted versus D (Ni) or D (Co) (Irvine and Kushiro, 1976; Irving 1978).

At least two types of clinopyroxenes can be considered to be xenocrysts abundant in alkali basalts and hawaiites. They occur always as corroded cores in larger phenocrysts and are Al-rich. They can, however, further be separated by contrasting amounts of Mg, Fe, Ti, Cr, Ni, Sc, Co, and the REE. Cr-rich clinopyroxenes (Mg-ratios 80-84; Cr 4600-3000 ppm, Ni ~200 ppm; Sc ~90 ppm; Co ~35 ppm) are colorless to light green and the altered remnants of peridotite disintegration. At best, they reflect clinopyroxenes from marginal parts of peridotite fragments close to, or already in contact with, the alkali basalt host. Some of the xenocrysts, however, reveal greater REE fractionation $[D_{La}(cpx/liquid):D_{Yb}(cpx/liquid) = 0.1$ to $0.12]$ and can be considered as megacrysts (that are phenocrysts) which crystallized when the magma was still at high pressure (Frey et al., 1978; Irving, 1978). Cores of fassaitic green to dark green clinopyroxenes are chemically identical with the clinopyroxenes of the clinopyroxenite fragments. Clinopyroxenes of that kind exhibit low Mg-ratios (56 to 65), low Cr (~ 70 ppm), Ni (~ 75 ppm), Sc (~ 44 ppm), and Co (~ 25 ppm) contents. They are rimmed by zones which have greater Mg-ratios (> 75 to 80) and greater abundances of these trace elements. REE fractionation $[D_{La}(cpx/liquid):D_{Yb}(cpx/liquid) = 0.15$ to $0.23]$ reveals these clinopyroxenes as phenocrysts. They apparently have crystallized prior to peridotite incorporation when basaltic liquids (as products of low degrees of metling) still have low contents of Cr, Ni, Sc, and Co.

The corroded remnants of kaersutite megacrysts are ubiquitous phases in primary alkali basalts, but occur also in derivative hawaiites. Their abundances, however, vary from a fraction of a per cent up to about 1% to 2% in general. The hawaiite from Hüstchen (South loop of the Nürburg racing track) and nepheline basanite from Alte Burg (near Reifferscheid) with 11% and 18% modal kaersutite, respectively, are exceptions.

All kaersutite megacrysts (maximum size up to 5 cm in diameter) are marginally or completely replaced by magnetite + clinopyroxene + rhönite intergrowths and containing olivine and biotite as a result of amphibole melt interaction. Skeletons of individual rhönite crystals appear to be final products of the kaersutite breakdown (at low pressure of H_2O). Major and trace element chemistry of kaersutite is consistent with amphibole which plays an important role (together with olivine and clinopyroxene) in the fractionation scheme of alkalibasaltic lavas toward mugearites and benmoreites (Irving and Price, 1981). Kaersutite megacrysts in alkali basalts of the Hocheifel area indicate the presence of H_2O sufficient to saturate the melt for the precipitation of amphibole when the magma was still at (high) pressure prior to eruption. Low Rb/Sr ratios (< 0.015) of kearsutite megacrysts give way to kb-enrichment in melts from which amphibole is subtracted by fractionation. The results are higher Rb/Sr-ratios as displayed for the phonolitic benmoreite from the Selberg near Quiddelbach. The initial $^{87}Sr/^{86}Sr$ ratio of this derivative volcanic (K-Ar age 30·Ma) is 0.70385 and well within the range of 0.7035 to 0.7039 derived from primary alkali basalts of the Hocheifel Area (4 samples). The variation in the $^{87}Sr/^{86}Sr$-ratios could be explained in terms of a slight mantle source heterogeneity because a crustal contamination can be ruled out for these rocks.

Hocheifel Alkali Basalt Petrogenesis

Hocheifel alkali basalt petrogenesis is related to a melting zone below the volcanic field. Specific petrologic and geochemical parameters of primary alkali basalts reflect physicochemical conditions within the upper mantle underneath the volcanic area. A N-S shaped body (40 km in N-S and 25 km in E-W direction) is projected on to the surface from within the upper mantle which has released (primary) alkali basaltic melts from (unknown) depths 46 to 18 Ma ago. Fragmental inclusions restrict the depths of the alkali basalt generation in the Hocheifel area below a cataclastic deformed but metasomatically altered spinel peridotite level. Ascending alkali basaltic melts (highly ne-normative) have accidentally collected fragments from that level at depths ranging from 50 to 30 km (about 10 to 17 kbar and temperatures of 890° to 1040°C (orthopyroxene-clinopyroxene equilibration). Major and trace elements of Hocheifel alkali basalts including the REE infer a fertile peridotite as common source. Geochemical constraints imply a 5% to 20% partial melting model with no or very small amounts of garnet left in mantle residue.

At the time being, only 31 potassium-argon data out of about 300 eruptive centers are available. Thus, a correlation of alkali basalt composition versus time of eruption is vague. The conditions of melting at a given mantle source is also a matter of nuance and cannot be defined with the (geochemical) data at hand. For a complete composition-age-PT record on Hocheifel alkali basalt petrogenesis more information particularly about the time sequence of the (over) 300 eruptions is required. To deduce and decipher the nature of the primary melts whether primitive (O'Hara, 1968) or incipient (Yoder, 1976) high-pressure laboratory experiment on major Hocheifel alkali basalts types and their peridotite inclusions must find out about the nature of liquidus phases in the melt (alkali basalt) - solid (mantle residue)-fluid (H_2O, CO_2, O_2) equilibria. This procedure is one effective way to define possible temperature and pressure conditions at which a partial melt could have been in coexistence with the solids of the mantle source prior to eruption.

References

Cantarel, P. and Lippolt, H.J., 1977. Alter und Abfolge des Vulkanismus in der Hocheifel. Neues Jahrb. Geol. Paläontol. Monatsh., 600-612.

Eggler, D.H., 1978. The effect of CO_2 upon partial melting of peridotite in the system $Na_2O-CaO-Al_2O_3-MgO-SiO_2$ CO_2 to 35 kbar, with an analysis of melting in a peridotite H_2O-CO_2 system. Am. J. Sci., 278:305-343.

Finnerty, A.A. and Boyd, F.R., 1978. Pressure-dependent solubility of calcium in forsterite coexisting with diopside and enstatite. Carnegie Inst. Yearb., 77:713-717.

Frey, F.A., Green, D.H., and Roy, S.D., 1978. Integrated models of basalt petrogenesis: A study of quartz tholeiites to olivine melilitites from South Eastern Australia utilizing geochemical and experimental petrological data. J. Petrol., 19:463-513.

Huckenholz, H.G., 1973. The origin of fassaitic augite in the alkali basalt suite of the Hocheifel Area, Western Germany. Contr. Mineral. Petrol., 40:315-326.

Huckenholz, H.G. and Puchelt, H., 1982. Trace element and REE data on alkalibasalts from the Hocheifel Area, Western Germany (in preparation).

Irvine, T.N. and Kushiro, I., 1976. Partitioning of Ni and Mg between olivine in silicate liquids. Carnegie Inst. Yearb., 75:668-675.

Irving, A.J., 1978. A review of experimental studies of crystal/liquid trace element partitioning. Geochim. Cosmochim. Acta, 42:743-770.

Irving, A.J. and Price, R.C., 1981. Geochemistry and evolution of lherzolite-bearing phonolitic lavas from Nigeria, Australia, East Germany, and New Zealand. Geochim. Cosmochim. Acta, 45:1309-1320.

Lindsley, D.H. and Dixon, S.A., 1975. Diopside-enstatite equilibria at 850°C to 1400°C, 5 to 35 kbar. Am. J. Sci., 276:1282-1301.

Lippolt, H.J. and Fuhrmann, U., 1980. Vulkanismus der Nordeifel: Datierung von Gang- und Schlotbasalten. Aufschluss, 31:540-547.

Mysen, B.O., 1976. The role of volatiles in silicate melts: Solubility of carbon dioxide and water in feldspar, pyroxene, and feldspathoid melts to 30 kbar and 1625°C. Am. J. Sci., 276:969-996.

Mysen, B.O., 1977. The solubility of H_2O and CO_2 under predicted magma genesis conditions and some petrological and geophysical implications. Rev. Geophys. Spec. Phys., 15, 3:351-361.

Mysen, B.O. and Kushiro, I., 1977. Compositional variations of coexisting phases with degree of melting of peridotite in the upper mantle. Am. Mineral., 62:843-865.

Mysen, B.O., Arculus, R.J., and Eggler, D.H., 1975. Solubility of carbon dioxide in melts of andesite, tholeiite, and olivine nephelinite compositions to 30 kbar pressure. Contrib. Mineral. Petrol., 53:227-239.

O'Hara, M.J., 1968. The bearing of phase equilibria in synthetic and natural systems on the origin and evolution of basic and ultrabasic rocks. Earth Sci. Rev., 4:69-133.

Roeder, P.L. and Emslie, R.F., 1970. Olivine-liquid equilibrium. Contrib. Mineral. Petrol., 29:275-289.

Vieten, K., 1980. Chemismus der Olivin-Idiokristen und Olivin-Xenocrysten in Alkalibasalten des Siebengebirges. Protokoll über das 5. Kolloquium im Schwerpunkt "Vertikalbewegungen und ihre Ursachen am Beispiel des Rheinischen Schildes". Dtsch. Forschungsgem., November 1980, pp. 161-163.

Wood, B.J. and Banno, S., 1973. Garnet-orthopyroxene and orthopyroxene-clinopyroxene relationships in simple and complex systems. Contrib. Mineral. Petrol., 42:109-124.

Yoder, H.S., Jr., 1976. Generation of Basaltic Magma. Natl. Acad. Sci., USA

5.3 Volcanism in the Southern Part of the Hocheifel

E. Bussmann and V. Lorenz[1]

Abstract

In the Hocheifel Area, preexisting fractures oriented N-S and E-W were obviously of great importance in the rise of the Tertiary magmas.

In the southern part of the Tertiary basaltic Hocheifel volcanic field the structures of the Devonian bed rocks, as well as the highly eroded volcanoes themselves, were studied with the aim of recognizing relationships between magma rise and bed rock structures.

On aerial photographs about 90% of the basalts have been recognized as lying on or very close to principal lineations, frequently on their intersections (Fig. 1). These lineations usually coincide with faults mapped by Fuchs (1974) and others. With their maximum diameter many of the basalt feeders are oriented parallel to the lineation on which they are localized. We therefore assume that magma rise made use of preexisting fractures and especially of those oriented N-S and E-W. In contrast to the Quarternary Westeifel volcanic field where NW-SE structures were of great importance to magma rise, this direction was of minor relevance in the Hocheifel. Possibly this points to a different orientation of the Tertiary stress-field related to the Hocheifel when compared with that of the Quarternary stress-field related to the Westeifel.

In the southern part of the field 18 of the 43 Tertiary basaltic plugs investigated in some detail are mantled by tuff, i.e., 42%. Therefore, phreatomagmatic activity is assumed to have been rather important. The large diameter of several basaltic "plugs", e.g., Höchstberg, Kastelberg, Steineberger Lei, is the result of a two-phase activity. An initial phreatomagmatic activity caused ejection of respective quantities of fragmented bed rock material and thus formation of an initial maar. In an immediately following second phase, magma rose and emplaced itself in the initial maar as a thick lava lake. In the southern Hocheifel G. Büchel also recognized several Tertiary maars.

The great relevance of phreatomagmatic activity in the Quarternary and Tertiary volcanic fields of the Eifel and the principles observed suggest the following: Owing to their structural and hydrogeological setting a great many of the volcanoes of the Rhenish Massif and surrounding areas erupted phreatomagmatically, thus causing formation of the many maars, diatremes, and plugs mantled by tuff.

Reference

Fuchs, G., 1974. Das Unterdevon am Ostrand der Eifeler Nordsüd-Zone. Beitr. Naturkd. Forsch. Südwestdtschl. Beih., 2:3-163.

---

[1] Institut für Geowissenschaften der Universität Mainz, Saarstraße 21, D-6500 Mainz, Fed. Rep. of Germany

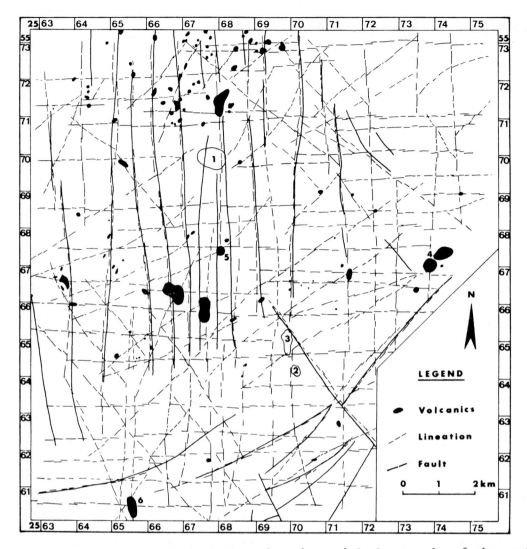

Fig. 1. Map of the principal lineations in aerial photographs of the southern Hocheifel. Faults after Fuchs (1974) and others. 1 Mosbruch Maar; 2 Ulmen Maar; 3 Jungferweiher; 3 Höchstberg; 5 Kastelberg; 6 Steineberger Lei

5.4 Tertiary Volcanism in the Siebengebirge Mountains

K. Vieten[1]

Abstract

In the Siebengebirge Mountains, a moderately alkaline and a strongly alkaline basalt-trachyte series can be distinguished. Both, however, are assumed to be derived from a single parent magma. The vast majority of the acid and intermediate differentiates is restricted to the centre of the region. A close relationship between faults and volcanism is indicated.

The Tertiary volcanic region extending over about 40 km in NW-SE and 30 km in NE-SW direction at the lower Middle Rhine is named here after the geographic Siebengebirge. This district, 7 by 5 km in area, is situated excentrically in the NE of the volcanic region near the southern margin of the Lower Rhine basin. It is considered as the centre because of the comparatively large number of closely associated volcanic edifices, the great volume of volcanic products, and the vast variety of volcanic rocks.

The alkali basalt-trachyte association of the Siebengebirge is divided into two series with different alkaline affinities. The moderately alkaline series comprises latitebasalts [i.e., olivine-basalts according to compositional parameters (cf. Huckenholz, 1983)], latites (i.e., basic mugearites), quartz-latites (i.e., basic benmoreites), trachytes (i.e., benmoreites), and quartz-trachytes (i.e., trachytes), whereas the strongly alkaline series consists of feldspathoid-latite-basalts (i.e., nepheline-basanites and alkali olivine-basalts), feldspathoid-latites (i.e., hawaiites), feldspathoid-trachytes (i.e., hawaiites and basic mugearites), and alkali trachytes.

Each of the two series is typified by a systematic variation of the modal and chemical composition of the rocks, and of the chemistry of the major minerals. There are significant mineralogical and petrochemical differences between both series with regard to acid and intermediate members. However, the basic volcanites, which are altogether alkali basalts in virtue of the contents of SiO_2 and $(Na_2O + K_2O)$, show gradual transitions. Inasmuch as these two spatially and temporally associated rock series converge within the basic range, they are assumed to be cogenetic and derivatives from a single starting magma.

According to area measurements, quartz-trachytic and trachytic pyroclastics (9.9 km^2 in total/ 7.1 km^2 of which in the centre) and quartz-trachytes and trachytes (2.03/2.02) on the one hand, and alkali basalts (6.3/1.5) together with alkali basaltic pyroclastics (0.76/0.2) on the other hand predominate by far over quartz-latites and latites (0.82/0.72) and their pyroclastics (0.04/0.04) as well as over alkali trachytes (0.3/0.01), feldspathoid-trachytes (0.16/0.16), and feldspathoid-latites (0.02/0.02). As these data show, the vast majority of the acid and intermediate differentiates is restricted

[1] Mineralogisch-Petrologisches Institut und Museum der Universität Bonn, Poppelsdorfer Schloß, D-5300 Bonn, Fed. Rep. of Germany.

to the centre of the volcanic region. Moreover, strongly alkaline rocks with higher contents of SiO_2 are relatively rare. Towards the outer parts of the region, a few latites and one alkali trachyte and trachyte are encountered in addition to prevalent alkali basalts. In the outermost parts alkali basalts are the only volcanic rocks.

Current investigations concerning the areal and compositional pattern of the approximately 150 alkali basalts of the Siebengebirge allow the preliminary conclusion that alkali basalts associated spatially with latites and/or trachytes differ from isolated alkali basalts with respect to the contents of certain major, minor, and trace elements, indicating a somewhat more original nature of the latter rocks. However, there are also petrochemical differences (although less significant), between the alkali basalts of different parts of the region. Hence, it appears that the whole volcanic region can be divided into subregions, each with an independent volcanic history.

Within the centre, volcanic activity started with the eruption of trachytic pyroclasts. A wide-spread cover, some hundred meters thick, was piled up into which successively trachytic, latitic, and alkali basaltic melts intruded, forming domes, funnel fillings (Trichter-Kuppen), necks, dykes and sills. At such volcanic centres where this sequence is incomplete as only two rock types are encountered, trachytes always prove to be older than latites and alkali basalts respectively, and latites older than alkali basalts.

As can be inferred from petrographical and petrochemical differences between rocks of subregions, volcanism occurred with certain temporal shifting. This is supported by K-Ar age determinations. Unfortunately, there are not enough data to clear up the temporal distribution pattern of the volcanic events in more detail.

The faults along which the Devonian rocks of the northern Eifel and the Bergisches Land are displaced against the Lower Rhine basin, converge towards the centre of the volcanic region which is dissected in horsts and grabens. The orientation of dykes and the strict alignment of central volcanoes in about the same direction as the faults indicate a close relationship between tectonics and volcanism. Moreover, it follows from the differences of the areal and compositional pattern of the rocks (e.g., most of the strongly alkaline acid and intermediate rocks are confined to a narrow district within the Siebengebirge Graben) that even the genesis of the volcanic rocks may be linked to the tectonic movements.

References

Frechen, J., 1976. Siebengebirge am Rhein - Laacher Vulkangebiet - Maargebiet der Westeifel. Sammlung Geologischer Führer, Vol. 59. Borntraeger, Berlin, Stuttgart, 209 pp.

Frechen, J. and Vieten, K., 1970. Petrographie der Vulkanite des Siebengebirges. Decheniana, 122:337-377.

Stindl, H. and Vieten, K., 1982. The minerals of the volcanic rock association of the Siebengebirge. II. Olivines. 1. Idiocrysts and xenocrysts. Neues Jahrb. Mineral. Abh., 145:183-199.

Todt, W. and Lippolt, H.J., 1980. K-Ar-age determination on Tertiary volcanic rocks: V. Siebengebirge, Siebengebirge-graben. J. Geophys., 48:18-27.

Vieten, K., 1972. Über die Heteromorphie-Beziehungen in der Vulkanit-Assoziation des Siebengebirges und ihre petrogenetische Bedeutung. Neues Jahrb. Mineral. Abh., 117:282-323.

Vieten, K., 1979. The minerals of the volcanic rock association of the Siebengebirge. I. Clinopyroxenes. 1. Variation of chemical composition of Ca-rich clinopyroxenes (salites) in dependence of the degree of magma differentiation. Neues Jahrb. Mineral. Abh., 135:270-286.

Vieten, K., 1980. The minerals of the volcanic rock association of the Siebengebirge. I. Clinopyroxenes. 2. Variation of chemical composition of Ca-rich clinopyroxenes (salites) in the course of crystallization. Neues Jahrb. Mineral. Abh., 140:54-88.

Vieten, K. and Stindl, H., 1983. The alkali basalts of the Tertiary volcanic region at the lower Middle Rhine (Siebengebirge) (in preparation).

5.5 Tertiary Volcanism in the Westerwald Mountains

K. von Gehlen and W. Forkel[1]

Abstract

In the Westerwald, nepheline basanites predominate. Mantle xenoliths are very rare. There is again an excentric cluster of trachytes.

The almost circular Westerwald volcanic province of less than 1000 km^2 has not yet been studied in the same detail as that of the Eifel or of the Northern Hessian Depression. As far as volcanic rocks have been dated in the Westerwald (Lippolt, this Vol.), basaltic and trachytic rocks about 25 Ma old (Oligocene/Miocene) predominate, followed by some eruptions of about 5 Ma (Miocene/Pliocene) in the western part, and finally only a few younger than 1 Ma (Quaternary) close to the Rhine. These last basaltic rocks must be considered as belonging to the Quaternary Eifel province.

As far as has been investigated, about 5% of the rocks are trachytic. All other rocks are basaltic in character. Half of them are nepheline basanites (in the sense of Green and Ringwood, 1967), and some 20% are alkali olivine basalts. There are moreover nearly 15% olivine nephelinites and a few tholeiitic and hawaiitic rocks. The group of trachytes which belong to the 25 Ma series occurs excentrically in a rather small area north of Montabaur. For the 5 Ma basalts, so far no remarkable differences in composition from the older series have been found.

In contrast to the approximately N-S trending distribution of basalt types found in the Tertiary Hocheifel volcanics (Huckenholz, 1983) and those of the Northern Hessian Depression (Wedepohl, 1983), no such trend was found so far in the Westerwald. In SiO_2 and MgO trend surfaces and plots of various major element ratios, a tendency towards a W-E to SW-NE strike seems to show up. The eruption points of the trachytes and the few tholeiites seem to underline this tendency.

Mantle xenoliths are very rare in the Westerwald basalts, among them only four spinel lherzolites so far. Their investigation by electron microprobe has produced results on pyroxenes similar to those from the Eifel (Seck, 1983), but neither amphiboles nor phlogopites have been observed.

References

Green, D.H. and Ringwood, A.E., 1967. The genesis of basaltic magmas. Contrib. Mineral. Petrol., 15:103-190.
Huckenholz, H.-G., 1983. Tertiary volcanism of the Hocheifel area. This Vol.
Seck, H.A., 1983. Eocene to Recent Volcanism Within the Rhenish Massif and the Northern Hessian Depression. This Vol.
Wedepohl, K.H., 1983. Tertiary volcanism in the Northern Hessian Depression. This Vol.

---

[1] Institut für Geochemie, Petrologie und Lagerstättenkunde, Universität Frankfurt, Senckenberganlage 28, D-6000 Frankfurt 11, Fed. Rep. of Germany.

5.6 Tertiary Volcanism in the Northern Hessian Depression

K. H. Wedepohl[1]

Abstract

About 250 of the larger basalt exposures of the volcanic field east of the Rhenish Massif (5000 km² in size) have been investigated petrographically, chemically and geochronologically. Magma of the predominating alkali olivine basalts, as well as that of basanites and nephelinites, was probably formed by partial melting of metasomatically altered spinel lherzolites. Quartz tholeiites can be explained as products of shallow partial melting of depleted mantle rocks.

Areal Distribution and Time Sequence of Basaltic Species

For general conclusions about the genesis of basaltic magmas under certain tectonic conditions, volcanic rocks in a reasonably large area should be investigated by petrological and geochemical methods. The present investigation covers the volcanic province north of the Vogelsberg Mountains, which is more than 5000 km² in size. The map of Fig. 1 gives the position of almost 250 larger volcanic necks and flows out of a total of more than 2000 separate basaltic masses.

On the base of their areal coverage, alkali olivine basalts represent by far the most common type (73%). Nepheline basanites (as well as chemically equivalent limburgites) and olivine nephelinites share a proportion of 12% and 9% respectively. Quartz tholeiites cover only 6% of the basaltic area. This petrographic specification is based on the investigation of several hundred thin sections and detailed chemical investigations (Wedepohl, 1983). It is used to recognize the areal distribution of basaltic species and the relations between tectonism and magmatism. The map demonstrates that olivine nephelinites and nepheline basanites are restricted to the NW and SE part of the area. The NNW-SSE trending long dashed line in Fig. 1 connects more than ten volcanoes with nepheline-rich basalt over a distance of 100 km (Sandebeck-Eisenberg Zone). Nepheline basalts are the youngest volcanic products of our area (Wedepohl, 1982). The genetically related melilite-containing olivine nephelinites occur in two belts in the NW part of our area: one north of Fritzlar, the other one close to Hofgeismar (dashed line in Fig. 1). Many olivine nephelinites and nepheline basanites contain lherzolite-harzburgite xenoliths. The volcanic activity in our area started with quartz tholeiites 20 Ma ago. Flows and necks of tholeiitic composition are restricted to two rather small belts: one occurs E and NE of the town Treysa and the other one crosses the Weser river close to the northern edge of the map. Both belts trend almost in N-S direction. Tholeiitic basalts never contain peridotite xenoliths.

---

[1] Geochemisches Institut, Goldschmidtstraße 1, D-3400 Göttingen, Fed. Rep. of Germany.

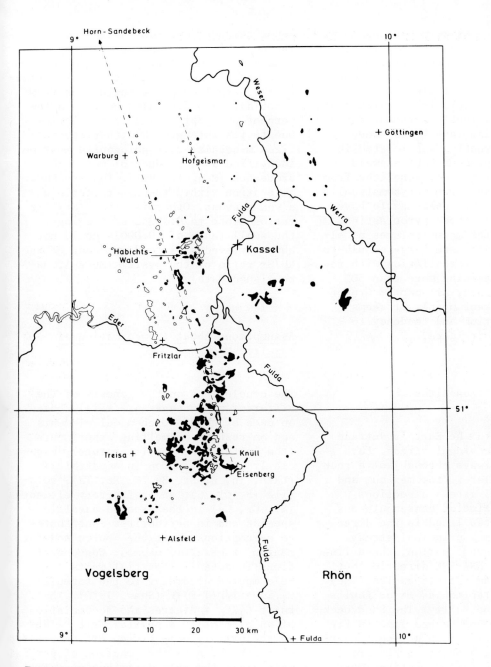

■ Alkali olivine basalt   □ Olivine nephelinite, nepheline basanite, limburgite   ▣ Quartz tholeiite

<u>Fig. 1.</u> Areal distribution of species of basaltic rocks in the northern Hessian Depression. For definitions, modal and norm compositions of basaltic rocks see Wedepohl (1983). *Long dashed line* connects localities with nepheline-rich basalts of the Eisenberg Sandebeck Zone. *Short dashed line* connects exposures of melilite (larnite)-bearing olivine nephelinites close to Hofgeismar

The large number of alkali olivine basalts is produced during a relatively short period of time (10-15 Ma ago). Less than a quarter of the basalt occurrences of this type contain lherzolite-harzburgite xenoliths from the upper mantle. Rock compositions intermediate between alkali olivine basalt and nepheline basanite or between nepheline basanite and olivine nephelinite have been abundantly observed. Links between alkali olivine basalt and tholeiite are extremely rare

(Bühl near Weimar, Kassel-West), probably because of genetic reasons. Voluminous beds of pyroclastics with a preerosional volume of several cubic kilometers were predominantly formed of alkali olivine basaltic magma (Mengel, 1981) which probably contained slightly more water than the melts producing lava flows. One center of tuff eruption was located in the area of the present Habichtswald Mountains west of Kassel. Xenoliths from the lower crust and from metasomatically altered parts of the upper mantle have been discovered in these pyroclastics. At a few localities magma volumes of alkali olivine basalt were large enough to undergo in situ differentiation with residual melts containing more than 50% SiO_2 (Hoher Meissner, Allendorf; Großer Belgerkopf, Oberkaufungen). It seems that differentiation and tendency to form sills are correlated.

Tectonics, Mantle Xenoliths

Magma transport to the Earth's surface needs tectonic tensional stress. Lava forms its channelways through solid rock units perpendicular to the maximum and minimum principal stress directions (Shaw, 1980). Chains of consanguineous volcanoes are often lined in the direction of the maximum principal stress.
In the area of investigation, these lines usually trend in NNW-SSE direction but are not exposed as visible faults. They have the same direction as major faults in the Upper Rhine Valley. The Sandebeck-Eisenberg Zone (long dashed line in Fig. 1) is identical with such a line. Several alkali olivine basaltic dykes are intrusions with the same direction. Quartz tholeiites as the oldest volcanic products of our area are lined in N-S direction.

Because of the difference in density between peridotites and basaltic melts, all basaltic species containing xenoliths cannot have resided in magma chambers above the source of xenoliths. They must originate from the upper mantle. According to their average mineral composition (73% olivine, 18% orthopyroxene, 6.7% clinopyroxene, ≤1.8% spinel) abundant xenoliths are depleted mantle rocks. The depletion of clinopyroxene etc. was caused by former partial melting. We have never observed garnet in our peridotites. Rare lherzolite inclusions from pyroclastics and basalts contain a few percent phlogopite (and or amphibole). Due to its intergrowth with recrystallized minerals, this phlogopite must be identified as a product of metasomatic processes (Mengel, 1981). This secondary alteration within the upper mantle was a young event. Otherwise the $^{87}Sr/^{86}Sr$ ratios close to 0.70362 scattering in the small range of ±0.00014 could not have survived a sixfold increase of the Rb/Sr ratio of the system (Mengel, unpublished data).

Assumptions About the Formation of Basaltic Magmas

The conclusions on the genesis of the different magma types are mainly based on data of about 40 chemical elements and on results of melting experiments (the latter from the literature). Especially information on incompatible (LREE, Th, U, Sr, Ba, K, Rb, Ta), volatile (F, Cl, S), and refractory elements (Ni, Cr, Co) in the exposed basaltic species and in peridotites contributes to assumptions about the source and formation of basaltic magmas. Concentrations in potential partial melts can be computed according to the formula: $C_1/C_o = 1/D(1-F)+F$ (Shaw, 1970). The ratio C_1/C_o indicates the accumulation of an element in a partial melt relative to the primary composition of the related mantle rock. F is the fraction of partial melt and D the bulk distribution coefficient for partition of an element between mantle rock and melt.

Highly incompatible elements like La, Ce, U, Th have bulk distribution coefficients D~0.01 for partition between mantle rocks containing 10% clinopyroxene and 90% olivine plus orthopyroxene and melt. At a minimum degree of partial melting (F = 0.01) and with D = 0.01, the C_1/C_o ratio is about 50. This figure indicates that incompatible elements can approach a 50-fold accumulation in melts

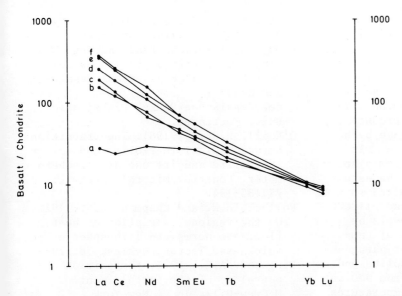

Fig. 2. Concentrations of REE in six basaltic species from the northern Hessian Depression relative to abundances in ordinary chondrites (the latter according to Mason, 1979). Number of samples of INA analyses used for averages (a) 10; (b) 18; (c) 10; (d) 3; (e) 5; (f) 7. (For individual data see Wedepohl, 1983)

relative to mantle source rocks as a maximum. Concentrations in source peridotite probably range from 0.30 ppm La and 0.90 ppm Ce to 0.50 ppm La and 1.5 ppm Ce (chondrites, depleted lherzolite Stöpfling etc.). Figure 2 indicates that a ratio C_1/C_0 = [basalt]/[chondrite] $\leq$ 50 for La and Ce is restricted to tholeiitic magmas. The remainder of the magma species from alkali olivine basalt to olivine nephelinite cannot be explained by partial melting of depleted lherzolites represented by common xenoliths. They must originate from mantle rocks which have gained La, Ce and other incompatible elements from reactions with metasomatic volatiles. A potential metasomatically altered mantle rock probably contains 2%-3% phlogopite, ~0.3% phosphate, some Ca carbonate etc. Spera (1981) has estimated the conditions for metasomatic transport of materials and heat from lower to higher mantle levels. The same author contributed data to demonstrate that melt fractions as low as a few percent have very low velocities of flow in a porous rock to be separated from restite peridotite into magma chambers within a reasonable time ($\leq 10^7$ years).

The heavy REE have distinctly higher bulk distribution coefficients (D) than the light REE especially for systems with garnet peridotite. A $D_{Yb} \geq 0.2$ has been estimated for melts equilibrated with lherzolite containing 5% garnet, 10% clinopyroxene, 85% olivine, orthopyroxene (Shaw, 1972; Harrison, 1981). This means that the following ratios can be computed: C_1/C_0 = [basalt]/[peridotite] < 5. We have observed for all basaltic species: $C_1/C_0 \sim 10$ (Fig. 2). This indicates that the different types of basaltic magmas occurring in our area were probably equilibrated with spinel peridotites. The maximum depth of stability of spinel (containing about 50% chromite molecule) is close to 90 km (O'Neill, 1981).

Our assumptions on compositions of basaltic magmas and upper mantle rocks can be related to melting experiments from the literature (cf. Ringwood, 1975), with the following results: The variety of basaltic magma species produced in the northern Hessian Depression was formed by partial melting of spinel lherzolites in the upper mantle. The gradient in composition from alkali olivine basalts to melilite-containing olivine nephelinites was probably caused by increasing pressure (depth). Quartz tholeiites can be explained as products of 5% to 10% partial melting of depleted spinel peridotite which was lifted adiabatically into a pressure level of about 15 kb in small uprising diapirs. The alkali olivine basaltic and the nepheline basaltic magmas required a spinel lherzolite mantle which had gained some phlogopite, Ca phosphate, Ca carbonate etc. through metasomatic reactions. The degree of partial melting to produce alkali olivine basalt could have been comparable to that of tholeiite. Olivine nephelinite originated from a smaller fraction of partial melt. We assume that the additional heat required to melt spinel peridotite at ≤ 30 kb and $>1180°C$ was introduced from deeper levels by metasomatic fluids. Estimates of temperatures of equilibration for paragenetic pyroxenes in spinel lherzolite and harzburgite xenoliths range from 900 to 1100°C (Oehm, 1980). These temperatures can be correlated with pressures from 20 to 24 kb (Pollack and Chapman, 1977). A general adiabatic uplift of mantle material under the Hessian Depression should have caused large scale rifting. This and abundant cataclastic structures of peridotites from shearing in diapirs cannot be observed. Therefore partial melting to form alkali olivine basalt and olivine nephelinite must be possible in a mechanically almost stable mantle after introduction of metasomatic fluids.

References

Harrison, W.J., 1981. Partitioning of REE between minerals and coexisting melts during partial melting of a garnet lherzolite. Am. Mineral., 66:242-259.

Mason, B., 1979. Chapter B, cosmochemistry, part 1. Meteorites. In: Fleischer, M. (ed.) Data of geochemistry. U.S. Geol. Surv. Prof. Pap., 440-B-1: 1-132.

Mengel, K., 1981. Petrographische und geochemische Untersuchungen an Tuffen des Habichtwaldes und seiner Umgebung und an deren Einschlüssen aus der tieferen Kruste und dem oberen Mantel. Dissertation, Univ. Göttingen.

Oehm, J., 1980. Untersuchungen zu Equilibrierungsbedingungen von Spinell-Peridotit-Einschlüssen aus Basalten der Hess der Hessischen Senke. Dissertation, Univ. Göttingen.

O'Neill, H.St.C., 1981. The transition between spinel lherzolite and garnet lherzolite and its use as a geobarometer. Contrib. Mineral. Petrol., 77:185-194.

Pollack, H.N. and Chapman, D.S., 1977. On the regional variation of heat flow, geotherms and lithospheric thickness. Tectonophysics, 38:279-296.

Ringwood, A.E., 1975. Composition and petrology of the earth's mantle. McGraw-Hill, Inc., New York.

Shaw, D.M., 1972. Development of the early continental crust. Part 1. Use of trace element distribution coefficient models for the Protoarchean crust. Can. J. Earth Sci., 9:1577-1595.

Shaw, D.M., 1970. Trace element fractionation during anatexis. Geochim. Cosmochim. Acta, 34:237-243.

Shaw, H.R., 1980. The fracture mechanisms of magma transport from the mantle to the surface. In: Hargraves, R.B. (ed.) Physics of magmatic processes. Princeton Univ. Press, Princeton.

Spera, F.J., 1980. Aspects of magma transport. In: Hargraves, R.B. (ed.) Physics of magmatic processes. Princeton Univ. Press, Princeton.

Spera, F.J., 1981. Carbon dioxide in igneous petrogenesis: II. Fluid dynamics of mantle metasomatism. Contrib. Mineral. Petrol., 77:56-65.

Wedepohl, K.H., 1982. K-Ar-Altersbestimmungen an basaltischen Vulkaniten der nördlichen Hessischen Senke und ihr Beitrag zur Diskussion der Magmengenese. Neues Jahrb. Mineral. Abh., 144:172-196.

Wedepohl, K.H., 1983. Die chemische Zusammensetzung der basaltischen Gesteine der nördlichen Hessischen Senke und ihrer Umgebung. Geol. Jahrb. Hessen, 111 (in press).

5.7 The Quaternary Eifel Volcanic Fields

H.-U. Schmincke[1], V. Lorenz[2], and H. A. Seck[3]

Abstract

The Quaternary (ca. 0.7 Ma) volcanic fields in the western central part of the Rhenish Massif (West Eifel and East Eifel) have formed roughly synchronously with the main Quaternary phase of uplift. The fields are 50 and 30 km long, elongated in NW-SE direction, contain ca. 240 and 90 volcanoes and are dominantly made of K-rich nephelinitic-leucititic-basanitic scoria cones. The larger West Eifel differs from the East Eifel field by more mafic and silica-undersaturated magmas, greater abundance and larger size of peridotite xenoliths and near absence of highly differentiated magmas contrasted with the occurrence of four highly differentiated phonolite volcanoes in the smaller East Eifel field. Two major groups of primitive magmas generated in different mantle reservoirs were erupted in both fields, basanites and nephelinites-leucitites. Differentiation was accomplished dominantly by fractionation of olivine, clinopyroxene, phlogopite and, in more differentiated magmas, sphene, amphibole, apatite, magnetite, joined by feldspar and feldspathoids in highly derivative magmas. Diffusion-controlled processes may have led to the formation of the extremely LIL-element-enriched phonolite magmas.

The magmas appear to have been derived from metasomatically modified mantle sources. NW-SE-oriented pathways may have been largely controlled by the present tensional lithospheric stress field, local N-S and NE-SW orientations possibly being influenced by pre-existing zones of structural weakness in the crust. Volcanic activity coincides broadly with increasing rates of uplift during the Quaternary. Data are insufficient to correlate distinct volcanic and tectonic phases. Quaternary magmas may have been generated in a low velocity anomaly in the mantle beneath the area (Raikes).

Several distinct mantle domains differing in isotope and trace element ratios were affected during partial melting. Concentric zonation in volume erupted, Mg-numbers and, in part, type of xenoliths in the West Eifel might reflect the detachment level of the magmas in the mantle. Uplift and volcanism may be related via ascent of hotter mantle material beneath the area, partial melting being due to decompression.

---

1 Institut für Mineralogie, Ruhr-Universität Bochum, Postfach 102148, D-4630 Bochum, Fed. Rep. of Germany

2 Institut für Geowissenschaften, Johannes Gutenberg-Universität, Saarstraße 21, D-6500 Mainz, Fed. Rep. of Germany

3 Institut für Mineralogie und Petrographie der Universität zu Köln, Zülpicher Str. 49, D-5000 Köln, Fed. Rep. of Germany.

Introduction

The formation and evolution of magmas beneath the Quaternary Eifel volcanic fields (Fig. 1), has proceeded in complex ways in both mantle and crustal domains. Their eruption concurrent with uplift of

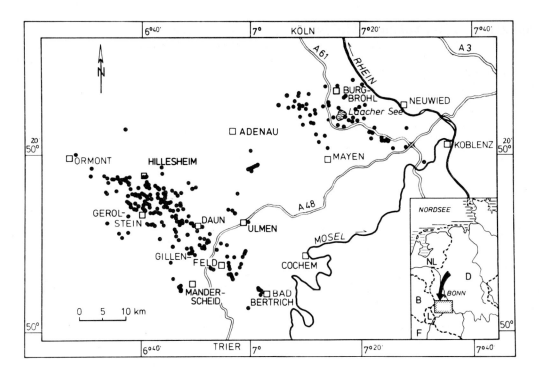

Fig. 1. Distribution of eruptive centers of Quaternary West and East Eifel volcanic fields. (Mertes and Schmincke, 1983)

the Rhenish Massif suggests a close relationship to mantle processes which, judging from the chronological pattern of past eruptions and current uplift, are still continuing.

Many aspects of these volcanic fields have been restudied during the course of the present research project: in the West Eifel, periodotite nodules and mineralogical and chemical aspects of lavas by Sachtleben, Seck and Stosch; maars, ^{14}C ages and tectonic features by Büchel and Lorenz; $^{40/39}Ar$ ages by Fuhrmann and Lippolt; a general assessment of the geochemical and volcanological evolution of the West and East Eifel fields by Mertes, Schmincke, Bogaard, Duda, Viereck, and Wörner; Sr-, Nd- and Pb-isotope ratios by Zindler and Staudigel and by Kramers et al., and Sr-isotopes by Webb; REE concentrations by Hertogen, Gijbels and Stosch; volcanic stratigraphy in the East Eifel based on soil and loess stratigraphy by Brunnacker and Windheuser.

In order to keep the following summary to a reasonable length publications by these groups are only listed at the end of the report and are not referred to in the text. Reference to previous work may be found in these publications as well as in recent summaries on the same general subject (Illies et al.,1979; Schmincke, 1982a).

Spatial and Volcanic Evolution

The western field (Fig. 2) covers approximately 600 km^2, extends for about 50 km from Ormont in the northwest to Bad Bertrich in the southeast and consists of about 240 eruptive centers (66% scoria cones, half of them with lava flows; 30% maars and tuffrings; 2% scoria rings and 2% pyroclastic vents). The total amount of magma (dense rock) erupted is about 1.5 km^3.

About 12 of these eruptive centers occur in an intervening zone about 20 km wide between the eastern and western volcanic fields. While spatially closer to the West Eifel, there are stronger geochemical and tectonic similarities to the eastern province. The eastern

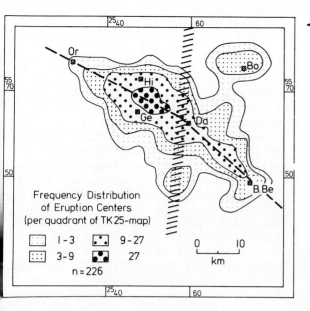

Fig. 2. Frequency distribution of eruptive centers identified in West Eifel volcanic field. Cross-hatched area marks the eastern limit of the Eifel N-S zone. (Mertes, 1982)

Fig. 3. Zonation of MgO-content of West Eifel volcanic rocks. (Mertes, 1982) ▼

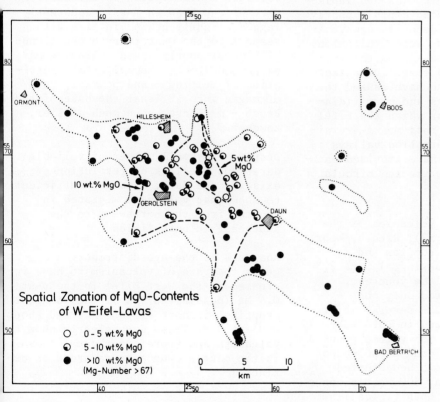

field extends for about 35 km roughly NW-SE around the centrally located Laacher See, is about 10-20 km wide, covers about 400 km² and consists of about 70 to 100 major eruptive centers. Quantitative assessment is not completed, but it is estimated that about 80% of the volcanoes are scoria cones. Maars are rare while phonolitic volcanoes are represented by four small calderas, which produced exogeneous and endogeneous domes, ash flow tuffs, vulcanian base surge and Plinian fall-out deposits. Total magma erupted may approach 10 km³ with some 90% of highly differentiated composition.

Number of volcanoes, in part size of volcanoes, spacing of volcanoes and the relative abundance of differentiated compared to primitive magmas is highest in the central part of both volcanic fields (Figs. 1-3).

The well-exposed interiors of many volcanoes show them to be complex. Many are composed of several major and minor adjacent and partly superimposed craters representing several but mostly short-lived eruptive phases. The complex nature is due to an unusually broad spectrum of eruptive styles rather than prolonged duration of volcanic activity.

Scoria cones (up to 100 m high) are made up largely of strombolian/hawaiian type agglutinates, scoria breccias and lapilli deposits but most also contain poorly sorted, xenolith-rich deposits apparently formed by phreatomagmatic (vulcanian) activity, usually representing the initial phase of a scoria cone. Vulcanian activity is believed to have caused the formation of tuff-rings and maars (diameter up to 1.7 km) for which the Eifel is the type locality. Most maars are situated in fracture-controlled valleys and the fractures may have provided pathways for both rising magma and circulating groundwater.

The larger phonolitic volcanoes of the East Eifel (Kempenich (?), Rieden, Wehr, Laacher See) are more complex and consist of caldera-like basins, 1.5 to 2.5 km in diameter. While the youngest, Laacher See, has erupted only once - about 11000 years B.P. - several major phases can be recognized in both Wehr and Rieden volcanoes. The volume erupted by the Laacher See volcano alone - about 5 km^3 - exceeds the combined erupted magma volume of all ca. 300 scoria cones and maar volcanoes in both fields. The erupted phonolite magma may represent only about 10% of chemically zoned high-level magma chambers. Most of the phonolite magma was erupted in high Plinian eruption columns. Tephra from the Laacher See eruption forms a widespread ash layer throughout Europe, from Northern Italy to central Sweden.

Temporal Evolution

Dating of volcanoes in both fields is still rather sketchy. However, application of several different methods ($^{40}Ar/^{39}Ar$, K/Ar, ^{14}C, morphological preservation, relationship to river terraces, stratigraphic correlation with soil horizons and loess deposits) has established a general framework of the temporal evolution of Quaternary Eifel volcanism (see also Lippolt, 1983).

The exact time of initiation of surface activity is still unknown in both fields. In the West Eifel, volcanism began at least as early as 0.6-0.7 m.y. and similar ages also appear to hold for the East Eifel. Thus, the first Quaternary volcanic manifestation in the western part of the Rhenish Massif apparently lagged slightly behind in time with respect to the Quaternary phase of uplift whose poorly dated beginning may be around 1.5 Ma (Semmel, 1983). Initiation of melting - and possibly rise of magma into at least the base of the crust - may have begun much earlier, however.

Despite the small number of reliable dates and the lack of resolution of existing age data, data are sufficient to show that volcanism migrated in general from the northwest to the southeast in both fields.

Analysis of the age distribution suggests the rate of volcanism to have increased in the West Eifel in the past 0.4 Ma (possibly 0.1 Ma). The youngest eruptions in both fields occurred about 0.01 Ma B.P. These data clearly show that volcanism and therefore thermal anomalies in the mantle cannot be regarded as extinct.

Mineralogical and Chemical Composition

The Quaternary Eifel lavas are phyric to highly phyric, total phenocrysts ranging up to 30%. The mafic lavas are dominated by clinopyroxene (Ti-augite) with generally less than 5% of olivine, phlogopite

Table 1. Representative chemical analyses of Quaternary Eifel volcanic rocks

	West Eifel				East Eifel		
	ME 90 Bad Bertrich Basanite	Me 77 Mosenberg Olivine nephelinite	ME 41 Beuel (Kirchweiler) Melilite nephelinite	ME 31 Steinrausch Nepheline-leucitite	E 314 Kunkskopf Basanite	E 337 Krufter Ofen Tephrite	1002-1 Laacher See Highly evolved phonolite
SiO_2	42.70	40.60	41.30	41.80	43.70	44.7	54.60
TiO_2	2.38	2.58	2.60	2.72	2.75	2.67	0.12
Al_2O_3	11.70	11.20	10.40	12.10	13.65	15.97	20.80
Fe_2O_3	3.00	3.85	6.89	7.29	5.20	6.54	1.34
FeO	8.19	7.05	3.99	3.70	5.14	3.86	0.47
MnO	0.13	0.19	0.19	0.19	0.16	0.23	0.55
MgO	14.10	14.50	11.40	8.95	9.70	5.72	0.07
CaO	12.10	12.70	16.10	14.70	12.03	10.51	0.42
Na_2O	3.15	3.41	2.62	3.22	2.65	4.22	11.40
K_2O	1.51	1.91	2.58	3.25	3.35	3.62	5.04
P_2O_5	0.91	0.90	0.71	0.60	0.53	0.76	0.02
H_2O	0.22	0.20	0.49	0.53	0.40	0.57	2.10
CO_2	0.05	0.05	0.07	0.05	0.08	0.21	0.04
	100.30	99.18	99.37	99.14	99.34	99.63	97.36
Cr	654	604	464	257	204	63	5
Co	62	60	62	49	69	44	4
Ni	330	368	174	120	115	44	7
Cu	64	62	125	112	68	61	49
Zn	83	87	75	80	78	98	309
Rb	37	51	73	99	61	107	712
Sr	936	1045	789	980	723	1285	5
Y	28	28	22	26	25	31	43
Zr	176	223	205	247	232	378	2614
Nb	66	100	77	94	61	122	413
Ba	721	795	1045	1204	931	1166	359
MgNo	72.37	73.61	69.36	63.83	66.94	51.00	7.79

ME 31, 41, 77, 90 from Mertes (1982).

Laacher See from Wörner (1982).

Kunkskopf, Krufter Ofen Schmincke (unpublished).

and titanomagnetite except for the more olivine-rich basanites and olivine nephelinites in the West Eifel. Amphibole, apatite and sphene appear as liquidus phases in the intermediate rocks with phenocrystic plagioclase, sanidine, hauyne, nosean, leucite and biotite being restricted to the phonolitic and rare trachytic (Wehr) magmas.

Volcanic rocks of the Eifel volcanoes are grouped into two major suites which show some geochemical, spatial and temporal coherency, some comprising a range in compositions from primitive to highly differentiated (Table 1): a leucitite-(melilite)-nephelinite suite (LMN) which is highly enriched and a basanite (B) and local olivine nephelinite (O) suite in the West Eifel which is moderately enriched in large-ion lithophile (LIL) elements. The suites are referred to subsequently as LMNW (west), LMNE (east), BW (west) and BE (east) (Fig. 4a,b). In figures 4a and 4b, the leucitite-nephelinite and melilite nephelinite fields are shown separately.

LMNW Suite

The volcanoes of the West Eifel volcanic field are dominated by nephelinite, leucitite and melilite nephelinite compositions, characterized by high K_2O (mostly 2.5%-4%) and related elements (Rb, Ba, LREE) as well as high Ca and low Al concentrations and high MgO contents, ranging from 7% to 14% (mean Mg-values: 63-67). Rocks of intermediate to highly differentiated compositions are rare (only five volcanoes). Chemically, and, in part, petrographically two major distinct subgroups occur. For example, melilite nephelinites are characterized by very high CaO (up to 16.5%) and Nb, higher P, Fe and CO_2, and by slightly lower Al, Si and K-concentrations compared to the leucitites and nephelinites. Carbonate, believed to be primary, possibly due to unmixing of a carbonatitic liquid, occurs in the groundmass of all melilite nephelinites of the West and East Eifel fields, but only rarely in rocks of other composition.

LMNE Suite

Leucitites, nephelinites and melilite nephelinites and their derivative magmas make up the older western subfield in the East Eifel, grouped around the centrally located larger Rieden and Kempenich complex phonolite volcanoes. The mafic magmas can be further subdivided based on differences in major (Na, Ti, Fe) and trace (Sr, Nb, Rb, Ba) element concentrations. The suite as a whole chemically resembles the LMNW magmas of the West Eifel except for some significant differences such as their higher Al_2O_3-content (more than 12.5%) with very little overlap with the West Eifel magmas (less than 13.5%). Among the trace elements, compositions are similar to the LMNW suite except for the much higher Sr, Ba and Nb in some or all of the Hochstein/Herchenberg/Leilenkopf magmas.

BW Suite

About 11% of the West Eifel volcanoes form a young chemically distinct suite mainly in the southeastern part of the field which consists of two major groups: groundmass plagioclase-bearing basanites and olivine nephelinites. All rocks of this suite are very mafic with high Cr, Ni, Co and MgO > 11% (mean Mg-values 71-72). They represent the most primitive suite of all Quaternary Eifel magmas and approximate primary magmas assuming current models of mantle compositions. However, the interpretation of the high Cr, Ni and Mg concentrations is complicated by the fact that xenocrystic nodule debris is common in these rocks. They are distinguished chemically from the LMNW-suite by lower Ba, Rb, Ca, Nb and Zr among the trace and K and Ca among the major elements. However, their P, Sr and Y contents are among the highest in the West Eifel lavas.

BE Suite

The majority of the East Eifel magmas form a basanite-tephrite-phonolite suite occupying the younger eastern part of the field around, E and S of the

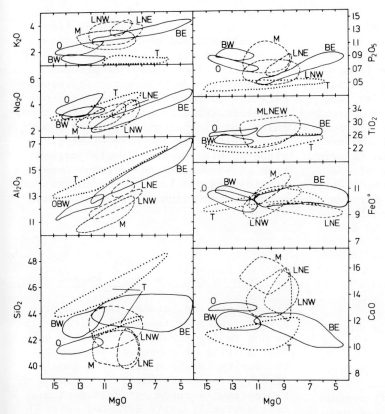

Fig. 4a

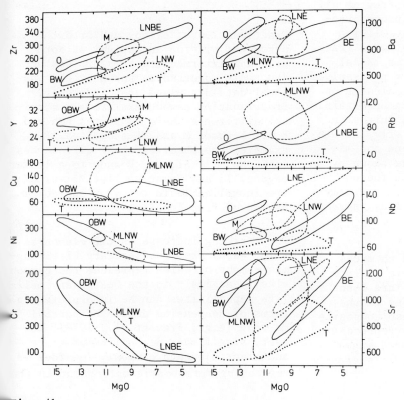

Fig. 4b

Fig. 4a,b. Composition of different magmatic suites of West and East Eifel volcanic fields. West Eifel (W) LN leucitites, nephelinites; M melilite nephelinites; O olivine nephelinites; B basanites. East Eifel (E): LN leucitites, nephelinites; B basanites; T Tertiary alkali basalts from the Hocheifel. (Data from Duda and Schmincke, 1978; Mertes, 1982, and Schmincke, unpubl.). a) Major elements versus MgO; b) Trace elements versus MgO

Laacher See, overlapping in part with the LMNE suite in the western part.

The basanite magmas can be clearly distinguished from the nephelinite/leucitites by their higher SiO_2, and, in part, FeO and lower Na_2O, K_2O, P_2O_5, CaO, NB, Sr, Ba, Cu, and, in part TiO_2. Al_2O_3 concentrations are similar between the suites except for the Leilenkopf and Hochsimmer lavas which have less Al_2O_3. The BE primitive magmas differ from those of the BW-suite by being less mafic - MgO never exceeding 11%, the lower limit for the BW magmas - and by the significantly higher K, Rb, Ba and, in part, Al and lower Sr.

Geochemical Evolution

Differentiation Within Suites

Some 10% of the West Eifel contain large (up to 50 cm) and some 10% of the East Eifel volcanoes small (cm-sized) peridotite nodules, but only the OBW and some of the LMNW magmas qualify as primary using criteria summarized by Frey et al. (1978) which assume a homogeneous pyrolite-type mantle composition. Olivine (and/or enstatite) and spinel were fractionated in these mafic OBW magmas although nodule debris possibly redissolved in these magmas makes exact calculation of quantitative fractionation uncertain. Most primitive magmas within suites in both fields have evolved chiefly through fractionation of clinopyroxene and minor olivine and phlogopite as shown by decrease in Mg, Cr, Ni and Ca (below 10% MgO), only minor decrease initially in Si, Fe, Ti and general steady increase in Al and Na.

Systematic mineralogical and chemical changes from initial tephritic to late stage basanitic compositions are found in several larger scoria cones in the East Eifel (Rothenberg). These composite volcanoes are interpreted as representing successive eruptions from zoned magma columns developed at intermediate crustal levels (Fig. 5).

Processes in addition to crystal fractionation have led to the evolution of extremely evolved phonolite magmas such as erupted during the initial stage of the Laacher See volcano eruption. In the BE suite two trends are represented: basanite to tephrite and basanite to phonolite. Fractionation in large chemically and mineralogically zoned phonolitic systems in the East Eifel took place at high crustal levels judging from geobarometry and basement xenolith studies and the large magma columns that probably had volumes approaching tens of km^3.

Differences Between Parent Magmas

The differences between the primitive parent magmas of the Eifel suites might be due to fractionation, variable partial melting of a homogeneous mantle and derivation from different domains of a heterogeneous mantle. Crustal contamination is ruled out as a mechanism having significantly affected these highly silica-undersaturated mafic magmas.

Fractionation of clino- and/or orthopyroxene, olivine and spinel at both high and intermediate pressure has undoubtedly occurred as shown by the chemical data discussed above. However, fractionation cannot explain all chemical differences between suites LMN and B nor between M and NL subsuites. For example, the observed clear differences in K, Na, Al, Si, Cr, P and, to a lesser degree, Ti and Fe at similar Mg values are inexplicable in terms of any combination of fractionating phases.

The striking increase in LIL elements (Fig. 4a,b) from the Tertiary Eifel magmas - many of which represent the common alkali basalts - to the Quaternary melilite nephelinites can be interpreted in terms of decreasing degree of partial melting (e.g., Frey et al., 1978). However, significant decoupling of trace and minor elements commonly regarded as reflecting metasomatic enrichment in the

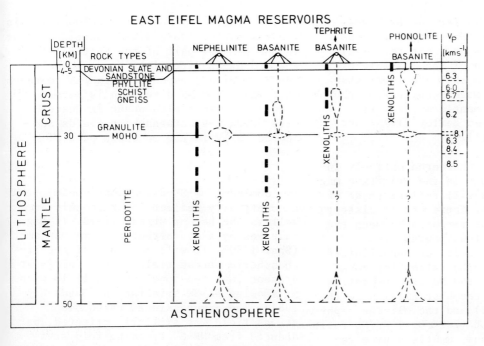

Fig. 5. Schematic crustal section through East Eifel showing possible magma reservoirs at various crustal levels as deduced from xenolith studies. Seismic subdivision after Mechie et al. (1982). (Schmincke 1982a)

mantle, can be recognized between the Quaternary and many other alkalic provinces as well as between different Quaternary Eifel magma groups. For example, the Quaternary Eifel magmas are not especially enriched in Ti, P and Y compared to similar rocks elsewhere while showing unusually strong enrichment in Nb, Rb, Ba and K. The higher Ba, Rb, Nb, Zr and K concentrations in the LMN and olivine nephelinite suites compared to the BW magmas are not correlated with P_2O_5 concentrations. In fact, the "less enriched" BW magmas have the highest P contents. If P is a reliable indicator of the degree of partial melting, assuming a homogeneous source with respect to P (Sun and Hanson 1975; Frey et al. 1978), then the Quaternary Eifel magmas do not represent unusually low degrees of partial melting (perhaps 5-10%?) but have been derived from a heterogeneous mantle source enriched by metasomatic fluids of variable composition.

Among the major elements, Al contents strongly contrast with the LIL concentrations, the Tertiary alkali basalts showing the highest and the melilite nephelinites the lowest Al concentrations. This feature, coupled with the very high Ca-concentrations, steep REE patterns, constant concentrations in Y, Sc, Yb, Lu and high La/Y ratios (exceeding 37) is most easily explained if garnet is a residual phase during low degrees of partial melting producing the LMN suite magmas. The fact that garnet has not been found in xenoliths in the Eifel - as in most nodule occurrences - is not considered significant because the area of melt generation is probably located below the zone of rapid uprise in which mantle fragments become incorporated in the magma en route to the crust.

The suggestion of two different mantle source reservoirs is also supported by differences in $^{87}Sr/^{86}Sr$ and $^{143}Nd/^{144}Nd$-isotope ratios which are below 0.704 and above 0.5128 for the B suite and higher resp. lower for the LMNW suite magmas. The LMN magmas are evidently derived from a mantle reservoir isotopically close to bulk Earth evolution while the B suite magmas are derived from a mantle depleted in incompatible elements relative to the source of the LMN magmas but similar to that from which the Hocheifel alkali basalt magmas were derived. This contrast holds strictly only for the West Eifel.

The depleted source for the BW magmas as reflected in their Sr- and Nd-isotope ratios contrasts with their relatively high concentration of incompatible elements, even though they are lower than in the LMN suite magmas suggesting geologically "recent" mantle metasomatism [see also Lloyd and Bailey (1975) and Frey et al. (1971)].

If Zr and Nb are considered entirely incompatible elements in the primitive magmas, Zr/Nb ratios will reflect those of the source region. There are consistent and significant differences between the East Eifel basanites showing high (3.55 ± 0.3) and leucitites/nephelinites low (2.20 ± 0.3) Zr/Nb ratios for rocks having similar Sr and Nd isotope ratios. The basanites of the West Eifel show the highest (2.9 ± 0.3) and the olivine nephelinites the lowest (2.2 ± 0.1) Zr/Nb ratios in that field. Thus, mantle source regions beneath the fields are chemically distinct on different scales, larger domains differing in isotopic and smaller-scale domains by trace element ratio signatures.

Eruption of BW magmas partly overlaps in time and space with that of LMN suite magmas. This is most easily explained in terms of a layered mantle, the more enriched reservoirs possibly underlying the more depleted ones. In both fields, the more enriched magmas were erupted chiefly during the initial phase prior to appearance of the less enriched magmas.

Tectonic Evolution

Both volcanic fields show abundant evidence for a dominant NW-SE trend. This is reflected chiefly in (a) the shape of the entire West Eifel and the two East Eifel subfields, (b) dominant directions of dikes, chiefly in the West Eifel, (c) elongate shape of volcanoes and alignment of volcanoes of similar composition believed to reflect a dike system, (d) zones of peridotite xenoliths of different degrees of deformation and recrystallization, and (e) magmas of similar Mg-values (Figs. 2, 3, 6). In both fields, there are major exceptions to this general pattern. N-S dike directions occur in the central part of the West Eifel. E-W and NW-SE directions are common in the East Eifel, especially in its southern part where the NW direction is rare. NE-SW directions also occur in Quaternary volcanoes between both fields and sporadically in the southern part of the West Eifel.

The dominant NW-SE directions could theoretically reflect the Paleozoic as well as the Recent stress field. If magma rises by a hydrofracturing mechanism (Shaw 1980), it will make use of the lithospheric stress field. The Eifel is at present in a tensional stress field with σ_3 showing NE-SW orientation (Baumann, 1981). The compressional Paleozoic stress field was similarly oriented, as deduced from NE-SW trending fold axes and NW-SE transverse fractures.

These data suggest the overriding influence of the present lithospheric stress field, although the pronounced NW-SE alignment of volcanoes and dike directions may be due to enhancement of both paleo and present-stress fields. The near-surface influence of the paleo-stress field may be reflected in N-S dike directions of volcanoes located in the Paleozoic/Mesozoic structural N-S zone in the central part of the West Eifel.

The fact that NW-SE directions are less well developed in the East Eifel may in part be explained by the more common establishment of high-level magma reservoirs within the crust. Magma rising from shallow reservoirs may more easily use pre-existing lines of weakness such as the Siegen overthrust, a major NE-SW tectonic structure in the area. Moreover, the collapse of the Tertiary Neuwied Basin overlapping in space with the East Eifel volcanic field may reflect a more pronounced tensional regime in the East than in the West Eifel. The more common development of high-level magma reservoirs in this area may likewise be due to the development of the volcanic field in the crest of the uplifted shield.

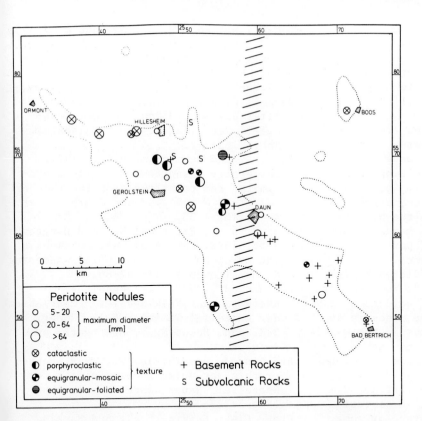

Fig. 6. Spatial distribution of peridotite (texture: maximum grade of deformation) and crustal xenoliths. Crosshatched area marks the eastern limit of the Eifel N-S zone (Mertes, 1982)

Discussion

Many intraplate volcanic fields occur in uplifted regions related to rift structures which are characterized by thinner crust and lithosphere thickness, tensional stress fields, and high heat flow or high geotherms as reflected in nodule mineral geothermometers (see Seck et al., 1983). These features are commonly interpreted as resulting from the rise of hot mantle material generating both uplift and magmas - resulting from decompression. The Wilson/Morgan hot spot/plume hypothesis, a specific model for rising mantle streams, has been applied to the Quaternary Eifel volcanic fields, believed to be presently situated above a stationary sublithospheric hot spot which earlier generated most of the Tertiary volcanic provinces in Central Europe (Duncan et al., 1972). Although age data available do not support the progressive younging of Tertiary volcanic fields to the west, the area of Quaternary volcanism in the Eifel is underlain by a region of teleseismic attenuation at a depth of ca. 40 to 150 km whose lower density may be due to up to 1%-2% partial melting (Raikes, 1980) (see also Raikes and Bonjer, 1983). The magma erupted at the surface (ca. 10 km^3) would represent only about 0.1% of a 2% melt volume. The melt fraction would be even higher if the compressibility of magmas at high pressure is taken into account (Stolper et al., 1981). Such areas of wave attenuation are now known to underlie many large active volcanic areas and are generally situated within broader lithospheric swells. The generation of melt would tend to enhance the buoyancy of the hot mantle, especially if the melt is not drained away rapidly.

Metasomatism of more or less depleted mantle by LIL-rich fluids and magmas is an additional mechanism to decrease mantle density and also to generate compositional heterogeneity in the mantle. Some authors have invoked such processes to generate mantle expansion, uplift and partial melting in part using evidence from the Eifel (Bailey, 1972; Lloyd and Bailey, 1975). Metasomatism was widespread prior to generation of the Quaternary magmas. Although metasomatism may be more widespread in the zones of magma generation underlying the mantle sources for the nodules, density reduction due to thermal expansion of mantle material and trapped magma are believed to have a more pronounced effect on uplift.

If phosphorus concentrations are a reliable indicator of the degree of melting, the Quaternary magmas were derived by much lower degrees of melting than the Tertiary basalt magmas (Fig. 4a,b). Moreover, the more volatile-rich nature of the Quaternary compared to the bulk of the Tertiary alkali-basalt magmas could be interpreted in terms of partial melting at lower temperatures. Such a hypothesis might be contradicted by the widespread Quaternary uplift but it must be remembered that surface volcanoes only highlight much broader root zones in the mantle. Generation of magmas in the mantle beneath the Rhenish Massif may have contributed to recent uplift even though the large scale mantle upwelling (?) that may be responsible for the Tertiary doming, volcanism and rifting in central Europe may have been decaying for some time. The similarity of the general succession of voluminous less alkalic volcanism followed after an erosional interval of up to several million years by small volumes of highly alkalic primitive magmas formed in several oceanic intraplate settings with that in the Rhenish Massif suggests operation of similar processes, perhaps a decaying cycle of lithosphere heating, the bleeding-off of late stage magmas possibly facilitated by fractures in the cooling contracting lithosphere.

Acknowledgements. We thank G. Büchel, D.S. Chapman, H. Mertes and L. Viereck for helpful suggestions on the manuscript. Much unpublished information was kindly supplied to us by van den Bogaard, Büchel, Duda, Mertes, Viereck and G. Wörner.

References

Bailey, D.K., 1972. Uplift, rifting and magmatism in continental plates. J. Earth Sci., 8:225-239.
Baumann, H., 1981. Regional stress field and rifting in western Europe. Tectonophysics, 73:105-111.
Böhnel, H., Kohnen, H., Negendank, J., and Schmincke, H.-U., 1982. Paleomagnetism of Quaternary volcanics of the East Eifel, Germany. J. Geophys., 51:29-37.
Büchel, G. and Lorenz, V., 1982. Zum Alter des Maarvulkanismus der Westeifel. Neues Jahrb. Geol. Paläontol. Abh., 163:1-22.
Duda, A. and Schmincke, H.-U., 1978. Quaternary basanites, melilite nephelinites and tephrites from the Laacher See Area, Germany. Neues Jahrb. Miner. Abh., 132:1-33.
Duncan, R.A., Petersen, N., and Hargraves, R.B., 1972. Mantle plumes, movement of the European plate, and polar wandering. Nature (London), 239:82-86.
Frechen, J., 1976. Siebengebirge am Rhein-Laacher Vulkangebiet - Maargebiet der Westeifel. Samml. Geol. Führer, 3rd. edn. Bornträger, Stuttgart, pp. 1-195.
Frechen, J. and Thiele, W., 1979. Petrographie der vulkanischen Foidite der Westeifel. Neues Jahrb. Miner. Abh. 136, 3:227-237.
Frey, F.A., Haskin, L.A., and Haskin, M.A., 1971. Rare-earth abundances in some ultramafic rocks. J. Geophys. Res., 76:2057-2070.
Frey, F.A., Green, D.H., and Roy, S.D., 1978. Integrated models of basalt petrogenesis: a study of quartz-tholeiites to olivine melilitites from Southern Eastern Australia. J. Petrol., 19:463-513.
Illies, J.H., Prodehl, C., Schmincke, H.U., and Semmel, A., 1979. The Quaternary uplift of the Rhenish shield in Germany. Tectonophysics, 61:197-225.
Kramers, J.D. Betton, P.J., Cliff, R.A., Seck, H.A., and Sachtleben, Th., 1981. Sr and Nd isotopic variations in volcanic rocks from the West Eifel and their significance. Fortschr. Mineral. Beih., 59:246-247.
Lippolt, H.J., 1983. Distribution of volcanic activity in time and space. (this vol.).
Lloyd, F.E. and Bailey, D.K., 1975. Light element metasomatism of the continental mantle: the evidence and the consequences. Phys. chem. earth, 9:389-416.
Lorenz, V., 1973. On the formation of Maars. Bull. Volcanol., 37:183-204.
Lorenz, V. and Büchel, G., 1980a. Die Kesseltäler der vulkanischen Westeifel Nachweis ihrer Maargenese. Mainzer Geowiss. Mitt., 8:173-191.

Lorenz, V. and Büchel, G., 1980b. Zur Vulkanologie der Maare und Schlackenkegel der Westeifel. Mitt. Pollichia. Pfaelz. Ver. Naturkd. Naturschutz, 68:29-100.

Mechie, J., Prodehl, C., and Fuchs, K., 1983. The long-range seismic refraction experiment in the Rhenish Massif. This vol.

Mertes, H., 1982. Aufbau und Genese des Westeifeler Vulkanfeldes. Dissertation, Ruhr-Univ. Bochum, pp.1-415.

Mertes, H. and Schmincke, H.-U., 1983. Age distribution of Quaternary volcanoes in the West Eifel Volcanic Field. Neues Jahrb. Geol. Palaeontol. Abh. 166:260-293.

Raikes, S., 1980. Teleseismic evidence for velocity heterogeneity beneath the Rhenish Massiv. J. Geophys., 48:80-83.

Raikes, S. and Bonjer, K.P., 1983. Large scale mantle heterogeneities beneath the Rhenish Massif and its vicinity from teleseismic P-residuals measurements. This vol.

Sachtleben, T., 1980. Petrologie ultrabasischer Auswürflinge aus der Westeifel. Unveröffentl. Disseration, Univ. Köln, pp.1-160.

Sachtleben, T. and Seck, H.A. 1981. Chemical control of Al-Solubility in orthopyroxene and its implications on pyroxene geothermometry. Contrib. Mineral. Petrol., 78:157-165.

Schmincke, H.-U., 1977a. Eifel-Vulkanismus östlich des Gebietes Rieden-Mayen. Fortschr. Mineral., 55, 2:1-31.

Schmincke, H.-U., 1977b. Phreatomagmatische Phasen in quartären Vulkanen der Osteifel. Geol. Jahrb. Reihe A, 39:3-45.

Schmincke, H.-U., 1981. Volcanic activity away from plate margins. In: Smith, D.G. (ed.) The Cambridge encyclopaedia of earth sciences. Cambridge Univ. Press, London, pp.201-209.

Schmincke, H.-U., 1982a. Vulkane und ihre Wurzeln. Rheinisch. Westfael. Akad. Wiss. Nat. Ing. Wirtschaftswiss. Vortr. Westdtsch. Verlag, Opladen, 315:35-78.

Schmincke, H.-U., 1982b. Volcanic and chemical evolution of the Canary Islands. In: v. Rad et al. (eds.) Geology of the Northwest African continental margin. Springer, Berlin, Heidelberg, New York, pp.273-308.

Semmel, A., 1983. Early Pleistocene terraces of the Upper Middle Rhine and its southern foreland-questions concerning their tectonic interpretations. This vol.

Shaw, H.R., 1980. The fracture mechanisms of magma transport from the mantle to the surface. In: Hargraves, R.B. (ed.) Physics of magmatic processes. Princeton Univ. Press, Princeton, pp.201-264.

Staudigel, H. and Zindler, A., 1978. Nd and Sr isotope compositions of potassic volcanics from the East Eifel, Germany: Implications for mantle source region. Geol. Soc. Am. Abstr. W. Progr., 10:497.

Stolper, E., Walker, D., Hager, B.H., and Hays, J.F., 1981. Melt segregation from partially molten source regions: the importance of melt density and source region size. J. Geophys. Res., 86:6261-6271.

Stosch, H.G., 1980. Zur Geochemie der ultrabasischen Auswürflinge des Dreiser Weihers in der Westeifel: Hinweise auf die Evolution des kontinentalen oberen Erdmantels. Dissertation, Univ. Köln, pp.1-233.

Stosch, H.G. and Seck, H.A., 1980. Geochemistry and Mineralogy of the two Spinel Peridotite Suites from Dreiser Weiher, West Germany. Geochim. Cosmochim. Acta, 44:457-470.

Stosch, H.G., Carlson, R.W., and Lugmair, G.W., 1980. Episodic Mantle Differentiation: Nd and Sr Isotopic Evidence. Earth Planet. Sci. Lett., 47:263-271.

Sun, S.S. and Hanson, G.N., 1975. Evolution of the mantle: Geochemical evidence from alkali basalt. Geology 3:297-302.

Windheuser, H., 1977. Die Stellung des Laacher Vulkanismus (Osteifel) im Quartär. Sonderveröff. Geol. Inst. Univ. Köln, 31:1-223.

Windheuser, H. and Brunnacker, H., 1978. Zeitstellung und Tephrostratigraphie des quartären Osteifel-Vulkanismus. Geol. Jahrb. Hessen, 106:261-271.

Wörner, G., 1982. Geochemisch-mineralogische Entwicklung der Laacher See Magmakammer. Dissertation, Ruhr-Univ. Bochum, pp.1-331.

Wörner, G., Schmincke, H.-U., and Schreyer, W., 1982. Crustal xenoliths from the Quaternary Wehr volcano (East Eifel). Neues Jahrb. Mineral. Abh., 144:29-55.

Zindler, A., 1980. Geochemical processes in the earth's mantle and crust-mantle interactions: evidence from studies of Nd and Sr isotopic ratios in mantle derived igneous rocks and lherzolite nodules. Ph.D. thesis, M.I.T., Cambridge, Mass., pp.1-238.

5.8 Carbon Dioxide in the Rhenish Massif

H. Puchelt[1]

Abstract

Carbon dioxide in the order of $0.5-1 \times 10^6$ t yr^{-1} emerges from the whole Eifel region. Carbon isotope studies indicate a magmatic source. Locally, isotopic alteration is caused by kinetic fractionatin in unsaturated water bodies.

In the whole Eifel area, approximately 0.5 to 1×10^6 metric tons of carbon dioxide emerge from the ground or from oversaturated waters per year. It can be assumed that this has been happening since the last volcanic activity, i.e., for more than 10.000 years. Consequently, 5 to 10×10^9 tons of CO_2 must have been produced since.

In order to define the source of the CO_2, carbon isotope analyses of gases and the co-existing hydrogen carbonate-containing waters have been performed on more than 250 samples (natural waters and drilled producing wells). Over a time span of 5 years, the $\delta^{13}C$ values in CO_2-gas cover a range from -0.7‰ to -7‰. Co-existing bicarbonate is isotopically heavier in most cases, but may sometimes even be lighter than CO_2-gas (Puchelt and Hubberten, 1980, 1983). The reason for this C-isotope fractionation, which cannot be explained by equilibrium distribution in the gas/water system, was found to be a kinetic process during which $^{12}CO_2$ enters faster into the aqueous phase than $^{13}CO_2$. Such processes occur when CO_2 penetrates through unsaturated waters, i.e., in the gases emerging in Lake Laach or in the Lahn river. This mechanism of kinetic C-isotope fractionation was investigated in laboratory experiments using water heights from 1 to 4 m. Carbon dioxide bubbling through these water columns was enriched in ^{13}C during the initial stages of the experiments by 3‰, 5‰, and 7‰, respectively (Puchelt, 1982). This effect explains the occurrence of "heavy" carbon dioxide with $\delta^{13}C$ values between -4‰ and -0.7‰.

Thus one general source for the CO_2 occurring in the Eifel can be assumed: magmatic carbon dioxide, which may locally be isotopically altered by kinetic fractionation involving unsaturated water bodies.

References

Puchelt, H., 1982. Kinetische Kohlenstoff-Isotopenfraktionierung im CO_2-H_2O-System des Laacher Sees. Ber. Bunsenges. Phys. Chem., 86:1041-1043.
Puchelt, H. and Hubberten, H.-W., 1980. Vulkanogenes Kohlendioxid: Aussagen zur Herkunft aufgrund von Isotopenuntersuchungen. ZFI-Mitt., Berichte des Zentralinstituts für Isotopen- und Strahlenforschung, Leipzig, 29:198-210.
Puchelt, H. and Hubberten, H.-W., 1983. Kohlenstoffisotopenuntersuchungen an Gasen und Wässern der Eifel. Neues Jahrb. Mineral. Monatsh. (submitted).

---

[1] Institut für Petrographie und Geochemie, Universität Karlsruhe, Kaiserstraße 12, D-7500 Karlsruhe, Fed. Rep. of Germany.

5.9 Eocene to Recent Volcanism Within the Rhenish Massif and the Northern Hessian Depression – Summary

H. A. Seck[1]

Abstract

The volcanism of the Rhenish Massif is associated with tectonic events related to the formation of the Rhine rift system. During the Tertiary, the beginning of volcanic activity shifts from west (Hocheifel, 46 Ma) to east (Northern Hessian Depression, 20 Ma). The Quaternary volcanism is restricted to two fields west of the Rhine. Except for the Siebengebirge, SiO_2-undersaturated basaltoids predominate throughout the Tertiary. Their uniform major and trace element chemistry as well as their Sr ($^{87}Sr/^{86}Sr$ = 0.7037 ± 0.0003) and Nd ($^{143}Nd/^{144}Nd$ = 0.512858 ± 0.000055) isotope similarities are indicative of homogeneous mantle sources enriched in LIL elements prior to partial melting. Magma generation is believed to be a result of asthenosphere updoming for which evidence exists from geophysical observations and xenolith studies. Quaternary magmas presumably are related to a region of P-wave attenuation whose maximum elevation at 50 km depth is located beneath the West Eifel. From geochemical and isotopic evidence it can be concluded that these magmas in their majority come from mantle sources whose Sr ($^{87}Sr/^{86}Sr$ = 0.7042 ± 0.0003) and Nd ($^{143}Nd/^{144}Nd$ = 0.512695 ± 0.000045) isotopies and LIL element geochemistry are distinctly different from the Tertiary sources.

Evolution of Rhenish Massif Volcanism in Space and Time

The study of volcanism associated with the uplift of the Rhenish Massif cannot be expected to give direct answers to the question of the forces causing the uplift. However, the evolution of magmatism in this area through the interpretation of mantle xenoliths and geophysical data may be used to derive constraints on the nature of thermal processes in the lower lithosphere below the massif. The geochemistry of the magmas may further provide insight into the chemical and petrological constitution of the lower lithosphere and into chemical processes involved in magma generation. Under this aspect, the data reported in the previous articles are summarized and discussed here with main emphasis on the basaltoids.

For the purpose of a synopsis of volcanic activity, data of more geologic nature pertinent to the volcanic fields are compiled in Table 1, and geochemical data are listed in Table 2. It should be kept in mind, however, that the volcanic field of the Northern Hessian Depression is not located within the massif but occupies an old depression adjacent to it which existed since the uppermost Palaeozoic.

Volcanic activity in the Rhenish Massif ranged from mid-Eocene (46 Ma) until about 10.000 years before now. Although periods of volcanic activity in the different volcanic fields overlap as shown in Fig. 1, there is an inter-relationship of the spatial distribution of volcanism in the Massif with time, in that

---

[1] Mineralogisch-Petrographisches Institut, Zülpicher Str. 47, D-5000 Köln, Fed. Rep. of Germany.

Table 1. Geologic data of volcanic fields

	Tertiary				Quaternary	
	Hocheifel	Siebengebirge	Westerwald	Northern Hessian Depression	East Eifel	West Eifel
Periods of volcanic activities						
a) basaltic rocks	46-18 Ma	26-18 Ma	30-18 Ma (10)-5 Ma	20-7 Ma	0.7 Ma to 10.000 a	0.7 Ma to 10.000 a
b) trachytes, benmoreites, phonolites	44-38 Ma	26-24 Ma	28-23 Ma			
Eruptive styles	Stocks, necks, tuffites less abundant	Explosive, effusive, no lava flows	Effusive, stocks, necks, lava flows, tuffites less abundant	Effusive, explosive, necks, stocks, lava flows, pyroclastics	~80% scoria cones 20% endogene and exogene domes, maars	66% scoria cones 30% maars
Number of eruption centers	300	200	x00	x00 (x > 5)	~80	240
Lateral extension of volcanic fields	1000 km²	1200 km²	700 km²	6000 km²	400 km²	600 km²
Xenoliths	Spinel peridotites of mantle origin (4 locations), spinel peridotite debris very abundant	Spinel peridotites of mantle origin, comagmatic peridotites and pyroxenites, gneisses and schists	Spinel peridotites of mantle origin (very rare), pyroxenites (only few locations)	Spinel peridotites of mantle origin, garnet websterite (1 location), eclogite (1 location), granulites and other crustal rocks	Spinel peridotites of mantle origin, comagmatic xenoliths, complete suite of crustal rocks	Spinel peridotites of mantle origin, comagmatic xenoliths, crustal rocks

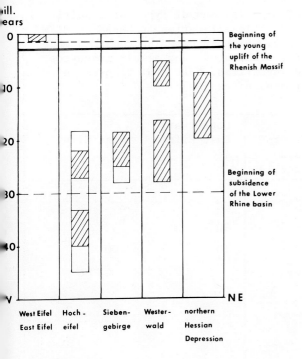

Fig. 1. Periods of volcanism in the Rhenish Massif and the Northern Hessian Depression in their relation to important tectonic events indicated by *dashed lines*. *Hatched fields* mark the periods of main volcanic activities

the beginning of magmatic activity shifts to younger ages from SW to NE. The Tertiary/Quaternary boundary in two aspects marks an important turning point in this scheme. First, the spatial evolution during the Tertiary outlined above is interrupted and volcanism is reactivated in two rather narrow fields in the western part of the Massif again. Second, the Quaternary magmatics are distinctly different from the Tertiary ones in several aspects of their major element geochemistries. Further, the Quaternary volcanism appears to be very intense in terms of either the mass of material erupted (East Eifel) or the number of eruption centers (West Eifel) per square unit (Table 1).

There are two important tectonic events marked in Fig. 1. The first is the beginning of subsidence of the Lower Rhine Basin associated with a phase of uplift of the Rhenish Massif which is not very well documented (Meyer et al. 1983) and a reorientation of the stress field.

Whereas the Hocheifel field as a whole and the majority of its volcanic edifices are aligned in N-S direction, magma ascent in the Siebengebirge becoming active after the subsidence of the Lower Rhine Basin started, occurred along NW-SE directed paths parallel to the rift structures of the Lower Rhine basin. In the Northern Hessian Depression, NNW-SSE orientation predominates for eruption centers less than 12 m.y. in age, however, also the NW-SE direction is observed.

The second tectonic event is a young phase of uplift of the Rhenish Massif whose beginning is dated at approximately 1.5 m.y. (Semmel, 1983), preceding the Quaternary volcanism by about 1 m.y.

Geochemical Evolution

Percentages of rock types encountered in the Tertiary volcanic fields are listed in Table 2. The figures reported are not strictly comparable, though, for two reasons. First, the subdivisions chosen for the various fields are not entirely compatible, because they are based on CIPW-norms for Hocheifel, Siebengebirge, and Westerwald but on combined modal compositions and CIPW-norms for the Northern Hessian Depression. Second, the percentages given partly refer to numbers of eruption centers and partly to square dimensions mapped. Nevertheless, the dominance of SiO_2-undersaturated magmas among the basaltoids showing a transition from olivine nephelinites to alkali olivine basalts and olivine basalts is evident in all fields. Considering the Tertiary magmatism as a whole, there is, on the basis of data available, no evidence for a chemical evolution trend in terms of an increase or decline of the average degree of SiO_2-undersaturation with space and time. However, a trend from alkali olivine basalt to more nepheline-rich basaltic rocks with decreasing age is observed in the Northern Hessian Depression. It is of importance that the tholeiites which amount to less than 5% on average, are concentrated in a small area in the NE of the Rhenish Massif and in two others in the Northern Hessian De-

Table 2. Percentages of trachytoids and basaltoids (a) and of types of basaltic rocks (b) of the Rhenish Massif volcanic fields. Classification is based on CIPW-norm except for the Northern Hessian Depression, which is based on combined mode and CIPW-norm. Trachytoids include benmoreites, trachytes, and alkali trachytes. Alkali olivine basalts include hawaiites and mugearites. Note that figures refer either to number of eruption centers (EC) or area units (AU)

	Tertiary				Quaternary	
	Hocheifel	Siebengebirge	Westerwald	Northern Hessian Depression	East Eifel	West Eifel
a)	EC	AU	EC	AU	EC	EC
Trachytoids	6	68	5	-	90[a]	[b]
Basaltoids	94	32	95	100	10	100
b)						
Basaltoids normalized to 100						
Olivine nephelinite	9	-	11	9.3		
Nepheline basanite	51	70	58	11.8		
Alkali olivine-basalt + olivine basalt	40	28	26	73		
Tholeiite	-	2	5	6		

[a] Phonolites and trachytes.
[b] Very rare phonolites.

pression. According to experimental evidence (e.g., Jaques and Green, 1980), this requires higher temperatures in a shallow mantle depth in these zones. It is characteristic that the trachytes, if present, are confined to the center of maximum activity in each field and erupted during a limited time span at the beginning of the individual periods of activity. In contrast to Hocheifel and Westerwald where trachytes are subordinate, in the Siebengebirge volume production of qz-trachytes outweighs that of alkali basalts.

Except for potassium which appears to increase in a W-E direction across the Massif, major element compositions of identical magma types assumed to be primitive (Table 3) are surprisingly similar. Also $^{87}Sr/^{86}Sr$- and $^{143}Nd/^{144}Nd$-ratios fall into a narrow range (Table 4). In an ε_{Juv}-versus $^{87}Sr/^{86}Sr$-diagram they plot on the Mantle Array as defined by oceanic and continental volcanics (e.g., O'Nions et al., 1979) at $^{87}Sr/^{86}Sr$-ratios lower and ε_{Juv}-values higher than those of the Chondritic Uniform Reservoir (CHUR). This high degree of homogeneity in terms of major element chemistey and Sr- and Nd-isotopes indicated for that part of the mantle underneath the massif which produced the Tertiary magmas, contrasts markedly with the major element and isotope heterogeneity of the uppermost mantle as

Table 3. Major element compositions (wt.%) of representative Tertiary basaltoids of the Rhenish Massif

	Hocheifel			Siebengebirge		Westerwald		Northern Hessian Depression			
	1	2	3	4	5	6	7	8	9	10	11
SiO_2	41.32	41.37	45.78	42.7	44.3	44.1	44.8	40.3	42.3	45.2	48.3
TiO_2	2.68	2.59	2.39	2.4	2.4	2.52	2.30	2.7	2.7	2.2	2.2
Al_2O_3	13.51	13.42	13.93	14.2	14.3	12.2	11.6	11.4	11.8	12.5	13.3
Fe_2O_3	4.67	4.47	4.18	4.0	3.8	3.32	3.89	4.1	4.0	3.0	3.1
FeO	5.96	5.67	6.47	7.5	7.7	7.86	6.89	6.8	6.9	7.5	7.2
MnO	0.20	0.17	0.19	0.2	0.2	n.d.	n.d.	0.19	0.17	0.18	0.17
MgO	10.83	10.64	10.05	11.8	9.6	11.6	12.5	12.2	11.5	10.9	8.7
CaO	13.61	13.62	10.95	11.0	10.4	11.0	11.5	12.6	11.2	10.2	8.7
Na_2O	3.34	2.98	3.08	3.0	3.1	2.8	2.4	3.3	3.4	2.9	3.3
K_2O	1.34	0.89	0.93	1.4	1.2	1.4	1.1	1.7	1.8	1.9	1.8
P_2O_5	0.82	0.80	0.59	0.5	0.6	n.d.	n.d.	1.1	0.88	0.73	0.56
CO_2	0.5	0.4	0.20	n.d.	n.d.	n.d.	n.d.				
H_2O	1.29[a]	2.98[a]	1.39[a]	0.8	1.2	n.d.	n.d.				
	100.07	100.0	100.13	99.5	98.8	96.8	96.98				
$\frac{100\ Mg}{Mg + Fe}$	65.5	66.2	63.7	66	61	65.6	68.2	67.5	66.1	65.6	60.8

[a] Total H_2O.

1) Olivine nephelinite, Nürburg, Nürburg racing course. 2) Nepheline basanite, Nitzbach Steinchen, near Adenau. 3) Alkali olivine basalt, Scharfe Kopf, Nürburg racing course. 4) Nepheline basanite (N=45). 5) Alkali olivine basalt (N=18). 6) Nepheline basanite, Käfernberg near Ahlbach. 7) Alkali olivine basalt, Schmalburg near Beilstein. 8) Olivine nephelinite (N=9). 9) Nepheline basanite (N=11). 10) Basanitic alkali olivine basalt (N=13). 11) Alkali olivine basalt (N=32). Data sources: 1-3) Huckenholz (unpubl.) 4-5) Vieten (unpubl.) 6-7) Junk (unpubl.) 8-11) Wedepohl (1982)

Table 4. $^{87}Sr/^{86}Sr$- and $^{143}Nd/^{144}Nd$-isotope ratios of Quaternary and Tertiary basaltoids

	Quaternary			Tertiary		
	East Eifel	West Eifel Nephelinites	Basanites	Hocheifel	Westerwald	Northern Hessian Depression
	(N = 6)[1]	(N=10)[2]	(N=6)[3]	(N=7)[4]	(N=3)[5]	(N=18)[6]
$^{87}Sr/^{86}Sr$	0.7040	0.7042	0.7037	0.7035	0.7034	0.7033 $\bar{x}_{18}$=0.70357±0.0003
	0.7045	0.7045	0.7039	0.7037	0.7039	0.7043
$^{143}Nd/^{144}Nd$	0.51260	0.512650	0.512780	0.512803	0.512867	
	0.51265	0.512739	0.512882	0.512858	0.512912	

Data sources:

1) Staudigel and Zindler (1978)
2) Staudigel and Zindler (1978), Kramers et al. (1981)
3) Kramers et al. (1981)
4) Staudigel and Zindler (1978), Kramers et al. (1981), Huckenholz and von Drach (unpubl.)
5) Stosch and Lugmair (unpubl.)
6) Mengel (unpubl.).

deduced from spinel peridotite xenoliths (Seck and Wedepohl, 1983).

To comply with the LIL element geochemistry of the lavas observed at the surface (except tholeiites), it is required that their mantle sources were enriched in incompatible elements prior to partial melting. For lanthanum of olivine nephelinites throughout the massif, enrichment factors observed range from 280 to 340 times chondritic abundances. This requires melting rates of less than 0.5% in a mantle containing non-volatile elements in chondritic ratios. Apart from the fact that geochemical and petrological arguments stand against such small melting rates (Frey et al., 1978), it was shown by Spera (1980) that such small portions of melt are unlikely to segregate. There is additional evidence for young mantle metasomatism from xenoliths of the Northern Hessian Depression, Hocheifel, and West Eifel. However, if or in what way xenolith metasomatism is related to that of the magma source regions is still uncertain. That such a relation may exist is indicated by the Sr- and Nd-isotopies of the Quaternary magmas and xenoliths of the West Eifel. LIL-element-enriched xenoliths approach $^{87}Sr/^{86}Sr$- and $^{143}Nd/^{144}Nd$-ratios similar to those of the Quaternary magma source regions (see below) whereas the dry xenoliths from the Rhenish Massif not affected by metasomatism reveal $^{87}Sr/^{86}Sr$- and $^{143}Nd/^{144}Nd$-ratios typical for recent MORB (Stosch et al., 1980).

Pleistocene to recent volcanism which is associated with a very young phase of shield uplift from 1.5 Ma on (Semmel, 1983 and Chap. 4 this volume) is restricted to the western part of the massif. In spite of differences in their magmatic evolution patterns, the East Eifel and West Eifel fields have some features in common which distinguish them from Tertiary volcanism. Representative analyses of Quaternary and Hocheifel Tertiary basaltic rocks listed in Table 5 for comparative purposes show this. The marked geochemical contrast is of particular interest, because both fields slightly overlap. Apart from the dominance of various types of K-rich nephelinites, which amount to about 90%, it is the lower Al_2O_3, the higher Fe_2O_3 and CaO contents, and the strong enrichment of highly incompatible elements (Rb, K, Ba) which are of importance and presumably are the result of both differences in the processes of magma generation and in the mantle source geochemistry. The assumption of a different magma source for the nephelinites of the West Eifel - and the basaltic magmas of the East Eifel - finds support from their $^{87}Sr/^{86}Sr$- and $^{143}Nd/^{144}Nd$-ratios, which in contrast to those of the Tertiary magmas are very similar to CHUR. Whether this difference in the mantle source chemistry is the corollary of a pre-melting enrichment of the mantle which produced the Tertiary magmas, as suggested by Staudigel and Zindler (1978), or if mantle sources rising from greater mantle depths were involved in the generation of the Quaternary magmas is not clear yet. The fact that a rather small portion of basanitic lavas in the West Eifel which are similar in terms of major and trace element geochemistry to the Tertiary volcanics, also reveal Sr- and Nd-isotopes typical of the Tertiary magmas lends some support to the latter possibility.

Conditions of Magma Generation

It is the distinct chemical similarity of the West Eifel volcanics to the nephelinites of the central volcanoes of the East African Rift system that has led to the idea that the Quaternary volcanism of the Rhenish Massif is related to the Upper Rhine rifting (e.g., Lloyd and Bailey, 1975). It has further been suggested that the Upper Rhine Graben continues underneath the Rhenish Massif to the Lower Rhine Basin (Illies et al., 1981), which finds its surface expression in a NW-SE-trending fault system. This has served since the Miocene as paths of magma ascent. Eruption centers in the Siebengebirge and of the Quaternary fields are aligned along this direction. In situ stress measurements (Baumann, 1983) and the study of focal mechanisms of recent earthquakes (Ahorner, 1983) demonstrate that tensional forces perpendicular to this fault system are still active. In a reconnaissance study of the lithosphere structure in Europe (Panza et al., 1980), a zone of anomalously thin lithosphere was shown to ex-

Table 5. Representative analyses of Quaternary and Tertiary lavas from the Eifel

	Quaternary			Tertiary	
	1	2	3	4	5
SiO_2	42.54	40.90	42.79	43.53	42.54
TiO_2	2.88	2.83	2.49	2.47	2.38
Al_2O_3	11.78	11.43	11.82	13.72	14.37
Fe_2O_3	5.62	7.47	4.40	4.24	4.37
FeO	5.25	3.55	6.63	6.32	6.03
MnO	0.19	0.21	0.18	0.20	0.18
MgO	10.45	8.93	13.36	10.17	8.46
CaO	14.24	16.10	11.66	12.04	12.56
K_2O	3.88	2.55	3.01	0.85	0.75
Na_2O	2.55	2.8	1.50	3.06	4.20
P_2O_5	0.68	0.84	0.89	0.75	0.85
CO_2	0.11	0.29	0.11	0.30	0.72
H_2O	0.59	2.36	1.00	1.68	3.02
	100.76	100.26	99.84	99.33	100.43

1) Leucite nephelinite, Rudersbüsch, West Eifel; 2) Melilite nephelinite, Hohenfels-Essingen, West Eifel; 3) Basanite, Wartgesberg, West Eifel; 4) Nepheline basanite, Bränkekopf, Nürburg racing course (Huckenholz unpubl.); 5) Basanite, Arensberg, West Eifel.

tend from the Upper Rhine Graben underneath the Rhenish Massif into the Lower Rhine Basin. Average shear wave velocities in the subcrustal lithosphere, which are low relative to adjacent areas, indicate anomalous physical properties of the lithospheric upper mantle in this zone. Both observations are in agreement with the widespread occurrence of strongly deformed mantle xenoliths in the Tertiary volcanics. The structure and temperature history of these xenoliths provide evidence for shearing and transport which could be related to upwelling of hot mantle material prior to the Tertiary volcanism in the Rhenish Massif (Seck and Wedepohl, 1983).

Temperatures and depths of magma segregation from the mantle residua are difficult to assess. Production of basanitic melts which are widespread in the Rhenish Massif requires pressures of 18 to 23 kbar in the absence of volatiles (Bultitude and Green, 1971).

Under CO_2-pressure, the SiO_2-undersaturated basic magmas encountered in the Rhenish Massif may be produced in much shallower depth compared to dry melting (Eggler, 1978). On the other hand, there is agreement today that water-saturated melts are distinctly more siliceous than dry melts at moderate degrees of partial melting (Jaques and Green, 1980). It is still a matter of debate, however, whether this is also valid for small degrees of partial melting under H_2O-pressures. In addition to this uncertaincy, the available data pertinent to the role volatiles have played in the partial melting processes of the Tertiary volcanism is very poor. That the magma source regions were affected by the influx of H_2O as a corollary of LIL-element enrichment may be inferred from the presence of hydrous silicates in the LIL-element-enriched xenoliths. However, if the amount of amphibole and/or phlogopite was not higher than that of the xenoliths, the total H_2O concentrations in the magma source rocks did not exceed a few tens of a percent. Thus, the amount of melt that could be saturated with H_2O is very small and the primary H_2O content of alkaline olivine basalts resulting from about 10% partial melting of a pyrolite source would be less than 1%. Nevertheless, this is sufficient for alkali olivine basalt melts to be formed at temperatures well below those of dry melting and the Tertiary magmas may have formed at depths higher than those equivalent to 15 to 20 kb obtained for dry partial melting (Jaques and Green, 1980; Thompson, 1974).

In summary, it is concluded that Tertiary magmas of the Rhenish Massif were generated in a zone of asthenospheric upwelling which according to Panza et al. (1980) exists underneath the Rhenish Massif as part of an "astenosphere channel" with an upper surface at about 80 km depth following the Rhine Rift System. The fact that volcanism is restricted to a number of narrow fields requires either the rise of local diapirs to shallower mantle depths beneath these fields or special chemical and physical conditions in those parts of the mantle from which the magmas come. The uniform major element and isotope geochemistries imply that the magma source regions as well as the magma-generating processes were similar throughout the Rhenish Massif.

In the absence of rifting phenomena and of both geophysical and petrological evidence for diapiric uprise, magma generation in the adjacent Northern Hessian Depression must have occurred in a mechanically nearly stable upper mantle. The heat required for partial melting is assumed to have been provided by the same fluid phases which caused the LIL-element enrichment of the magma source region. From the study of mantle xenoliths it is inferred that the magmas come from a depth equivalent to a pressure range of 24 to 27 kbar.

Quaternary volcanism appears to be related to a body of P-wave attenuation whose maximum elevation at 50 km depth was recorded underneath the West Eifel (Raikes, 1980). The fact that an abundant group of mantle xenoliths assumed to come from a depth range of less than 60 km (Seck and Wedepohl, 1983) reveals equilibrium temperatures of about $1150°C$, points to higher temperatures within the low-velocity body and gives support to the assumption of Raikes (1980) that the P-wave attenuation within this body may be due to the presence of small amounts of partial melt. This and the temperature history of strongly deformed mantle xenoliths from the West Eifel provide evidence that the body rose up diapirically from greater depth. Whether it is a young local diapir piercing the asthenosphere-lithosphere boundary or an older dome-like elevation of this boundary is an open question. However, the following facts and observations are in favor of the first possibility. The Quaternary volcanism is in several aspects entirely different from the Tertiary one. This is not restricted to differences in LIL-element geochemistries and Sr- and Nd-isotopies but applies to the whole magma type association. Melilite-bearing basic magmas, which except for the Northern Hessian Depression are missing in the Tertiary volcanism, are prominent representatives of the Quaternary magmatism. According to our present knowledge (Brey and Green, 1977; Eggler, 1978), this requires the presence of CO_2 and H_2O in the partial melting processes, and indicates that volatiles played a more important role.

References

Ahorner, L., 1983. Historical seismicity and present-day micro-earthquake activity of the Rhenish Massif, Central Europe. This Vol.
Baumann, H. and Illies, J.H., 1983. Stress field and strain release in the Rhenish Massif. This Vol.
Brey, G. and Green, H.D., 1977. Systematic study of liquidus phase relations in olivine melilitite + H_2O + CO_2 at high pressures and petrogenesis of an olivine melilitite magma. Contrib. Mineral. Petrol., 61:141-162.
Bultitude, R.J. and Green, H.D., 1971. Experimental study of crystal liquid relationships at high pressures in olivine nephelinite and basanite compositions. J. Petrol., 12:121-147.
Eggler, D.H., 1978. The effect of CO_2 upon partial melting of peridotite in the system $Na_2O-CaO-Al_2O_3-MgO-SiO_2-CO_2$

to 35 kb, with an analysis of melting in a peridotite -H_2O-CO_2 system. Am. J. Sci., 278:305-343.

Frey, F.A., Green, H.D., and Roy, S.D., 1978. Integrated models of basalt petrogenesis: A study of quartz tholeiites to olivine melilitites from South Eastern Australia utilizing experimental and geochemical data. J. Petrol., 19:463-513.

Illies, H., Baumann, H., and Hoffers, B., 1981. Stress pattern and strain release in the Alpine foreland. Tectonophysics, 71:157-172.

Jaques, A.L. and Green, D.H., 1980. Anhydrous melting of peridotite at 0 to 15 kb pressure and the genesis of tholeiitic basalts. Contrib. Mineral. Petrol., 73:287-310.

Kramers, J.D. Betton, J.P., Cliff, R.A., Seck, H.A., and Sachtleben, Th., 1981. Sr and Nd isotopic variations in volcanic rocks from the West Eifel and their significance. Fortschr. Miner. Beih. 1, 59:246-247.

Lloyd, F.E. and Bailey, D.K., 1975. Light element metasomatism of the continental mantle: the evidence and the consequences. Phys. Chem. Earth, 9:389-416.

Meyer, W., Albers, H.J., Berners, H.P., v. Gehlen, K., Glatthaar, D., Löhnertz, W., Pfeffer, K.-H., Schnüttgen, A., Wienecke, K., and Zakosek, R., 1983. Pre-Quaternary uplift in the central part of the Rhenish Massif. This Vol.

O'Nions, R.K., Carter, S.R., Evensen, N.M., and Hamilton, J.P., 1979. Geochemical and chosmochemical applications of Nd isotope analysis. Annu. Rev. Earth Planet. Sci., 7:11-31.

Panza, G.F., Mueller, St., and Calcagnile, G., 1980. The gross features of the lithosphere-asthenosphere system in Europe from seismic surface waves. Pageoph., 118:1209-1213.

Raikes, S., 1980. Teleseismic evidence for velocity heterogeneity beneath the Rhenish Massif. J. Geophys., 48:80-83.

Seck, H.A. and Wedepohl, K.H., 1983. Mantle xenoliths from the Tertiary and Quaternary volcanics of the Rhenish Massif and the Tertiary basalts of the Northern Hessian Depression. This Vol.

Semmel, A., 1983. Plateau uplift during Pleistocene time-Preface. This Vol.

Spera, F.J., 1980. Aspects of magma transport. In: Hargraves, R.B. (ed.) Physics of magmatic processes. Princeton Univ. Press.

Staudigel, H. and Zindler, A., 1978. Nd and Sr isotope composition of the potassic volcanics from the East Eifel, Germany: Implications for mantle source regions. Geol. Soc. Am. Annu. Meet., Toronto, pp.497.

Stosch, H.-G., Carlson, R.W., and Lugmair, G.W., 1980. Episodic mantle differentiation: Nd and Sr isotopic evidence. Earth Planet. Sci. Lett., 47:263-271.

Thompson, R.N., 1974. Primary basalts and magma genesis I. Skye, North-West Scotland. Contrib. Mineral. Petrol., 45:317-341.

Wedepohl, K.H., 1982. K-Ar Altersbestimmungen an basaltischen Vulkaniten der nördlichen Hessischen Senke und ihr Beitrag zur Diskussion der Magmengenese. Neues Jahrb. Miner. Abh., 144:172-196.

6 Present-Day Features of the Rhenish Massif

6.1 Height Changes in the Rhenish Massif: Determination and Analysis

H. Mälzer[1], G. Hein[2], and K. Zippelt[1]

Abstract

For the Rhenish Massif and its marginal surroundings, relative recent height changes are determined using the results of the precise levellings of the Departments of Ordnance Survey. The modelling and the calculation methods are described. The massif is not lifting as a rigid block. At the present time the uplift is concentrated to limited areas. The western part of the massif is characterized by uplift and stability, the eastern part by tendencies for alternating uplift and subsidence. Trend analysis of recent vertical crustal movements for the Rhenish Massif and the Federal Republic of Germany show an east-west tilt. An analysis of the recent vertical height changes in terms of surface deformation parameters is discussed.

Introduction

Geodetic studies were included in the scientific research program investigating the plateau uplift of the Rhenish Massif. At the start of the program it was supposed that the massif has been uplifted by about 300 m since the Pliocene. This regional up-arching was associated with widespread Quaternary volcanic activity mainly in the western part about 600,000 years ago (Windheuser, 1977). The youngest volcanic episode culminated about 11,000 years b.p. with the eruption of the Maar type volcanoes of the Eifel, e.g., Laacher See (Duda and Schmincke, 1978). The massif is bordered at its southern and northern ends by the active rift segments of the Rhine Graben und Lower Rhine Embayment which are areas of subsidence. Physiographically, the uplifted block of the Rhenish Massif thus crosscuts the belt of active rifting and the uplift is indicated by the terraces of the river Rhine. The question of whether or not an area has been influenced by relatively recent tectonic movements can often be answered to a great extent by the study of fluvial terraces. This answer will concern many thousand years, but the answer to "what happens today" can only be given by evaluation of geodectic measurements.

Data-Base

In the Federal Republic of Germany the repeated precise levellings can be used since the year 1912 for the evaluation of recent height changes. From this time the accuracy of the measurements of the former Reichsamt für Landesaufnahme is sufficiently good for geodynamic purposes. In the northern part of the Rhenish Massif the earliest levellings were made around 1920, in the southern part about 15 years later. Because of the loss of many junction points, it is difficult to join the earlier measure-

---

[1] Geodetic Institute, University Karlsruhe, D-7500 Karlsruhe, Fed. Rep. of Germany

[2] Institute of Physical Geodesy, The Technical University, D-6100 Darmstadt, Fed. Rep. of Germany.

Plateau Uplift, ed. by K. Fuchs et al.
© Springer-Verlag Berlin Heidelberg 1983

Table 1. In the different departments different numbers of first-order levellings at different times have been carried out

Department	First	Last	Number of levellings
	measurements		
Baden-Württemberg	1922-1939	1952-1962	2
Rheinland-Pfalz	1934-1957	1968-1973	2-3
Hessen:			
Levelling network	1936-1957	1968-1973	2-3
Part of former subnet IV	1936-1938	1955-1958	2
Nordrhein-Westfalen:			
Southern part	1949-1954	-1973	2-6
Northern part	1921-1925	-1975	2-8

ments to the later levellings provided by the Departments of Ordnance Survey (Landesvermessungsämter) of Baden-Württemberg, Rheinland-Pfalz, Hessen and Nordrhein-Westfalen (Table 1).

Not only for the first levellings but also for the relevellings the standard deviations of the observations amount to $m_l = (0.2-0.5)$ mm/$\sqrt{km}$, computed from the differences of the double (forward and reverse) measurements of levelling sections. The network available is bounded to the north by the Ruhrdistrict and the Lower Rhine Embayment, to the west and southwest by the border of the Federal Republic of Germany, and to the east by a line running from Siegen through Frankfurt to Karlsruhe (Fig. 1).

Special investigations were first done in the northern part. Quitzow and Vahlensieck (1955) determined in the Lower Rhine Embayment near Düren a relative subsidence of up to 2.7 mm a^{-1}. Later the surrounding region of the Ruhrdistrict was investigated to fix the limits of subsidence caused by mining (Jacobs, 1965; Ghitau, 1970; Jacobs and Lindstrot, (1980).

In the south, all the levellings in the northern Rhine Graben were used to estimate height changes and the influence of the subsided groundwater level. The significance of height changes discovered from the first order levelling network of Hessen and Rheinland-Pfalz was tested by applying multivariate analysis (Hein, 1978). A first sketch overlaying most parts of the massif is visible in the *Map of Height Changes in the Federal Republic of Germany - Status 1979* (DGK-Arbeitskreis, 1979). Height changes of about 200 junction points in the Rhenish Massif are the base of this map.

In the first step of data-processing the levelling lines were combined to form a connected network, based on those junction points which were identical for the first levelling and the relevellings. Along a levelling line, a point, where the measurement was interrupted during a period of observation, was used like a junction point.

In the second step each levelling line was prepared pointwise, considering only points where at least two levelling measurements had been made. This data could be used to check the total height differences of the first step, and thus errors in the data of both steps could be eliminated.

Modelling and Calculation Method

Because of the inhomogeneity in temporal distribution and the large amount of data, only a velocity model was suited to model the point movements. Ghitau (1970) and Holdahl (1975) describe general methods, of which the single point model was chosen (Mälzer et al., 1979), yielding adjusted time-dependent heights, point velocities, and, if necessary and possible, accelerations corresponding to the reference time, $t_o = 1960$. The observation equation for the height differences, Δh_{ij}^k, at time, t_k, between the junction points, P_i and P_j, is given by:

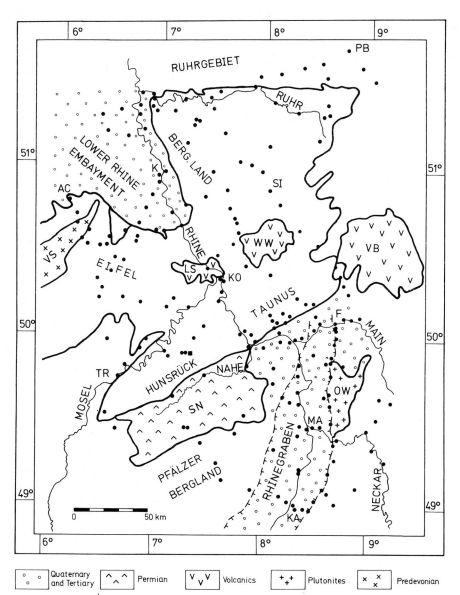

Fig. 1. Junction points of the used levelling network in association with geological pattern

○ ○ Quaternary and Tertiary   ∧ ∧ Permian   v v Volcanics   + + Plutonites   × × Predevonian

● Junction point   ■ Fundamental bench mark (reference point)

AC = Aachen; F = Frankfurt; K = Köln; KA = Karlsruhe; KO = Koblenz; MA = Mannheim; PB = Paderborn; SI = Siegen; TR = Trier. LS = Laacher sea area; SN = Saar-Nahe trough; OW = Odenwald; VB = Vogelsberg; VS = Venn Sattel; WW = Westerwald.

$$\Delta h_{ij}^k = H_j^o - H_i^o + \sum_{n=0}^{m_j} (\alpha_{jn} \cdot \Delta t_k^n)$$

$$- \sum_{n=0}^{m_i} (\alpha_{in} \cdot \Delta t_k^n) \quad , \quad (1)$$

where H_i^o, H_j^o are heights at the reference time, t_o; α_{in}, α_{jn} are coefficients of one-dimensional polynomials of maximum degree m_i, m_j, and $\Delta t_k^n = (t_k - t_o)^n$. The functional model for an adjustment of assumed accelerated movements is given by:

$$E(1) = Ax + B\alpha_1 + C\alpha_2 = \bar{A}y \quad (2)$$

where

x = heights referring to the time, t_o
α_1 = linear polynomial coefficients (velocities)
α_2 = quadratic polynomial coefficients (accelerations).

This model has the great advantage that, for each junction point of the network, the degree of the polynomial can be fixed depending on the number of measurements and the assumed movements. Therefore the model will allow for some points moving linearly as well as for other points which are characterized by no movements or by third-order polynomial behavior. Introducing the stochastic model:

$$C_{11} = \sigma_o^2 \, Q_{11} \qquad (3)$$

and neglecting the covariances between the measurements, the matrix of the normal equations is:

$$N = \bar{A}^T \, Q_{11}^{-1} \, \bar{A} \quad . \qquad (4)$$

As a consequence of only the relative height changes being known, this matrix has a rank deficiency caused by the lacking datum information on absolute height and the velocity reference. Therefore one junction point, called the reference point, must serve as the basis for the height and velocity system, such that the unknowns belonging to this point P_o, satisfy the conditions:

$$x_{P_o}, \quad \alpha_{1_{P_o}} \rightarrow \text{const.}$$

By adding these conditions to the normal equations, a uniquely solvable system is obtained, and the solution leads to the estimated values of the unknowns:

$$\hat{y} = N^{-1} \, \bar{A}^T \, Q_{11}^{-1} \, 1 \quad . \qquad (5)$$

By solving for the parameters using a least-squares fit, the residuals:

$$v = \bar{A}\hat{y} - 1 \quad , \qquad (6)$$

the (a posteriori) variance factor of unit weight:

$$\hat{\sigma}_0^2 = \frac{v^T Q_{11}^{-1} v}{r} \quad (r = \text{number of redundant observations}) \qquad (7)$$

and the variance-covariance matrix of the unknowns:

$$C_{xx} = \hat{\sigma}^2 N^{-1} \qquad (8)$$

are computed. The observations are tested for outliers by the method of d a t a - s n o o p i n g (e.g., Förstner, 1979), and the significance of the estimated polynomial coefficients is tested by a pointwise t-test. The results of the adjustment are depending on the chosen reference point and will be influenced by individual movements of this point. These movements, which can never be determined by levellings alone, have an effect similar to a displacement of the computed velocities. Therefore the reference point has to be chosen carefully.

Not only the reference point, but also all junction points can be subjected to disturbed single point movements which may not be congruent with the behavior of the surrounding of the point. The evaluation of the neighborhood of junction points is necessary, as well as the selection of representative linepoints. For this purpose, each relevelled bench mark in a relevelling line is used to compute detailed height changes. The information connected with the bench marks along the levelling lines can be combined with the results of the junction point solution by a stepwise approach. Beside the observation equations of type (1) for the line points, additional observation equations for estimated heights, $\hat{x}_f$, and polynomial coefficients, $\hat{\alpha}_{1f}$ and $\hat{\alpha}_{2f}$ referring to the junction points must be considered:

$$\begin{vmatrix} 1_x \\ 1_{\alpha_1} \\ 1_{\alpha_2} \end{vmatrix} = I \, y_f \quad (I = \text{unit matrix}) \quad . \qquad (9)$$

The stochastic model is completed by the variance-covariance matrix of the junction point parameters contained in the inverse of the matrix N occurring in Eq. (4):

$$C = \hat{\sigma}_0^2 \begin{vmatrix} Q_{11} & 0 \\ 0 & Q_{y_f y_f} \end{vmatrix} \quad . \qquad (10)$$

The resulting normal equations have full rank and are therefore directly solvable. The now estimated values are still related to the reference point. The "observed" parameters of the accounted junction points obtain small modifications, based on a larger number of measurements (information about movements), compared to the junction point adjustment. By investigating the behavior of neighbored points the representativity of the junction points can be evaluated. Furthermore, representative line points are chosen to describe the height changes.

The Map of Height Changes

The connected levelling network is about 4000 km long and contains 4800 relevelled points. Nearly 800 representative points, judged by neighborhood and local realization, have been selected for the contouring process. The bound of subsidence area caused by mining of the Ruhrdistrict is quoted by published data (Jacobs, 1965). The prediction of the interpolation grid is realized by the multiquadratic method. The contour lines in the map of height changes (Fig. 2, inside the pocket of the backside cover) are plotted at an interval of 0.2 mm a^{-1} using an integrated interpolation and plotting program. The basic data of boundary-lines and river-lines are provided by the Institut für Angewandte Geodäsie (IfAG), Frankfurt am Main.

The resulting map of height changes includes remarkable effects. The height changes show that the Rhenish Massif is not lifting as a rigid block but that the present uplift is concentrated in limited areas. West of the river Rhine the massif is predominantly either uplifting or is stable while east of the Rhine tendencies of alternating subsidence and uplift occur.

Northeast of the massif, near Paderborn, uplift of up to 0.6 mm a^{-1} occurs. This area of uplift is bounded to the north by the influences of mining in the Ruhrdistrict and to the south by a tendency of subsidences correlated to the Wittgensteiner Mulde, situated in the northeastern part of the Rhenish Massif. In the west, towards Bergisches Land between the Lüdenscheider Mulde and the Elsper Mulde a rapid change from uplift to subsidence up to 0.6 mm a^{-1} takes place.

Entering Bergisches Land, uplift of up to 0.6-0.9 mm a^{-1} occurs northeast of the Lower Rhine Embayment. The uplift occurs not only in the Devonian rocks but also in the eastern Quaternary sedimentary series of the Kölner Scholle. In the Lower Rhine Embayment itself, opencast working is widespread and its influences, especially the lowering of groundwater level, overlay the tectonic movements. Nevertheless, the recent height changes in this area can still be correlated to the tectonic pattern as already shown by Quitzow and Vahlensieck (1955).

In the Nordeifel and Venn Sattel, the greatest rates of uplift of up to 1.6 mm a^{-1} can be correlated with the presence of a strong seismic reflector at 3-4 km depth indicating a prominent fault zone near Aachen (Meissner et al., 1981). These movements, determined from measurements made later than 1950, decrease after crossing the Aachen Thrust. This area is characterized by seismic activity, which is not only restricted to the thrust zone. On the other hand, it seems possible that the uplift in the Nordeifel and Venn Sattel and in the Bergisches Land, surrounding the subsiding Lower Rhine Embayment, indicate a continuous taphrogenesis (Ahorner, 1983).

On both sides of the river Rhine near Koblenz an area of uplift of up to 0.6 mm a^{-1} trends ENE-WSW. From the geodetic point of view a strong correlation between heights and height changes, especially in the eastern part of this uplifting area, points to an incorrect scale in one of the measurements. This possible effect can never be eliminated by later computations without knowledge of the real and the entered scale. On the other hand, geomorphological studies also indicate a recent uplift in this area.

In general, the southern border of the Rhenish Massif, marked by the Hunsrück and Taunus fault systems, is not recognizable by height changes. Only in the southwestern part of the massif, near Trier, uplift is found in a small area of the Hunsrück. West of the river Rhine near the border fault system, a trend of recent subsidence, of up to 0.3 mm a^{-1}, occurs in the massif. This is the only zone where negative height changes are found west of the river Rhine. However, these rates are very small and not reliable although the subsidence is continued in the Taunus, east of the river Rhine. Further, this area is situated at the northern end of the Rhine Graben and it may be that extension and shear in the Rhine Graben causes subsidence in this part of the Rhenish Massif, especially at the southern end of the Middle Rhine Valley. The map of height changes shows that the Rhine Graben is not only an area of subsidence but also an area of uplift. Subsidence is greatest in the northern part of the graben. In this area, shear controlled movement causes an extension, which results in an average subsidence of 0.6 mm a^{-1}. Maximum values amounting to 1.4 mm a^{-1} near Mannheim and Frankfurt am Main are certainly influenced by a subsiding groundwater level. Accordingly, the recent compaction of Quaternary and Tertiary sediment fill has to be taken into account.

In the central segment of the Rhine Graben, uplift between 0.3 and 0.5 mm a^{-1} (Zippelt and Mälzer, 1981) reflects the effects of a compressional shear. This is surprising but in agreement with geological conceptions (Illies and Baumann, 1982). It has been shown that the western border fault system does not interrupt these motions in the direction of the Pfälzer Bergland (Palatinate mountains). Here, uplift amounts to 0.9 mm a^{-1} in the south and about 0.7 mm a^{-1} in the north.

Critical Remarks

The reference point for the computed height changes is the fundamental bench mark of Rheinland-Pfalz located on a geological stable plateau of the Hunsrück mountains. The stability and hence the qualification of this point as the reference point could be criticized for geological reasons. A geodetic examination, whether spontaneous movements of this point are occurring or not, is carried out by a comparison with additional fundamental bench marks, e.g., Freudenstadt situated in the Black Forest. These investigations establish a strong agreement of relative stability between several suitable reference points.

For the determination of absolute movements, additional absolute vertical motions with respect to a reference system must be known from space methods in at least one point.

Furthermore, the single point model permits to compute only apparent height changes and not the real recent crustal movements. The real crustal movements are combined from surface deformations obtained by levellings and changes of geoid derived from gravity changes. Also, temporal changes of orthometric heights do not describe real crustal deformations as the reference surface used for this height system changes its shape in the course of time. Different models have been presented in the past years to describe the connection between levelled height differences and gravity measurements (Biro, 1980; Strang van Hees, 1977; Heck and Mälzer, 1982). Due to the lack of repeated gravity measurements of high precision, which can indicate changes in gravity, the present investigation has been done using orthometric heights. Numerical values of the influence of secular variations of gravity on relevelling data are estimated by use of a time-dependent orthometric correction (Kistermann and Hein, 1979). It is pointed out that only gravity changes ≥ 50 µgal cause deformations in height greater than 0.2 mm. Even in the area of the Rhine Graben, no secular variations are found in this order of magnitude with the exception of areas with man-made influences (Deichl et al., 1978). This observation can be extended to the whole area of the Rhenish Massif, with the exception of the areas influenced by mining in the Lower Rhine Embayment and the Ruhrdistrict.

The contour interval of the present map is 0.2 mm a^{-1} and gives a clear impression of the described motions. By this way it is very simple to classify the movements in connection to their significance. Starting from a standard deviation of about $\hat{\sigma}_0 \approx 0.9$ mm/$\sqrt{km}$, obtained from Eq.(7), the velocity values (i.e., linear polynomial coefficients) obtain a standard deviation of about $\hat{\sigma}_{\alpha_1} \approx 0.4$ mm a^{-1} on average. The presented motions, which are not greater than these estimated values, are certainly insignificant and very unreliable. Increasing up towards $\alpha_1 \approx 0.8$ mm a^{-1}, the motions exceed their standard deviations expressing a trend of height changes without statistical significance. Only those rates of uplift and subsidence exceeding their double standard deviation can be considered as reliable by statistical standards. The test statistic for the velocities, due to the hypothesis, H_o: $\alpha_1 = 0$, given by:

$$T_{\alpha_1} = \frac{\alpha_1^2}{\hat{\sigma}_0^2 \, Q_{\alpha_1 \alpha_1}}$$

is compared to the critical value of a Fisher's $F(S,1,r)$-distribution, or $\sqrt{T_{\alpha_1}}$ is compared to a Student's $t(S,r)$-distribution (Koch, 1975). Significance of motions, based on the 95% confidence level, is given only in a few parts of the map, e.g., Rhine Graben, southern Pfälzer Bergland, Nordeifel and Bergisches Land.

Geodetic Interpretation

Whereas in the preceding sections a detailed map of recent height changes was presented, the present section is concerned with the analysis and interpretation of the results which is done in a mathematical framework provided by geodetic least-squares collocation and a subsequent differential-geometrical analysis.

In order to understand the philosophy behind the analysis, the reader is reminded that geodetic relevelling observations are conventionally regarded as having three parts: the more or less regional height changes due to a trend in recent crustal movements, the local field of non-tectonic influences caused by the artificial height changes of bench marks, subsidence due to groundwater variations or long-period variations in the physical parameters of the soil in which the investigated points are implanted, and the measurement noise. The non-tectonic influences can be modelled either as a deterministic function or a random process. Inasmuch as both of these models are inevitably idealizations of the real world, it is a matter of individual judgment as to which model is more appropriate in a particular situation.

Obviously, any information regarding the sources of the non-tectonic changes is helpful. Consequently, areas with groundwater subsidence and mining industry have already been outlined and discussed. However, since no detailed numerical information is available for the Rhenish Massif area, it is impossible to treat the non-tectonic influences in a deterministic way, as in the Upper Rhine Graben where the influence of groundwater subsidence on the relevelling data was calculated using functional relationships provided by the theory of soil mechanics (Fahlbusch et al., 1980; Hein, 1980).

When detailed information is missing, as in the Rhenish Massif area, the non-tectonic influences on the results can be treated in a pseudo-stochastic way using the method of geodetic least-squares collocation for which the general observation equation is given by Moritz (1980):

$$x = A X + B s + n \qquad (11)$$

where x is the vector of the observations, X is the vector of the systematic parameters, s is the signal vector, and n is the vector of the measuring errors (noise). A, B are appropriate coefficient matrices. A priori knowledge of the covariances C_{ss} and C_{nn} is needed.

By introducing the hybrid minimum condition:

$$\Omega = n^T C_{nn}^{-1} n + s^T C_{ss}^{-1} s = \min \qquad (12)$$

the solutions for $\hat{X}$ and $\hat{s}$ are obtained.

The sensitivity analysis and computation of error estimates are found in Moritz (1980), and not repeated here.

In the model described by Eq.(11), the regional changes of recent crustal movements can be characterized as the deterministic part whereas the local field of non-tectonic influences and the noise represent the stochastic parameters. Considering that Δh_{ij}^k are the measured height differences between points, P_i and P_j, at the time, t_k, the observation equation is:

$$\Delta h_{ij}^k = H_j^o + v_j(t_k - t_o) - H_i^o - v_i(t_k - t_o) + s_j(t_k - t_o) - s_i(t_k - t_o) + n_{ij} \quad (13)$$

where H_i^o, H_j^o are heights referring to a central time, t_o; v_i, v_j are point velocities, $\partial h/\partial t$, which can be associated with recent crustal movements, and s_i, s_j are the changes of points, P_i and P_j, with time in terms of velocities due to local non-tectonics. The appropriate matrices fitting Eq.(11) are:

$$x^T = [\Delta h_{ij}^k \ldots]^T$$

$$X^T = [H_i H_j \ldots v_i v_j \ldots]^T$$

$$\begin{bmatrix} 1 & -1 & \ldots & | & (t_k-t_o) & -(t_k-t_o) & \ldots \\ \ldots & 1 & -1 & | \ldots & & -(t_1-t_o) & (t_1-t_o) \ldots \\ \ldots & \ldots & & | \ldots & & \ldots & (t_m-t_o) \ldots \end{bmatrix} = [A_1 | T]$$

$$B = T \quad (14)$$

where t_k, t_l and t_m are the times of the different observation epochs.

Such an approach leads to a simultaneous adjustment of heights referring to a certain time, t_o, point velocities due to recent vertical crustal movements, and non-tectonic point velocities. The latter can also be obtained as residual heights by introducing $B = I$ where I is the identity matrix. By setting:

$$H_o = \text{const.} \quad \text{and} \quad v_o = 0 \quad (\text{or const.}) \quad (15)$$

for a reference point, P_o, the defect in the normal equation matrix $(A^T \bar{C}^{-1} A)$ is removed. The subvector of X containing the velocity quantities can be transformed into coefficients of a trend polynomial describing analytically the vertical crustal motion.

Interpreting the above-mentioned approach geometrically, it turns out that by filtering the non-tectonic movements by Eq.(12) the derived recent crustal movements result as a smooth trend function. One could object to this method by arguing that important information from the geological point of view is lost. The counter argument, however, is that geodetic levelling can only record certain instantaneous snapshots of the discontinuous process of crustal movements. Considering further, that, for example, in most parts of the Rhenish Massif only two observations were carried out, one easily comes to the conclusion that besides the introduction of a linear model any assumption of more than a smooth function describing recent crustal movements is not justified. The model described by Eq.(11) was applied to representative observation stations in the Rhenish Massif area, already used by Hein (1978), as well as to the data of the map of recent height changes of the Fed. Rep. of Germany (DGK-Arbeitskreis, 1979). The corresponding results, the trend of recent crustal movements in the Rhenish Massif area and the Federal Republic of Germany, respectively, are shown in Figs. 3 and 4. The distance, s, dependent autocovariance function of the form:

$$C(s) = \frac{C_o}{1 + (s/d)} \quad (16)$$

was determined as having the following parameters:

the variance, $C_o = 0.1705 \, [\text{mm a}^{-1}]^2$ and

the correlation length, $d = 29.4 \, [\text{km}]$.

By inspecting the graphs of the autocorrelation function for data from the German Democratic Republic, a good agreement for the correlation length, within 2-3 km, is found (Harnisch 1977).

All computed trends of recent vertical crustal movements (Figs. 3 and 4) show an east-west tilt, the zero line of which lies nearly in the middle of West Germany. Towards the west uplift takes place whereas towards the east subsidence takes place. In addition, the area near the North Sea

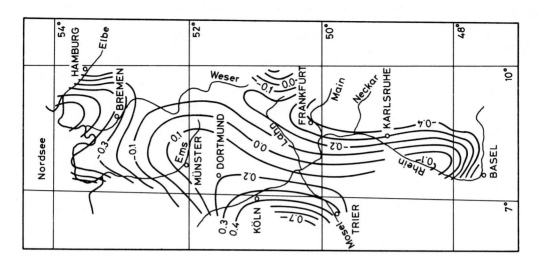

Fig. 4. Trend of recent crustal movements in the Federal Republic of Germany. Isolines in [mm a⁻¹]

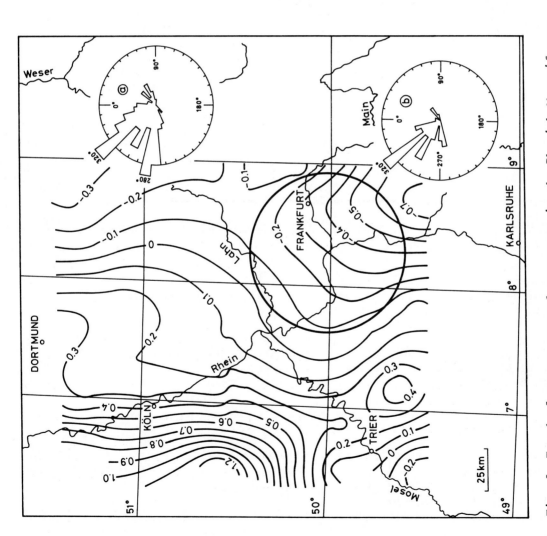

Fig. 3. Trend of recent crustal movements in the Rhenish Massif area. Isolines in [mm a⁻¹]. *Circles* distribution of the directions of the horizontal gradients (see Fig. 5) for the whole area (*a*) as well as for the Mainzer Basin (*b*)

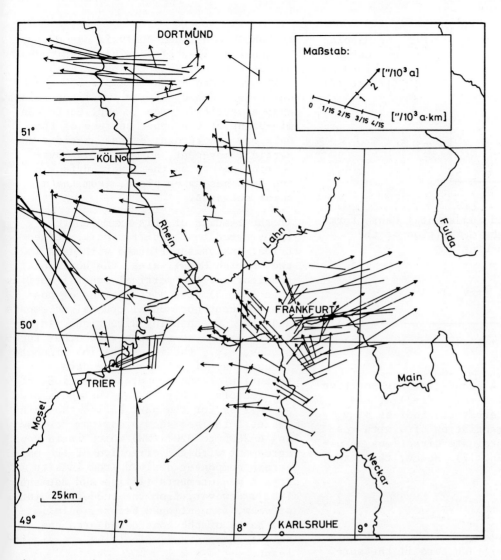

Fig. 5. Horizontal gradients of recent vertical crustal movements and related curvatures as derived by Eqs.(17) to (21). For the mapping of the related quantities see Fig. 6

also shows subsidence which increases towards the coast.

Analysis of the recent vertical crustal movements by surface deformation parameters: It is obvious that for a full description of the dynamic features of the earth's crust the quantities bending, translation, and rotation need to be considered since the latter two alone do not cause any deformation of the earth's surface. This only happens if there is a bending. Therefore, as a quantity describing the tilt, the horizontal gradient of the vertical crustal motions, v, is defined, the total value of which is given by:

$$|\text{grad } v| = [(\partial v/\partial x)^2 + (\partial v/\partial y)^2]^{\frac{1}{2}}$$
$$= (v_x^2 + v_y^2)^{\frac{1}{2}} \quad (17)$$

and the direction of which is given by:

$$\alpha = \text{arc tan}(v_x/v_y) \quad (18)$$

where (x,y) represent a planar mathematical coordinate system. The components v_x, v_y as well as the horizontal gradient, Eq.(17), have a formal similarity with the strain-rate of a one-dimensional deformation process.

For the bending, use is made of differential geometry and, in particular, of

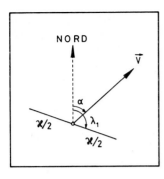

Fig. 6. Mapping of the quantities derived by Eqs. (17) to (21)

first and second fundamental Gauss forms which lead to the definition of the principal curvatures:

$$k_i = \frac{1}{R_i} = \frac{1}{2}(v_{xx} + v_{yy}) \\ \pm \sqrt{\frac{1}{4}(v_{xx} + v_{yy})^2 - (v_{xx}v_{yy} - v_{xy}^2)} \quad . \tag{19}$$

The index i of k stands for the positive (i=1) or negative value (i=2) of the square root, whereas the indices of v indicate differentiation of v with respect to the coordinate directions as already used in Eq. (17). Hence, the mean curvature is:

$$\kappa = k_1 + k_2 = v_{xx} + v_{yy} \tag{20}$$

which, from a view point above the surface, is greater than zero if the surface is concave and less than zero if the surface is convex. The direction of the main principal curvature, k_1, is given by:

$$\lambda_1 = \arctan[v_{xy}/(k_1 - v_{xx})] \quad . \tag{21}$$

The results of the described analysis applied to the unsmoothed data of the vertical crustal movements in the Rhenish Massif area are shown in Fig. 5, in which the different quantities and their directions are mapped as in torsion balance gradiometry (see Fig. 6). Hein and Kistermann (1981) give corresponding results for the whole Federal Republic of Germany.

Although geodesy is not yet able to detect significant horizontal movements at the submillimetre level, the results of the above mentioned analysis can give at least an idea of possible magnitudes and directions. This can be mathematically verified by the example of a balk which is spanned between two fixed points. When a vertical force is applied to the mid-point of the upper surface of the balk one can recognize that, besides the vertical movement, a *horizontal* movement of points on the lower surface of the balk has taken place, from the mid-point outwards.

If the results of this example are transferred to the differential-geometrical analysis performed with data in the Rhenish Massif area, the horizontal gradients of the vertical crustal motions (see Fig. 5) give a rough idea of possible recent horizontal crustal movements and the corresponding directions. In order to have a comparison with other techniques, e.g., the investigation of photo lineations, Fig. 3 shows the data distribution of the computed directions of the vectors in Fig. 5 for the whole area as well as for the part called the Mainz Basin. Both maps show a maximum for the angle $280°$ to $320°$ (NW) which is in good agreement with the direction of the main strain component derived from in situ strain measurements (Illies and Baumann, 1982). However, from the geodetic point of view, as mentioned before, only in a few parts of the considered area the height changes and velocities are significant.

In conclusion it can be stated that the above mentioned techniques - trend and differential-geometrical analysis - are mathematically based interpretations of geodetic results (see also Fahlbusch et al., 1982). It is obvious that they cannot replace the detailed results derived before. However, they supply and complete a careful treatment of recent crustal movements in the geodetic context.

Acknowledgements. The authors are indebted to the Departments of Ordnance Survey (Landesvermessungsämter) of Baden-Württemberg, Hessen, Nordrhein-Westfalen and Rheinland-Pfalz for providing the levelling data. The computations have been carried out at the Computing Centres of the University Karlsruhe and the Technical University Darmstadt. The financial support given by the German Research Society (Deutsche Forschungsgemeinschaft) is gratefully acknowledged.

References

Ahorner, L., 1983. Historical seismicity and present-day microearthquake activity of the Rhenish Massif, Central Europe. This Vol.

Biro, P., 1980. Dynamic aspects of repeated geodetic levellings. Periodica polytechnica, civil engineering. Geodesy. Tech. Univ. Budapest, 24:3-12.

Deichl, K., Groten, E., Kahle, H.-G., Kiviniemi, A., Klingele, E., and Weichl, B., 1978. Precise gravimetry in the Rhinegraben net. Meet. Int. Gravity Commiss., Paris 1978.

DGK-Arbeitskreis, 1979. On the "map of height changes in the Federal Republic of Germany - status 1979" 1:1000000. Allg. Vermessungsnachr., 86:362-363.

Duda, A. and Schmincke, H.-U., 1978. Quaternary basanites, melilites nephelinites and tephrites from the Laacher See area (Germany). Neues Jahrb. Mineral. Abh., 132:1-33.

Fahlbusch, K., Hein, G., and Kistermann, R., 1980. Rezente Bewegungen im nördlichen Oberrheingraben. Verknüpfung von Meßdaten aus Geodäsie, Geologie und Bodenmechanik. Neues Jahrb. Geol. Palaeontol. Monatsh.:460-476.

Fahlbusch, K., Hein, G., and Kistermann, R., 1982. Geologisch-geodätische Interpretation von Höhenänderungen in West- und Südwestdeutschland. Allg. Vermessungsnachr., 89:49-62.

Förstner, W., 1979. Das Programm TRINA zur Ausgleichung und Gütebeurteilung geodätischer Lagenetze. Z. Vermessungswes., 104:61-72.

Ghitau, D., 1970. Modellbildung und Rechenpraxis bei der nivellitischen Bestimmung säkularer Landhebungen. Dissertation, Inst. Theor. Geodäs., Univ., Bonn.

Harnisch, G., 1977. Zur Autokorrelationsfunktion der rezenten vertikalen Krustenbewegungen im Gebiet der Deutschen Demokratischen Republik. In: Proc. 3rd Int. Symp. Geodes. Phys. Earth. Weimar, Oct. 25-31, pp.611-616.

Heck, B. and Mälzer, H., 1982. Determination of vertical recent crustal movements by levelling and gravity data. Gener. Meet. IAG, Tokyo, 1982.

Hein, G., 1978. Multivariate Analyse der Nivellementsdaten im Oberrheingraben und Rheinischen Schild. Z. Vermessungswes., 103:430-435.

Hein, G., 1980. Influence of ground water variations on relevelling data. In: Proc. 2nd. Int. Symp. Problems related to the redefinition of North American vertical geodetic networks, May 26-30. Ottawa (Canada) 1980, pp.603-622.

Hein, G. and Kistermann, R., 1981. Mathematical foundation of non-tectonic effects in geodetic recent crustal Movement Models. Tectonophysics, 71: 315-334.

Holdahl, S.R., 1975. Models and strategies for computing vertical crustal movements in the United States. Int. Symp. Recent crustal movements. Grenoble, France.

Illies, J.H. and Baumann, H., 1982. Crustal dynamics and morphodynamics of the Western European Rift System. Z. Geomorphol. N.F., 42:135-165.

Jacobs, E., 1965. Vertikale Bodenbewegungen im weiteren Bereich des niederrheinisch-westfälischen Steinkohlengebietes. Dissertation, Inst. Theor. Geodäs., Univ. Bonn.

Jacobs, E. and Lindstrot, W., 1980. Geodätische Präzisionsmessungen zur Untersuchung rezenter Bodenbewegungen am Nordrand des Ruhrgebietes. Forschungsber. d. Landes Nordrhein-Westfalen, Nr. 2934, Fachgruppe Bergbau/Energie.

Kistermann, R. and Hein, G., 1979. Der Einfluß säkularer Schwereänderungen auf das Wiederholungsnivellement. Z. Vermessungswes., 104:471-475.

Koch, K.R., 1975. Ein allgemeiner Hypothesentest für Ausgleichungsergebnisse. Allg. Vermessungsnachr., 82: 339-345.

Mälzer, H., Schmitt, G., and Zippelt, K., 1979. Recent vertical movements and their determination in the Rhenish Massif. Tectonophysics, 52: 167-176.

Meissner, R., Bartelsen, H., and Murawski, H., 1981. Thin-skinned tectonics in the northern Rhenish Massif, Germany. Nature (London), 290, 5805: 309-401.

Moritz, H., 1980. Advanced Physical Geodesy. Wichmann, Karlsruhe.

Quitzow, H.W., and Vahlensieck, O., 1955. Über pleistozäne Gebirgsbildung und rezente Krustenbewegungen in der Niederrheinischen Bucht. Geol. Rundsch., 43:56-67.

Strang van Hees, G.L., 1977. Zur zeitlichen Änderung von Schwere und Höhe. Z. Vermessungswes., 102:444-450.

Windheuser, H., 1977. Die Stellung des Laacher Vulkanismus (Osteifel) im Quartär. Sonderveröff. Geol. Inst. Köln, 31:223 pp.

Zippelt, K. and Mälzer, H., 1981. Recent height changes in the central segment of the Rhinegraben and its adjacent shoulders. Tectonophysics, 73:119-123.

6.2 Stress Field and Strain Release in the Rhenish Massif

H. Baumann[1] and J. H. Illies†

Abstract

The stress conditions prevailing in the Rhenish Massif were determined at 14 sites by means of the doorstopper method. The mean direction of the maximum horizontal principal stress is 140° to 150°, thus not differing from the rest of the western Alpine foreland. Tensile stresses are the characteristic differences of the stress field of the Rhenish Massif with respect to its surroundings. These tensile stresses are due to the diverging σ_1-trajectories, but besides that they are probably promoted by thermal effects at depth or by diapiric doming. The special kind of strain release found in the Rhenish Massif, if compared with that in the bordering rift segments of the Rhine Graben and the Lower Rhine Embayment, is mainly due to differing mechanical behavior of the crust in these areas.

Method

The plateau uplift of the Rhenish Massif should be reflected in the present stress field. Seismic investigations of earthquake mechanisms provide information on the orientation of the stress tensor and the stress drop of depth at those locations where earthquakes occur. The methods of in situ stress determination yield orientation and magnitude at shallow depth and also at places where no earthquakes are available.

The method applied throughout this research program covering the Rhenish Massif is the doorstopper method developed by Leeman (1964) in South Africa. It is an overcoring techniques; the principle of this borehole relief technique is illustrated in Fig. 1.

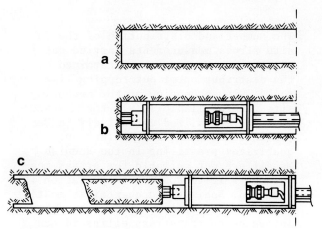

Fig. 1. *a* A borehole (diameter 76 mm) is drilled into the rock and its bottom is smoothed, polished and dried up. *b* An oriented CSIR doorstopper is glued to the base of the borehole and an initial reading is taken. Then the installing tool is separated. *c* Then the doorstopper is overcored and straining of the cores is determined by strain readings

[1] Geologisches Institut, Universität Karlsruhe, Kaiserstr. 12, D-7500 Karlsruhe, Fed. Rep. of Germany.

By applying Hooke's law $\sigma = E \cdot \varepsilon$ (σ = stress, E = Young's modulus, ε = measured strains), the directions and amounts of the horizontal principal normal stresses are calculated from the measured strains. The resulting indexes are corrected with the stress concentration factors of Van Heerden (1969) for the "falsifying" strains induced through the borehole. In order to compare the results of measurements carried out at differing depths, the measured horizontal stresses are reduced by the horizontal strain portion, which is caused by the overburden of the rock. This corrected horizontal stress is termed "excess stress". (The amounts of stress indicated in the figures and in the table are excess stress data.) The elastic moduli, on which these calculations are based, were measured in situ with a Goodman jack (Goodman et al., 1968). The Poisson's ratios were obtained by unconfined compression tests in the laboratory.

The Recent Stress Field in the Rhenish Massif

The number and distribution of the in situ stress measurements carried out in the Rhenish Massif were determined by the distribution of outcropping limestone rocks. With exception of the measuring site Nenning (Table 1, 9), the measurements were realized exclusively in Devonian limestones. The schists and quartzites prevailing in the Rhenish Massif are not suitable for stress measurements if tectonic interpretation is intended. Schists and quartzites mostly have a strong elastic anisotropy, whose influence on the stress ellipsoid can be calculated only approximately. Also Rahn (1981) in his theoretical study comes to the result that the influence of elastic anisotropies on the deformation measurements can only be estimated. Due to this distribution of rocks, no stress determination sites could be placed into the central part of the eastern Rhenish Massif (Westerwald and Siegerland).

In total 15 stress measurements are available in the region of the Rhenish Massif (see Table 1). Apart from the site Waldeck all measuring points were analysed with respect to the directions of the principal stresses, i.e., 14 directional values for parameter σ_1 can be used for statistical evaluation.

The 14 directional values are:

$146°$	(02)
$184°$	(03)
$129°$	(04)
$167°$	(05)
$58°$	(06)
$215°$	(07)
$139°$	(08)
$126°$	(09)
$150°$	(10)
$150°$	(11)
$160°$	(12)
$150°$	(13)
$145°$	(14)
$136°$	(15)

(see Table 1)

When these stress measurements are considered as measurements on an identical object (the Rhenish Massif), the statistic evaluation of the results is rendered possible. Applying an outliers test (Nalimov-test in: Kaiser and Gottschalk, 1972), the directions of σ_1 of $58°$, $215°$ and $184°$ result to be singularities. From the remaining 11 directional values a mean σ_1 can be estimated as $145.3° \pm 12.3°$ (95%; 11 samples); i.e., with a probability of 95% the mean values of the σ_1 directions would be between $157.6°$ and $133.0°$.

From these data a uniform anisotropy of the stress field with respect to the directions of the principal normal stress can be derived. The statistical investigation of the mean directions and of the standard deviations of the mean σ_1 directions determined in the Rhenish Massif and in the regions shown in Fig. 2 shows that these directions do not differ, and consequently it is possible to indicate a common mean value for the direction of σ_1 in the investigated (for location see Fig. 2) western Alpine foreland: $146.1° \pm 14.2°$ (95%; 25 samples). The confidence interval of the mean value is $\pm 5.9°$.

Considering the amplitudes of the stress values measured in the western Alpine foreland (see Fig. 2, Table 1), we find that the Rhenish Massif differs signifi-

Table 1. Test sites of in situ stress determinations and results. Negative "excess" stress values mean tension

Location	N-Latitude	E-Longitude	Direction σ_1	Magnitude σ_1 (MPa)	τ_{max} (MPa)	Magnitude σ_2 (MPa)	Reference	Lithology	System
01 Eldagsen	52°09'	07°17'	135°	+ 0.2	+ 0.1	− 0.03	Baumann 1982	Limestone	Jurassic
02 Gressenich	50°46'	06°18'	146°	− 1.3	+ 1.6	− 4.5	Schmitt 1981	Limestone	Devonian
03 Keldenich	50°32'	06°35'	004°	− 0.5	+ 0.5	− 1.4	Baumann 1981	Limestone	Devonian
04 Pelm	50°14'	06°41'	129°	− 0.3	+ 0.3	− 0.8	Baumann 1981	Limestone	Devonian
05 Dornap	51°15'	07°03'	167°	+ 0.6	+ 0.3	− 0.02	Baumann 1982	Limestone	Devonian
06 Warstein	51°26'	08°23'	058°	− 0.2	+ 0.3	− 0.9	Baumann 1982	Limestone	Devonian
07 Brilon	51°23'	08°37'	035°	− 0.4	+ 0.1	− 0.6	Baumann 1982	Limestone	Devonian
08 Düstertal	51°28'	08°42'	139°	+ 0.2	+ 0.2	− 0.1	Baumann 1982	Limestone	Devonian
09 Nennig	49°31'	06°23'	126°	+ 2.1	+ 0.5	+ 1.2	Greiner 1978	Limestone	Triassic
10 Stromberg	49°57'	07°46'	150°	+ 0.2	+ 0.2	− 0.2	Schmitt 1981	Limestone	Devonian
11 Hahnstätten	50°19'	08°04'	150°	+ 0.02	+ 0.2	− 0.3	Schmitt 1981	Limestone	Devonian
12 Fachingen	50°22'	07°54'	160°	− 0.01	+ 0.1	− 0.3	Schmitt 1981	Limestone	Devonian
13 Wirbelau	50°26'	08°13'	150°	− 1.3	+ 0.5	− 2.3	Illies and Greiner 1979	Limestone	Devonian
14 Villmar	50°24'	08°11'	145°	− 0.8	+ 0.5	− 1.7	Illies and Greiner 1979	Limestone	Devonian
15 Wetzlar	50°33'	08°38'	136°	+ 1.4	+ 0.3	+ 0.9	Elmohandes pers. commun.	Limestone	Devonian
16 Waldeck	51°12'	09°05'	---	+ 8.0	---	---	Jagsch 1974 / Lögters 1974	Graywacke	---
17 Oppenheim	49°52'	08°21'	126°	− 0.3	+ 0.1	− 0.5	Greiner 1978	Limestone	Tertiary
18 Auerbach	49°43'	08°39'	125°	+ 3.3	+ 0.6	+ 2.1	Greiner 1978	Diorite	---
19 Albersweiler	49°13'	08°01'	075°	+ 0.7	+ 0.6	− 0.5	Greiner 1978	Gneiss	---
20 Wössingen	49°01'	08°37'	140°	+ 2.2	+ 0.6	+ 1.0	Greiner 1978	Limestone	Triassic
21 Baden-Baden	48°42'	08°15'	138°	+ 5.4	---	---	Schirmer 1979	Rhyolite	Permian
22 Freudenstadt	48°27'	08°29'	170°	---	---	---	Schmitt 1981	Sandstone	Traissic
23 Onstmettingen	48°17'	09°01'	130°	+ 1.9	+ 1.2	− 0.5	Greiner 1978	Limestone	Jurassic
24 Straßberg	48°10'	09°05'	152°	+ 1.9	+ 1.1	− 0.2	Greiner 1978	Limestone	Jurassic
25 Kaiserstuhl	48°09'	07°45'	078°	+ 2.4	+ 0.7	+ 1.1	Leopoldt 1979	Limestone	Jurassic
26 Bollschweil	47°55'	07°47'	153°	+ 0.8	+ 0.2	+ 0.4	Greiner 1978	Limestone	Jurassic
27 Kleinkems	47°41'	07°32'	176°	+ 1.9	+ 0.3	+ 1.4	Greiner 1978	Limestone	Jurassic
28 Choignes	48°07'	05°10'	148°	+ 1.1	+ 0.7	− 0.2	Paquin et al. 1978	Limestone	Jurassic
29 Etrochey	47°53'	04°31'	149°	+ 1.2	+ 0.6	+ 0.03	Paquin et al. 1978	Limestone	Jurassic
30 Ravières	47°43'	04°13'	152°	+ 2.2	+ 0.9	+ 0.5	Paquin et al. 1978	Limestone	Jurassic

Table 1 (cont.)

Location	N-Latitude	E-Longitude	Direction σ_1	Magnitude σ_1 (MPa)	τ_{max} (MPa)	Magnitude σ_2 (MPa)	Reference	Lithology	System
31 Massangis	47°38'	03°58'	167°	+ 2.4	+ 1.3	− 0.2	Paquin et al. 1978	Limestone	Jurassic
32 Luzern	47°04'	08°17'	104°	+ 5.4	+ 1.9	+ 1.6	Gysel 1975	Sandstone	Tertiary
33 Grimsel	46°34'	08°19'	171°	+ 17.5	+ 1.2	+ 15.1	Greiner 1978	Granodiorite	---
34 Gotthard	46°35'	08°35'	---	+ 15.0	---	---	Kovári et al. 1972	Granite	---
35 Piedilago	46°17'	08°21'	030°	---	---	---	Ribacchi and Martinetti 1979	Gneiss	---
36 Roncovalgrande	46°04'	08°44'	150°	+ 24.9	+ 7.4	+ 10.2	Ribacchi and Martinetti 1979	Gneiss	---
37 Mont Blanc	45°53'	06°54'	040°	+ 36.0	---	---	Hast 1973	Gneiss	---
38 S. Fiorano	46°03'	09°50'	175°	+ 14.8	+ 1.6	+ 11.6	Ribacchi and Martinetti 1979	Phyllite	---
39 Edolo	46°11'	09°49'	147°	+ 35.6	+ 10.8	+ 14.0	Ribacchi and Martinetti 1979	Phyllite	---

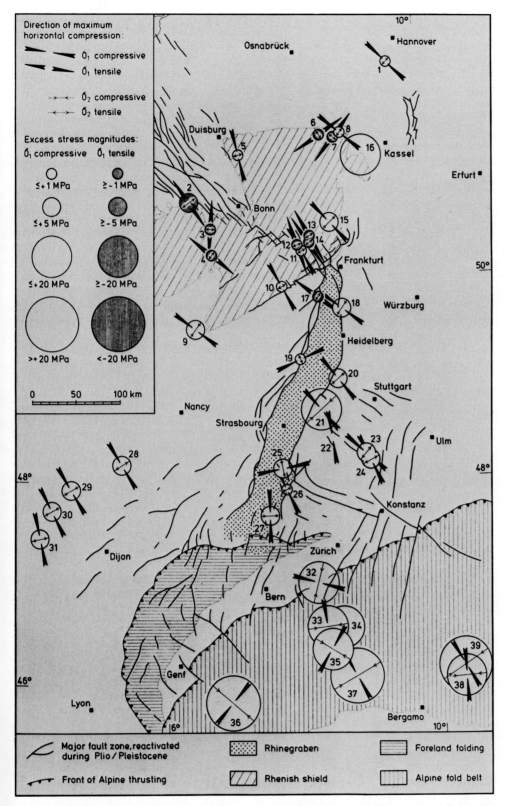

Fig. 2. Direction and magnitude of maximum horizontal compression create a characteristic pattern in Central Europe: an approximately SE-NW-tending stress flux, which is locally deflected by active shear strain along the Rhine Graben rift. High stress values are revealed in the Central Alps, which act as the stress "generator"; due to the decreasing stress gradient in the northern foreland, this region becomes the stress "consumer" (Rhenish shield = Rhenish Massif)

cantly from its surroundings. Not only the smallest horizontal normal principal stress (σ_2), but also the largest horizontal normal principal stress (σ_1) show tensile stress characteristics. The transition from tensile stresses in σ_1-direction in the Rhenish Massif to small compressive stresses in σ_1-direction in the surrounding are seems to be continuous, at least at the southern and eastern borders of the Rhenish Massif.

The amount of the highest shear stress is of particular relevance with respect to the tectonic effect of a stress field. The largest amount of horizontal shear stress was determined for each of our measuring points from the equation

$$\tau_{max.horiz.} = \frac{\sigma_{max.horiz.} - \sigma_{min.horiz.}}{2}$$

From these shear stress values (Fig. 3) four shear stress provinces can be deduced (see Fig. 4). The increase of the shear stresses towards the Alps indicates that the shear stresses are likely to be caused by the Alpine region.

Discussion of the In Situ Stress Measurements

The causes of the existing stress field of the Rhenish Massif can only be understood when the stress distribution of the entire Alpine foreland is regarded.

Not only according to the stress measurements, but also according to the seismic data (Ahorner, 1975), we can suppose a uniform direction of the largest horizontal normal principal stress ($\sigma_1 \sim 140°$-$150°$) to exist in Central Europe. This direction is maintained in the Rhenish Massif, although the compressive stress field is transformed into a tensional stress field. To N and E this tensional stress field changes in the marginal regions and beyond the Rhenish Massif into a stress field with low compressive stresses in σ_1, the tension in σ_2 direction being maintained. To the east (Harz Mountains) σ_1 turns to an approximate N-S direction (Knoll et al., 1977). To the west, particularly in the region of the Paris basin, the fault plane solutions (Veinante-Delhaye and Santoire, 1980) and the axes of neotectonic deformations (Fourniguet et al., 1981) show a flattening of the strike direction from σ_1 to about $120°$. This divergence of the σ_1 trajectories in the Alpine foreland was recognised and described for the first time by Illies (1974). The Alps, acting as stress generator (Illies and Greiner, 1978), are considered to be the cause of this divergence. The fan-like image of the σ_1-trajectories in the Alpine foreland corresponds to the pattern of principal stress directions deduced from fault plane solutions in the Alps (Pavoni, 1980). Following the σ_1 trajectories in radial direction away from the western Alps, we find a decrease of stress magnitudes in this direction. At a certain distance from the Alps, the direction of the stress trajectories changes (Illies et al., 1981): the largest horizontal principal stress (σ_1) again trends from NW-SE to NNW-SSE. This direction was found also in Southern France (Fourniguet et al. 1981) and in Scandinavia (Slunga, 1981). This pattern of stress distribution is regarded as the result of two superimposed stress fields (Illies and Baumann, 1982): on the one hand the plate tectonic stress field (SW-NE orientation), induced by the collision of Eurasia and the African plate (Ahorner, 1978), and on the other hand the diverging "force-arc" pattern perpendicular to the Alps. The stresses induced by the body of the Alps result from the topographic stresses, which are caused by the fact that the Alps overlook their foreland by about 2000 m (mean value), and by the buoyancy of their roots.

The tensile stresses in σ_2-direction can be explained by the diverging σ_1-trajectories in the western Alpine foreland, but this explanation is not valid for the tensile stresses in σ_1-direction. Those tensile stresses typical of the stress field in the Rhenish Massif must have their origin in a phenomenon, which keeps the regional principal stress directions constant and induces an isotrope tensile stress field. As causes for such a tensile stress field geothermic effects (Gay 1980) and a diapiric doming are discussed (Neugebauer et al., 1983).

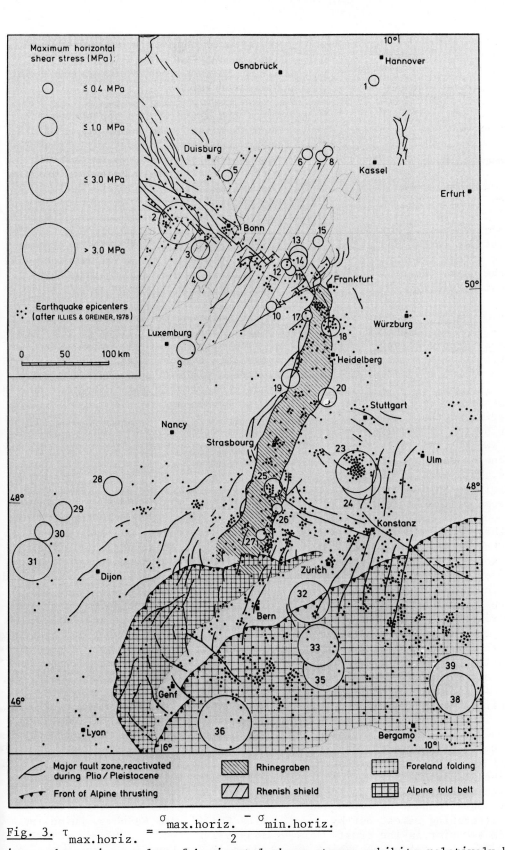

Fig. 3. $\tau_{max.horiz.} = \dfrac{\sigma_{max.horiz.} - \sigma_{min.horiz.}}{2}$

i.e., the maximum value of horizontal shear stress exhibits relatively high values in those foreland areas, where seismic activity is observed. The low indeces in the Rhine Graben may be explained by high crustal weakness along the repeatedly activated disruption zone (Rhenish shield = Rhenish Massif)

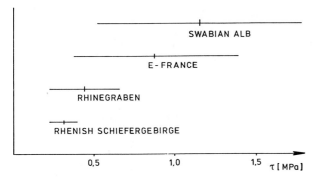

Fig. 4. The mean values of maximum horizontal shear stress in the indicated areas reveal a stress gradient decreasing towards the north (Rhenish Schiefergebirge = Rhenish Massif)

The Mechanism of Strain Release in the Rhenish Massif

The pattern of recent tectonics observed in the key region of the Neuwied basin and the distribution of the micro-earthquakes along NW-SE trend lines show that a weakness zone crosses the Rhenish Massif laterally. This zone connects the northern end of the Rhine Graben with the active rift zone of the Lower Rhine Embayment. Although the tectonic investigations and the stress measurements showed that the stress field in the Rhenish Massif is a tensional stress field providing the direction of the relatively largest horizontal principal normal stress of about 145°, no far-reaching dip-slip faultings or rift-like structures are developed (with the exception of the Neuwied Basin). This is a surprising finding and particularly when we start from the fact that the active stretching rate in the Rhenish Massif ranges in the same order of magnitude as the rate found in the flanking rift systems. The Rhenish Massif responded to the recent stress field not by developing discrete faults, but by uplifting. The strain release observed in the Rhenish Massif, which differs from the strain release in the adjacent rift zones, cannot be explained by differing stretching rates, but by different rock behavior in the crust (see Fig. 5). The basement of the Rhine Graben consists mainly of granites and gneisses of the Hercynian orogeny, which due to their mechanically competent behavior respond with faulting to the tectonic stress.

Where the Rhine Graben approaches the Rhenish Massif, the composition of the crust changes and consequently also the response of the rocks to stress. The rocks of the southern basement, reacting tectonically competent, are replaced in the Rhenish Massif by incompetently reacting folded Devonian and Lower Carboniferous rocks, which are 5,000 to 10,000 m thick. These rocks are able to consume stress by ductile behavior and show rifting only at higher deformation rates. The change from tendency to uplift in the Rhenish Massif to tectonic subsidence in the Lower Rhine Embayment is also accompanied by modifications of the mechanical properties of the crust. Here, we enter into a crustal zone, which is influenced by the Brabant Massif and responds tectonically more competent due to strain hardening.

References

Ahorner, L., 1978. Horizontal compressive stresses in Central Europe. In: Closs, H., Roeder, D., and Schmidt, K. (eds.) Alps, Apennines, Hellenides. Schweizerbart, Stuttgart, pp.17-19.

Baumann, H., 1981. Regional stress field and rifting in Western Europe. In: Illies, H. (ed.) Mechanism of graben formation. Tectonophysics, 73:105-111.

Baumann, H., 1982. Spannung und Spannungsumwandlung im Rheinischen Schiefergebirge. Forneck, Koblenz, 232 pp.

Fourniguet, J., Vogt, J., and Weber, C., 1981. Seismicity and recent crustal movements in France. In: Vyskocil, P., Green, R., and Mälzer, H. (eds.) Recent crustal movements, 1979. Tectonophysics 71:195-216.

Gay, N., 1980. The state of stress in the plates. In: Bally, A., Bender, P., McGetchin, T., and Walcott, R. (eds.) Dynamics of plate interiors. Geodynamics Ser. 1, Am. Geophys. Union, pp. 145-153.

Goodman, R.E., Van, T.K., and Heuze, F. E., 1968. The measurement of rock de-

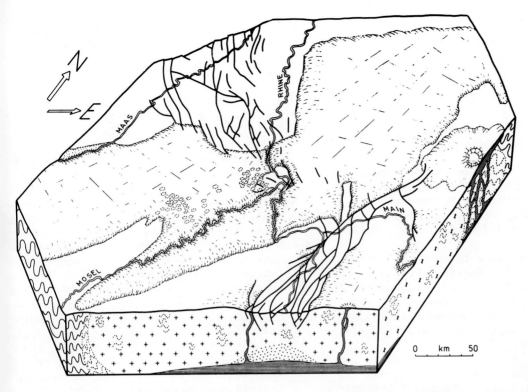

Fig. 5. Block diagram illustrating how the Rhenish Massif acts as a hinge between shear rifting along the Rhine Graben and extensional rifting at the Lower Rhine Embayment. Due to an incompetent behavior of the affected rocks, physiographic rift features are absent over wide parts of the Rhenish Massif

formability in boreholes. 10th Symp. Rock Mechanics. Univ. Texas, pp.1-45.

Greiner, G., 1978. Spannungen in der Erdkruste - Bestimmung und Interpretation am Beispiel von in situ-Messungen im süddeutschen Raum. Dissertation, Univ. Karlsruhe, 198 pp.

Gysel, M., 1975. In-situ stress measurements of the primary stress state in the Sonnenberg Tunnel in Lucerne, Switzerland. Tectonophysics, 29:301-314.

Hast, N., 1973. Global measurements of absolute stress. Philos. Trans. R. Soc. London Ser. A, 274:409-419.

Heerden, van, W.L., 1969. Stress concentration factors for the flat borehole end for use in rock stress measurements. Eng. Geol., 3:307-323.

Illies, J.H., 1974. Taphrogenesis and plate tectonics. In: Illies, H. and Fuchs, K. (eds.) Approaches to taphrogenesis. Schweizerbart, Stuttgart, pp.433-460.

Illies, J.H. and Baumann, H., 1982. Crustal dynamics and morphodynamics of the Western European Rift System. Z. Geomorphol. N.F., Suppl. Bd., 42:135-165.

Illies, J.H. and Greiner, G., 1978. Rhinegraben and the Alpine system. Geol. Soc. Am. Bull., 89:770-782.

Illies, J.H. and Greiner, G., 1979. Holocene movements and state of stress in the Rhinegraben rift system. Tectonophysics, 52:349-359.

Illies, J.H., Baumann, H., and Hoffers, B., 1981. Stress pattern and strain release in the Alpine foreland. In: Vyskočil, P., Green, R., and Mälzer, H. (eds.) Recent crustal movements, 1979. Tectonophysics, 71:157-172.

Jagsch, D., 1974. Ermittlung von Gebirgsspannungen in einem Großbohrloch. Festschrift Leopold Müller, Salzburg, pp.261-269.

Knoll, P., Vogler, G., and Schmidt, M., 1977. Bisherige Ergebnisse von Spannungsmessungen mit Hilfe der Bohrlochentlastungsmethode. Freiberg. Forschungsh., A 569:29-45.

Kovári, K., Amstad, Ch., and Grob, H., 1972. Ein Beitrag zum Problem der Spannungsmessungen im Fels: Int. Symp.

Untertagebau, Luzern, 11-14 Sept. 1972, pp.502-512.

Leeman, E.R., 1964. Borehole rock stress measuring instruments, Part 2. J.S. Afr. Inst. Min. Metall., 65:82-114.

Leopoldt, W., 1979. In situ Spannungsmessungen am Michaelsberg bei Riegel, Nordöstlicher Kaiserstuhl, Südwestdeutschland. Oberrhein. Geol. Abh., 28:17-28.

Lögters, G., 1974. Interfels - Mess-stern zur Bestimmung der Spannungsverteilung in situ. Interfels Messtechnik Inf., 1974:33-35.

Neugebauer, H.J., Woidt, W.-D., Wallner, H., 1983. Uplift, volcanism and tectonics: evidence for mantle diapirs at the Rhenish Massif. This Vol.

Paquin, Ch., Froidevaux, C., and Souriau, M., 1978. Mesures directes des contraintes tectoniques en France septentrionale. Bull. Soc. Geol. Fr., (7), XX:727-731.

Pavoni, N., 1980. Crustal stresses inferred from faultplane solutions of earthquakes and neotectonic deformation in Switzerland. In: Scheidegger, A. (ed.) Tectonic stresses in the Alpine-Mediterranean region. Rock Mech. Suppl, 9:63-68.

Rahn, W., 1981. Zum Einfluß der Gesteinsanisotropie und des bruchbedingten nichtlinearen Materialverhaltens auf die Ergebnisse von Spannungsmessungen im Bohrloch. Bochumer Geol. Geotechn. Arb., 5:209 pp.

Ribacchi, R. and Martinetti, S., 1979. In situ stress determinations in Italy. Paper submitted to the Symposium on "Stresses in the Alpine-Mediterranean region". Vienna, September 11-14, 1979, 14 pp.

Schirmer, P., 1979. In situ Spannungsmessungen in Baden-Baden. Oberrhein. Geol. Abh., 28:7-15.

Schmitt, T.J., 1981. The West European stress field: new data and interpretation. J. Struct. Geol., 3:309-315.

Slunga, R., 1981. Fault mechanisms of Fennoscandian earthquakes and regional crustal stresses. Geol. Föeren. Stockholm Föerh., 103:27-31.

Veinante-Delhaye, A. and Santoire, J., 1980. Sismicité récente de l'Arc Sud-Armoricain et du Nord-Quest du Massif Central. Mécanismes au foyer et tectonique. Bull. Soc. Geol. Fr., XXII: 93-102.

6.3 General Pattern of Seismotectonic Dislocation and the Earthquake - Generating Stress Field in Central Europe Between the Alps and the North Sea

L. Ahorner[1], B. Baier[2], and K.-P. Bonjer[3]

Abstract

We have used 30 reliable fault-plane solutions from earthquakes which occured during 1975 to 1982 in Central Europe between the Alps and the North Sea to derive the general pattern of seismotectonic dislocations and the characteristic features of the earthquake generating stress field in the Rhenish Massif area and surrounding region. Stress directions from fault-plane solutions are in good agreement with the results of in situ stress measurements. Although the maximum compressive stress shows an average direction of about 145° with respect to north in the whole area, regional differences in the stress directions and in the seismotectonic regime are recognizable. In the southern part of the Upper Rhine Graben, the Swabian Jura, and in the western Vosges mountains, strike slip mechanisms are predominantly, whereas in the northernmost part of the Upper Rhine Graben, in the Rhenish Massif, and the Lower Rhine Embayment tensional dip-slip dislocations along NW-SE trending fault planes are typical. Seismotectonic evidence strongly suggests that an active rift zone separates the Rhenish Massif along a line between the Lower Rhine Embayment and the Upper Rhine Graben.

Introduction

To investigate the nature and cause of the Late Tertiary and Quaternary plateau uplift of the Rhenish Massif in Central Europe was the main subject of a multidisciplinary research program sponsored by the Deutsche Forschungsgemeinschaft during the years 1976-1982 (Illies et al., 1979, 1983). Geodynamical models, which try to explain the phenomenon of the plateau uplift, require, besides many other basic data, reliable information on the large-scale crustal stress field and the general pattern of seismotectonic dislocations, not only in the Massif area proper, but also in the surrounding regions. Therefore an attempt is made in this paper to compile an up-to-date map showing the pattern of seismotectonic block movements and the regional characteristics of the earthquake-generating stress field in the Central European area between the Alps and the North Sea. Maps of this kind have been published previously by Ahorner et al. (1972), and Ahorner (1975). But since 1976, the number of seismic stations has increased considerably, and detailed seismicity studies have been carried out (Bonjer, 1979; Ahorner, 1983; Baier and Wernig, 1983). These recent seismological investigations provide us with high precision source information, which allows reliable determination of fault-plane solutions even for small magnitude earthquakes

---

[1] Abteilung für Erdbebengeologie, Geologisches Institut, Universität zu Köln, Vinzenz-Pallotti-Str. 26, D-5060 Bergisch Gladbach 1 (Bensberg), Fed. Rep. of Germany

[2] Institut für Meterologie and Geophysik, Universität Frankfurt, Feldbergstr. 47, D-6000 Frankfurt, Fed. Rep. of Germany

[3] Geophysikalisches Institut, Universität Karlsruhe, Hertzstr. 16, D-7500 Karlsruhe 21, Fed. Rep. of Germany.

down to about $M_L = 2$. In addition, new techniques using amplitude data of seismic body waves were successfully applied to improve the quality of the solutions (Kisslinger, 1980; Koch, 1982). Thus the data base for the compilation of a seismotectonic map for the Central European area is today much better than some years ago.

Data

Our study is based on earthquakes occurring in the Central European area between the Alps and the North Sea during the years 1975-1982. The epicenters of events with magnitude $M_L \geq 2$ are shown in Fig. 1. Focal parameters for most of these shocks are published in the yearly bulletins of the seismological observatories of the Federal Republic of Germany (edited by the Gräfenberg observatory).

The source region with the highest seismicity is the Swabian Jura, south of Stuttgart, where recently a larger earthquake with $M_L = 5.7$ occurred on September 3, 1978, near the town of Albstadt (Haessler et al., 1980; Turnovsky and Schneider, 1982). A rather diffuse pattern of seismicity with respect to $M_L \geq 2$ has been observed in the southern part of the Upper Rhine Graben and its adjacent mountain ranges (Black Forest, Vosges Mountains). One of the most prominent events in the southern Rhine Graben occurred near Sierentz on July 15, 1980 ($M_L = 4.9$; Rouland et al., 1980) during the course of this study. The central part of the graben around Karlsruhe was quiet with respect to events with $M_L \geq 2$ during the research period. The northermost part of the graben, however, shows a remarkable seismicity, which is concentrated along a NW-SE trending zone between Heidelberg and Mainz. This seismically active zone continues into the Rhenish Massif, where epicenters are clearly aligned along the Middle Rhine zone between Mainz and Bonn (see Ahorner, 1983). The seismicity of the Rhenish Massif outside the Middle Rhine zone is more spreadout and is concentrated in the western half of the Massif (Hohes Venn, Ardennes, Eifel). A local source region in the eastern half, however, released the strongest shock of the Rhenish Massif during the study period near the village of Bad Marienberg, Westerwald, on June 28, 1982 ($M_L = 4.7$; Ahorner, 1983). The seismicity of the Lower Rhine Graben between Bonn and Nijmegen shows nearly the same activity level as the Middle Rhine zone and the northermost part of the Upper Rhine Graben. The large events in the Lower Rhine area occurred near Roermond on June 5, 1980 ($M_L = 3.8$) and in the region of Waldfeucht on May 22, 1982 ($M_L = 3.8$). Both sources are situated within the Roer Valley Graben.

For 30 events, distributed over the whole area of investigation, reliable fault-plane solutions could be derived. The parameters of the solutions are given in Table 1. The geographical distribution of different types of focal mechanisms and of stress data is shown in the seismotectonic map (Fig. 2). Examples of well defined fault plane solutions are displayed in Fig. 3. Figures 4 and 5 show the spatial orientation of P-, B- and T-axes of fault plane solutions, and in addition the strike of σ_1-axes from in situ stress measurements.

Interpretation

From the geographical distribution of focal mechanisms it reveals that different seismotectonic regimes exist in the area of investigation (Fig. 2). Each regime is characterized by the predominance of a particular type of focal mechanism or by a specific combination of mechanism types. Regional differences in the general pattern of seismotectonic dislocations have been found already by Ahorner (1975) and Bonjer (1979) from previous focal mechanism studies.

The Lower Rhine Embayment, the Middle Rhine zone (central part of the Rhenish Massif), and the northermost part of the Upper Rhine Graben are characterized by tensional dip-slip dislocations (normal faults) along NW-SE trending fault-plane. This implies an extensional crustal deformation in SW-NE direction. Focal

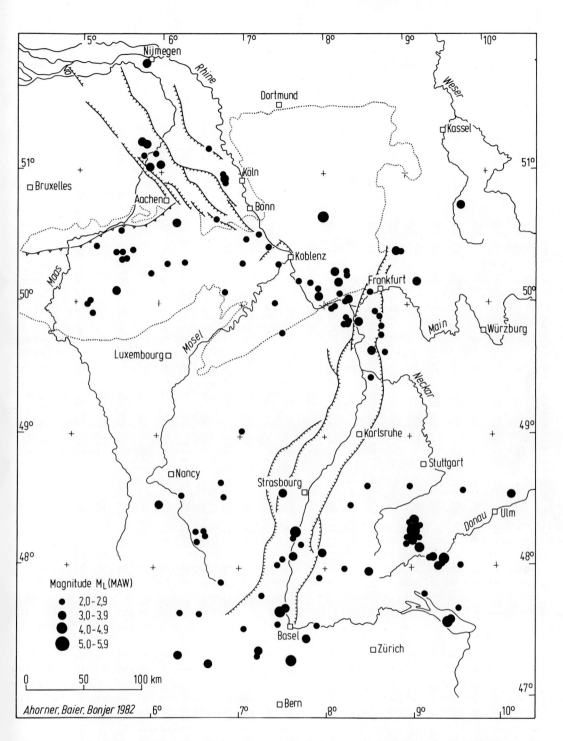

Fig. 1. Epicenter map of earthquakes 1976–1982 with local magnitudes $M_L \geq 2$. The contourline of the Rhenish Massif is *dotted*

Table 1. Fault plane solutions of earthquakes 1975–1982

No.	Region	Date	Lat N	Long E	h km	M_L	Plane 1 Strike	Plane 1 Dip	Plane 2 Strike	Plane 2 Dip	P-Axis Az	P-Axis Plunge	T-Axis Az	T-Axis Plunge	B-Axis Az	B-Axis Plunge	Type	Q
1	Roermond	Jun.5,1980	51.24	5.75	14	3.8	134	72 NE	64	45 SE	267	46 SW	12	18 NE	128	40 SE	N	B
2	Waldfeucht	May 22,1982	51.06	5.97	13	3.8	121	52 SW	169	49 NE	328	64 NW	54	1 NE	144	26 SE	N	B
3	Brauweiler	Nov.6,1977	50.96	6.78	14	3.6	17	48 NW	34	42 SE	185	81 SW	296	3 NW	26	8 NE	N	A
4	Roetgen	Jun.10,1978	50.62	6.20	6	3.0	64	80 NW	64	10 SE	334	35 NW	154	55 SE	64	1 NE	R	C
5	Euskirchen	Mar.29,1978	50.65	6.69	9	2.5	122	42 SW	132	48 NE	300	86 NW	39	4 NE	128	2 SE	N	A
6	Sinzig	Apr.30,1978	50.53	7.21	6	2.6	163	65 SW	613	25 NE	73	70 NE	252	20 SW	163	1 SE	N	B
7	Marienberg	Jun.28,1982	50.68	7.99	13	4.7	83	58 SE	152	60 NE	296	46 NW	28	1 NE	152	44 SE	S	A
8	Ochtendung	1977–1981	50.35	7.38	–	–	124	60 NE	109	30 SW	230	74 SW	28	15 NE	120	6 SE	N	A+
9	Braubach	1978–1981	50.27	7.64	–	–	130	60 NE	75	45 SE	272	59 SW	20	8 NE	120	32 SE	N	B+
10	St. Goar	Dec.31,1980	50.18	7.70	6	2.8	113	64 SW	170	42 NE	336	57 NW	227	13 SW	130	30 SE	N	A
11	Geroldstein	May 8,1977	50.07	7.94	11	3.1	45	45 NW	45	45 SE	137	0 SE	0	90	45	0 NE	R	B
12	Echzell	Nov.4,1975	50.41	8.87	11	3.6	63	80 NE	162	50 NW	300	20 NW	195	34 SW	51	57 NE	S	C
13	Hanau	Sep.21,1981	50.17	9.10	8	3.2	13	80 SE	112	50 SW	324	36 NW	68	20 NE	182	48 SW	S	C
14	Wiesbaden	Nov.4,1979	50.04	8.30	5	3.2	140	35 NE	153	56 SW	90	78 E	237	12 SW	338	6 NW	N	A
15	Oppenheim	Jul.31,1979	49.87	8.40	10	2.9	2	85 NE	94	75 SE	135	8 SE	228	14 SW	25	75 NE	S	C
16	Darmstadt	Jun.7,1979	49.91	8.66	6	2.3	100	42 NE	145	58 SW	106	65 SE	215	10 SW	308	24 NW	N	B
17	Lorsch	Apr.9,1977	49.65	8.56	15	2.6	105	50 NE	153	49 SW	130	63 SE	41	0 NE	310	25 NW	N	B
18	Schwetzingen	Apr.3,1979	49.44	8.55	5	2.1	155	30 NE	162	60 SW	80	74 NE	249	18 SW	340	2 NW	N	C
19	Enzklösterle	Oct.12,1980	48.65	8.50	21	3.3	5	70 SE	100	30 SW	318	40 NW	56	7 NE	152	55 SE	S	B
20	Epinal	Mar.3,1981	48.28	6.58	5	2.9	109	71 SW	26	70 NW	158	1 SE	67	28 NE	250	61 SW	S	B
21	Rhinau	Oct.27,1979	48.29	7.65	7	3.9	12	90	102	80 SW	326	3 NW	57	8 NE	192	80 SW	S	A
22	Kaiserstuhl	Apr.30,1978	48.10	7.63	11	2.8	4	80 NW	105	45 NE	133	39 SE	242	22 SW	355	45 NW	S	B

Table 1 (cont.)

No.	Region	Date	Lat N	Long E	h km	M_L	Plane 1 Strike	Plane 1 Dip	Plane 2 Strike	Plane 2 Dip	P-Axis Az	P-Axis Plunge	T-Axis Az	T-Axis Plunge	B-Axis Az	B-Axis Plunge	Type	Q
23	Waldkirch	Jan.27,1979	48.12	7.97	17	3.1	60	82 SE	153	60 SW	110	15 SE	12	26 NE	225	58 SW	S	A
24	Waldau	Jan.15,1976	48.00	8.22	10	2.1	13	80 SE	106	60 SW	328	23 NW	62	5 NE	170	68 SE	S	A
25	Oberried	Mar.29,1976	47.93	7.93	16	2.2	15	45 NW	95	80 SW	160	31 SE	53	26 NE	265	43 SW	S	B
26	Wolterdingen	Febr.29,1976	47.98	8.51	11	3.1	12	60 SE	133	50 SW	340	55 NW	77	6 NE	168	33 SE	S	A
27	Sierentz	Jul.15,1980	47.68	7.48	12	4.9	40	90	130	80 SE	354	8 N	86	8 E	220	80 SW	S	A
28	Delle	Febr.11,1978	47.55	7.06	8	2.4	128	60 SW	62	54 NW	186	4 SW	92	50 SE	279	41 NW	S	C
29	Albstadt	Sept.3,1978	48.29	9.03	7	5.7	20	75 NW	113	77 NE	157	19 SE	66	2 NE	333	71 NW	S	A
30	Bodensee	Mar.2,1976	47.6	9.4	20	3.7	30	0	120	0	346	0 NW	76	0 NE	–	90	S	C

Quality of solution: A = very good, B = good, C = fair, + = composite solution.
Dislocation type: N = normal faulting, S = strike slip faulting, R = reverse (thrust) faulting.
References of fault plane solutions:
1-11 (Ahorner), 12 (Neugebauer and Tobias 1977), 13-14 (Baier and Wernig), 15-20, 22-26, 28 (Bonjer), 21 (this paper),
27 (Bonjer, Rouland et al. 1980, this paper), 29 (Turnovsky and Schneider 1982), 30 (Mayer-Rosa and Pavoni 1977).

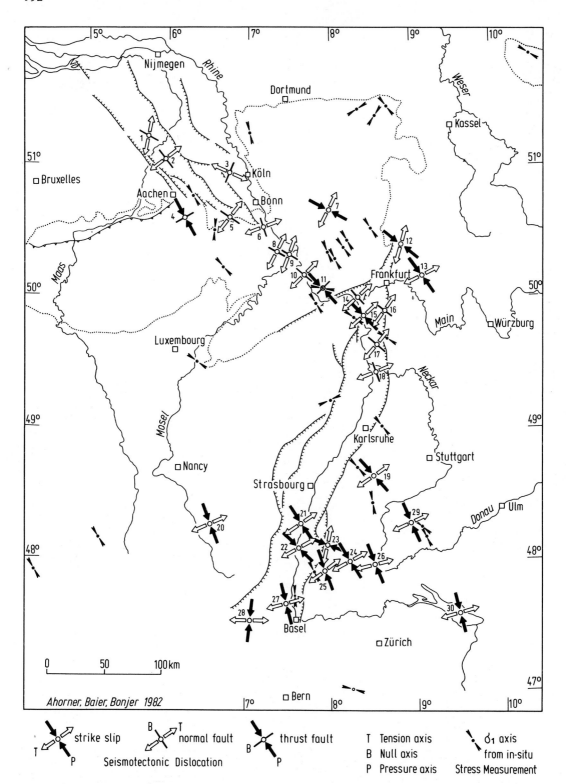

Fig. 2. Seismotectonic map for the Central European area based on fault plane solutions of earthquakes 1975-1982. In situ stress data are added (Baumann, 1982)

Fig. 3. Typical fault plane solutions. Equal-area projection of the lower focal hemisphere. Compressional quadrants are *hatched*. P-, B- and T-axes are marked by *squares*. Examples *1,3,5,10* and *14* are tensional dip-slip dislocations. Example *11* shows a compressional dip-slip dislocation (thrust fault). Examples *13,21* and *27* reveal strike-slip dislocations

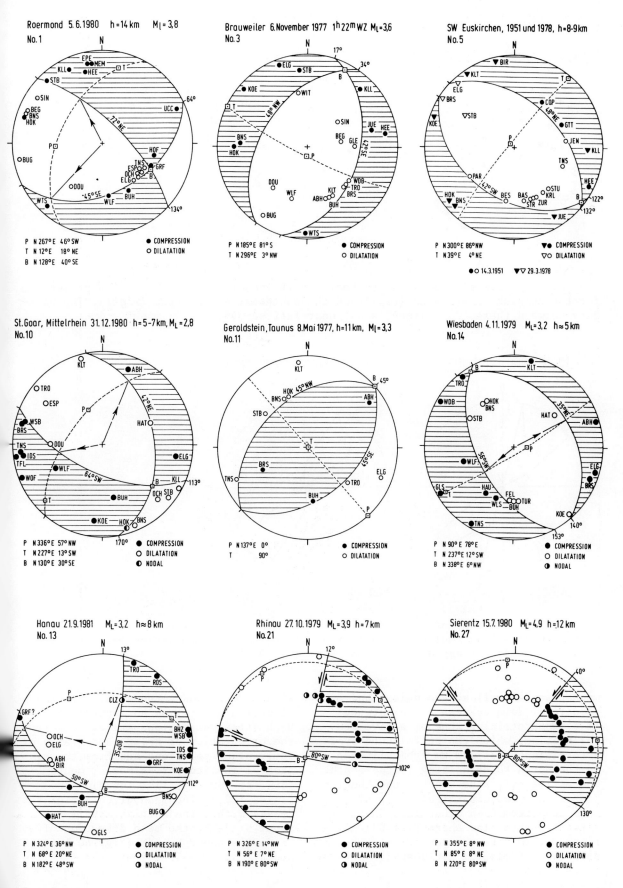

Fig. 3

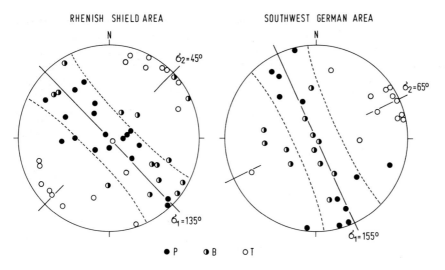

Fig. 4. Composite plot of seismotectonic P-, B- and T-axes in equal-area projection for the Rhenish Massif area (*left*) and the Southwest German area (*right*). The boundary between these two regions is set near latitude 49.3°N

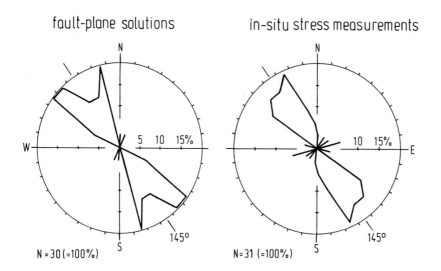

Fig. 5. Rose diagrams of the direction of maximum horizontal compressive stress in Central Europe from fault-plane solutions (*left*) and in situ stress measurements (*right*). In situ stress data are from Baumann (1982)

mechanisms as well as geological and geodetical data suggest the existence of a NW-SE striking zone of active crustal rifting, which can be followed over a total length of 350 km from Nijmegen in the North to Heidelberg in the South. The active rift zone bisects the Rhenish Massif into two crustal blocks, which move away from each other in a SW-NE direction (see also Ahorner, 1983). The rift zone is arranged parallel to the direction of maximum horizontal compressive stress, and perpendicular to the direction of minimum horizontal stress. Both stress directions are evident from the orientation of P-, B- and T-axes of fault-plane solutions and from the σ_1-axes of in-situ stress measurements (Fig. 2; Baumann, 1982, Baumann and Illies, 1983). The mean direction of the P-axes within the Rhenish Massif is about N 135°E, the mean direction of T-axes is about N 45°E (Fig. 4, left side).

In addition to the tensional dip-slip dislocations two compressional dip-slip dislocations were observed in the area of the Hohes Venn (solution 4) and in the Taunus mountains (solution 11). Fault planes of those mechanisms are arranged SW-NE, which is parallel to the Hercynian fold axes of the Rhenish Massif. The P-axes are orientated NW-SE, parallel to the direction of the maximum horizontal compressive stress. In the eastern part of the Rhenish Massif and in the adjoining Hessian Depression some strike-slip mechanisms with a minor tensional dip-slip component were observed. One of these fault-plane solutions is of the Echzell, Wetterau, earthquake (solution 12), published by Neugebauer and Tobias (1977). The Bad Marienberg, Westerwald, earthquake 1982 (solution 7) is another well defined example. Strike-slip mechanisms were found in previous studies by Ahorner (1975) also in the northwestern border region of the Rhenish Massif. Again the P-axes and T-axes of strike-slip mechanisms are arranged generally NW-SE or SW-NE, respectively, in agreement with the regional stress field.

In contrast to the mostly tensional dip-slip dislocations of the northernmost part of the Upper Rhine Graben (Bonjer, 1979), the southern Upper Rhine Graben with its adjacent mountain ranges, the Swabian Jura region and the Lake Constance region display predominantly strike-slip dislocations with fault planes striking NNE-SSW or WNW-ESE, respectively (Bonjer and Fuchs, 1979; Bonjer, 1979; Mayer-Rosa and Pavoni, 1977; Turnovsky and Schneider, 1982). This has been mentioned already by Ahorner and Schneider (1974) on the basis of older focal mechanisms. As a result of the new data the average direction of P-axes in the southwest German area is found to be N 155°E, the direction of T-axes is N 65°E (see Fig. 4, right side). With respect to the Rhenish Massif a clockwise rotation of the principal stress axes by about 20° is evident from focal mechanism data. This trend has already been pointed out by Bonjer and Gelbke (1976). The σ_1-axes of in situ stress measurements in both areas exhibit no significant difference of the stress directions. The slight disagreement between the results of both methods might be explained either by a dependency of the stress direction with depth, or more likely by regional differences in the prevailing strike direction of pre-existing fault zones within the crust, which are controlling the shear-plane orientation and consequently the direction of P-, B- and T-axes of fault plane solutions.

The predominance of strike-slip mechanisms in the Southwest-German area in contrary to the Rhenish Massif is probably due, first to the prevailing NNE and WNW directions of pre-existing fault zones in this area (e.g., the Upper Rhine Graben), which are arranged parallel to the direction of the maximum shear stress of the actual stress field, and second to the increasing stress gradient of the maximum horizontal compression from the Rhenish Massif area in the North to the Alps in the South. In situ stress measurements show that the excess stress magnitude is much higher within the Alps than in the Rhenish Massif area (Illies et al., 1981; Illies and Baumann, 1982; Baumann, 1982; Baumann and Illies, 1983). Seismotectonic strike-slip dislocations require a maximum compression in an approximately horizontal plane, whereas tensional dip-slip dislocations require an approximately vertical orientation of the maximum compressive stress. The vertical compression is in many cases not of tectonic, but of gravitational origin.

Stress results gain in accuracy if we use not a single fault plane solution or in situ stress measurement but as many data as possible from a certain region in order to get an average stress direction. The nearly perfect agreement of rose diagrams showing P-axes from 30 fault-plane solutions and σ_1-axes from 31 in situ stress measurements is displayed in Fig. 5. Both data sets result in an average direction of the maximum horizontal stress in Central Europe of about 145° with respect to North. Nearly the same stress direction has been found by Ahorner (1975) from focal mechanism data prior to 1975. The high density of in situ stress measuring points in Central Europe makes this region a key area for testing the reliability of stress field results from fault plane solutions of local earthquakes or vice versa.

The practical results disprove some critical objections coming from theoretical considerations against the applicability of the relation between fault-plane solutions and the directions of principal stresses, if earthquakes occur on pre-existing faults in heterogeneous rocks, and not in homogeneous material (e.g., McKenzie, 1969).

Summary

On the basis of high precision determinations of focal parameters it has been found that the area of the Rhenish Massif (including the Lower Rhine Embayment and the northernmost part of the Upper Rhine Graben) is experiencing active rifting today. This is corroborated by fault plane solutions of earthquakes between Nijmegen and Heidelberg. In addition the fault plane solutions show a clockwise rotation of the principle stress directions from North to South. The average direction of the regional tectonic field is in concordance with the results obtained by former seismological studies (e.g., Ahorner, 1975) and are in surprising good agreement with the results deduced by in situ stress measurements (Baumann and Illies, 1983).

Acknowledgments. The authors enjoyed countless discussions with H. Illies (†), G. Greiner, K. Fuchs, H. Neugebauer, N. Pavoni, and D. Rouland and thank all colleagues who kindly made data of their station network available. The computations have been carried out at the computing centers of the universities of Köln, Frankfurt and Karlsruhe. K. Fuchs and J. Zucca read the manuscript and made valuable suggestions to improve the quality of the manuscript, which was typed by H. Gadau.

References

Ahorner, L., 1975. Present-day stress field and seismotectonic block movements along major fault zones in Central Europe. In: Pavoni, N. and Green, R. (eds.) Recent crustal movements. Tectonophysics, 29:233-249.

Ahorner, L., 1983. Historical seismicity and present-day micro-earthquake activity of the Rhenish Massif, Central Europe. This Vol.

Ahorner, L. and Schneider, G., 1974. Herdmechanismen von Erdbeben im Oberrheingraben und in seinen Randgebirgen. In: Illies, H.J. and Fuchs, K. (eds.) Approaches to taphrogenesis. Schweizerbart, Stuttgart, pp.104-117.

Ahorner, L., Murawski, H., and Schneider, G., 1972. Seismotektonische Traverse von der Nordsee bis zum Apennin. Geol. Rundsch., 61:915-942.

Baumann, H., 1982. Spannung und Spannungsumwandlung im Rheinischen Schiefergebirge. Forneck, Koblenz, 232 pp.

Baumann, H. and Illies, H.J., 1983. Stress field and strain release in the Rhenish Massif. This Vol.

Baier, B. and Wernig, J., 1983. Microearthquake activity near the southern border of the Rhenish Massif. This Vol.

Bonjer, K.-P., 1979. The seismicitiy of the Upper Rhinegraben. A continental rift system. In: Petrovski, J. and Allen, C. (eds.) Proc. Int. Res. Conf. Intra-Continental Earthquakes. Ohrid, Yugoslavia, 1979, pp.107-115.

Bonjer, K.-P. and Fuchs, K., 1979. Real time monitoring of seismic activity and earthquake mechanism in the Rhine graben area as a basis for prediction. Proc. ESA-Counc. Eur. Sem. Earthquake Prediction, Strasbourg, pp.383-386.

Bonjer, K.-P. and Gelbke, C., 1976. Seismizität und Dynamik im Bereich des Oberrheingrabens. Proc. 36. Jahrestag. DGG, Bochum, p.31.

Haessler, H., Hoang Trong, P., Schick, R., Schneider, G., and Strobach, K., 1980. The September 1, 1978, Swabian Jura earthquake. Tectonophysics, 68: 1-14.

Illies, H.J. and Baumann, H., 1982. Crustal dynamics and morphodynamics of the Western European rift system. Z. Geomorphol. N.F., Suppl.-Bd., 42: 135-165.

Illies, H.J., Prodehl, C., Schmincke, H.-U., and Semmel, A., 1979. The Quaternary uplift of the Rhenish shield in Germany. In: McGetchin, T.R. and Merril, R.B. (eds.) Plateau uplift: mode and mechanism. Tectonophysics, 61:197-225.

Illies, H.J., Baumann, H., and Hoffers, B., 1981. Stress pattern and strain release in the Alpine foreland. Tectonophysics, 71:157-172.

Kisslinger, C., 1980. Evaluation of S and P amplitude ratios for determining focal mechanism from regional network observations. Bull. Seismol. Soc. Am., 53:1039-1074.

Koch, K., 1982. Bestimmung von Herdmechanismen aus SV/P - Amplitudenverhältnissen für Mikroerdbeben im Bereich des Oberrheingrabens. Diplom-Thesis, Geophys. Inst. Univ. Karlsruhe, 116 pp.

Mayer-Rosa, D. and Pavoni, N., 1977. Fault plane solutions of earthquakes in Switzerland from 1971 to 1976. Proc. XV. Gen. Assem. E.S.C., Krakow, 1976. Publ. Inst. Geophys. Pol. Acad. Sci., A-5, 116:321-326.

McKenzie, D.P., 1969. The relations between fault plane solutions for earthquakes and the directions of the principal stresses. Bull. Seismol. Soc. Am., 59:591-601.

Neugebauer, H.J. and Tobias, E., 1977. A study of the Echzell-Wetterau earthquake of November 4, 1975. J. Geophys., 43:751-760.

Rouland, D., Haessler, H. Bonjer, K.-P., Gilg, B., Mayer-Rosa, D., and Pavoni, N., 1980. The Sierentz Southern Rhinegraben earthquake of Juli 15, 1980 - Preliminary results. Proc. ESC-EGS Gen. Assem., Budapest 1980.

Turnovsky, J. and Schneider, G., 1982. The seismotectonic character of the September 3, 1978, Swabian Jura earthquake series. Tectonophysics, 83: 151-162.

6.4 Historical Seismicity and Present-Day Microearthquake Activity of the Rhenish Massif, Central Europe

L. Ahorner[1]

Abstract

The earthquake activity of the Rhenish Massif, which has undergone some 300 m of plateau uplift since the Late Tertiary, is analyzed. Seismotectonic implications on geodynamic processes controlling crustal uplift are discussed. The study is based on a detailed investigation of the historical seismicity pattern since 1500 A.D., and on a special microearthquake survey carried out during the period 1976-1982. A local station network consisting of high-sensitive analog or digital recording seismographs has been installed in an area of about 200 × 200 km^2 with an average station spacing of 40 km. More than 800 local earthquakes of tectonic origin, ranging in magnitude from $M_L = 0$ to 4.7, were recorded during the 6-year study. Most of them could be located with reasonable accuracy (±1 km horizontally and ±2 km vertically). Fault-plane solutions were derived, in general, for shocks with magnitudes greater than $M_L = 3$.

The main features of the historical seismicity pattern are reproduced by the short-term microseismicity, although differences exist in detail. Focal regions with large historical earthquakes show, for instance, in general an unusual low level of microseismicity. From S-wave spectra of digitally recorded earthquakes the source characteristics (seismic moment, source radius, dislocation and stress drop) have been determined using the Brune (1970) source model. Between local magnitude M_L and seismic moment M_0 (dyne cm) the following relation holds (in the magnitude range $1 \leq M_L \leq 5$): $\log M_0 = 17.4 + 1.1\ M_L$. Stress drops vary between several bar and more than 70 bar; they are loosely related to magnitude (small magnitude shocks have in general smaller stress drops). A local anomaly of the stress drop regime correlating with an anomaly of the focal depth distribution is probably existing in the central part of the Rhenish Massif near the Neuwied basin and the Late-Quaternary volcanic field of the Laacher See.

The overall pattern of seismicity within the Rhenish Massif and its vicinity is clearly related to stress-field controlled block movements along major crustal fracture zones, like the Rhine Graben system, and shows no direct relationship to the massif uplift. Only in some special regions, e.g., the Hohes Venn in the northwestern part of the massif, where an exceptional large present-day uplift rate of 1.6 mm yr^{-1} has been found by geodetical means, some evidence for a causal correlation between seismicity and plateau uplift is achieved.

Introduction

The earthquake activity of a region gives insight into the nature and spatial distribution of seismotectonic processes down to the base of the brittle crust.

---

[1] Erdbebenstation Bensberg, Abteilung für Erdbebengeologie, Geologisches Institut, Universität Köln, Vinzenz-Palotti-Str. 26, D-5060 Bensberg, Fed. Rep. of Germany.

Plateau Uplift, ed. by K. Fuchs et al.
© Springer-Verlag Berlin Heidelberg 1983

Thus the depth range of geodynamic observations is greater by seismological investigations than with geological and geodetical methods, which are restricted in their findings mainly to near-surface phenomena. Under these general aspects the analysis and seismotectonic interpretation of the local seismicity was of special interest within the scope of the Rhenish Massif Project, a multidisciplinary research program sponsored by the German Research Society (DFG) in the years 1976-1982. The main subject of this program was to investigate nature and causes of the epirogenic plateau uplift of the Rhenish Massif in the northern foreland of the Alps (Illies et al., 1979). The large outcropping basement complex of the Rhenish Massif consists of a series of predominantly Devonian slates, quartzites and limestones, which were strongly folded during the Hercynian orogeny and have subsequently undergone some 300 m of plateau uplift since the Pliocene. Young volcanism, still active until the Late Quaternary, accompanied the uplift.

This paper is concerned mainly with the description and interpretation of the regional seismicity pattern of the Rhenish Massif, as deduced from earthquakes observed since 1500 A.D., and with the present-day microearthquake activity of the central part of the Rhenish Massif and its northern foreland. Seismotectonic implications, especially with respect to the shield uplift, are discussed.

Historical Seismicity Pattern

The historical seismicity of the Rhenish Massif and its vicinity has been previously investigated under different aspects by Hiller et al. (1967), Ahorner (1968, 1970, 1975), Ahorner et al. (1970), Van Gils and Zaczek (1978), Bonjer (1979), and others, on the basis of numerous older publications.

An overview of the seismicity pattern, as deduced from earthquakes observed since 1500, and of the geological features of the shield area is given in Fig. 1. The map shows the locations of earthquake epicenters classified according to local magnitude, focal depth and accuracy of epicenter determination. Seismological data are based on historical earthquake catalogues of Sieberg (1940), Sponheuer (1952) and others, and, beginning with shocks later than 1900, on a re-interpretation of all original available macroseismic and microseismic information. For the period 1500-1749 only shocks with magnitudes $M_L \geq 4.5$ were plotted, for 1750-1899 those with $M_L \geq 3.5$ and for 1900-1982 those with $M_L \geq 2.0$.

It is obvious from the distribution of epicenters that nearly all foci active since 1500 are situated within four seismicity zones:

1. The Rhenish earthquake zone, which follows approximately the course of the river Rhine from Karlsruhe in the South to Nijmegen in the North. This zone represents the most active seismogenetic feature of the Rhenish Massif and its vicinity. An almost continuous line of epicenters connects the neotectonic fracture zones of the Upper Rhine Graben in the South with the Lower Rhine Graben in the North

2. The Belgian earthquake zone, which trends in a West-East direction across Belgium from Ostend to Aachen, following the border line between the Caledonian Brabant Massif and the Hercynian Rhenish Massif.

3. The earthquake zone of the Stavelot-Venn Massif, which follows the SW-NE trending axis of a major anticline in the northwestern part of the Rhenish Massif south of Aachen.

4. The Hunsrück-Taunus earthquake zone, which covers the southern border region of the Rhenish Massif between Luxembourg and Frankfurt. The trend of this zone is also SW-NE, parallel to the Hercynian fold axes.

The seismic activity outside the above-mentioned earthquake zones is quite low. Large parts of the Rhenish Massif appear to be even aseismic in the light of historical earthquake observations. The highest level of seismicity has been observed in the western part of the Lower Rhine Graben near the intersection point between the Rhenish and the Belgian earthquake zones. In this region, the

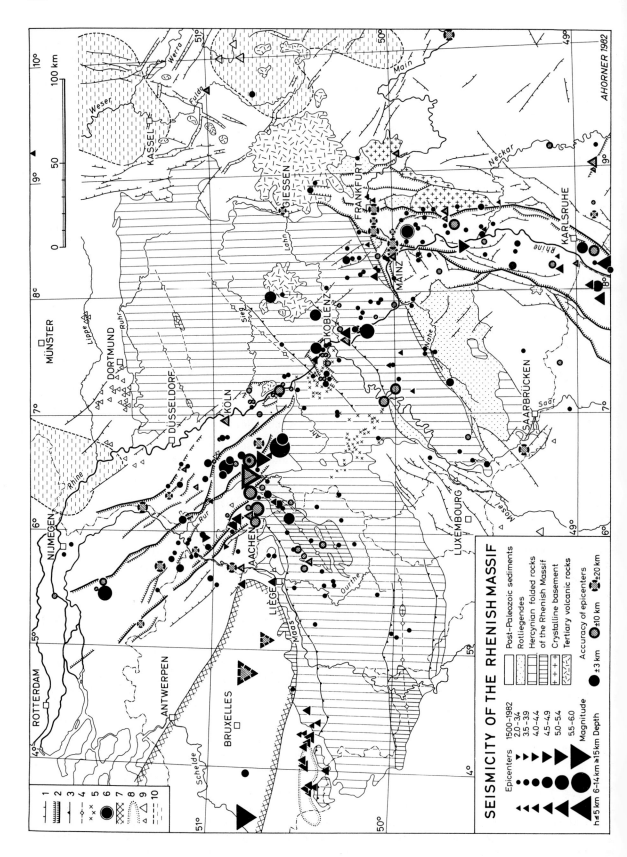

Fig. 1 (figure caption see opposite page)

strongest earthquake of the Rhenish Massif area took place near the town of Düren 1756 (magnitude $M_L = 6$, epicentral intensity I_0 = VIII MSK-scale). Another major shock occurred in 1938 in the Belgian zone southwest of Brussells ($M_L = 5.8$, I_0 = VII). The maximum observed event in the Middle Rhine zone is the St. Goar earthquake of 1846 ($M_L = 5.0$, I_0 = VII). The northern part of the Upper Rhine Graben has experienced shocks up to magnitude $M_L = 4.8$, in particular during the earthquake swarm of Großgerau 1869-1871.

The great majority of earthquakes felt within the Rhenish Massif area are clearly of tectonic origin. Focal depth varies between several km and 25 km as a maximum. The predominant depth range of tectonic events is 6 to 14 km. Only within some special regions, situated mostly outside the Rhenish Massif, non-tectonic shocks do occur. This is true primarily for regions with extensive underground mining, e.g., the Ruhr Coal district and the Werra Potash district, where rockbursts and other shocks induced by mining are frequent (Casten and Cete, 1980; Leydecker, 1976). Natural collapse earthquakes due to the outwashing of evaporitic and carbonatic rocks by underground water are to be expected mainly in the region east of the Rhenish Massif, where favorable geological conditions exist for this kind of seismicity. The extreme shallow-focus events of the Hainaut zone, between Mons and Charleroi at the northwestern border of the Rhenish Massif, are probably also of non-tectonic origin. Recent geological and geophysical investigations in this area have proven Paleozoic evaporites at a depth of a few km, partly overriden by the low-angle thrust faults of the Faille du Midi system (Bouckaert et al., 1977). Collapsing cavities within these soluble rocks and accompanying shear dislocations in the overlying material might be a plausible explanation for the specific seismic activity of the Hainaut zone, with focal depths of only 1 to 3 km and magnitudes up to $M_L = 4.6$.

From focal mechanism studies of earthquakes prior to 1976 - the beginning of the Rhenish Massif Project - the general pattern of seismotectonic block movements along major fracture zones in Central Europe was already known to some extent from papers of Schneider et al. (1966), Ahorner and Schneider (1974), Ahorner (1975), and others. Data available at that time suggest that strike-slip mechanisms dominat in and around the Upper Rhine Graben and within the Belgium earthquake zone, whereas in the Lower Rhine Graben tensional dip-slip mechanisms prevail. From a synoptical interpretation of focal mechanism data and in situ stress measurements, it was clear already in this early stage of research that the regionally different regimes of seismotectonic dislocations are controlled by the present-day large-scale stress field within the Earth's crust in Central Europe (Ahorner, 1975). Improved fault-plane solutions of earthquakes since 1976 have confirmed and modified these previous findings, as will be described in the following text (see also Ahorner et al., 1983).

Microearthquake Activity

Station Network and Instrumentation

Small magnitude microearthquakes ($0 < M_L < 3$) are comparatively frequent even in regions with minor seismicity. This favors a worldwide application of microearthquake data to many seismological problems, such as detailed studies of regional seismicity patterns and focal

---

Fig. 1. Historical seismicity pattern and geological features of the Rhenish Massif. *1* fault zone active during Tertiary time; *2* fault zone active until Quaternary and Recent time; *3* thrust fault of Hercynian age; *4* major Hercynian anticline; *5* Quaternary volcano; *6* earthquake swarm; *7* border line of the Caledonian Brabant massif; *8* Hainaut basin; *9* seismic events due to mining activity; *10* border line of underground salt deposits

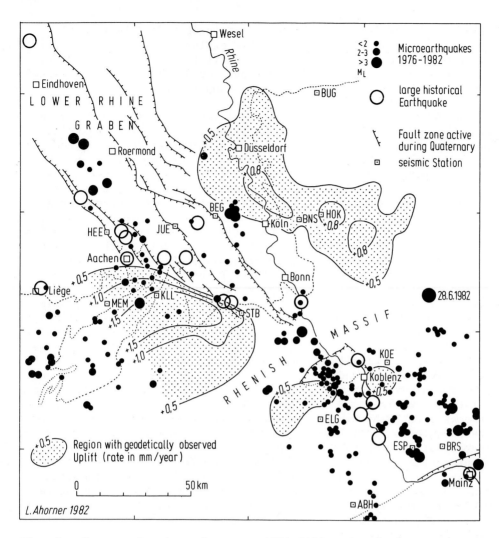

Fig. 2. Microearthquake epicenters 1976-1982 and seismic station network in the Rhenish Massif and its northern foreland. Regions with crustal uplift as deduced from geodetical measurements are marked by *points* (Mälzer et al., 1983). Isolines give uplift rates in mm yr^{-1}

mechanisms, identifying and mapping of active faults, etc. (for an overview see Lee and Steward, 1981). In Central Europe a permanent microearthquake network has been successfully in operation since 1971 in the Upper Rhine Graben area (Bonjer and Fuchs, 1974; Bonjer, 1979).

In order to improve the detection capability for small earthquakes within the Rhenish Massif a network of high-sensitive seismograph stations has been installed since 1976 by the Geological Institute of the University of Cologne in the central part of the Rhenish Massif and its northern foreland. Additional stations were set up in the Hunsrück-Taunus region and in the southern foreland of the Rhenish Massif by the Geophysical Institutes of the Universities of Frankfurt and Karlsruhe. The locations of these stations are plotted in Fig. 2. Station coordinates are listed in Table 1.

The instrumental equipment of the stations operated by the Cologne team consists of short-period seismometers of the Geotech S-13 or the Geospace HS-10 type with natural periods between 1 and 1.5 s, combined either with analog ink recorders (paper speed 120 or 240 mm min^{-1}) or with digital tape recorders using pulse code modulation (Lennartz

Table 1. High-sensitive seismograph stations operating in 1976-1982 in the Rhenish Massif

Code	Name	Latitude North	Longitude East	Height	Component	Institution
HOK	Hohkeppel	50.980°	7.313°	230 m	Vertical	Cologne
KLL	Kalltalsperre	50.647°	6.311°	390 m	Triaxial	Cologne
STB	Steinbachtal	50.595°	6.840°	270 m	Vertical	Cologne
KOE	Köppel	50.425°	7.732°	540 m	Vertical	Cologne
OCH	Ochtendung	50.371°	7.376°	120 m	Vertical	Cologne
ELG	Burg Eltz	50.206°	7.337°	140 m	Triaxial	Cologne
ESP	Espenschied	50.099°	7.893°	400 m	Triaxial	Cologne
BRS	Bärstadt	50.105°	8.066°	490 m	Vertical	Frankfurt
IDS	Idstein	50.217°	8.271°	–	Vertical	Frankfurt
ABH	Alteburg	49.882°	7.548°	620 m	Vertical	Karlsruhe
BIR	Birkenfeld	49.702°	7.060°	755 m	Vertical	Karlsruhe

PCM-system S-5000). The tape recorders are not in continuous operation but are triggered by seismic events. The dynamic range of the PCM-system is 72 db (12 bit). High-frequency cut-off filters have a corner frequency of 20 Hz. The magnification of the analog recording seismographs is normally set to about 3×10^5 at 10 Hz depending on local noise conditions. The timing precision is ±0.05 s, in case of digital tape recording even better.

Calculation of Hypocenter Parameters

Seismograms of microearthquakes typically show sharp onsets of P- and S-waves. This facilitates the hypocenter determination. From error analysis it is assumed that well-recorded shocks within the network can be located with an accuracy of about ±1 km horizontally and ±2 km vertically. Location tests with quarry blasts confirm this accuracy. The routine hypocenter determination has been performed with the computer program HYPO-ITERATION, which has been developed at the Cologne institute and works with a Commodore CBM 4032 microcomputer. Both, P- and S-phase arrivals are taken as input data. The average crustal model used for computing travel times in the iterative location procedure is listed in Table 2. The model is based on seismic refraction data for the Rhenish Massif (Mooney and Prodehl, 1978; Mechie et al., 1983). A velocity ratio V_p/V_s = 1.69 has been derived as mean value for events distributed over the whole area of investigations, but single shock data suggest that there are significant local variations of the V_p/V_s ratio ranging from 1.65 to 1.73.

Local magnitudes M_L have been calculated from S-wave amplitudes A_h (horizontal component of maximum ground displacement in 10^{-6} m) and hypocenter distance R (in km) with the following formula deduced from Richter's original log A_0-values by converting Wood-Anderson amplitudes to ground amplitudes:

$$M_L = \log_{10}(A_h) + 1.90 \log_{10}(R) - 0.35$$

Magnitudes determined in the distance range of R = 10-300 km are fairly consistent with standard Wood-Anderson magnitudes MAW as reported by the Gräfenberg observatory (GRF) for Central European earthquakes.

Seismicity Pattern from Microearthquakes

More than 800 microearthquakes of tectonic origin were recorded by the Cologne station network during the period 1976-1982 within the Rhenish Massif and its vicinity. Of these about 600 could be located with reasonable precision, using additional readings from the station networks operated by the colleagues of Frankfurt and Karlsruhe. The total number of microearthquakes observed during the 6-year study is of the same order as of all known historical earth-

Table 2. Crustal velocity model

Layer No.	Thickness H (km)	Velocity V_p (km s^{-1})	V_s (km s^{-1})
1	1	5.00	2.96
2	1	5.50	3.25
3	1	5.80	3.43
4	7	6.00	3.55
5	9	6.25	3.70
6	12	6.90	4.08
7	–	8.10	4.76

MOHO-depth 31 km (between layer 6 and 7).

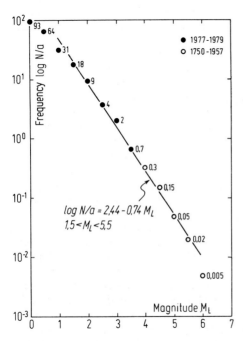

Fig. 3. Cumulative magnitude-frequency diagram for the Rhenish Massif and its northern foreland. The covered area is about 40,000 km^2. *Open circles* are data points from historical earthquakes, *black dots* are from microearthquakes 1977-1979

quakes since 1500. Epicenters of microearthquakes are plotted in Fig. 2. Magnitudes vary between $M_L = 0$ and 4.7. The strongest shock, causing slight damage in the epicentral region, occurred on June 28, 1982, in the Bad Marienberg, Westerwald, area ($M_L = 4.7$, I_o = V-VI MSK-scale, macroseismic radius 150 to 200 km). The main shock was accompanied with more than 100 foreshocks and aftershocks, which have been recorded in the near-field by a mobile station network. The focal region of Bad Marienberg is situated at the eastern periphery of the seismoactive Middle Rhine zone and has never before in historical time experienced such a large earthquake. The majority of microearthquakes located in the period 1976-1982, however, fall within zones that have been active already in previous times. This becomes clear, if we compare Fig. 1 with Fig. 2. The seismicity pattern as deduced from historical events is reproduced in its main features by the short-term observation of microactivity. In detail, of course, there are some differences, which will be discussed later.

A magnitude-frequency diagram derived for the Rhenish Massif and its northern foreland (area of Fig. 2) by combining microearthquake data 1977-1979 with historial data of larger earthquakes is shown in Fig. 3. Both data sets give nearly the same b-value (slope of the regression line) and the same yearly seismic rate, thus indicating that short-term observation of microactivity can be a reliable tool for estimating long-term recurrence rates of larger events. In the magnitude range of 1.5 to 5.5, the relationship between magnitude and the logarithm of frequency of occurrence log N yr^{-1} seems to be quite linear with a b-value of 0.74. This b-value has also been found by Bonjer (1979) for the Upper Rhine Graben. Seismotectonic implications of such similarities in b-values between the Rhenish Massif and the Upper Rhine Graben are not fully understood at this time. Bonjer (1979) has pointed out that not too much weight should be given to b-values being only an average for larger areas, because smaller focal regions within those areas often show significantly different b-values.

This seems to be true, indeed, for the Rhenish Massif and its northern foreland. Within the main seismoactive zones there are obviously regions with comparatively high microearthquake activity and minor macroseismicity, e.g., the southeastern part of the Middle Rhine zone between Koblenz and Mainz, whereas in other regions, in contrast, large historical earthquakes are frequent and the present day microearthquake activity is comparatively low, e.g., in the western part of the Lower Rhine Graben between Aachen and Cologne. This implies local variations of b-values. Macro- and microseismicity seem to substitute each other to some extent. Two separate maps with isolines of earthquake frequency for smaller and larger events demonstrate this fact (Fig. 4, see pages 206 and 207), which is of practical importance for the evaluation of local seismic risk. A high rate of microseismicity implies a more continuous strain release and acts like a "safety valve", which prevents a focal region from high stress concentrations and therefore from larger earthquakes.

Source Characteristics

In order to get an impression of size and amount of seismotectonic processes, the source characteristics of local earthquakes within the Rhenish Massif and its vicinity have been estimated using S-wave displacement spectra from near-field digital recordings. For 27 events ranging in magnitude from 0.9 to 4.7 the focal parameters seismic moment M_o, source radius r_o, stress drop $\Delta\sigma$, and average dislocation d_o were calculated from spectral parameters (e.g., long-period spectral level Ω_o, spectral corner frequency f_o) with formulas published by Brune (1970) for his physical source model (see also Lee and Steward, 1981). Spectra were corrected for instrumental response and anelastic energy absorption ($Q = 500$). The results are listed in Table 3 (see page 208).

The relationship between seismic moment M_o (in dyne cm) and local magnitude M_L derived from our results is well described by the formula (Fig. 5, see page 207):

$$\log M_o = 17.4 + 1.1 M_L$$

The standard deviation for $\log M_o$ is 0.21. The relation holds for magnitudes from 0.9 to 4.7, or even higher, because the data point of the Albstadt earthquake 1978 ($M_L = 5.7$; Haessler et al., 1980), occurring in southwestern Germany, is near the calculated regression line. Similar relationships have been found by Durst (1981) for earthquakes in the southern part of the Upper Rhine Graben, and by Johnson and McEvilly (1974) for earthquakes along the San Andreas fault zone in Central California.

The relationship between average source dislocation d_o and local magnitude M_L seems to be more complex and cannot be fitted by a straight regression line, but there is no doubt that source dislocation increases with magnitudes from a few tenths of a mm in case of $M_L = 1.0$ to about 70 mm in case of $M_L = 4.7$ (Fig. 6, see page 209). The same is true, in general, for the radius r_o of the assumed circular source, which ranges from 100 m to 700 m and increases with increasing magnitude, but there is also a dependency on stress drop.

Stress drops $\Delta\sigma$ vary between a few tenths of a bar and more than 70 bar. The correlation to magnitude is not very clear due to scattering of data, but smaller shocks show on an average smaller stress drops than larger shocks. This has been found also by Durst (1981) for events in the southern Upper Rhine Graben, and by Tucker and Brune (1973) for aftershocks of the San Fernando earthquake 1971 in California. Stress drop data for the Rhenish Massif available up to now suggest probably some regional differences in stress drop regimes, although conclusions like this have to be drawn with caution, considering the magnitude dependency of stress drop. Unusual low stress drops in the order of 1 bar have been obtained from (mostly small magnitude) earthquakes in the central part of the Rhenish Massif west of Koblenz (Neuwied basin and neovolcanic zone of the Eastern Eifel and the Laacher See region), whereas in the southeastern parts of the massif (Taunus mountains) and also in the northwestern parts including the Lower Rhine Graben, stress drops between 10 and 70 bar are prevailing (Fig. 7, see page 209). A slight tendency of decreasing horizontal shear stress towards the neovolcanic zone in the central part of the Rhenish Massif is also evident from in situ measurements (Baumann, 1982; Baumann and Illies, 1983).

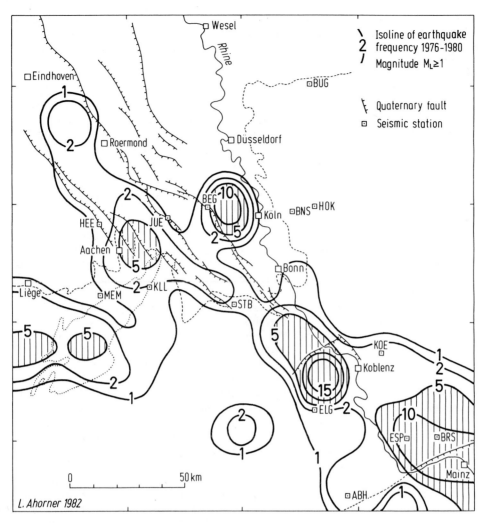

Fig. 4a,b. Maps showing isolines of earthquake frequency per unit area of 20×20 km^2 for microearthquakes 1976-1980 with magnitudes $M_L \geq 1$ (*above*), and for large historical earthquakes 1500-1975 with magnitudes $M_L \geq 4$ (*opposite page*). Regions with increased microearthquake activity are *hatched*

Focal Depth Distribution

Additional evidence for the existing of a crustal zone with anomalous geomechanical behavior in the central part of the Rhenish Massif comes from the focal depth distribution of earthquakes. A cross-section through the Rhenish Massif showing hypocenters of microearthquakes and larger historical earthquakes is given in Fig. 7. Focal depths of historical events have been determined from macroseismic data using the method of Sponheuer (1960). Both data sets agree in indicating a broad seismogenetic band within the upper part of the Earth's crust, above the Conrad discontinuity, which marks the boundary between the granitic and gabbroidic layers of the crust. The active band is about 10 to 15 km thick and has its highest earthquake frequency at a depth of about 10 km. Beneath the Neuwied basin and the nearby neovolcanic zone the seismogenetic crustal band is shifted in its entirety towards shallower depths by the amount of about 2 to 3 km. The contourline of the maximum depth of hypocenters follows this upwarping, which coincides in its regional position with the zone of low stress drops mentioned above. The most likely reason for the anomalous

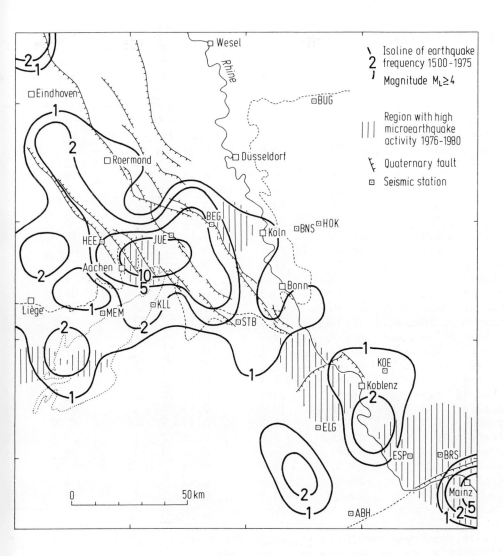

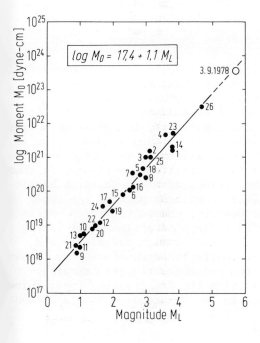

Fig. 5. Relation between seismic moment M_0 and local magnitude M_L for earthquakes occurring 1977-1982 in the Rhenish Massif area. The data point of the Albstadt, Swabian Jura, earthquake 1978 (*open circle*) is near the calculated regression line

Table 3. Source characteristics of local earthquakes within the Rhenish Massif

Date	Region	Magnitude	Moment log M_0 (dyne cm)	Corner frequency f_0 (Hz)	Source radius r_0 (km)	Stress Drop $\Delta\sigma$ (bar)	Average dislocation d_0 (mm)
07.03.77	Hünstätten	3.8	21.20	3.8	0.37	14	8
08.05.77	Geroldstein	3.1	21.18	6.4	0.21	71	27
02.11.77	Brauweiler	3.0	21.00	4.5	0.31	15	9
06.11.77	Brauweiler	3.6	21.66	3.1	0.44	24	22
06.11.77	Brauweiler	2.9	20.67	5.6	0.23	17	9
29.03.78	Euskirchen	2.5	20.04	4.6	0.30	2	1
04.09.78	Andernach	0.9	18.18	10.2	0.13	0.3	0.1
30.04.78	Sinzig	2.6	20.54	4.2	0.31	5	3
22.08.78	Neuhof	3.0	20.40	10.7	0.12	63	16
29.09.78	Kruft	1.1	18.74	11.5	0.11	1.8	0.4
12.10.78	Plaidt	1.0	18.34	11.3	0.12	0.6	0.2
24.12.78	Eveshausen	1.6	19.09	9.3	0.14	1.9	1
10.04.80	Ochtendung	1.0	18.72	10.8	0.12	1.3	0.3
05.06.80	Roermond	3.8	21.30	4.6	0.30	32	22
28.07.80	Ingelheim	2.3	19.90	10.7	0.12	20	5
30.07.80	Ingelheim	2.6	20.11	9.0	0.15	17	6
31.12.80	St. Goar	2.8	20.52	9.2	0.14	53	17
01.01.81	St. Goar	2.0	19.45	11.1	0.12	8	2
15.12.80	Großgerau	1.9	19.72	9.1	0.14	8	3
05.01.81	Ochtendung	1.4	18.89	10.4	0.13	1.6	0.5
01.03.81	Lahnstein	1.5	18.98	10.6	0.13	2	0.6
23.02.81	Ochtendung	0.9	18.38	10.2	0.13	0.5	0.02
02.03.82	Selfkant	3.6	21.26	4.8	0.28	41	23
22.05.82	Waldfeucht	3.8	21.71	2.3	0.59	13	16
20.06.82	Marienberg	1.7	19.57	5.0	0.26	0.9	0.5
26.06.82	Marienberg	3.1	20.99	2.9	0.45	5	5
28.06.82	Marienberg	4.7	22.50	1.9	0.69	42	69

geomechanical behavior of the crust in this part of the Rhenish Massif is lateral variations of crustal temperature, related to the Late Quaternary volcanism of the Eifel and the Laacher See area (Schmincke et al., 1983). Brace and Byerlee (1970) and others have emphasized the important role of temperature for the transition from stick-slip to stable-sliding in tectonic processes and therefore for the depth control of crustal earthquakes. Positive geothermal anomalies within the crust are in many cases correlated with comparatively shallow-focus seismicity. Alternatively or in addition, the anomalous crustal behavior might be explained by material inhomogeneities controlling the frictional properties of fault zones, by the presence of pore fluids under high pressure, and by large-scale heterogeneities of the tectonic strain-stress field. Further investigations are needed to resolve those questions raised by the present study.

Seismoactive Fault Lines

The spatial distribution of microearthquake hypocenters, which are more frequent and more precisely located than historical events, offers for the first time the opportunity to map seismoactive faults within the Rhenish Massif. As an example the alignment of epicenters along NW-SE striking fault zones in the southeastern part of the Rhenish Massif and in the northern part of the Upper Rhine Graben is shown in Fig. 8. An en echelon pattern of active fault lines, characterized by microseismicity and in some parts by macroseismicity, transfers the seismogenetic structural activity from the NNE-trending Upper Rhine Graben to the NW-trending Middle Rhine

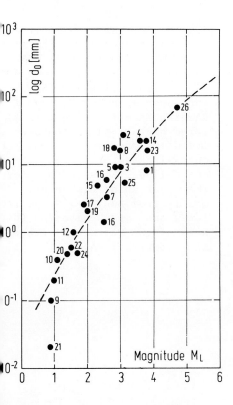

Fig. 6. Relation between focal dislocation d_0 and local magnitude M_L for earthquake occurring 1977-1982 in the Rhenish Massif area

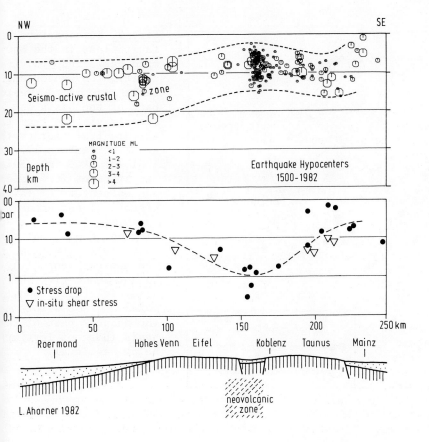

Fig. 7. Cross section through the Rhenish Massif along the Rhine valley with focal depth distribution of earthquakes (*above*) and stress data (*below*). *Black dots* denote stress drops within seismic sources, *triangles* give the maximum horizontal shear stress from in situ stress measurements (Baumann, 1982). Geological section is schematic

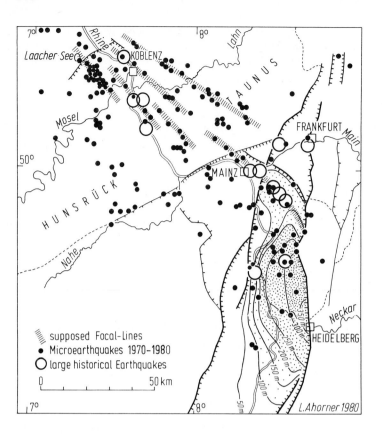

Fig. 8. Seismoactive fault lines as indicated by the alignment of microearthquake epicenters 1970-1980 (*black dots*) in the transition zone between the Upper Rhine Graben and the Rhenish Massif. Epicenters of large historical earthquakes (*open circles*) are in many cases located on microearthquake lines. Seismic data for the Upper Rhine Graben are mostly from Bonjer (1979). *Isolines* and *points* give the thickness of the Quaternary graben fill (Bartz, 1974).

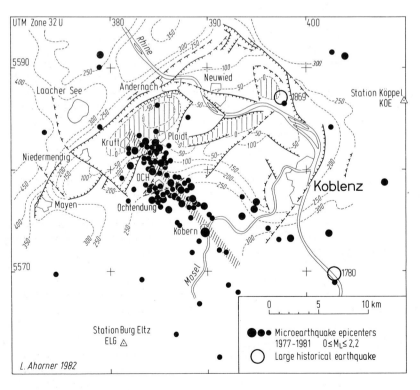

Fig. 9. Seismoactive fault lines in the Neuwied basin and its vicinity. The Ochtendung fault zone in the western part of the basin is clearly visible from the alignment of microearthquake epicenters 1977-1981 (*black dots*). Seismic stations (*triangles*) and epicenters of large historical earthquakes (*open circles*) are also plotted. Geological surface faults are from N. and Ahorner (unpubl.). *Height lines* give the base of Tertiary beds. *Hatched areas* denote the deepest part of the basin (base of Tertiary below sea level)

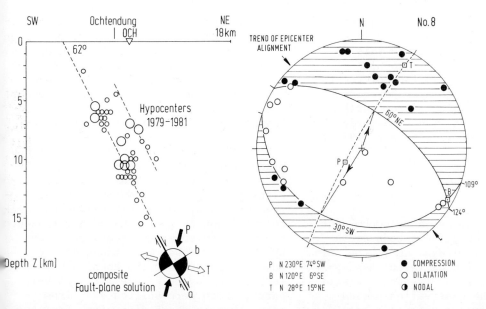

Fig. 10. Cross-section through the Ochtendung fault zone with the spatial distribution of hypocenters 1979-1981 (*left*), and a composite fault plane solution of seismic events occurring along the fault zone (*right*). Equal-area projection of the lower focal hemisphere. Azimuth and plunge of pressure axis *P*, tension axis *T*, and null axis *B* are given

zone. Most of the active faults identified show no clear surface features and can therefore not be mapped with geological or geodetical methods.

An interesting seismoactive structure, which has been investigated in detail, is the Ochtendung fault zone in the western part of the Neuwied basin (Fig. 9). This zone represents the most productive microearthquake source within the Rhenish Massif, generating quite regulary seismic events in the magnitude range from $M_L = 0$ to 2. During the period 1977-1981 more than 150 shocks have been located, mostly with a good depth control, because the recording station OCH is situated near the focal line. The epicenters are clearly aligned along a NW-SE trending zone of 15 km length. Focal depths vary between several km and 15 km. From the spatial distribution of hypocenters along a cross-section perpendicular to the Ochtendung fault zone it becomes clear that the moving zone dips by about $60°$ to the NE (Fig. 10). This is in agreement with the results of a composite fault-plane solution, which indicate a pure tensional dip-slip dislocation along a NW-SE striking and $60°$ towards NE dipping plane (Fig. 10). It is evident from these findings that the Ochtendung fault zone represents an active normal fault, being one of the southwestern border faults of the Cenozoic Neuwied basin. The outcrop of the fault zone at the surface is most likely identical with a fault line known from geological field work in the region southeast of Niedermendig (see Fig. 9). The post-Oligocene throw of this surface fault is about 50 to 100 m, which gives an estimate for the long-term dislocation rate in the order of 0.005 mm yr^{-1}. It is of interest to compare the long-term geological rate with short-term seismotectonic slip-rates calculated for the Ochtendung fault zone from seismic moments of related earthquakes using a method suggested by Brune (1968).

The seismic moment M_o of a single shock, determined either from spectra of seismic waves or from the magnitude M_L using a generalized $M_o(M_L)$-relation as given above, is defined as:

$$M_o = \mu F_o d_o$$

where μ = modulus of rigidity, F_o = slip surface (focal area), and d_o = average slip on F_o (focal displacement). If we sum up the moments of all earthquakes occurring along a fracture zone in a given time period, and divide this moment sum by an appropriate modulus of rigidity and the slip-surface of the total fracture zone, the average seismotectonic slip D resulting from earthquake activity during the time period considered is (see also Ahorner, 1975):

$$D = \frac{\sum M_o}{\mu A}.$$

Assuming a slip surface $A = 160$ km^2 for the total Ochtendung fault zone and a rigidity $\mu = 3 \times 10^{11}$ dyne cm^{-2}, the calculation yields with different assumptions on the magnitude-frequency-distribution during the observation period 1977-1981 a seismotectonic slip rate D/t of approximately 0.004 to 0.027 mm yr^{-1} (4 to 27 m per 10^6 years). This is about the same order as the long-term geological slip-rate mentioned above, although averaging over geological time periods is certainly problematic. In any case we have to conclude that the seismotectonic slip-rates along active structures like the Ochtendung fault zone are very small. Even if we take into account a strong portion of aseismic creep movement, the resulting total slip-rate is smaller than one tenth of a mm per year, which means that recent block movements along discrete fault lines within the Rhenish Massif might be proven by geodetical methods only, if the time interval between repeated measurements is at least a few decades.

General Pattern of Seismotectonic Dislocations

Normal block faulting requires an extensional crustal deformation and an adequate tensional stress field in the direction perpendicular to the strike of the fault. This is indicated by the orientation of the T-axis (tension axis) in corresponding fault-plane solutions. In case of the Ochtendung fault zone the T-axis is oriented nearly horizontally with a strike direction of N 28° E. This implies a local crustal extension in SW-NE direction (Fig. 10). Quite similar orientations of T-axes have been found from fault-plane solutions of many earthquakes occurring within the Middle Rhine zone (Fig. 11). Focal mechanisms in this area are predominantly of tensional dip-slip type (normal fault), sometimes combined with a small strike-slip component. Examples are given in Fig. 12 (see also Ahorner et al., 1983).

The general seismotectonic regime of the Middle Rhine zone between Mainz and Bonn is clearly characterized by extensional rifting along a broad NW-SE trending fracture zone, which follows approximately the course of the Rhine and connects the major seismoactive features of the Upper Rhine Graben with those of the Lower Rhine Graben. The Rhenish Massif is bisected by this zone into two half parts, which were shifted away from each other in SW-NE direction. From the geological point of view, the hidden zone of active rifting consists most probably - even in the deeper underground - not of large pervading fault lines, but of numerous smaller faults, oriented sub-parallel or in en-echelon configuration. This implies a rather diffuse picture of the ruptural deformation of the crust, which might be caused by the slaty composition of the majority of rocks in the upper part of the crust of the Rhenish Massif. Illies and Baumann (1982) and others have mentioned that such rocks reveal a ductile, mechanically incompetent behavior and sustain horizontal stretching without responding by major brittle faulting.

In the northermost part of the Upper Rhine Graben, between Heidelberg and Mainz, earthquake focal mechanisms are mostly of the same type as in the Middle Rhine zone. Tensional dip-slip dislocations, partly with a minor strike-slip component, are clearly predominating in this area, too (Bonjer, 1980; Ahorner et al., 1983). Active fault planes have a prevailing strike direction NW-SE, which is oblique to the general NNE-trend of the graben axis (Bonjer, 1980). Quite different from that of the northermost part is the seismotectonic regime of the central and southern parts of the Upper Rhine Graben, between Heidelberg and Basel. Seismotectonic block movements in this larger graben segment are characterized nearly exclusively by

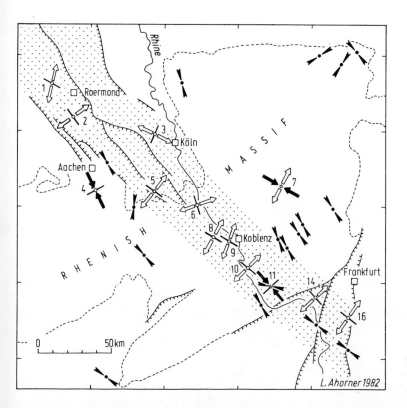

Fig. 11. Focal mechanisms of earthquakes 1976-1982 within the Rhenish Massif, and σ_1-directions (*small block arrows*) of maximum compressive stress from in situ stress measurements (Baumann, 1982). Tension axes (*white arrows*), pressure axes (*large black arrows*), and null axes (*lines*) of earthquake focal mechanisms are plotted, if the plunge of these axes is smaller than 45°

strike-slip mechanisms with left-lateral moving sense along fault-planes striking parallel to the graben axis (Ahorner and Schneider, 1974; Bonjer, 1979; Ahorner et al., 1983).

The Lower Rhine Graben, north of the Rhenish Massif, shows again a clear predominance of tensional dip-slip dislocations along NW-SE trending fault-planes (Fig. 11, and Ahorner, 1975). We can therefore state a close agreement between seismotectonic regimes in the Lower Rhine Graben, the Middle Rhine zone, and the northermost part of the Upper Rhine Graben. Reliable fault-plane solutions in all these regions speak for a pervading zone of active crustal rifting, which can be followed over a total length of 350 km from Nijmegen in the north to Heidelberg in the south (see also Ahorner et al., 1983).

Quaternary structural activity is evident along this zone especially in its northern and southern parts, whereas the central part within the Rhenish Massif shows only minor, or locally moderate, faulting activity during the Quaternary epoch. In the Lower Rhine Graben, however, Quaternary deposits are dislocated along still active normal faults, e.g., the Erft fault and the Peel boundary fault, up to 175 m (Ahorner, 1962). Recent slip-rates of about 1 mm yr^{-1} have been proven by high-precision levellings (Quitzow and Vahlensieck, 1955). The Upper Rhine Graben, forming the southern part of the active rift zone, shows also a very distinct Quaternary and recent structural activity (Bartz, 1974; Illies and Greiner, 1979). Tectonic subsidence of the graben floor near Heidelberg amounts to more than 350 m during Quaternary times (see Fig. 8). Geodetically observed present-day slip-rates are in the order of 0.5 mm yr^{-1} (Prinz and Schwarz, 1970; Mälzer et al., 1983).

Available data from focal mechanism studies as well as from geological and geodetical field work leave no doubt on the existence of a prominent still active rift zone, which crosses the Rhenish Massif in NW-SE direction nearly parallel to the Rhine valley. The general pattern of the plateau uplift of the Rhenish Massif is certainly influenced by the interaction of forces causing the crustal rifting, with those causing the uplift. One of the open

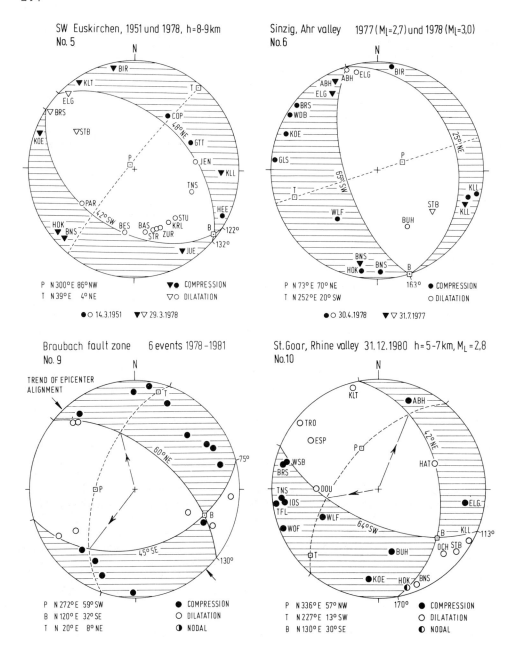

Fig. 12. Examples of fault-plane solutions of the tensional dip-slip type (normal fault). Equal-area projection of the lower focal hemisphere. Azimuth and plunge of pressure axis P, tension axis T, and null axis B are given

questions is why the surface features of the rift zone are so inconspicuous within the Rhenish Massif. Is this phenomenon due only to the slaty (incompetent) rock behavior of the upper crust, which enables crustal stretching without the formation of discrete faults, as Illies and Baumann (1982) have pointed out, or is the total amount of crustal rifting a priori smaller in the central segment, perhaps due to a progressive propagation of the initial rifting process from both ends of the zone to its center? If the latter is true, we may presume an intensification of the rifting process and the formation of more outstanding surface features along the rift zone within the Rhenish Massif in the geological future.

Besides the prevailing tensional dip-slip dislocations, which indicate the crustal extension in SW-NE direction, some other types of focal mechanisms have been found within the Rhenish Massif (Fig. 11). The well-defined fault plane solution of the Bad Marienberg, Westerwald, earthquake on June 28, 1982 (M_L = 4.7), for instance, reveals a strike-slip dislocation along a N 83° E striking fault plane, combined with a minor tensional dip-slip component. Furthermore, focal mechanisms of the thrust-fault type (compressional dip-slip dislocation) along SW-NE-trending fault planes have been proven in the Taunus mountains and in the Hohes Venn. P-axes of these fault plane solutions are oriented NW-SE, thus indicating crustal compression and a corresponding stress field in this direction, which is transverse to the Hercynian fold axes of the Rhenish Massif.

Both stress directions resulting from focal mechanisms within the Rhenish Massif, the compressive stress in NW-SE direction, and the tensile stress in SW-NE direction, belong to the same general stress regime and are not contradicting each other. In situ stress measurements have proven a quite uniform regional stress field for the Rhenish Massif and its vicinity, characterized by mean directions (of horizontal projections) of the maximum compressive stress (σ_1) in NW-SE, and of the least compressive stress (σ_2) in SW-NE (Illies and Greiner, 1979; Illies and Baumann, 1982; Baumann, 1982; Baumann and Illies, 1983). In general, σ_1-directions from in situ stress measurements agree quite well with the orientation of P-axes from fault-plane solutions of earthquakes occurring in the same region (Fig. 11; Ahorner, 1975; Bonjer, 1980; Ahorner et al., 1983). The mode of seismotectonic dislocations and the overall pattern of seismicity are obviously controlled by the present-day regional stress field within the Earth's crust in Central Europe (Ahorner, 1975).

Seismicity and Plateau Uplift

From the foregoing chapters it becomes clear that the main features of the historical and present-day seismicity pattern of the Rhenish Massif are related to block movements along major crustal fracture zones, like the Rhine Graben system. Such deep-reaching zones of crustal weakness border crustal units which are comparatively stable in their interior from the seismotectonic point of view. Some minor seismicity within those "stable" blocks is possibly caused by local stress concentrations due to plateau uplift.

This might be true, for instance, for the Hunsrück-Taunus zone at the southern border of the Rhenish Massif, where epicenters are aligned along SW-NE-trending structural features, which are arranged parallel to the Hercynian fold axes. The most prominent of these structures is the deep-reaching fault zone at the southern border of the Hunsrück mountains, which has released earthquakes up to magnitude M_L = 4.1 (Ahorner and Murawski, 1975). Focal mechanisms of shocks prior to our microearthquake study seem to be of tensional dip-slip type with fault planes striking SW-NE. Because of the very low seismicity of this part of the Rhenish Massif during the study, well-defined fault plane solutions are lacking. Dip-slip mechanisms along the Hunsrück border fault would be consistent with a relative uplift of the Hunsrück block with respect to its southern foreland, although the amount of vertical movement is certainly very small, considering the low level of seismicity. Geodetical measurements do not show any significant upheaval of the Hunsrück-Taunus region during the last decades, if one refers to the "normal" southern foreland (Saar-Nahe region, Palatinate forest), outside of the active rift zone of the Upper Rhine Graben (Mälzer et al., 1979; Mälzer et al., 1983).

Maps of geodetical height changes for the Rhenish shield area, as compiled by Mälzer et al. (1983), and others, make clear that there is no en-block uplift of the Rhenish Massif as a whole in recent time. Large portions of the Massif, mainly in its southern and southeastern half part, seem to be rather stable or show even slight tendencies of relative subsidence. Only in a few regions, situated nearly exclusively in the northwestern and northern part of the Rhenish Massif, on both sides of the Lower Rhine Graben, significant positive height changes have been proven (Fig. 2; Mälzer and Zippelt, 1979; Mälzer et al., 1983). We find there the typical configuration of a subsiding active rift zone, flanked by two rising graben shoulders, as suggested by geodynamical models explaining taphrogenesis. The maximum uplift with present-day rates up to 1.6 mm yr^{-1} has been observed in the northwestern Eifel mountains and in the Hohes Venn, at the southwestern side of the Lower Rhine Graben. At the northeastern side, smaller uplift rates of some 0.8 mm yr^{-1} were determined, mainly in the western part of the Bergisches Land, but also in the Rhine valley between Cologne and Düsseldorf. The latter region belongs already to the northeastern part of the ancient Lower Rhine Graben, with a thick cover of Tertiary sediments overlying the Paleozoic basement. From the distribution of Quaternary structural activity and seismicity it becomes clear that this northeastern part of the Lower Rhine Graben is inactive in Quaternary and recent times and belongs not to the present-day active rift zone (Ahorner 1962, 1968).

The southwestern border of the active rift zone of the Lower Rhine Graben against the rising Eifel mountains and the Hohes Venn is formed by a NW-SE-trending structural zone showing strong differential crustal movements. The average vertical strain rate as deduced from geodetical measurements is in the order of 1 to 2 mm yr^{-1} (Quitzow and Vahlensieck, 1955; Ahorner, 1968). Major normal faults of Quaternary age are connected with this zone, e.g., the Feldbiß fault and Sandgewand fault in the region northeast of Aachen. The zone represents in fact the most prominent seismogenetic feature in the northern Rhine area. The majority of large historical earthquakes (e.g., Düren 1640, Gressenich 1755, Düren 1756, Herzogenrath 1877, Euskirchen 1951) has been released within this zone (Fig. 2). Judging from the fault-plane solution of the Euskirchen earthquake (M_L = 5.7, I_o = VII-VIII MSK; Fig. 12) and from smaller shocks in the region north of Aachen, tensional dip-slip mechanisms with NW-SE-striking fault-planes are typical for the zone, which is in agreement with the observed surface features. An estimation of the average seismotectonic slip rate resulting from earthquake activity since 1500 leads to about 0.1 mm yr^{-1}. This is only 5% to 10% of the geodetically observed strain rate. The seismotectonic slip rate has been calculated with the method of Brune (1968), as described earlier in this paper. If one accepts the reliability of geodetical and seismological results, then a large amount (about 90% to 95%) of the total strain, which is built up in a long-term process in the border zone between the graben and the rising Eifel mountains and the Hohes Venn, must be released with aseismic crustal movements (creep movement, etc.).

Questions left open up to now are: How far is the geodetically observed uplift of regions bordering the active rift zone of the Lower Rhine Graben caused by a crustal response to the nearby rifting process, and are there other primary causes of uplift? A definite answer is hardly to be given from seismological evidence alone. It is obvious, however, that the rising regions on both sides of the graben show a different seismotectonic behavior. The Bergisches Land at the northeastern side is practically aseismic, in terms both of the historical seismicity and of the present-day microseismicity. This might be an indication of a more "passive" crustal uplift in response to the nearby rifting. The northern Eifel mountains and the Hohes Venn at the southwestern side, on the contrary, exhibit remarkable historical and present-day seismicity with shocks up to magnitude M_L = 4.6 (Fig. 13). This suggests an active process of crustal uplift, which possibly superimposes a rifting-related component of uplift.

Most epicenters of the Hohes Venn are aligned along a broad, SW-NE trending zone, which follows approximately the axis of the Hercynian Stavelot-Venn anti-

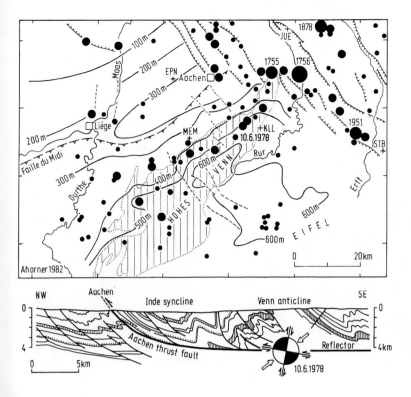

Fig. 13. Seismicity map 1500-1982 for the Hohes Venn and its vicinity (*above*) and a geological cross-section (*below*), showing the thrust fault as deduced from seismic reflection data (Meissner et al., 1981). The cross-section goes from Aachen to the SE. Earthquake epicenters are denoted in the map by *black dots*. *Dot size* varies with magnitude. Seismic stations are given by *crosses* and the *station code*. *Isolines* give the height of the base of Oligocene beds in meter above sea level as mapped by Albers (see Meyer et al., 1983). The hypocenter and focal mechanism of the Roetgen earthquake 1978 is plotted in the geological cross-section

cline, respectively a parallel striking hinge-zone of increased post-Oligocene uplift, coinciding with the northwestern flank of the anticline. Some secondary NW-SE-trending seismoactive features are poorly defined, e.g., the epicenter alignment along the Hockai fault zone between Verviers and Malmedy (south of station MEM in Fig. 13). The post-Oligocene uplift of the Hohes Venn is exceptionally large compared with other parts of the Rhenish Massif, and amounts more than 600 m with respect to the northern foreland as determined by Albers (see Meyer et al., 1983). If one corrects for eustatic sea-level fluctuations, at least 400 m true uplift remain.

The close correlation of the seismicity pattern of the Hohes Venn with geological surface features and geodetical results demonstrating the young uplift gives strong evidence for similarities of primary causes of both phenomena. The northern Eifel and the Hohes Venn might be a key area for a better understanding of geodynamic processes causing both plateau uplift and related seismicity. It is of interest in this connection that the large-scale mantle heterogeneity, found from teleseismic P-residuals be-

neath the western part of the Rhenish Massif (Raikes, 1980; Raikes and Bonjer, 1983), extends with its northernmost marginal portions beneath the Hohes Venn.

Due to the peripheral position of the Hohes Venn with respect to our station network, and the lacking of larger shocks during the observation period, only a few fault plane solutions are available for this interesting area. Two examples are shown in Fig. 14. The focal mechanism of the Roetgen earthquake on June 10, 1978 (M_L = 3.0; focal depth about 5 km) is a compressional dip-slip dislocation along a N 64° E-striking faultplane, which dips either with 10° towards SSE, or with 80° towards NNW. A composite fault-plane solution for six events occurring in 1975-1978 in the Hohes Venn shows practically the same type of focal mechanism. Thus compressional dip-slip mechanisms (thrust faults) seem to prevail in the region of increased uplift. The hypocenter of the Roetgen earthquake is located near a strong reflector in the upper crust, which has been found by Meissner et al. (1981, 1983) and interpreted as a prominent thrust fault, dipping gently towards SSE (Fig. 13). It is assumed

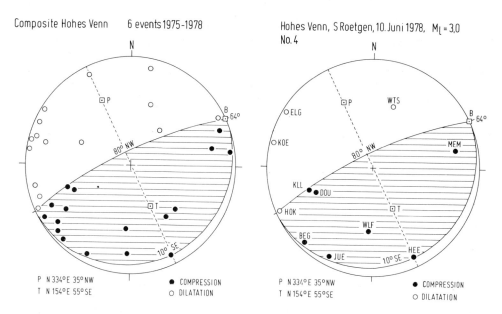

Fig. 14. Fault plane solutions of earthquakes in the Hohes Venn indicating compressional dip-slip dislocations (thrust faults). Equal-area projection of the lower focal hemisphere. Azimuth and plunge of pressure axis P, tension axis T, and null axis B are given

that a nearly horizontal nappe-displacement took place along this thrust-plane during the last stages of the Hercynian orogeny, shifting the overlying rocks of the Hohes Venn towards NNW. The outcrop of the main fault plane corresponds most probably to the Aachen thrust fault. Judging from the geological situation in the hypocentral region of the Roetgen earthquake, the NNE-dipping plane is probably the true fault plane. This interpretation implies a thrust dislocation of the overlying rocks towards NNW, as in case of the Aachen thrust fault. Hence an old fault zone seems to be reactivate during the focal process. This is not unlikely from the geomechanical point of view, because the present-day stress field within the Rhenish Massif is characterized by a NW-SE orientation of the maximum compressive stress, as described earlier, and this is the same direction as during the Hercynian orogeny, when the Aachen thrust fault was primarily formed.

A locally increased compressive deformation of the crust accompanied with seismotectonic thrust fault mechanisms would be able to explain, at least in parts, the increased uplift of the Hohes Venn. The adequate geomechanical model has to take into account that secondary thrust faults with curved fault planes (planes becoming steeper towards higher crustal levels) branch off from the main shear planes at depth, as shown in the geological cross-section (Fig. 13). Horizontal shear dislocations are in this way transformed to vertical movements near the surface. Of course, the few fault plane solutions available up to now, and the scarce information about the exact spatial distribution of hypocenters in the Hohes Venn area are not sufficient to verify the above model beyond doubt. Further investigations, based on an improved station network especially within the Hohes Venn, are needed to solve those problems.

Conclusions

The overall seismicity pattern of the Rhenish Massif, as investigated in this chapter, is clearly related to stress-field-controlled large-scale block movements along major crustal fracture zones like the Rhine Graben system, and shows no direct relationship to the plateau uplift of the Rhenish Massif. Only in some special regions, e.g., in the Hohes Venn in the northwestern part, where the uplift since Oligocene times has amounted more than 600 m and the present-day uplift rate is in the order of 1 mm yr^{-1}, evidence for a causal correlation between seismicity and crustal uplift is achieved.

From maps of geodetical height changes it becomes clear that the Rhenish Massif has undergone no en-bloc uplift during the last decades. Only smaller parts are significantly rising, whereas larger portions of the Rhenish Massif seem to be rather stable with respect to the environment. This is an important fact, if we look for an uplift-related seismicity. If the plateau uplift has come to a stand still in large parts of the Rhenish Massif, as indicated by geodetical measurements, then stress concentrations due to the uplift process and consequently uplift-related seismicity can not reasonably be expected in those regions. Geological and geomorphological methods, as applied today, are in general not able to decide whether the uplift of the shield as a whole or in parts is continuing during the Late Quaternary and Recent epoch or not. So the question is left open how long the geodetically suggested stand still of the uplift process has already lasted.

In summary, the following final conclusion can be drawn: Although the analysis of historical and present-day seismicity of the Rhenish Massif, as described in this paper, has failed in the aim to find out primary causes of a wide-spread plateau uplift, many new results concerning the neotectonic deformation of the shield area have been achieved, which give a detailed picture of geodynamic processes being still alive in Central Europe.

Acknowledgements. I dedicate this article to my academic teacher Professor Martin Schwarzbach (Bensberg) on the occasion of his 75th birthday. He tought me, many years ago, the basic principles of earthquake geology.

I am indebted to all those who have supported this research. In particular I mention Manfred Budny, Horst Loosen and Klaus Steuerwald from our institute, who helped to operate the microearthquake network and gave valuable assistance in routine data processing. Klaus-Peter Bonjer (Karlsruhe), Bodo Baier (Frankfurt), and other colleagues, kindly made data of their station networks available.

The research was supported by the German Research Society (DFG) within the priority program *Vertikalbewegungen und ihre Ursachen am Beispiel des Rheinischen Schildes* (Ah 13/3-13/9).

References

Ahorner, L., 1962. Untersuchungen zur quartären Bruchtektonik der Niederrheinischen Bucht. Eiszeitalter Ggw., 13:24-105.

Ahorner, L., 1968. Erdbeben und jüngste Tektonik im Braunkohlenrevier der Niederrheinischen Bucht. Z. Dtsch. Geol. Ges., 118:150-160.

Ahorner, L., 1970. Seismotectonic relations between the graben zones of the Upper and Lower Rhine valley. In: Illies, H.J. and Müller, St. (eds.) Graben problems. Schweizerbart, Stuttgart, pp.155-166.

Ahorner, L., 1975. Present-day stress field and seismotectonic block movements along major fault zones in Central Europe. In: Pavoni, N. and Green, R. (eds.) Recent crustal movements. Tectonophysics, 29:233-249.

Ahorner, L., Baier, B., and Bonjer, K.P., 1983. General pattern of seismotectonic dislocations and the earthquake generating stress field in the Rhenish Massif and its vicinity. This Vol.

Ahorner, L. and Murawski, H., 1975. Erdbebentätigkeit und geologischer Werdegang der Hunsrück-Südrand-Störung. Z. Dtsch. Geol. Ges., 126:63-82.

Ahorner, L. and Schneider, G., 1974. Herdmechanismen von Erdbeben im Oberrhein-Graben und in seinen Randgebirgen. In: Illies, H.J. and Fuchs, K. (eds.) Approaches to taphrogenesis. Schweizerbart, Stuttgart, pp.104-117.

Ahorner, L., Murawski, H., and Schneider, G., 1970. Die Verbreitung von schadenverursachenden Erdbeben auf dem Gebiet der Bundesrepublik Deutschland. Z. Geophys., 36:313-343.

Bartz, J., 1974. Die Mächtigkeit des Quartärs im Oberrheingraben. In: Illies, H.J. and Fuchs, K. (eds.) Approaches to taphrogenesis. Schweizerbart, Stuttgart, pp.78-87.

Baumann, H., 1982. Spannung und Spannungsumwandlung im Rheinischen Schiefergebirge. Forneck, Koblenz, 232 pp.

Baumann, H. and Illies, H.J., 1983. Stress field and strain release in the Rhenish Massif. This Vol.

Bonjer, K.-P., 1979. The seismicity of the Upper Rhinegraben. A continental rift system. Proc. Int. Res. Conf. Intra-Cont. Earthqu. Ohrid, Yugoslavia, 1979, pp.107-115.

Bonjer, K.-P. and Fuchs, K., 1974. Microearthquake activity observed by a seismic network in the Rhinegraben region. In: Illies, H.J. and Fuchs, K. (eds.) Approaches to taphrogenesis. Schweizerbart, Stuttgart, pp.99-104.

Bouckaert, J., Delmer, A., and Graulich, J.M., 1977. La structure varisque de l'Ardenne: essai d'interpretation. Meded. Rijks Geol. Dienst, 28:133-134.

Brace, W.F. and Byerlee, J.D., 1970. California earthquakes: Why only shallow focus? Science, 168:1573-1576.

Brune, J., 1968. Seismic moment, seismicity and rate of slip along major fault zones. J. Geophys. Res., 73:777-784.

Brune, J., 1970. Tectonic stress and spectra of seismic shear waves from earthquakes. J. Geophys. Res., 75:4997-5009.

Casten, U. and Cete, A., 1980. Induzierte Seismizität im Bereich des Steinkohlenbergbaus des Ruhrreviers. Glückauf-Forschungsh., 41:12-16.

Durst, H., 1981. Digitale Erfassung und Analyse der physikalischen Prozesse in den Erdbebenherden aus dem Bereich des Oberrheingrabens. Diploma Thesis, Univ. Karlsruhe, 123 pp.

Gils, van, J.-M. and Zaczek, Y., 1978. La séismicité de la Belgique et son application en génie parasismique. Ann. Trav. Publ. Belg., 6:1-38.

Haessler, H., Hoang Trong, P., Schick, R., Schneider, G., and Strobach, K., 1980. The September 3, 1978, Swabian Jura earthquake. Tectonophysics, 68:1-14.

Hiller, W., Rothé, J., and Schneider, G., 1967. La séismicité du fossé Rhénan. Abh. Geol. Landesamtes Baden-Württemb., 6:98-100.

Illies, H.J. and Baumann, H., 1982. Crustal dynamics and morphodynamics of the Western European rift system. Z. Geomorphol. N.F., Suppl. Bd., 42:135-165.

Illies, H.J. and Greiner, G., 1979. Holocene movements and the state of stress in the Rhinegraben rift system. Tectonophysics, 52:349-359.

Illies, H.J., Prodehl, C., Schmincke, H.-U., and Semmel, A., 1979. The Quaternary uplift of the Rhenish shield in Germany. In: McGetchin, T.R. and Merril, R.B. (eds.) Plateau uplift: mode and mechanism. Tectonophysics, 61:197-225.

Johnson, L.R. and McEvilly, T.V., 1974. Near-field observations and source parameters of Central California earthquakes. Bull. Seismol. Soc. Am., 64:1855-1886.

Lee, W.H.K. and Steward, S.W., 1981. Principles and applications of microearthquake networks. Academic Press, London, New York, 293 pp.

Leydecker, G., 1976. Der Gebirgsschlag vom 23.6.1975 im Kalibergbaugebiet des Werratales. Geol. Jahrb. Hess. Geol. Landesamt, 104:271-277.

Mälzer, H. and Zippelt, K., 1979. Local height changes in the Rhenish massif area. Allg. Vermessungsnachr., 86:402-405.

Mälzer, H., Schmitt, G., and Zippelt, K., 1979. Recent vertical movements and their determination in the Rhenish massif. Tectonophysics, 52:167-176.

Mälzer, H., Hein, H., and Zippelt, G., 1983. Height changes in the Rhenish Massif: Determination and analysis. This Vol.

Mechie, J., Prodehl, C., and Fuchs, K., 1983. The long-range seismic refraction experiment in the Rhenish Massif. This Vol.

Meissner, R., Bartelsen, H., and Murawski, H., 1981. Thin-skinned tectonics in the northern Rhenish Massif, Germany. Nature (London), 290:399-401.

Meissner, R., Springer, M., Murawski, H., Bartelsen, H., Fluh, E.R., and Dürschner, H., 1983. Combined seismic reflection-refraction investigations in the Rhenish Massif and their relation to recent tectonic movements. This Vol.

Meyer, W., 1979. Influence of the Hercynian structures on Cainozoic movements in the Rhenish Massif. Allg. Vermessungsnachr., 86:375-379.

Meyer, W., Albers, H.J., Berner, H.P., Gehlen, von, K., Glatthaar, D., Löhnertz, W., Pfeffer, K.H., Schnüttgen, A., Wienecke, K., and Zakosek, H., 1983. Pre-Quaternary uplift in the central part of the Rhenish Massif. This Vol.

Mooney, W.D. and Prodehl, C., 1978. Crustal structure of the Rhenish Massif and adjacent areas; a reinterpretation of existing seismic refraction data. J. Geophys., 44:573-601.

Prinz, H. and Schwarz, E., 1970. Nivellement and rezente tektonische Bewegungen im nördlichen Oberrheingraben. In: Illies, H.J. and Müller, St. (eds.) Graben problems. Schweizerbart, Stuttgart, pp.177-183.

Quitzow, H.W. and Vahlensieck, O., 1955. Über pleistozäne Gebirgsbildung und rezente Krustenbewegungen in der Niederrheinischen Bucht. Geol. Rundsch., 43:56-67.

Raikes, S., 1980. Teleseismic evidence for velocity heterogeneity beneath the Rhenish Massif. J. Geophys., 48: 80-83.

Raikes, S. and Bonjer, K.P., 1983. Large scale mantle heterogeneities beneath the Rhenish Massif and its vicinity from teleseismic P-residuals measurements. This Vol.

Sieberg, A., 1940. Beiträge zum Erdbebenkatalog Deutschlands und der angrenzenden Gebiete für die Jahre 58-1799. Mitt. Dtsch. Reichserdbebendienst., 2:1-111.

Schneider, G., Schick, R., and Berckhemer, H., 1966. Fault-plane solutions of earthquakes in Baden-Württemberg. Z. Geophys., 32:383-393.

Sponheuer, W., 1952. Erdbebenkatalog Deutschlands und der angrenzenden Gebiete für die Jahre 1800-1899. Mitt. Dtsch. Erdbebendienst, 3:1-195

Sponheuer, W., 1960. Methoden zur Herdtiefenbestimmung in der Makroseismik. Freiberg. Forschungsh., C 88:1-120.

Tucker, B.E. and Brune, J.N., 1973. Seismograms, S-wave spectra, and source parameters for aftershocks of the San Fernando earthquake. In: Murphy, L.M. (ed.) San Fernando, California, earthquake of February 9, 1971, Vol. III. Geological and geophysical studies. pp.69-121. U.S. Govt. Printing Office, Washington, D.C.

6.5 Microearthquake Activity near the Southern Border of the Rhenish Massif

B. Baier and J. Wernig[1]

Abstract

Microearthquakes in the southern part of the Rhenish Massif and the adjacent areas have been observed during a period of 5 years. Epicentres show that historic activity is being continued into the present. New microearthquake activity seems to start in the Hessian Depression. Hypocentre sections show different maximum focal depths near the southern border of the Rhenish Massif and beneath it. Stress directions from fault plane solutions of well-recorded events are partly in good agreement with the results of in situ stress measurements (max. horizontal compressiv stress: average direction 145°).

The different types of fault plane solutions (Fig. 2) reflect the particular seismotectonic situation and indicate the region at a zone of shear blockade (Illies et al., 1981) and neotectonic fault action (Semmel, 1978).

Introduction

The microearthquake activity near the southern border of the Rhenish Massif was studied in the period between 1977 and 1981. The data used were recorded by the Taunus observatory and three temporary stations with ink recording systems on folded paper developed in the Frankfurt Institute of Geophysics. Since 1980 the network has been supplemented by Longterm Magnetic Recording instruments (MLR 78, unpubl.). During these 5 years, more than 100 microearthquakes were observed. We recorded 63 events on 4 or more stations and we were able to locate these events with reasonable (± 4 km) accuracy. Figure 1 shows the seismic network and epicentres of the recorded events. The localizations were made with the computer program HYPGRD (Gelbke, 1977). We used crustal models derived from reinterpretation of explosion seismic data (Mooney and Prodehl, 1978). For identification of the various phases in the seismograms, travel time curves were plotted for the two best crustal models, which were tested by observed travel times from a quarry blast.

Data Interpretation

A comparison of calculated epicentres of the present activity and the locations of historical events (Leydecker, unpubl. in this region shows that areas active in historical time are obviously still active (Fig. 1). An exception is the northernmost part of the Upper Rhine Graben with its NNE-striking branch, the Hessian Depression. It was long considered to be a stable (aseismic) region (Ahorner, 1970), since no earthquake activity was recorded until November 4, 1975, when the Echzell earthquak occurred (Neugebauer and Tobias, 1977).

---

[1] Institut für Meteorologie und Geophysik, Universität Frankfurt, Feldbergstraße 47, D-6000 Frankfurt, Fed. Rep. of Germany.

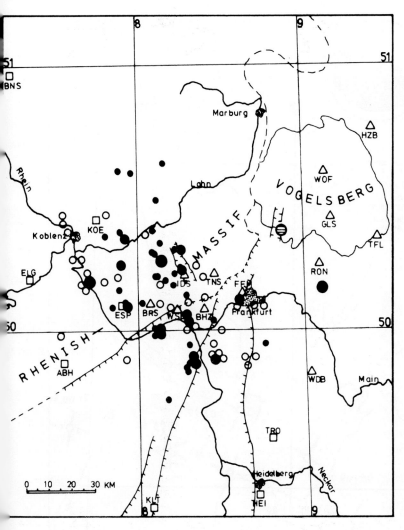

Fig. 1. Local seismic station network with the distribution of earthquake epicenters observed within 1977-1981; and historical events within 1400-1977

In the period of observation only a few earthquakes with intensities $I_o > 4$ ($M_L > 3$) were recorded by our network, as well as at the adjoining seismic stations. Locations and fault plane solutions of these events are shown in Fig. 2. As the result of fault plane solutions the horizontal projection of B-axes, T-(tension) or P-(pressure) directions are drawn respectively. The figure shows that not all of the fault plane solutions are in accordance with the general stress field. The microearthquakes in the southern part of the Rhenish Massif (Taunus mountains) have pressure axes oriented within the range of 139° to 162°. In general they follow the 142° (±20°) direction, obtained in earlier fault plane solutions (Ahorner, 1975) and the 138° (±16°) direction found by in situ stress measurements (Baumann, 1982), which are almost the same direction (average 145°, Ahorner et al., 1983). Events on the boundary between the Rhenish Massif and Upper Rhine Graben do not agree with the regional stress pattern. This is not surprising, because the complicated tectonic system splits the Upper Rhine Graben - starting at its northern end - in three directions: the lower Rhine Embayment, the Hessian Depression and in between a small rupture zone called the Idstein Depression. The deviation of the local pressure axes of maximum compression from the trend of the recent stress field indicates that this area may be an interference zone (zone of "shear blockade" Illies et al., 1981), where the seismic belt of the Upper Rhine Graben deflects in NW direction and where also a new seismic activity starts in the southern Hessian Depression. The

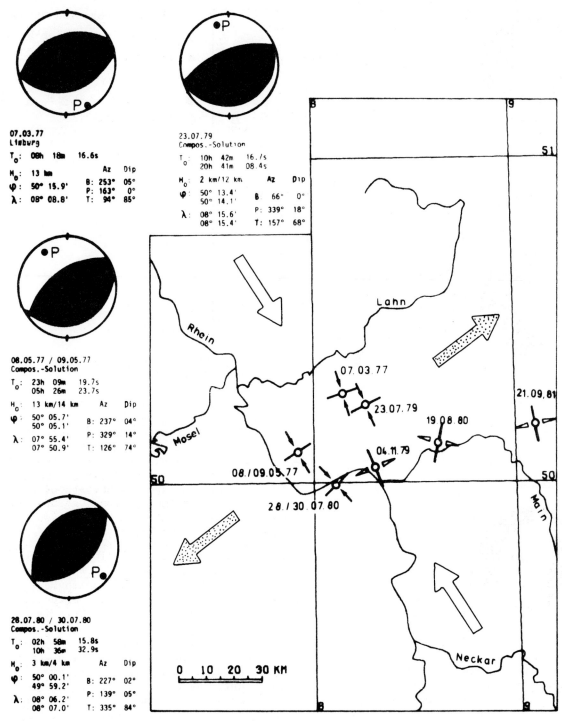

Fig. 2. Focal mechanisms of earthquakes $M_L \geq 3$ stereographic projections of the lower focal hemisphere. *White quadrants* dilatational first motion of p-wave; *black quadrants* compressional motion.
Orientation of B-(null) axes with max. horizontal stress directions

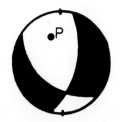

21.09.81
Hanau

T_0: 21h 32m 18.2s
H_0: 8 km
φ: 50° 10.1'
λ: 09° 06.0'

	Az	Dip
B:	171°	39°
P:	135°	49°
T:	77°	14°

19.08.80
Frankfurt

T_0: 13h 25m 19.6s
H_0: 6 km
φ: 50° 07.0'
λ: 08° 36.9'

	Az	Dip
B:	202°	10°
P:	326°	72°
T:	110°	17°

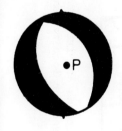

04.11.79
Wiesbaden

T_0: 02h 24m 58.5s
H_0: 7 km
φ: 50° 02.6'
λ: 08° 18.1'

	Az	Dip
B:	336°	08°
P:	106°	83°
T:	244°	07°

Fig. 2 (continued)

four fault plane solutions in the Taunus mountains with the NW oriented P-axes are reflecting well the general regional stress field. The mode of slip motion, however, is a thrust type with considerable updip component which might contribute to the general uplift movement of the Rhenish Massif in Neogene and present time. The calculated focal depths of analyzed earthquakes are shown in three vertical sections (Fig. 3). From section CD and EF it appears that the base of the hypocentre distribution is rising from a depth of 15-20 km in the middle part of the Upper Rhine Graben to a depth of 10 km near the southern border of the Taunus mountains and increasing again beneath the central part of the Rhenish Massif, where hypocentres are found at depth up to 25 km (section EF).

Conclusions

From the interpretation of microearthquake seismicity near the southern border of the Rhenish Massif we conclude:
Recent microearthquake activity is continuing in those areas where historical earthquakes have taken place. New seismic activity is observed in the southern Hessian Depression which was considered an inactive region in the recent past (Ahorner 1975; Bonjer and Gelbke, 1976). From the existance of a nearly homogeneous stress pattern in central Europe, and from the calculated, thrust fault plane solutions in the Taunus mountains, one can deduce that permanent pressure from the Southwest German Block must be present. The absence of hypocentre depths of more than 10 km in the southern border zone of the Rhenish Massif may reflect ductile behavior of tectonically incompetent material in the middle part of the crust and the inability to accumulate sufficient stress to produce earthquakes (Illies et al., 1979). This is also the zone of maximum relative vertical movements, i.e., uplift in the Rhenish Massif and subsidence at the northern end of the Upper Rhine Graben. Elevated temperature in the incompetent part of the middle crust is also suggested by the observation that the minimum in the hypocentre depth distribution coincides with the zone of strong hydrothermal activity along the southern rim of the Taunus mountains (Berckhemer, 1967; Jacob and Heintke, 1969).

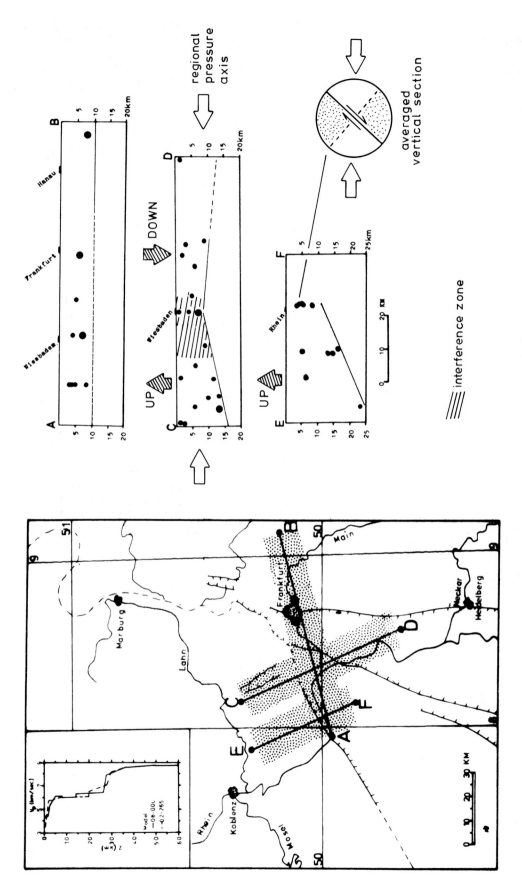

Fig. 3a. Profiles of vertical sections and its band-width for b and crustal models used for localization

Fig. 3b. Depth distribution of earthquake forci (lines in the sections reflect max. hypocentre depths)

Acknowledgements. The authors wish to thank all the collaborators who participated in data collection and the owners of the recording sites, who gave the permission to install the stations and partly assisted in the station service. Thanks are also expressed to Prof. Berckhemer for valuable assistance, Prof. Fuchs and Dr. Zucca for critical reading of the manuscript. The research work was supported by the Deutsche Forschungsgemeinschaft within the priority program *Vertikalbewegungen und ihre Ursachen am Beispiel des Rheinischen Schildes.*

References

Ahorner, L., 1970. Seismo-tectonic relations between the graben zones of the upper and lower Rhine Valley. In: Illies, J.H. and Mueller, St. (eds.) Graben problems. Schweizerbart, Stuttgart, pp.155-166.

Ahorner, L., 1975. Present-day stress-field and seismotectonic block movements along major fault zones in central Europe. Tectonophysics, 29: 233-249.

Ahorner, L., et al., 1983. General pattern of seismotectonic dislocations and the earthquake-generating stress field in Central Europe between the Alps and the North Sea. This Vol.

Baumann, H., 1982. Spannung und Spannungsumwandlung im Rheinischen Schiefergebirge. Numismatischer Verlag, Koblenz.

Berckhemer, H., 1967. Die Erdstöße in Wiesbaden am 04. Januar 1967. Notizbl. Hess. Landesamtes Bodenforsch., 95: 213-216.

Bonjer, K.-P. and Gelbke, C., 1976. Seismizität und Dynamik im Bereich des Oberrheingrabens (Abstract). 36. Jahrestag. Dt. Geophys. Ges., Bochum.

Gelbke, K.-P., 1977. Erweiterung des Erdbebenlokalisierungsprogramms HYPO71 (Lee and Lahr, US Geological Survey, 1971) um eine Laufzeitroutine für Mehrschichtenmodelle konstanter positiver und negativer Geschwindigkeitsgradienten. Geophys. Inst. Univ. Karlsruhe.

Illies, J.H., Prodehl, Cl., Schmincke, H.-U., and Semmel, A., 1979. The Quaternary uplift of the Rhenish shield in Germany. Tectonophysics, 61:197-225.

Illies, J.H., Baumann, H., and Hoffers, B., 1981. Stress pattern and strain release in the Alpine foreland. Tectonophysics, 71:157-172.

Jacob, K. and Heintke, H., 1969. Das Lorsbacher Beben vom 21. Juli 1968. Notizbl. Hess. Landesamtes Bodenforsch., 97:379-385.

Mooney, W.D. and Prodehl, C., 1978. Crustal structure of the Rhenish Massif and adjacent areas; a Reinterpretation of existing Seismic Refraction Data. J. Geophys., 44:573-601.

Neugebauer, H.J. and Tobias, E., 1977. A study of the Echzell/Wetterau Earthquake of November 4, 1975. J. Geophys., 43:751-760.

Semmel, A., 1978. Untersuchungen zur quartären Tektonik am Taunus-Südrand. Geol. Jahrb. Hessen, 106:291-302.

6.6 Geothermal Investigations in the Rhenish Massif

R. Haenel[1]

Abstract

The density, the specific heat capacity, the thermal diffusivity and the heat production have been determined for sedimentary, acid and basic rocks, and the thermal conductivity as a function of temperature.

In addition, 14 new heat-flow density determinations have been made in shallow boreholes. To eliminate the palaeoclimatic effect, palaeotemperatures have been reconstructed. This leads to an increase of the mean heat-flow density for Germany from 70 to 81 mW m^{-2}. A new heat-flow density map has been constructed which shows that the isolines have Hercynian trend within the Rhenish Massif.

Maps of temperature distribution at 10 km and 30 km, as well as a temperature graph down to 100 km depth were derived from the above data. The mean temperature at 30 km depth in the Rhenish Massif averages up to 800°C.

Model calculations show no significant increase in heat-flow density, either in the case of reasonable amounts of uplift, or over a large igneous body intruded more than 12 Ma ago, or for the remaining magma beneath the Laacher See volcano which erupted about 10,000 years ago. The heat of the latter has not yet reached the Earth's surface.

The calculated amount of uplift can be explained by thermal expansion of rock caused by intrusions.

Introduction

The geodynamics of the Rhenish Massif has been studied in an integrated research programme involving different geoscience disciplines including geothermics. The aims of the geothermal investigations were

- determination of geothermal rock parameters,
- heat-flow density determination in shallow boreholes,
- assessment of subsurface temperatures, and
- construction of mathematical models of the geothermal behavior of intrusive igneous rocks.

The results of the geothermal study presented here are a summary of a detailed report (Haenel 1983).

Geothermal Parameters

In order to make estimates of temperature at various depths and also to set up model calculations, a knowledge is required of the geothermal parameters of the area under consideration. The collection of relevant data for rocks in Germany (Kappelmeyer and Haenel, 1974) has been completed (see Tables 1 and 2). Furthermore, the thermal conductivity λ

---

[1] Niedersächsisches Landesamt für Bodenforschung, Postfach 510153, D-3000 Hannover 51, Fed. Rep. of Germany.

Table 1. Thermal parameters of rock samples collected in the Federal Republic of Germany; measurements made at 50°C, standard deviations given; λ = thermal conductivity, c = specific heat capacity, ρ = density, $\varkappa$ = thermal diffusivity

Formation	Number of Samples	λ (W m^{-1} K^{-1})	c (J kg^{-1} K^{-1})	ρ (10^3 kg m^{-3})	$\varkappa$ (10^{-6} m^2 s^{-1})
Tertiary	8	2.02 ± 0.79	816 ± 42	2.48 ± 0.06	1.00 ± 0.25
Cretaceous	5	2.16 ± 0.31	943 ± 42	2.56 ± 0.06	0.90 ± 0.15
Jurassic	42	2.74 ± 0.57	917 ± 50	2.62 ± 0.09	1.14 ± 0.21
Triassic	23	2.78 ± 0.44	883 ± 105	2.38 ± 0.17	1.35 ± 0.28
Permian	2	3.45	840[a]	2.56	1.60
Carboniferous	67	3.09 ± 0.85	871 ± 54	2.68 ± 0.10	1.40 ± 0.45
Devonian	32	2.83 ± 0.80	840[a]	2.59 ± 0.15	1.30
Middle	26	2.86 ± 0.84	840[a]	2.57 ± 0.15	1.32
Lower	6	2.70 ± 0.64	840[a]	2.69 ± 0.09	1.19
Silurian	4	3.02 ± 0.97	840[a]	2.67 ± 0.13	1.35

[a] Estimated value.

Table 2. Heat production and density of rocks in the Federal Republic of Germany with standard deviations

Material	Location	Number of Samples	ρ (10^3 kg m^{-3})	H (μW m^{-3})
Sediments:				
Triassic	Urach	25	2.64 ± 0.11	1.12 ± 0.87
Devonian	Rhenish Massif	5		1.02 ± 0.42
Gneiss	Urach	11	2.72 ± 0.04	2.70 ± 1.50
Syenite	Urach	4	2.76 ± 0.02	3.35 ± 0.21
Diorite	Odenwald	4	2.77 ± 0.12	1.54 ± 0.72
Granite	Odenwald	1	2.65	1.82
Gabbro	Odenwald	2	2.88	0.08

has been determined as a function of temperature (Table 3, Fig. 1). All rock samples which were measured during falling temperature gave different results (see Fig. 2) caused by changes in water content, by partial melting and by other factors. However, Fig. 1 shows some visually derived mean functions $\lambda = f(T)$ which can be expressed as it follows:

Rhenish sediments:

$$\lambda(T) = \frac{600}{350 + T} + 1.12 \quad 0 > T < 300°C \quad (1)$$

Acid rocks:

$$\lambda(T) = \frac{840}{350 + T} + 0.40 \quad 0 > T < 800°C \quad (2)$$

Basic rocks:

$$\lambda(T) = \frac{350}{350 + T} + 1.24 \quad 0 > T < 1000°C \quad (3)$$

Ultrabasic rocks:
(olivine)

$$\lambda(T) = \frac{1300}{350 + T} + 0.88 \quad 500 > T < 1600°C \quad (4)$$

where T is in °C and λ in W m^{-1} K^{-1}. Equation (4) is derived from Roy et al. (1981, Fig. 12.75).

Figure 3 gives parameters for the crust and upper mantle beneath the Federal Republic of Germany and the Rhenish Massif, which are used in the calculations.

Table 3. Temperature dependence of thermal conductivity of different rocks in $W\,m^{-1}\,K^{-1}$. Temperatures in $°C$.

No.	Location	Rock type	Density ($10^3\,kg\,m^{-3}$)	Thermal conductivity ($W\,m^{-1}\,K^{-1}$)								
				50	100	200	300	400	500	600	700	800
1	Hirzenhain, 137 m	Basalt	2.89		1.73	1.72	1.70	1.67	1.62	1.56	1.46	1.34
2	Rhen. Massif, No. 12	Slate	2.81	1.53	1.48	1.38	1.27	1.16	1.05	0.95	0.84	
3	Düppenweiler, Ia 35 m	Phyllite	2.50	3.17	2.94	2.50	2.18	1.96	1.86	1.20		
4	Düppenweiler, Ia 104 m	Graywacke	2.47	2.18	2.00	1.84	1.68	1.52	1.57	1.78		
5	Düppenweiler, Ia 165 m	Claystone	2.60		3.50	2.98	2.60	2.28	2.02			
6	Rhen. Massif, No. 17	Quarzitic Sandstone	2.70		3.51	2.92	2.51	2.17	1.87	1.61		
7	Schö T9, 2771,3 m	Marly clay	2.66		2.46	2.10	1.91	1.79	1.71	1.67		
8	Schö T9, 2769, 7 m	Marly clay	2.67		1.63	1.48	1.39	1.32	1.29			
9	Rhen. Massif, No. 9	Clay slate	2.70	3.20	2.90	2.48	2.24	2.04	1.89	1.77	1.66	1.58
10	Rhen. Massif, No. 4	Clay slate	2.74	2.80	2.71	2.53	2.36	2.28	2.15	2.01		
11	Gablingen	Granite	2.58	2.78	2.64	2.36	2.08	1.81	1.57	1.33	1.10	
12	Odenwald, No. 8	Granite	2.62	2.98	2.76	2.41	2.13	1.90	1.71	1.52		
13	Gaisbeuren	Metamorph. rock	2.62	2.25	2.15	1.93	1.73	1.50	1.29	1.10	1.26	1.48
14	Odenwald, No. 2	Gabbro	2.86	2.21	2.15	2.04	1.97	1.92	1.90	1.88		

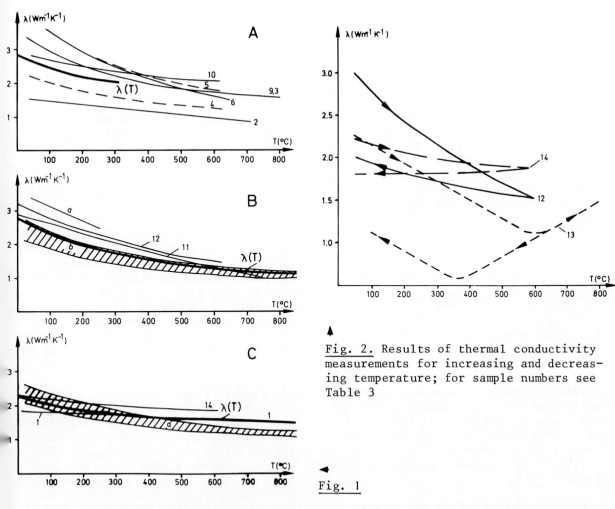

Fig. 2. Results of thermal conductivity measurements for increasing and decreasing temperature; for sample numbers see Table 3

Fig. 1. Temperature dependence of thermal conductivity; for sample numbers see Table 2. *A* Sediments of the Rhenish Massif; *B* Acid rocks; a Rockport granite (Roy et al. 1981); b Barre and Westerly granite (Roy et al. 1981); *C* Basic rocks; a basalt (Roy et al. 1981); $\lambda(T)$ visually estimated mean curve

	Germany excl. Rh. Massif	Rhenish Massif
Earth's surface	T_0	T_0
Sediments	λ from Tab.1	$\lambda = \frac{600}{350+T} + 1.12$
	H = 1.13	H = 1.02
Acid rock		$\lambda = \frac{840}{350+T} + 0.4$
		H = 2.92
Basic rock		$\lambda = \frac{350}{350+T} + 1.24$
		H = 0.46
Mohorovicic-Disc.		
Ultrabasic rock		$\lambda = \frac{1300}{350+T} + 0.88$
		H = 0.01

Fig. 3. Schematic model of the Earth's crust and upper mantle as used in this paper and the relevant geothermal parameters λ in W m^{-1} K^{-1} and H in µW m^{-3}

Voll (1983), Mengel and Wedepohl (1983), and Seck and Wedepohl (1983) point out that the crust below the sediments is predominantly acid in character, consisting of gneisses, phyllites and chlorite schists. For 2 or 3 km above the Mohorovičič discontinuity, the crust probably comprises gneisses, mainly granulites with bands of basic material. Since no geothermal data are available for these rocks in Germany, the model shown in Fig. 3 will be used. The heat-production values H are taken from Haenel (1979) and from Kappelmeyer and Haenel (1974) and Table 2.

Heat-Flow Density Determination and Temperature Representation

Heat-flow density determinations have been carried out in 14 shallow boreholes of about 50 m depth in the Rhenish Massif. The effect of the topography is eliminated using the method described by Bullard (1940) and Haenel (1971), and the necessary vertical components of thermal conductivity for heat-flow density calculations were determined according to the method described by Grubbe et al. (1983).

Another influence on the present-day heat-flow density which cannot be neglected is the temperature change at the Earth's surface during the past (palaeotemperature). The variation of palaeotemperature has been reconstructed for Germany and is given in Table 4 and Fig. 4 (Grube, 1981). It is based on Table 5.

The greatest influence of palaeotemperatures on the present-day heat-flow density can be observed down to about 25 m. The heat-flow density determinations have therefore only been carried out between 30 and 50 m depth in the above-mentioned shallow boreholes. If the correction from the data for 100,000 years ago are taken as being 100%, then the correction based on the last 380 years contributes about 60% and is reliable due to reasonably reliable temperature data. Assuming an accuracy in palaeotemperatures of only 60% between 380 and 100,000 years, then the total error of the correction is about 25%. From these estimates, a minimum depth of 100 m is recommended for future shallow boreholes, provided there is no subsurface water circulation.

The palaeoclimatic effect has been eliminated by using the formula (Kappelmeyer and Haenel, 1974):

$$q(z) = q_o - \sum_{n=1}^{n=\infty} \frac{\lambda \Delta T}{\sqrt{\pi \varkappa t_n}} \exp\left(-\frac{z^2}{4 \varkappa t_n}\right) \qquad (5)$$

where q_o = heat-flow density influenced by palaeotemperature, λ = thermal conductivity, $\varkappa$ = thermal diffusivity, ΔT = temperature difference caused by cold or warm periods at time t, and z = depth. By applying this correction the mean heat-flow density for the Federal Republic of Germany is increased from q_o = 69.8 mW m^{-2} to q = 81.2 mW m^{-2}; that is by about 17%.

All available heat-flow density values for Germany are listed in Table 5 and have been plotted in Fig. 6. This facilitates consideration and evaluation of the data from the Rhenish Massif. The isoline pattern within the massif has a more or less Hercynian trend. A distribution pattern exists which is also known from the Alps (Haenel and Zoth, 1982): high values in the northern and southern region and low values in the centre. As discussed in Haenel and Zoth (1982), the anomalous areas may be explained by upward migrating water, and/or by differences in the lateral distribution of thermal conductivity and of heat production by radioactive elements compared with the area surrounding the Rhenish Massif. Both causes must be sought in the crust. The numerous springs along the southern margin of the Rhenish Massif indicate water rising along the main fault system.

Subsurface temperatures can be calculated for a layered model of the Earth (Fig. 3) using the above heat-flow density values, assuming homogeneous layers. Thus at the base of the nth layer:

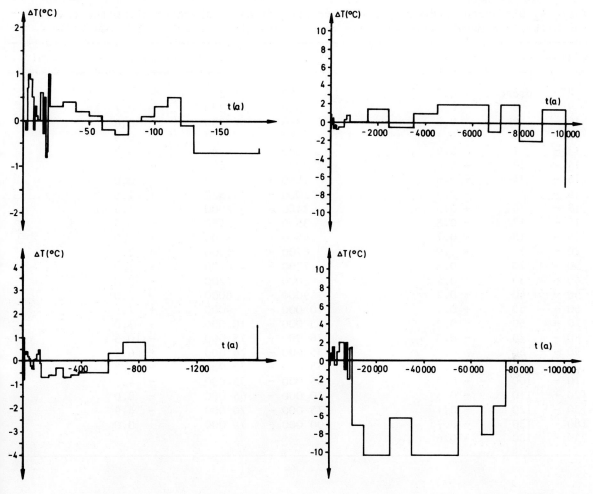

Fig. 4. Differences between palaeo- and present-day temperatures for the Federal Republic of Germany (see Table 4)

$$T_n(z) = T_o + \sum_{i=1}^{i=n} \left\{ \frac{z_i - z_{i-1}}{\lambda_i(T)} \left[q_o - \sum_{j=1}^{j=i-1} H_j(z_j - z_{j-1}) \right] - H_i \frac{(z_i - z_{i-1})^2}{2\lambda_i(T)} \right\} , \qquad (6)$$

where $z_o = H_o = 0$, and T_o = temperature and q_o = heat-flow density, both at the Earth's surface.

The crustal model employed here is based on data from deep boreholes, seismic work (Giese et al., 1976), and for the Rhenish Massif from Mechie et al. (1983) and from the geothermal parameters shown in Fig. 3.

In Figs. 6 and 7 the temperatures are represented at 10 km and 30 km depth. The isoline pattern is similar to that of the heat-flow density distribution (Fig. 5). The limits of error of the temperatures increase with depth. This is due to the inaccuracy of the geothermal parameters, which also increases with depth.

In Fig. 8 a temperature profile for the Rhenish Massif has been constructed down to 100 km. From this it follows that the rock may be partially molten at depths in excess of about 30 km. In the case

Table 4. Difference between palaeo- and present-day temperatures for the Federal Republic of Germany; t = time, ΔT = temperature difference (see Fig. 4)

t(a)		ΔT (°C)	t(a)		ΔT (°C)
8 –	present	+ 0.5	280 –	230	– 0.3
9 –	8	– 0.2	330 –	280	– 0.7
10 –	9	+ 0.3	380 –	330	– 0.6
11 –	10	+ 0.1	580 –	380	– 0.5
13 –	11	0.0	680 –	580	+ 0.3
14 –	13	+ 0.6	830 –	680	+ 0.8
15 –	14	+ 0.6	1580 –	830	0.0
16 –	15	– 0.3	2000 –	1580	+ 1.5
17 –	16	+ 0.5	2480 –	2000	+ 1.5
18 –	17	– 0.8	3500 –	2480	– 0.5
19 –	18	– 0.7	4500 –	3500	+ 1.0
20 –	19	+ 1.0	6700 –	4500	+ 2.0
30 –	20	+ 0.3	7200 –	6700	– 1.0
40 –	30	+ 0.4	8000 –	7200	+ 2.0
50 –	40	+ 0.2	9000 –	8000	– 2.0
60 –	50	+ 0.1	10 000 –	9000	+ 1.5
70 –	60	– 0.2	15 000 –	10 000	– 7.0
80 –	70	– 0.3	26 000 –	15 000	– 10.3
90 –	80	0.0	35 000 –	26 000	– 6.2
100 –	90	+ 0.1	55 000 –	35 000	– 10.3
110 –	100	+ 0.3	65 000 –	55 000	– 4.9
120 –	110	+ 0.5	70 000 –	65 000	– 8.0
130 –	120	– 0.1	75 000 –	70 000	– 4.9
180 –	130	– 0.7	100 000 –	75 000	0.0
230 –	180	– 0.6			

Table 5.

Reconstruction based on	For the period: present day to (in years)	Percent contribution towards correction of heat-flow density
Temperature measurements made by the meteorological station at Trier (Rhenish Massif)	– 150	44
Other temperature measurements, Rudloff (1967)	– 280	55
Analysis of tree rings of German oaks, Libby and Lamb (see Lamb 1977)	– 380	59
Various sources, e.g., Lamb (1977)	– 10 000	51
Data from Frenzel (1980)	– 100 000	100

Table 6. Heat-flow density data for the Federal Republic of Germany; q = determined heat-flow density including topographic correction, q_k = heat-flow density including topographic correction, also corrected for palaeoclimatic effect; the recently determined heat-flow density values are listed under numbers 76 and 78-90

No.	Latit.	Longit.	Depth	q	q_k	Location
1	52-11	10-24	1200	70.8	73.8	Konrad
2	50-45	7-56	830	57.8	64.5	Fuessenberg
3	50-34	7-31	800	57.8	65.1	Georg
4	49-24	12-10	230	72.0	85.2	Marienschacht
5	47-51	7-38	1010	69.9	79.1	Buggingen
6	47-47	11- 3	1000	79.0	79.3	Peissenberg
7	51-36	10- 0	750	55.7	67.1	Koenigshall
8	50-23	9- 7	610	70.4	81.3	Hirzenhain
9	48-52	12- 2	50	63.7	86.7	Waeldern
10	49-46	8-34	625	69.1	72.5	Haehnlein
11	49-35	11-37	600	62.4	75.2	Eschenfelden
12	49-26	12- 6	409	65.8	78.4	Naaburg
13	49- 6	9- 4	216	101.8	115.0	Heuchelberg
14	48-51	13-23	530	76.3	86.5	Grafenau
15	48-43	8-40	200	70.4	83.6	Oberreichenbach
16	48-40	8-39	110	72.5	96.0	Bad Teinach
17	48-29	10- 8	194	71.2	83.2	Musismuehle-Ulm
18	48-25	9-41	152	70.8	85.8	Tiefental-Ulm
19	48-20	9-56	445	88.0	97.2	Donaustetten-Ulm
20	47-45	8-52	600	69.1	79.7	Singen
21	49-15	8- 8	1600	110.6	109.8	Landau 1
22	49-13	8- 9	1600	139.2	139.9	Landau 2
23	49-15	8- 8	1300	116.4	117.8	Landau 3
24	50-20	7-22	700	62.2	68.3	Ochtendung
25	51- 5	6-25	350	60.0	67.8	Jackerather Horst
26	47-57	11-20	- 60	96.7	100.6	Starnberger See Nord
27	47-54	11-18	-115	81.7	91.0	Starnberger See Mitte
28	47-51	11-20	-100	81.7	86.6	Starnberger See Süd
29	47-35	11-21	-193	75.8	80.3	Walchensee
30	47-31	12-57	-602	76.2	80.8	Königssee
31	49-10	7- 5	707	59.0	64.8	Maybach
32	49-20	7- 4	900	65.8	72.0	Jungenwald
33	49-18	6-59	665	67.8	74.9	Netzbach
34	5-20	7-53	85	55.3	68.3	Hoever
35	48-36	10-18	175	82.1	96.2	Sachsenhausen
36	50-22	9-15	488	72.9	84.1	Roess Gesaess
37	47-33	10-43	- 38	88.0	91.5	Alpsee
38	47-59	11- 2	- 69	71.6	75.7	Ammersee
39	47-42	9-13	-138	64.0	67.8	Bodensee 1
40	47-39	9-16	-195	65.0	69.8	Bodensee 2
41	47-37	9-22	-250	80.0	84.8	Bodensee 3
42	47-32	9-29	-231	79.0	83.7	Bodensee 4
43	47-33	9-32	-230	77.0	81.6	Bodensee 5
44	47-28	9-38	-230	57.0	60.0	Bodensee 6
45	47-52	8- 2	- 32	64.9	68.8	Feldsee
46	51-39	10-20	- 28	58.6	64.0	Juessee

Table 6. (cont.)

No.	Latit.	Longit.	Depth	q	q_k	Location
47	50-24	7-16	- 48	72.9	75.7	Laacher See
48	50- 8	6-55	- 69	61.1	66.0	Pulver Maar
49	47-43	11-51	- 39	64.0	68.4	Schliersee
50	50-10	6-51	- 49	60.3	64.0	Weinfelder Maar
51	52- 0	7-20	5950	58.2	75.8	Münsterland
52	53-45	10- 5	319	46.9	53.9	Fahrenhorst
53	53-40	9-43	230	52.4	59.4	Appen
54	53-39	10- 1	505	48.8	56.1	Langenhorn
55	48-57	11-41	365	75.4	87.1	Riedenburg
56	49-50	6-38	190	63.7	78.5	Kylltal
57	50- 4	12- 9	115	59.5	77.3	Alexandersbad
58	50-47	8-45	500	69.3	78.2	Buchenau
59	50-48	8-53	730	72.4	79.4	Wehrhausen
60	49-50	12- 0	350	78.0	91.8	Falkenberg
61	48-50	10-38	100	80.7	100.7	Wörnitzostheim
62	48-53	10-30	1200	73.2	77.9	Nördlingen
63	50- 6	11-41	231	56.7	67.2	Weissenstein
64	48-30	9-22	3300	86.0	85.0	Urach 3
65	47-47	8-46	220	63.6	69.8	Hegau
66	49- 2	10-22	560	81.5	93.1	Dinkelsbühl
67	50-11	6-50	- 50	61.0	65.0	Gemündener Maar
68	48- 4	9- 2	52	70.2	88.2	Hausen im Tal
69	48 - 5	8-54	54	66.8	75.9	Baeratal
70	48-47	9-40	52	80.3	93.3	Lorch
71	48-22	9-37	81	78.0	95.2	Gundershofen
72	48- 8	9-29	1200	88.0	95.3	Saulgau
73	51-16	6-53	553	54.0	68.2	Schwarzbachtal
74	48-42	8- 7	2537	116.0	112.0	Bühl I
75	49-48	7- 6	50	54.0	85.0	Morbach
76	49-59	7-40	50	64.0	96.0	Rheinböllen
77	50- 8	8- 8	51	47.0	70.0	Bleidenstadt
78	50-18	8-35	46	31.0	56.0	Wehrheim
79	50-25	7-50	50	52.0	81.0	Montabaur
80	50-38	8-18	50	66.0	92.0	Fleischbach
81	50-32	6-23	51	32.0	61.0	Dreiborn
82	50-25	6-33	50	51.0	7.60	Schmidtheim
83	50-17	6-42	60	56.0	81.0	Hillesheim
84	50-36	6-56	48	59.0	89.0	Rheinbach
85	50-47	7-35	51	76.0	104.0	Helpenstell
86	50-53	8- 7	51	55.0	86.0	Deuz

both anomalous areas in the Rhenish Massif are not taken into consideration with regard to one of the above-mentioned reasons, the normal heat-flow density in the Rhenish Massif would be between 65-70 mW m^{-2}. This leads to the mean temperature graph A' in Fig. 8.

The variation of thermal conductivity with depth is derived from Fig. 8 and Eqs. (1-4) and is shown in Fig. 9.

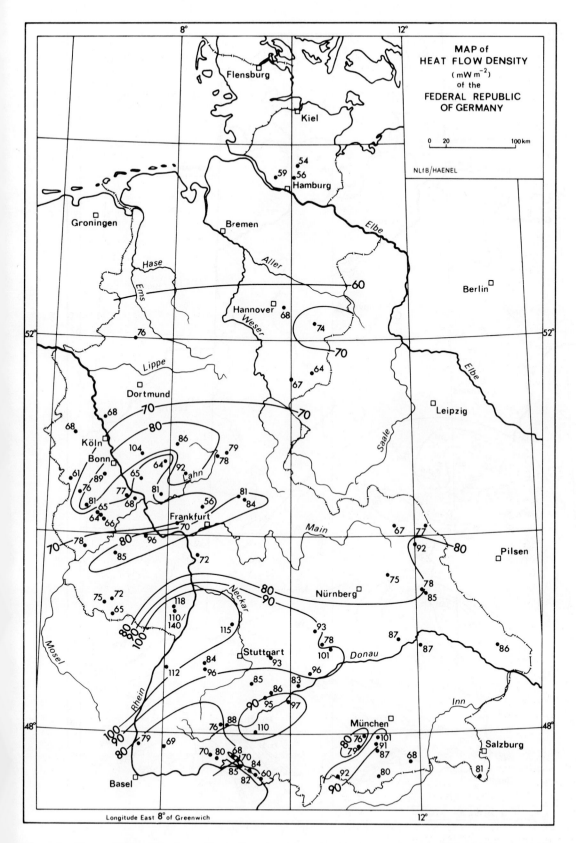

Fig. 5. Heat-flow density distribution in the Federal Republic of Germany

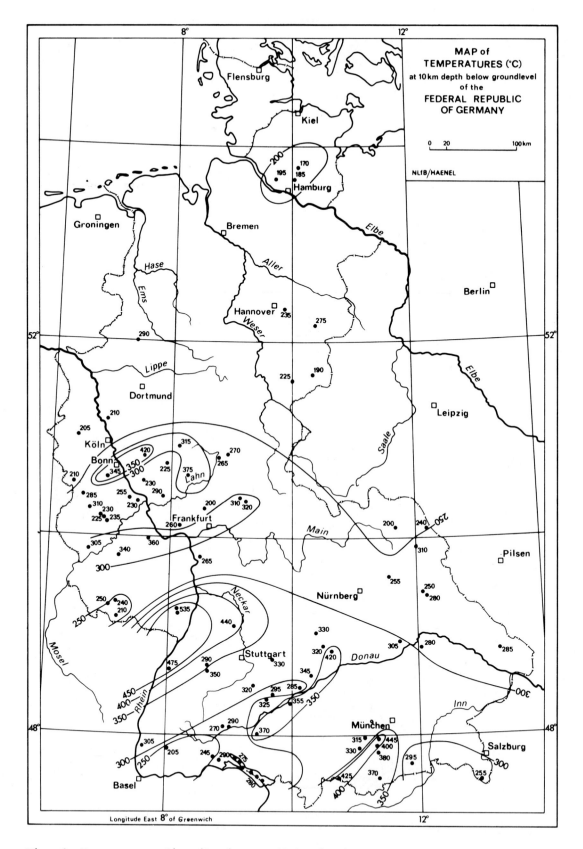

Fig. 6. Temperature distribution at 10 km depth

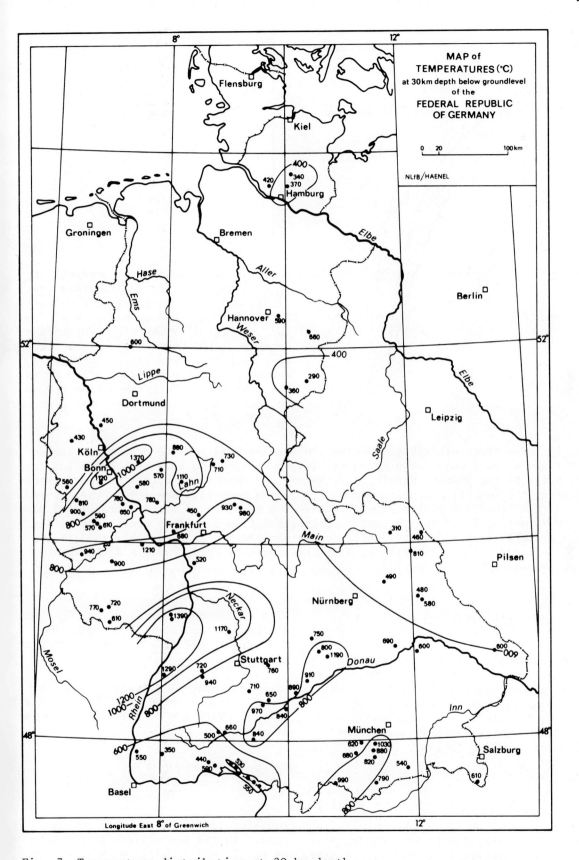

Fig. 7. Temperature distribution at 30 km depth

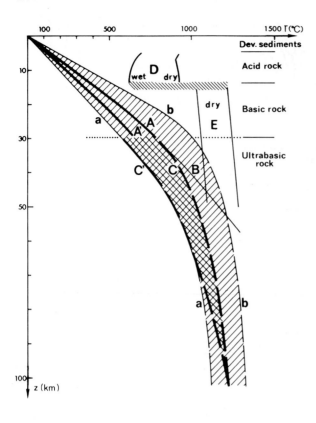
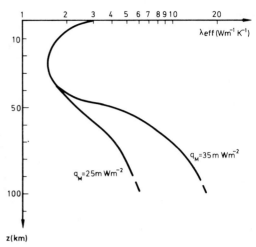

Fig. 9. Effective thermal conductivity λ_{eff} in the crust calculated from Eqs. (1-3), and in the upper mantle calculated from $q_o - H = q_M = 35$ m W m^{-2} and $\lambda_{eff} = q_M/\text{grad T}$, where q_M = heat-flow density at the Mohorovičić discontinuity, q_o = heat-flow density at the Earth's surface, H heat production of rocks in the crust; grad T is taken from Fig. 8. In addition λ_{eff} is represented also for $q_M = 25$ m W m^{-2}

Fig. 8. Temperature distribution below the Rhenish Massif. A, B Mean temperatures from Figs. 6 and 7 calculated using Eq. (6), corresponding to a mean heat-flow density value of the Rhenish Massif of 77 mW m^{-2}; A' Mean temperatures neglecting the effect of the heat-flow density anomaly: the mean heat-flow density is then 65-70 mW m^{-2}; a,b Lower and higher limits as given by Figs. 6 and 7 and extrapolated down to 100 km depth; C, C' Temperature reconstructed from stability conditions of igneous minerals erupted about 12 mill. years ago (Mengel and Wedepohl, 1983; Voll, 1983; Seck and Wedepohl, 1983); D, E Melting zones of acid, basic and ultrabasic rocks (Winkler, 1967; Wyllie, 1971)

Geodynamic Considerations

Birch (1950) gives a formula for the elimination of the effects of topography, uplift and denudation. In our case the topographic effect has already been eliminated; this term can therefore be neglected. The differentation yields the following equation (Grubbe, 1981):

$$G(z,t) = \frac{\partial T}{\partial z} = G' + b_o G' l t \quad (7)$$
$$+ (G_u - G')(1 + b_o d \cdot t - 2b_o z),$$

where l = uplift, d = denudation, $b_o = 2/\sqrt{\pi \varkappa t}$, $\varkappa$ = thermal diffusivity, t = time, G' = air temperature gradient, G_u = undisturbed temperature gradient, and G = actual (disturbed) temperature gradient.

Since the heat-flow density is determined and considered only near the Earth surface where $z = 0$, Eq. (7) can be rearranged:

$$\frac{G - G_u}{G_u} = \frac{2\sqrt{t}}{\sqrt{\pi \varkappa}} \left(d + \frac{G'}{G_u}(1 + d) \right), \quad (8)$$

Equation (8) yields the percentage change in the temperature gradient caused by uplift and erosion, as well as the

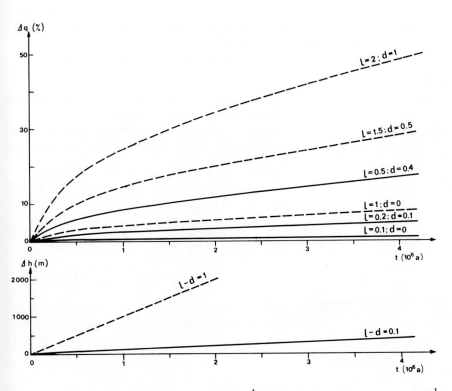

Fig. 10. Effect of uplift l (mm a^{-1}) and denudation d (mm a^{-1}) on heat-flow density and (*lower diagram*) the total amount of uplift $\Delta h = l - d$, l uplift, d denudation

corresponding heat-flow density using the well-known equation:

$$q = \lambda \frac{dT}{dz} = \lambda G \quad . \tag{9}$$

Some of the results are presented in Fig. 10. Since we are discussing an uplift of the Rhenish Massif of only about 0.1 mm a^{-1}, no remarkable change in heat-flow density can be expected there.

Assuming that a sudden increase of temperature ΔT takes place at an infinite surface (e.g., the upper surface of a laccolith) at depth d, then the temperature increase $T(z,t)$ at any level above that surface can be calculated by developing Eq. (1) of Carslaw and Jaeger (1959, p. 100):

$$T(z,t) = T_o + \frac{z(\Delta T - T_o)}{d}$$
$$+ \frac{2}{\pi} \sum_{n=1}^{n=\infty} \frac{T_o + (-1)^n \Delta T}{n}$$
$$\sin(\frac{n\pi z}{d}) \exp\left(-\frac{\varkappa n^2 \pi^2 t}{d^2}\right)$$
$$+ \frac{2}{\pi} \sum_{n=1}^{n=\infty} \sin(\frac{n\pi z}{d}) \exp\left(-\frac{\varkappa n^2 \pi^2 t}{d^2}\right)$$
$$\left[\frac{2(-1)^{2n+1} T_o}{n} + \frac{(-1)^{n+1} d}{n^2 \pi} \text{grad } T\right] . \tag{10}$$

The temperature gradient by differentiation of Eq. (10) is:

$$\frac{\partial T}{\partial z}(z,t) = \frac{\Delta T - T_o}{d} + \frac{2}{d} \sum_{n=1}^{n=\infty} (-T_o + (-1)^n \Delta T)$$
$$\cos(\frac{n\pi z}{d}) \exp\left(-\frac{\varkappa n^2 \pi^2 t}{d^2}\right) \tag{11}$$
$$+ \sum_{n=1}^{n=\infty} \cos(\frac{n\pi z}{d}) \exp\left(-\frac{\varkappa n^2 \pi^2 t}{d^2}\right)$$
$$\left[\frac{4(-1)^{2n+1} T_o}{d} + \frac{2(-1)^{n+1}}{n\pi} \text{grad } T\right] .$$

The heat-flow density at the Earth's surface ($z = 0$) is:

$$q_o(0,t) = \lambda \left[\frac{\Delta T - T_o}{d} + \frac{2}{d} \sum_{n=1}^{n=\infty} (-T_o + (-1)^n \Delta T) \exp\left(-\frac{\varkappa n^2 \pi^2 t}{d^2}\right) \right.$$

$$+ \sum_{n=1}^{n=\infty} \left(\frac{4(-1)^{2n+1} T_o}{d} \right. \quad (12)$$

$$\left. \left. + \frac{2(-1)^{n+1}}{n\pi} \text{ grad } T \right) \exp\left(-\frac{\varkappa n^2 \pi^2 t}{d^2}\right) \right),$$

where T_o = temperature at the Earth's surface and grad T = undisturbed temperature gradient at $t \leq 0$.

The uplift Δh at the Earth's surface is given by (thermal expansion coefficient $\beta = 3 \times 10^{-5}$ K^{-1}, Skinner, 1966):

$$\Delta h(t) = z\beta \frac{1}{z} \int_0^z T(z',t) dz' ,$$

$$\Delta h_{max}(t=\infty) = z\beta \left(\frac{z \text{ grad } T}{2}\right) = z\beta \frac{\Delta T}{2} , \quad (13)$$

and finally, the heat-flow density from Eq. (12) by:

$$q_o(t) = \lambda \frac{dT}{dz} , \quad q_{o,max}(t=\infty) = \lambda \left(\frac{dT}{dz}\right)_{max}. \quad (14)$$

The following examples show that uplifts in the range of 90-300 m can be explained by a temperature increase ΔT caused by large intrusive bodies at depth d ($\lambda = 2.0$ W m^{-1} K^{-1}):

	Δ = 200	400°C
d = 30 km : Δh_{max} =	90	180 m
Δq_{max} =	13	27 mW m^{-2}
d = 50 km : Δh_{max} =	150	300 m
Δq_{max} =	8	16 mW m^{-2} .

In Fig. 11, Δh and Δq have been related to Δh_{max} and Δq_{max}, respectively, and plotted against $F(t) = \pi^2 \varkappa t/d^2$. It follows that the uplift can be recognized earlier than the increase in heat-flow density.

Assuming an intrusive body of 50 × 50 × 30 km (length × width × thickness) at an intrusion temperature of about 1200°C, the temperature difference ΔT to the country rock can be taken again from Fig. 8. The formula is developed on the basis of Carslaw and Jaeger's equations (1959, pp. 56-62):

$$T(x,y,z,t) = \frac{\Delta T}{8} \left[\text{erf}\left(\frac{z-a}{A}\right) + \text{erf}\left(\frac{z+a}{A}\right) \right.$$

$$\left. - \text{erf}\left(\frac{z-b}{A}\right) - \text{erf}\left(\frac{z+b}{A}\right) \right] \quad (15)$$

$$\cdot \left[\text{erf}\left(\frac{c-y}{A}\right) + \text{erf}\left(\frac{c+y}{A}\right) \right]$$

$$\left[\text{erf}\left(\frac{d-x}{A}\right) + \text{erf}\left(\frac{d+x}{A}\right) \right] .$$

The heat-flow density at the Earth's surface can be derived from Eq. (15) by making $z = 0$:

$$q_o(x,y,t) = \lambda \frac{\partial T}{\partial z} = \frac{\lambda \Delta T}{8\sqrt{\pi \varkappa t}} \left[2 \exp\left(-\left(\frac{a}{A}\right)^2\right) \right.$$

$$\left. - 2 \exp\left(-\left(\frac{b}{A}\right)^2\right) \right]$$

$$\cdot \left[\text{erf}\left(\frac{c-y}{A}\right) + \text{erf}\left(\frac{c+y}{A}\right) \right]$$

$$\left[\text{erf}\left(\frac{d-x}{A}\right) + \text{erf}\left(\frac{d+x}{A}\right) \right] \quad (16)$$

where $A = 2\sqrt{\varkappa t}$, a = depth of the top surface of the intrusion, b = depth of its base, c and d = half axis of lateral extension, and x and y are the horizontal coordinates. The effect of latent heat may be included by making $\Delta T = \Delta T + L/c$, if crystallization commences at $T(x,y,z,t)$ or $q_o(x,y,t)$; where L = latent heat of crystallization and c = specific heat capacity of the magma.

The results of three examples are represented in Fig. 12. Since the Miocene volcanism took place about 12 Ma ago, it is unreasonable to expect a heat-flow

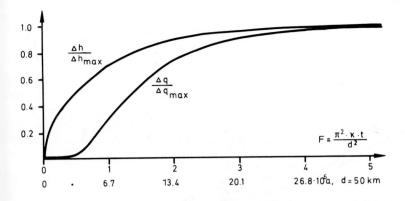

Fig. 11. Uplift Δh and heat-flow density change Δq related to Δh_{max} and Δq_{max}, respectively; suddenly increased temperature ΔT = 800°C, thermal conductivity λ = 2.7 W m^{-1} K^{-1}, thermal expansion coefficient β = 3 × 10^{-5} K^{-1}, t = time, ϰ = thermal diffusivity, d = depth, F = dimensionless number. Below the abscissa an example is given for F with d = 50 km, and ϰ = 37.8 km^2 a^{-1}

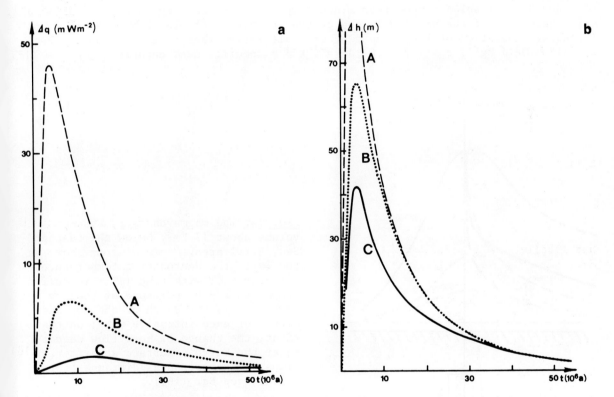

Fig. 12. Increase of heat-flow density Δq and uplift Δh at the Earth's surface caused by an intrusive body 50 × 50 × 30 km (length × width × thickness) at different depths [for symbols see Eq. (16)]:

	Model A	Model B	Model C
a (depth) =	20	30	50 km
b =	50	60	80 km
c = d =	25	25	25 km
ΔT =	700	400	200°C
λ =	2.2	2.2	3.5 W m^{-1} K^{-1}
ϰ =	25	25	35 m^2 a^{-1}

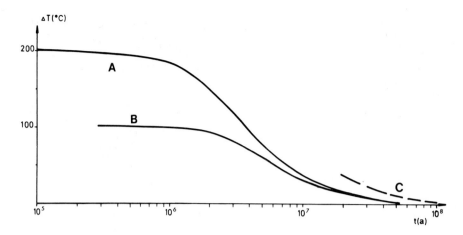

Fig. 13. Temperature decrease with time of an intrusive body, model C (Fig. 12); $\Delta T = 200°C$ = intrusion temperature minus temperature of country rocks, $\varkappa = 35$ m^2 a^{-1}. *A* In the centre of the body; *B* In the centre at the top of the body; *C* As in *A* but including liberation of latent heat of crystallization therefore $\Delta T = 200°C + L/c$, $L = 315 \times 10^3$ J kg^{-1}, $c = 1.25 \times 10^3$ J kg^{-1} K^{-1} = specific heat capacity

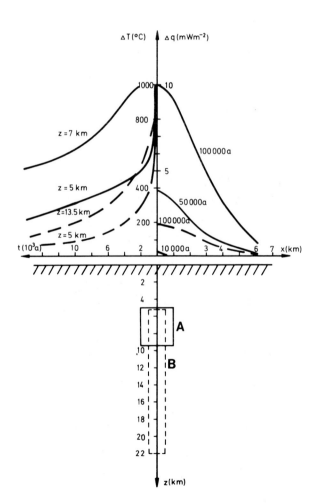

Fig. 14. The magma chamber, assumed volume about 17 km^3, below the Laacher See, Eifel area. *Upper diagram* shows on the *left*, the decrease of temperature difference ΔT with time (ΔT = intrusion temperature - temperature of country rocks), z is the depth of the top and centre of each intrusive body; on the right, the change of heat-flow density Δq at the Earth's surface at different times, x is horizontal distance [for symbols see Eq. (16)].

	Model A	Model B
Surface area =	4	1 km^2
a =	5	5 km
b =	9.25	22 km
c = d =	1	0.5 km
ΔT =	1000	800°C
λ =	2.35	2.15 W m^{-1} K^{-1}
$\varkappa$ =	32	29 m^2 a^{-1}

density which is significantly above normal. The uplift due to thermal expansion is relatively small, but it causes an increase in the heat-flow density at the Earth's surface (see Fig. 10, less than 10% of about 70 mW m^{-2}) which has to be added to the heat-flow density Δq shown in Fig. 12).

Because the exact size and the depth of the intrusive body are not known, the actual values (uplift and heat-flow density increase) must lie between those of an infinite plate [Eqs. (10)-(14), ΔT = constant for all times, which corresponds to a continuous supply of heat] and a body of finite size [Eqs. (15)-(16), ΔT decreases with time, as for a cooling body].

The temperature decrease for a deep lying body, model C, is shown in Fig. 13.

The possible existence of a magma chamber in the volcanic pipe of the Laacher See area is discussed by Schmincke (1978) and Viereck (pers. commun.). This is shown diagrammatically in Fig. 14. From Eqs. (15) and (16) one can calculate the present temperature in the centre and at the top of the magma chamber, as well as the heat-flow density at the Earth's surface (see Fig. 14). From these calculations it can be concluded that the heat has not yet reached the Earth's surface. The present heat-flow density anomaly due to the intrusion (10,000 years after its emplacement) is therefore zero.

Acknowledgements. The author gratefully acknowledges financial support from the Deutsche Forschungsgemeinschaft which enabled these geothermal studies to be carried out. The author also wishes to thank D. Brinkmann, K. Grubbe, St. Jobst, W. Reich, Ch. Reichert and G. Zoth for their kind cooperation, and Mr. Toms, who checked my English.

References

Birch, F., 1950. Flow of heat in the Front Range, Colorado.-Bull. Geol. Soc. Am., 61:567-630.

Bullard, E.C., 1940. The disturbance of the temperature gradient in the earth's crust by inequalities of height. - Month. Not. R. Astron. Soc., Geophys. Suppl., 4:300-362.

Carslaw, H.S. and Jaeger, J.C., 1959. Conduction of heat in solids. 2nd edn., Clarendon, Oxford.

Frenzel, B., 1980. Das Klima der letzten Eiszeit in Europa. In: Oescheer, H., Messerli, B., Svilar, M. (eds.) Das Klima. Springer, Berlin, Heidelberg, New York, pp 1-296.

Giese, P., Prodehl, C., and Stein, A., 1976. Explosion Seismology in Central Europe. Springer, Berlin, Heidelberg, New York.

Grubbe, K., 1981. Vertikalbewegungen und ihre Ursachen am Beispiel des Rheinischen Schildes, Teilprojekt Geothermik (1.4.80-30.4.81). Rep., NLfB Hannover, Arch. No. 89 517.

Grubbe, K., Haenel, R., and Zoth, G., 1983. Determination of the vertical component of thermal conductivity by line source methods. In: Haenel, R. and Gupta, M.L. (eds.) Results of the first workshop on standards in geothermics. Zentralbl. Geol. Paläontal., I, H. 1/2, Schweizerbart, Stuttgart, pp. 49-56.

Haenel, R., 1971. Determination of the terrestrial heat flow in Germany. Z. Geophys., 37:119-134.

Haenel, R., 1979. Determinations of subsurface temperatures in the Federal Republic of Germany on the basis of heat flow values. Geol. Jahrb. E15, Hannover, pp 41-49.

Haenel, R., 1983. Schlußbericht zum Forschungsvorhaben Vertikalbewegungen und ihre Ursachen am Beispiel des Rheinischen Schildes, Teilprojekt Geothermik (1.8.76-31.9.82). Rep. NLfB Hannover, Arch. No. 93 273.

Haenel, R. and Zoth, G., 1982. Heat flow density determination in shallow lakes along the geotraverse from München/Salzburg to Verona/Trieste. In: Cermak, V. and Haenel, R. (eds.) Geothermics and geothermal energy. Schweizerbart, Stuttgart, pp 1-299.

Kappelmeyer, O. and Haenel, R., 1974. Geothermics with special reference to application. Geoexpl. Monogr. S.1, No. 4. Borntraeger, Berlin, pp 1-238.

Lamb, H.H., 1977. Climate, present, past and future. Vol. 2, Climatic history and the future. Methuen, London.

Mechie, J., Prodehl, C., Fuchs, K., 1983. The long-range seismic refraction experiment in the Rhenish Massif. This Vol.

Mengel, K. and Wedepohl, K.H., 1983. Crustal xenoliths in Tertiary volcanics from the northern Hessian Depression. This Vol.

Roy, R.F., Beck, A.E., and Touloukian, Y.S., 1981. Thermophysical Properties of rocks. In: Touloukian, Y.S., Judd, W.R., and Roy, R.F. (eds.) Physical properties of rocks and minerals. McGraw Hill, New York, pp 409-502.

Rudloff, H., 1967. Die Schwankungen und Pendelungen des Klimas in Europa seit Beginn der regelmäßigen Instrumenten-Beobachtungen (1970). Die Wissenschaft, Vieweg, Braunschweig, p 122.

Schmincke, H.U., 1978. Report Number 2, Projekt 230/77/EGD. Inst. Miner. Univ. Bochum.

Voll, G., 1983. Crustal xenoliths and their evidence for crustal structure underneath the Eifel volcanic district. This Vol.

Skinner, B.J., 1966. Thermal expansion. In: Handbook of physical constant. Geol. Soc. Am. Mem. 97.

Seck, H.A. and Wedepohl, K.H., 1983. Mantle xenoliths from the tertiary and quaternary volcanics of the Rhenish Massif and the tertiary basalts of the northern Hessian Depression. This Vol.

Winkler, H.G., 1967. Die Genese der metamorphen Gesteine. Springer, Berlin, Heidelberg, New York.

Wyllie, P.J., 1971: Heacock, J.G. (ed.) The structure and physical properties of the earth's crust. Geophys. Monogr. Ser. Am. Geophys. Union, Vol. 14, AGU, Washington, DC, pp 279-300.

6.7 The Gravity Field of the Rhenish Massif

W. R. Jacoby[1], H. Joachimi[1,2], and C. Gerstenecker[3]

Abstract

The gravity field gives important information about the deep structure of the Rhenish Massif and is presented as the Bouguer anomaly, the free air anomaly, and the geoid. The Bouguer anomaly contour map is based on an incomplete older map and later published and unpublished material including our own measurements; it is still a preliminary map. The Bouguer anomaly is positive over the eastern Massif and negative over the western part. It shows a peculiar ring pattern about a center in the East Eifel/Neuwied Basin region and a pattern of narrow lows radiating into the western quadrants. Most of these features correspond to distinct tectonic units. Special attention is given to the problems of constructing a homogeneous gravity map from inhomogeneous data and on the errors and their propagation to the modelling. The free air anomaly, based on similar material and presented by smoothed contours of map averages (6' lat × 10' long), is, of course, more positive forming a broad high between the lows of the North German and Molasse Basins. There are similarities and dissimilarities with the Bouguer anomaly; both parts of the Massif are positive; in the east the two types of gravity anomalies are correlated, in the west they are anticorrelated. The geoid (or quasigeoid) is reproduced for completeness from recent work of geodesists. Its importance is that it stresses the longer wavelengths with a broad maximum over the eastern Massif falling off in all directions; it still shows narrow features as the Rhine Graben. Gravity in its different forms, particularly if compared to topography, clearly demonstrates mass excess nearer to the surface of the uplifted Massif; its nature and mechanism are obviously quite dissimilar to that of the Alps and their immediate foreland.

Introduction

The subject of this paper is the gravity anomaly of the Rhenish Massif. It is our aim to define the anomaly, to attempt a specification of what the gravity effect of the Massif actually is and to separate it from other effects. In this paper we shall try to recognize patterns and relationships with other parameters and discuss what may be considered the "anomaly" and what the norm or reference. The interpretation with the aid of quantitative models is deferred to Chapter 8.1 (Drisler and Jacoby, 1983).

Gravity is one of the geophysical fields that characterize the Rhenish Massif and help delineating its deeper structure. It must therefore be discussed

---

1 Institut für Meteorologie und Geophysik, Universität Frankfurt, Feldbergstraße 47, D-6000 Frankfurt am Main 1, Fed. Rep. of Germany

2 now at: Prakla-Seismos GmbH, D-3000 Hannover-Buchholz, Fed. Rep. of Germany

3 Institut für Physikalische Geodäsie, Technische Hochschule Darmstadt, Petersenstraße 13, D-6100 Darmstadt, Fed. Rep. of Germany.

here. It is presented in three different forms: The Bouguer anomaly (BA), the free air anomaly (FA), and the geoid (GE). These are viewed in conjunction with the relief or topography (TP) in a strongly smoothed version.

A special aspect and a reason for publishing this paper is the lack of complete and satisfactory published gravity maps for the region. This and the inaccessibility of new data has induced us to construct an improved, though still preliminary, BA contour map of the Rhenish Massif and its vicinity (which always must be taken into account in gravity studies), based on all material available to us.

The area of this study comprises parts of Germany, France, Belgium, the Netherlands, and Luxembourg. We had therefore to combine several national maps, but we used the BA map of Gerke (1957) as the base map since it covers the largest part of the Massif. The map had to be updated by results from more recent surveys. There are a number of difficulties in such a procedure of patching up a map: (1) we must piece together maps of different quality, accuracy, scale, and detail; (2) we need to adapt several maps to one common reference field and "gravity system" (as, e.g., a gravity base network which may be adjusted from time to time) as well as a common reduction density; (3) in some cases the documentation of the reduction parameters used is incomplete; (4) we had to use contoured maps instead of the original data; (5) there are problems of data exchange. The result is therefore not a BA map of homogeneous accuracy, but we believe it to be acceptable for the purpose of a regional study of the Rhenish Massif. We have also tested this quantitatively as will be demonstrated.

The problems were less serious for the free air anomaly, the geoid, and topography. We have mainly used published maps, only slightly updated and redrawn in a common format since they are used here merely in a qualitative manner anyway.

The four "fields", BA, FA, GE, and TP, are presented together with some geological information on the quadrangle from 48° to 53°N and from 3° to 12°E. The Rhenish Massif lies approximately in a central position of the map which extends sufficiently beyond the limits (at least as defined for the purposes of this project; see Murawski, Chapter 2, this volume). A broad margin is necessary as in any gravity study in order to distinguish what is anomalous from the reference background.

In the following sections BA, FA, GE, and TP are dealt with following the scheme: data sources; map construction; description of the fields and their characteristics, patterns, and cross correlations particularly with geology and tectonics; some qualitative remarks on the interpretation.

Establishing the Bouguer Anomaly (BA, Fig. 1)

The BA is the form of gravity representation most revealing about the density distribution at depth. If the terrain effect is neglected the result is termed the "simple Bouguer anomaly". In the Rhenish Massif the terrain effect may amount to -1.5 mgal in unfavorable situations as deep narrow valleys, but generally it is smaller in magnitude than -0.5 mgal (its value is always negative). The neglect will systematically bias the BA toward low values and, depending on the type of topography, may introduce spurious anomaly variations. In the Rhenish Massif such effects are not sufficiently large to affect the major conclusions (Joachimi, 1981).

The BA map in Fig. 1 is primarily based on Gerke's (1957) map. As the base map, also our map is one of the simple BA, although terrain-corrected values (maps) have been used here for updating beside uncorrected ones. A more sophisticated procedure was not warranted. The station density at the disposal of Gerke (1957) had been variable and, unfortunately, very low or nil in parts of the Rhenish Massif, particularly in the west. Since the results of a later regional survey by the Niedersächsisches Landesamt für Bodenforschung (NLfB) were unavailable to us we attempted to make use of, and

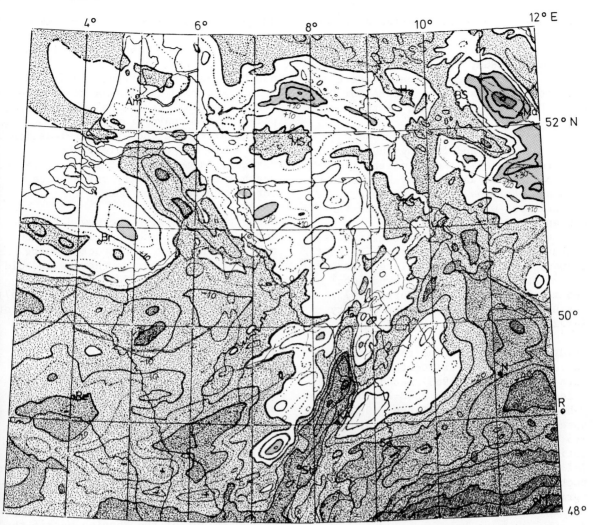

Fig. 1. Bouguer anomaly (BA) of the Rhenish Massif and its vicinity, based on Potsdam gravity System, Gravity Formula of 1930, and Bouguer density 2.67 g/cm^3. Compiled on the basis of Gerke (1957) and more recent data (see text). Contours at 10 mgal intervals (5 mgal *dotted*)

fit in, all availabe data, thus patching up the base map with pieces of varying size and detail, and at the same time assessing its errors (Joachimi, 1981). The new data include our own surveys in the Ruhr region (three small areas of together 300 km^2 with high point density: Miethke, 1978; Wabra, 1981; W.R. Jacoby, unpubl.), in the vicinity of Koblenz (three areas of together about 2000 km^2 with low point density along levelling lines: Joachimi, 1981), and in the Palatine (about 3500 km^2 with low point density along levelling lines: C. Gerstenecker, unpubl.) as well as unpublished results by others (local network Bad Weilbach, 1 km^2: K. Zippelt, unpubl.; 196 points from selected areas of the Massif previously not well surveyed: NLfB, S. Plaumann, unpubl.), and published maps (lower Rhine Graben: Dürbaum and Wolff (1958); eastern margin of the Brabant Massif: Bless et al. (1980); northeastern part of the Massif: Plaumann and Ulrich (1979a); Siegerland: Bosum et al. (1971); Limburg area: Plaumann and Ulrich (1979b); Hunsrück area: Plaumann (1978a); Taunus area: Plaumann (1974, unpubl.); Vogelsberg: v. Braunmühl (1973, 1975); West Harz: Plaumann (1978b); the gravity base network in most of the Rhenish Massif: Plaumann (1979).

The BA over the western extension of the Rhenish Massif, particularly the Ardennes, and beyond was taken from French and Dutch maps (BRGM, 1975; Atlas van Nederland, 1970). In a small-scale insert the Dutch map also gives the BA for Belgium, Luxembourg and parts of the North Sea.

The joining of, and fitting in, various maps requires adjustments of the reduction parameters unless they happen to be the same. For the present purpose it is not critical which gravity system is chosen, and for convenience we chose that used by Gerke (1957). Gravity was measured in the order I network of West Germany linked to the absolute Potsdam value. As reference the International Gravity Formula of 1930 (Torge, 1975) was used; the assumed Bouguer density was 2.67 g cm^{-3}.

Later maps were referred to updated gravity networks (DSN-62; Deutsches Schwerenetz 1962 (Kneißl, 1963) based on the Potsdam gravity value; DSGN-76: Deutsches Schweregrundnetz 1976 (Torge, 1980) based on a corrected absolute value, -15 mgal versus Potsdam; IGSN-71; International Gravity Standardization Net 1971 (Doergé et al., 1977) with the corrected value). The gravity reference formula was also updated from the 1930 one in 1967 and 1980 (Moritz, 1980):

$$\gamma = 978\ 049\ (1 + 0.005\ 288\ 4\ \sin^2\varphi$$
$$-\ 0.000\ 005\ 9\ \sin^2 2\varphi) \quad (1930)$$

$$\gamma = 978\ 031.8(1 + 0.005\ 302\ 4\ \sin^2\varphi$$
$$-\ 0.000\ 005\ 9\ \sin^2 2\varphi) \quad (1976)$$

$$\gamma = 978\ 032.7(1 + 0.005\ 302\ 4$$
$$-\ 0.000\ 005\ 8\ \sin^2 2\varphi) \quad (1980)$$

and the changes must be considered when matching the maps. At 50°N, crudely the following "corrections" must be applied to map values referred to System X when converting them to Gerke's (1957) map, i.e., the International gravity formula of 1930 and the Potsdam system:

From System X: to 1930/Potsdam:	Helmert 1901 − 12.3	Geod. Ref. Syst. 1967 + 5.8 (−9.2)	Geod. Ref. Syst. + 4.7 (−8.3)

where the value in brackets applies if in the new system the uncorrected Potsdam value had still been used. If the Bouguer density for gravity reduction had been assumed different from 2.67 g cm^{-3} (in some cases even depth-dependent) a crude "correction" δBA was applied in the conversion based on $\Delta\rho = 2.67 - \rho_{map}$ and average elevation $\bar{h}$ of small map quadrangles: $\delta BA = -2\pi G\bar{h}\Delta\rho$. As mentioned it was disregarded whether the new gravity values were terrain-corrected or not.

After the conversion the new data were incorporated into the old map and new contour lines were constructed where necessary. By comparing the new and old maps in various ways (Joachimi, 1981) the accuracy of the latter (Gerke, 1957) was tested in regions of different point densities. Although the results depend slightly on the procedure chosen they give a good idea of map accuracy. In sufficiently surveyed regions the mean discrepancies or errors in $\sim 10 \times 10$ km^2 areas of the old map are always smaller than ± 0.5 mgal and vary unsystematically, while the standard deviations based on 9 points in these areas are between 0.5 and 1.5 mgal, reflecting small-scale variations and reading error from the contours. Hence for our purposes we consider the old map satisfactory in such regions. In poorly surveyed regions of the Rhenish Massif the $\sim 10 \times 10$ km^2 mean deviations vary between −4 and +4 mgal while the 9-point standard deviations are similar to those in the surveyed regions; the mean deviations exhibit systematic undulations of typically 50 km wavelength, but for the whole Massif the average differs insignificantly. Our conclusion is that the errors of the old map in poorly surveyed regions are serious and will propagate into models of crustal structure while for the over-all deep structure they are not so serious. The new map, piecewise updated, is considerably improved and generally more reliable with the exception of the western Eifel and Hunsrück regions. Note finally that the absolute anomaly level, because referred to the 1930 gravity formula and

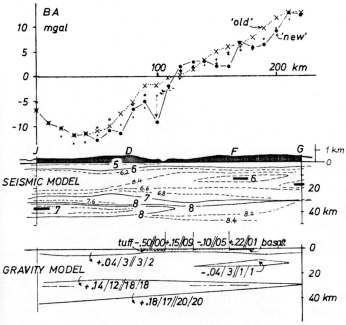

Fig. 2. Two-dimensional test model between shot points J and G of long-range seismic refraction profile I (Mechie et al., 1983). The purpose of the model is to demonstrate the effects on interpretation of the "old" inaccurate gravity data in comparison to the updated "new" ones (Fig. 1; for a more thorough interpretation see Drisler and Jacoby, 1983). *Top* BA profile sampled at 10 km spacing: "observed" *old* (x), *new* (•) and, where distinguishable, computed old (o), new without surface geology (·), with (↓); old with surface geology is indistinguishable from x. *Center* Vertically exaggerated topography (*black*) and seismic section: preliminary interpretation with P velocity contours (in km s^{-1}, low-velocity regions (—), shot points J, D, F, G) by C. Prodehl (pers. comm., 1980). *Bottom* density section; model geometry simplified from seismic section; density contrasts (shown in g cm^{-3} for model units in direction of *arrows*) computed by least-squares fit to the two "observed" BA's and for two models one including, the other not including, surface geology units. The density contrasts are given according to the scheme: new data with/without surface geology // old data with/without surface geology

the Potsdam system, is ~5 mgal high if compared to the most recent maps. For the anomaly variation and the interpretation of the Rhenish Massif deep structure this is fortunately irrelevant.

The above assertion concerning the propagation of map errors into models was substantiated by two-dimensional model calculations for a number of recent seismic refraction profiles (Mechie et al., 1983). As an example we show in Fig. 2 part of the long SW-NE profile between shot points J and G. On top the "old" and "new" gravity data are shown together with the computed values (least-squares, see Drisler and Jacoby, 1983). Below, a preliminary seismic model is presented (C. Prodehl, pers. comm., 1980), and beneath it, our simplified density model with the computed density contrasts. They differ depending on the model assumptions (surface geology included/not included) and on the data set used ("new" or "old") and thus give some indication of the "error propagation". As expected, the density contrasts computed for the small-scale surface geology elements, and to a lesser degree for the low-velocity lense in the upper crust, differ widely between the two gravity data sets fitted, while they are similar for the larger-scale model units as the 6 and 7 km s^{-1} interfaces and the low-velocity body below the M-discontinuity at about 30 km depth. This clearly demonstrates that the interpretation of Rhenish Massif gross structure is not much affected by the meagre "old" data set.

Special Features of Bouguer Anomaly

We shall now describe the Bouguer anomaly as presented by the "new" data shown in Fig. 1.

1. East-west asymmetry of the BA on the Rhenish Massif is manifest in more positive values east of the Rhine river than west of it. The zero contour (in

the gravity system chosen) closely follows the Rhine; the average BA in the east is ~-10 mgal, in the west ~+10 mgal. Beyond the eastern Massif limits, the positive BA extends with varying interruptions in all directions from N to E and S (Holland - Bramsche Massif - Lower Saxony Tectogen - Harz - Fläming - *not* Thuringian Forrest - Vogelsberg - Odenwald - Spessart); the negative interruptions will be discussed later. Beyond the limits of the western Massif the negative BA also extends to the NW and W through S (Lower Rhine Graben to North Sea - Ardennes - Somme - Paris Basin - Lorraine - Jura and further toward the Alps) with positive BA values on the Brabant Massif and the Palatine-Vosges shoulder of the Rhine Graben. The "negative interruptions" of the eastern positive BA are partly the expression of the NNE trending Hessian Depression-Leine Graben link between the Rhine Graben and the North German Basin, partly they follow an arc along Münsterland - SE Teutoburger Wald - Kassel region - Rhön - North Franconia, opening toward the Alps and branching from the Rhön to the Neckar/Rhine junction in the Rhine Graben. Another distinct narrow BA low follows the NW trending Lower Rhine Graben and marks the border between the positive east and the negative west.

2. Many of the above features can also be viewed as a ring pattern about a center near the Neuwied Basin (central Massif). The Massif is surrounded by lows of the North German Basin, the Paris Basin, and the Molasse Basin and Alps. A ~200 km radius ring of highs is formed by Bramsche Massif - Harz - Spessart, Kraichgau - Palatine, Vosges - Brabant Massif, and one of 140 km radius by lows of Münsterland - SE Teutoburger Wald - Kassel region - Rhön, linked with the Neckar/Rhine junction of the Rhine Graben - Lorraine - Ardennes - Lower Rhine Graben. The ring pattern is not very distinct in the west where a radiating pattern is more obvious: narrow lows radiate from the central to eastern Massif into the western quadrants (Lower Rhine Graben, West Vlanderen, Ardennes - Somme, Champagne - Paris Basin, Lorraine, Rhine Graben).

3. The significance of the above patterns is questionable. Some of the features correspond to surface geology, others not. Tertiary to Recent rift structures produce narrow gravity lows. Most of the horsts or Massifs of Hercynian and older rocks produce gravity highs with typically ~100 km dimension: Brabant, Bramsche, Harz, Fläming, but not Thuringian Forrest; similar positive features are the uplifted Rhine Graben shoulders: Palatine, Vosges and Odenwald, Black Forrest. Some gravity features are related to Hercynian basement structures: Hunsrück and Saar-Nahe Trough (gravity high), the narrow low connecting the Rhön and the Neckar/Rhine junction, and the adjoining gravity high of Franconia-Kraichgau. The basins outside the Rhenish Massif produce gravity lows although they differ strongly in crustal structure (thin crust in North Germany, thick crust in Molasse Basin). Most strikingly lacking a relationship to geology is the gravity asymmetry on the Rhenish Massif itself. Another feature with no clear geological relation is the North Franconian and Thuringian Forrest gravity low. Obviously the relationships between gravity and geology are varied and at places not evident. Hence a simple geological explanation of gravity is not possible. The patterns reflect a complex superposition of the effects of shallow and deep sources. In interpreting gravity it is necessary to incorporate crustal and upper-mantle structure (see Drisler and Jacoby, 1983).

4. It is also important to consider the relationships of gravity (BA: Fig. 1) with topography (TP: Fig. 5). Positive and negative correlations (anticorrelations) do occur in different areas and at different wavelengths. Again there is in this respect a clear difference between the eastern and western part of the Rhenish Massif. In the west the BA/TP correlation is largely negative except in the Hunsrück-Saar-Nahe Trough as well as in the Palatine-Vosges region. In the east the BA/TP correlation is clearly positive and this includes marginal features as the Lower and Upper Rhine Graben with shoulders, the Hessian Depression, Leine Graben, Vogelsberg, Harz, Odenwald; a few features are anti-

correlated: Thuringian Forrest, Rhön - Franconia - Kraichgau; and along the NE rim of the Rhenish Massif topography is highest in the east while BA is highest in the west (graben shoulder). A clear BA/TP anticorrelation is also evident with the approach to the Alps, interrupted only by the Rhine Graben - Vosges - Black Forest structure. Approaching the North German Basin the BA/TP correlation is positive.

In view of the varied relationships between gravity, geology, and topography, it is not surprising that the BA contours partly outline the Rhenish Massif boundaries, partly intersect them. Interpretation of gravity is complex. One important point is evident: the principal difference between the eastern and western Massif, suggesting differences at depth and also in evolution. On the other hand, the Eifel North-South Zone is not reflected in gravity although it has been an important tectonic element (Murawski et al., 1983).

The Free Air Anomaly (FA, Fig. 3)

The FA differs from the BA by the neglect of the effects of the topographic mass (as though its density were zero) between surface and geoid; specifically the Bouguer slab is neglected, the terrain effect is sometimes subtracted. The FA values particularly in high relief are much scattered. To remove the scatter one usually takes areal averages, as is done here. Since no mass reduction is applied to obtain the FA it gives mainly information on large-scale mass excess or deficiency and isostasy while the BA tells more about the mass distribution. But of course, the FA is not free of the latter effects.

The FA map of Fig. 3 is a contoured version of the Gravity Map of Western Germany, Mean Free Air Anomalies, Edition 1959 (Gerke and Watermann, 1959). The mean values refer to $6' \times 10'$ or approximately 10×10 km^2 quadrangles. For some quandrangles in the Rhenish Massif region updated mean values were taken and filled into the map (our own surveys; Institut für Angewandte Geodäsie). The map does not extend to the western neighborhood of Germany. No smoothing has been applied when contouring the mean values except that implied by the size of the areas. Such contouring has, of course, no high precision, but in view of the merely qualitative discussion of the freee air anomaly the errors are not serious.

On elevated land the FA is always more positive than the BA. The whole Rhenish Massif and its surroundings to more than 100 km distance beyond its limits form a broad FA dome. On the Massif the mean FA is between +30 and +40 mgal with local maxima of more than +60 mgal. Narrow lows along the Lower and Upper Rhine Graben cut into the positive dome, which is surrounded by low FA regions of the North German Basin, the Paris Basin, and the Molasse Basin and Alps. The FA/TP correlation is nearly in the whole map region clearly positive with the exception of the southwestern corner where the topography rises while the FA decreases toward the Alps and in regions in the north where topography is nearly nil.

On the whole the broad Rhenish Massif region is one of crust/upper mantle mass excess surrounded by mass deficiencies in the north, west and south.

The Geoid (GE, Fig. 4)

The geoid is the equipotential surface corresponding to the undisturbed sea level; gravity is the negative gradient of the potential. A geoid undulation is the distance between the geoid and the normal ellipsoid (exactly: the length of the plumbline between points P and Q of intersection with the two surfaces; Jung, 1956, pp. 534-548; Heiskanen and Moritz, 1967, p. 84). In a sense, the geoid undulations, briefly called the geoid GE here, are a low-pass filtered or smoothed form of FA, transformed to units of length.

The quasigeoid (which for practical purposes closely approximates the geoid)

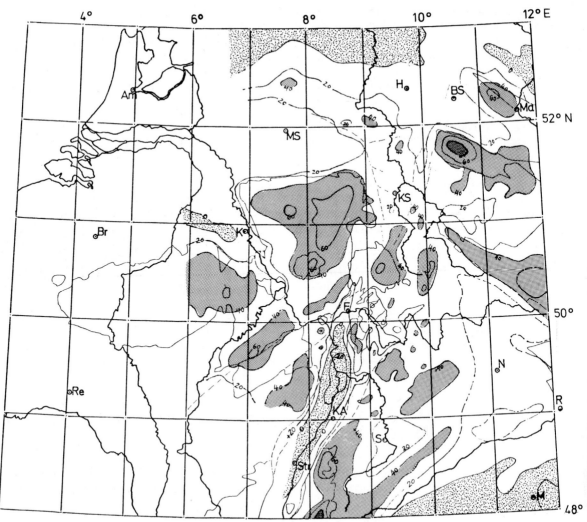

Fig. 3. Free air gravity anomaly (FA) of the Rhenish Massif and its vicinity, based on Gerke and Watermann (1959) and updated with more recent observations (see text)

presented in Fig. 4 is merely redrawn from Lelgemann et al. (1981), who give all the details of their solution. Generally we show only meter contours with a bit more detail in some regions.

GE has about 5 m relief in Central Europe. It is very similar to FA in a smoothed fashion, forming a broad dome centered on the eastern Massif and extending further east. The two main maxima in the map area (Fig. 4) are at the NE corner of the Rhenish Massif and in the Harz. The drop-off to the west is rather regular with only a weak "promontory" toward Eifel and Palatine and a distinct "embayment" from the Rhine Graben. Correlation with elevation (Fig. 5) is generally positive north of 49°N, but the geoid high is shifted east from the topography; south of 49°N with the approach to the Alps the long wavelength correlation with topography is negative. Thus, though a relationship of the geoid with the Rhenish Massif is indisputably present, it is not so clear what the mutual east-west shift means. Possibly the relation would become more direct if as a reference surface for the GE undulations one would use some low-order harmonic representation of the geopotential other than

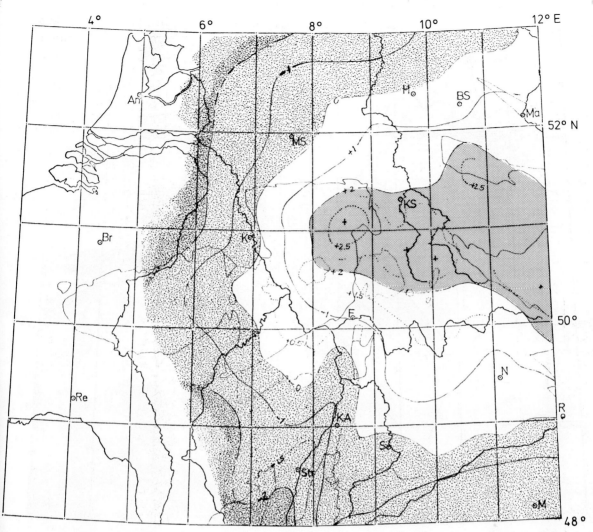

Fig. 4. Quasigeoid (GE) undulations of Rhenish Massif and vicinity, redrawn from Lelgemann et al. (1981). For reference system and other details see their paper

the spheroid. It is not, however, possible to separate the sources of the undulations. To date no quantitative modelling of the GE has yet been done.

Topography (TP, Fig. 5)

Topography must be taken into consideration in conjunction with gravity and uplift. The relationships with gravity have been pointed out in the above sections.

Our map is a contoured version of Schleusener's (1959) map of mean elevations for $12' \times 20'$ or $\geq 20 \times 20$ km^2 quadrangles. Water-covered quadrangles show the mean depth reduced by the water condensed from 1.027 to 2.67 g cm^{-3}. We acknowledge again the problems of contouring mean values, but the smooth topography map is perfectly adequate to give a broad impression of the regional features. Of course, because of the smoothing steep slopes even of regional extent will appear rather gentle.

The Rhenish Massif is a bounded, but not totally compact, feature on the TP map. It is disected by broad "furrows" along the Rhine and Moselle rivers. They partly follow sinking elements as that of the Lower Rhine Graben and the Neuwied Basin,

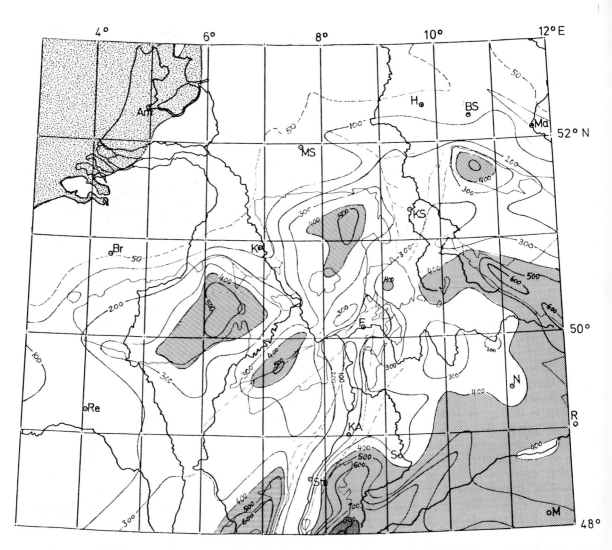

Fig. 5. Smoothed topography (TP), of Rhenish Massif and vicinity, contoured from mean elevations. (Schleusener, 1959)

partly they may be of erosional origin. The Massif therefore consists of three regional highs: (1) the eastern Massif with a small "outlyer" of the Taunus; (2) the Eifel radiating to the southern Ardennes (the northern Ardennes have no distinct elevation), and (3) the Hunsrück (south of the Moselle river) extending without interruption to the Palatine. TP is more nearly symmetric about the Rhine river than gravity (Figs. 1, 3, 4) as discussed above. The slight asymmetry is just opposite to that of gravity, the western Massif being the more elevated part than the eastern Massif.

In the north and west the Rhenish Massif is bounded by low country with a gentle transition in the west. The old Massifs (Brabant, Bramsche, Fläming) north of the Rhenish Massif have no, or hardly any topographic expression. In the east and south the Rhenish Massif is indistinctly bounded by the narrow Hessian Depression, the northern part of the Rhine Graben and the Lorraine. Beyond the boundaries TP varies in the same range as in the Massif itself; the east is therefore not very distinct from the Massif. Further south the TP map shows the rise toward the Alps.

Discussion

Mean elevation depends on several factors as the recent uplift or subsidence history, erosion (which in turn depends on rock type, climate, relief energy, etc.), or sedimentation (supply, subsidence rate, etc.), and volcanism. All these may play a role in the area. Thus the elevation of the Rhenish Massif itself is not sufficient evidence for uplift. It may, however, not be fortuitous that the mean TP relief is of the same order as the inferred Pleistocene uplift (Chapter 4, this vol.). This raises the question why similar regions as that to the east stand up topographically just like the Massif.

The more conspicuous uplift of the Alps is beyond the scope of this paper, but we note that relations with gravity and particularly the geoid are just the opposite. In the eastern Massif and to the east of it, topography and geoid are positive, while with the approach to the Alps topography rises and the geoid drops. This points to different causes for today's elevation.

A similarly different behavior of topography and gravity occurs between the eastern and western Massif. One is inclined to postulate a greater dynamic affinity of the eastern Massif with its eastern neighborhood than with the western Massif.

In conclusion we stress this asymmetry of the Rhenish Massif in gravity in contrast to elevation, exposed geology, and recent uplift on which the definition of the Massif rests. The asymmetry is also evident in crustal and upper-mantle structure and in volcanism. Therefore the Rhenish Massif may not be a homogeneous unit genetically; the causes of uplift may be different in the east and west. These aspects will be discussed further in Chapter 8 in connection with quantitative models.

Acknowledgements. We gratefully acknowledge all the help we received during this study. First of all, the valuable interdisciplinary discussion among the study group was prompted and greatly furthered by Henning Illies who had initiated and headed the project "Vertical motions and their causes, exemplified by the Rhenish Massif" which was organized and funded by Deutsche Forschungsgemeinschaft. We have been generously supplied with elevation data of benchmarks for our surveys by Landesvermessungsamt Rheinland-Pfalz, Koblenz, and Hessen, Wiesbaden as well as a number of institutions in the Ruhr region. We have used gravimeters of Institut für Geophysik, Universität Hamburg, Geologisches Landesamt Nordrhein-Westfalen, Krefeld, and Prakla-Seismos GmbH, Hannover. We had the help of many students in the surveys. We obtained unpublished gravity data from, and were directed to difficult-to-get data by, Afdeling der Geodesie, Technische Hogeschool Delft; K. Zippelt, Karlsruhe; Niedersächsisches Landesamt für Bodenforschung, Hannover; Hessisches Geologisches Landesamt, Wiesbaden, and Institut für Angewandte Geodäsie, Frankfurt. The latter institute also helped us in many other ways. We thank H. Closs (†), former President of Alfred Wegener Stiftung, for his long negotiations in our behalf with Niedersächsisches Landesamt für Bodenforschung about gravity data. Many individuals - too many to be named - helped with information on all aspects of data and interpretation. The computations were done at Hochschulrechenzentrum, University of Frankfurt. We especially acknowledge the financial support we received from Deutsche Forschungsgemeinschaft (grants Ja 258/5, 7, 8, 9, 12, 16-7 and Ge 381/2-3).

References

Atlas van Nederland, 1963-1977, Plate II-7. Staatsdrukkerej, Delft.
Bless, M.J.M., Bosum, W., Bouckaert, J., Dürbaum, H.J., Kockel, F., Paproth, H., Querfurth, H., and Rooyen, van, P., 1980. Geophysikalische Untersuchungen am Ostrand des Brabanter Massivs in Belgien, den Niederlanden und der Bundesrepublik Deutschland. Meded. Rejks Geol. Dienst, 32-17:313-343.

Bosum, W., Dürbaum, H.J., Fenchel, W., Fritsch, J., Lusznat, M., Nickel, H., Plaumann, S., Scharp, A., Stadler, G., and Vogler, H., 1971. Geologisch-lagerstättenkundliche Untersuchungen im Siegerländer-Wieder Spateisensteinbezirk. Beih. Geol. Jahrb., 90:139.

Braunmühl, W., 1973. Spezielle gravimetrische Untersuchungen im Vogelsberg. Diplomarb. Inst. Met. Geophys., Univ. Frankfurt.

Braunmühl, W., 1975. Gravimetrische Untersuchungen im Vogelsberg. Notizbl. Hess. Landesamt Bodenforsch., 103:327-338.

Bureau de Recherches Géologiques et minières, France, 1975. Carte Gravimétrique de la France - Anom. Bouguer. 1:1 000 000.

Doergé, W., Reinhardt, E., and Boedecker, G., 1977. Das "International Gravity Standardization Net 1971 (IGSN-71)" in der Bundesrepublik Deutschland. Dtsch. Geodät. Komm. B:225.

Drisler, J. and Jacoby, W.R., 1983. Gravity anomaly and density distribution of the Rhenish Massif. This Vol.

Dürbaum, H.J. and Wolff, W., 1958. Das Schwerebild des südlichen Teiles der Niederrheinischen Bucht. Fortschr. Geol. Rheinland Westfalen, 2:387-407.

Gerke, K., 1957. Die Karte der Bouguer-Isanomalen 1:1 000 000 von Westdeutschland. Dtsch. Geodät. Komm. B:46, I.

Gerke, K. and Watermann, H., 1959. Die Karte der mittleren Freiluftanomalien für Gradabteilungen 6' × 10' von Westdeutschland. Dtsch. Geodät. Komm. B:46, II.

Heiskanen, W.A. and Moritz, H., 1967. Physical Geodesy. Freeman, San Francisco.

Joachimi, H., 1981. Untersuchung der Schwereanomalie des Rheinischen Schildes. Diplomarb. Inst. Met. Geophys. Univ. Frankfurt.

Jung, K., 1956. Figur der Erde. In: Flügge, S. (ed.) Handbuch der Physik/Encyclopedia of Physics, 47, Geophysik I. Springer, Berlin, Heidelberg, New York, pp 535-639.

Kneißl, M., 1963. Deutscher Landebericht über die in den Jahren 1960 bis 1962 ausgeführten Arbeiten. Dtsch. Geodät. Komm. B:103.

Lelgemann, D., Ehlert, D., and Hauck, H., 1981. Eine astro-gravimetrische Berechnung des Quasigeoids für die Bundesrepublik Deutschland. Dtsch. Geodät. Komm. A:92, 67 pp.

Mechie, J., Prodehl, C., and Fuchs, K., 1983. The long-range seismic refraction experiment in the Rhenish Massif. This Vol.

Miethke, G., 1978. Anwendung eines Inversionsverfahrens auf eine Schweranomalie im Ruhrgebiet. Diplomarb. Inst. Met. Geophys. Univ. Frankfurt.

Moritz, H., 1980. Geodetic reference system 1980. Bull. Géodés., 54:395-405.

Plaumann, S., 1978a. Die Schweranomalie am Südrand des Hunsrück und ihre Interpretation. Mainz. Geowiss. Mitt., 7:171-181.

Plaumann, S., 1978b. Die Schwereanomalie des Westharzes. Geol. Jahrb., E12: 23-29.

Plaumann, S., 1979. Die Basisnetze für gravimetrische Regionalvermessungen im Bereich des Rheinischen Schildes. Geol. Jahrb., E16:5-18.

Plaumann, S. and Ulrich, H.J., 1979a. Geophysik. In: Erläut. Geol. Karten Hessen 1:25 000, Bl. 4618, 2. Aufl. Adorf, Wiesbaden, pp 97-101.

Plaumann, S. and Ulrich, H.J., 1979b. Geophysik. In: Erläut. Geol. Karten Hessen 1:25 000, Bl. 5514, 2. Aufl. Hadamar, Wiesbaden, pp 144-147.

Schleusener, A., 1959. Karte der mittleren Höhen von Zentraleuropa. Dtsch. Geodät. Komm. B:60.

Torge, W., 1975. Geodäsie. de Gruyter, Berlin.

Torge, W., 1980. Das Schweregrundnetz 1976 der Bundesrepublik Deutschland (DSGN76). Z. Vermessungswes., 9: 454-457.

Wabra, T., 1981. Auswertung von Schweremessungen im nordwestlichen Vorland des Ruhrgebietes. Diplomarbeit, Inst. Met. Geophys. Univ. Frankfurt.

7 Crust and Mantle Structure, Physical Properties and Composition

7.1 The Long-Range Seismic Refraction Experiment in the Rhenish Massif*

J. Mechie, C. Prodehl, and K. Fuchs[1]

Abstract

The long-range seismic refraction experiment to investigate the crust and uppermost mantle structure beneath the Rhenish Massif was completed during 1978 and 1979. It included a 600-km-long main profile extending from the Paris Basin in the southwest, across the Massif itself, to the Hessian Depression in the northeast, and three cross-profiles, up to 170 km long, almost wholly located in the Massif itself. Interpretation of the data, including ray-tracing through laterally inhomogeneous media and calculation of synthetic seismograms, has yielded the following velocity-depth information.

The P_g phase represents the refracted wave, with velocity 6.0-6.4 km s^{-1}, from the upper crust below an average depth of 3-5 km. Other intracrustal phases, of variable lateral extent and occurring in the depth range 9-24 km, can also be recognized.

The crust-mantle boundary structure beneath the profiles also displays marked lateral variability. Between shot-points A and B, beneath the Paris Basin, there exists a sharp crust-mantle boundary at 32 km depth. Along the main profile, between shot-points B and D, beneath the Southern Ardennes, an upper-mantle velocity of 8.1 km s^{-1} is reached at about 37 km depth, at the base of a 7.5 km thick transition zone. Northeast of shot-point D, beneath the Massif itself, there is a thin (<1 km) high velocity layer (V = 8.1 km s^{-1}) beneath a sharp crust-mantle boundary at 29-31 km depth. Below the thin high-velocity layer, there is a 4-6 km-thick transition zone, in which the velocity increases from low values (6.3-7.0 km s^{-1}) to upper-mantle values (8.4 km s^{-1}). Along the cross-profiles, mainly towards the northwest of the main profile, it appears that the crust-mantle boundary structure is a 2-4 km-thick transition zone, with an upper mantle velocity of 8.1 km s^{-1} being reached at 30-33 km depth.

An attempt is made to explain the results obtained from the seismic experiment in terms of a petrological model. In particular, the transition zone from low velocities (6.3-7.0 km s^{-1}) to high velocities (8.4 km s^{-1}) in the depth range 30-35 km, along the main profile northeast of shot-point D, may be explained in terms of magmatic material rising from greater depths and being trapped at the level of the crust-mantle boundary structure.

Geological Location and Description of the Experiment

As part of the geophysical investigation into the uplift of the Rhenish Massif, a long-range seismic refraction experiment with areal coverage was carried out to

---

[1] Geophysikalisches Institut, Universität Karlsruhe, D-7500 Karlsruhe 21, Hertzstraße 16, Fed. Rep. of Germany

* Contribution no. 251 Geophysical Institute, University of Karlsruhe

Plateau Uplift, ed. by K. Fuchs et al.
© Springer-Verlag Berlin Heidelberg 1983

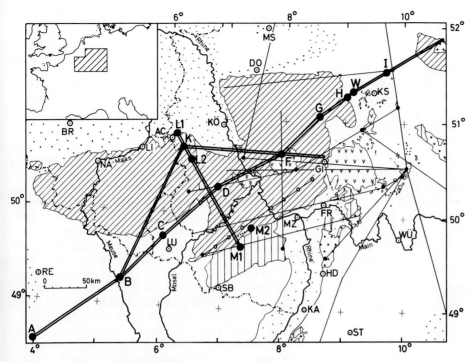

Fig. 1. Map of Rhenish Massif and adjacent areas, showing location of seismic profiles and simplified geology. Key: [·.·] Quaternary and Tertiary, [☐] Mesozoic, [⫼⫼⫼⫼] Permian, [▨▨] Carboniferous and Devonian, [∨∨∨] mainly Quaternary trachytes and phonolites, [∨∨] Tertiary volcanics, [+·+] Plutonic rocks. ●——— Shot-points and recording lines completed before 1978. ○———○ Common-depth point profile completed before 1978. ═══●═══ Shot-points and recording lines completed during 1978-1979. ⊙ Cities: AC Aachen; BR Brussels; DO Dortmund; FR Frankfurt; GI Giessen; HD Heidelberg; KA Karlsruhe; KÖ Cologne; KS Kassel; LI Liège; LU Luxembourg; MS Münster; MZ Mainz; NA Namur; RE Rheims; SB Saarbrücken; ST Stuttgart; WÜ Würzburg

determine the velocity structure of the crust and uppermost mantle beneath the Massif. Figure 1 shows the location of the profiles completed during the 1978-1979 experiments and interpreted in this chapter and also preliminarily by Mechie et al. (1982), together with the profiles completed between 1958 and 1974 and interpreted by Mooney and Prodehl (1978).

The 600-km-long main profile, completed during May 1979, runs from the Paris Basin in the southwest through the Rhenish Massif and into the Hessian Depression in the northeast (Fig. 1). Shot-points A, B, and C were sited on Mesozoic rocks southwest of the Massif, shot-points B and C on Jurassic rocks and shot-point A on the Cretaceous rocks of the eastern flank of the Paris Basin. Shot-points D, F, and G were located on the folded and slightly metamorphosed Carboniferous and Devonian sediments, of the Massif itself. In addition, shot-point D was situated about midway between the West and East Eifel Quaternary volcanic fields and shot-point F was located just north of the Tertiary volcanics of the Westerwald. Shot-points H, W, and I were sited northeast of the Massif, on the Triassic rocks of the Hessian Depression. The cross-profiles (see Fig. 1), up to 170 km long, were located mainly within the Massif, with only shot-points B and M1 being sited outside the Massif, the latter on Permian rocks 10 km south of the Massif's southern boundary fault zone. The pre-1978 profiles (see Fig. 1) were located mainly in the eastern part of the Rhenish Massif and in the Hessian Depression. Together with the new data, they have been used in compiling a map of depths to Moho (see Fig. 8).

A full description of the 1978-1979 experiments can be found in Mechie et al. (1982). Briefly, however, a series of (1982). Briefly, a series of reversed and overlapping profiles was completed. Recordings with an average spacing of 2km were made all along the main profile from the shot-points located on the main profile, while recordings with a spacing typically of 3 km were made along the cross-profiles from the shot-poins situated on these profiles. Thus, a large number of over 3000 three-component recordings was obtained, with maximum recording distances ranging from 50-600 km for the different profiles.

Data and Methods of Interpretation

Figures 2a-q and 3a-j (see fold-outs 1 and 2, inside the backcover) show 27 normalized and band-pass filtered P-wave record sections, plotted with a reduction velocity of 6 km s^{-1}. The widest band-pass filter used was 1.3-33 Hz, while the narrowest was 6-16 Hz. In the record sections, the maximum recording distance shown is 250 km, which seems to be the maximum distance at which the signal/noise ratio is good enough for phases to be recognized and correlated. Full details of the correlation and interpretation of the phases shown in the record sections are given below.

In interpreting the data, velocity-depth (V-Z) functions (see, e.g., Fig. 6a-d and Mechie et al., 1982, Fig. 4) were first of all fitted to the correlated phases on each record section, assuming that each record section is independent and that velocity varies only with depth. However, because of the close station spacing, good traveltime and, in particular, amplitude correlations could be made. Thus synthetic seismograms, using the reflectivity method (Fuchs and Müller, 1971) with a source signal of one cycle at 5 Hz dominant frequency, were calculated (see Figs. 6a-d and 7) in order to improve the preliminary V-Z functions. In addition, the large number of shots, giving rise to reversed and overlapping coverage, provided a superb opportunity for using the ray-tracing method for laterally inhomogeneous media (Červený et al., 1974) to improve upon and combine the one-dimensional velocity models to make a reliable two-dimensional velocity model (see Figs. 4 and 5).

In fact, except for the a_s and g phases, the phases derived from ray-tracing through the two-dimensional velocity model (Figs. 4 and 5) are those drawn in on 24 of the 27 record sections, i.e., on all the record sections except L1, M2, and F_{w-sw}, the latter being the record section composed of recordings along profile B-K from shot-point F (see Figs. 2a-q and 3a-e and h-i). It should be noted, however, that, in the ray-tracing analysis, record sections H_{sw} and H_{ne} have not been fully incorporated and that only phase a from shot-point H has been fitted. For profiles L1 and M2 (Fig. 3g and j) the observed phases have been drawn in. The average absolute differences between the theoretical arrivals derived from ray-tracing and the observed onsets on the seismograms for the different phases are shown in Table 1. The average absolute difference for the whole model is 0.08 s from 517 observations. In addition, the theoretical critical distances of the reflected (retrograde) phases are in 85% of the cases within ±10 km of the observed critical distances on the record sections (see Table 1).

The Upper Crustal Phases, a_s and a

The a_s phases, observed over the first few tens of kilometres on the record sections, are the direct and refracted phases through the sedimentary sequence. Their velocities, found by least-squares fitting, range from 2.6-5.6 km s$^-$

Phase a (P_g) is the refracted wave, with an apparent velocity nearly always greater than 6 km s^{-1} (see Figs. 2a-q and 3a-j) from the upper crust below depths in excess of 1.5 km (see Figs. 4 and 5). It was the best-observed phase

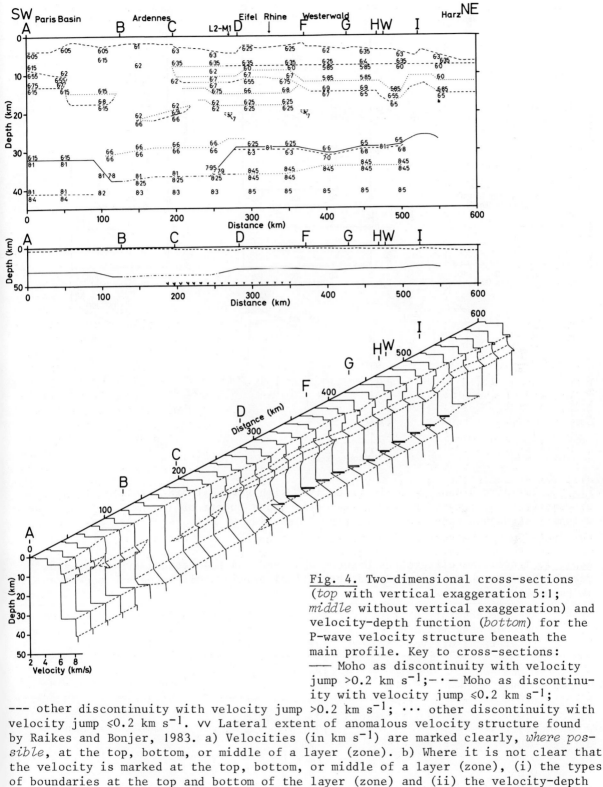

Fig. 4. Two-dimensional cross-sections (*top* with vertical exaggeration 5:1; *middle* without vertical exaggeration) and velocity-depth function (*bottom*) for the P-wave velocity structure beneath the main profile. Key to cross-sections: —— Moho as discontinuity with velocity jump >0.2 km s^{-1}; —·— Moho as discontinuity with velocity jump ≤0.2 km s^{-1}; --- other discontinuity with velocity jump >0.2 km s^{-1}; ··· other discontinuity with velocity jump ≤0.2 km s^{-1}. vv Lateral extent of anomalous velocity structure found by Raikes and Bonjer, 1983. a) Velocities (in km s^{-1}) are marked clearly, *where possible*, at the top, bottom, or middle of a layer (zone). b) Where it is not clear that the velocity is marked at the top, bottom, or middle of a layer (zone), (i) the types of boundaries at the top and bottom of the layer (zone) and (ii) the velocity-depth functions shown beneath the cross-sections, can be used in helping to determine whether it is the velocity at the top, bottom, or middle of the layer (zone) that is marked. c) To a good first approximation, one may assume a linear gradient between the velocity values given at the top and bottom of a layer (zone)

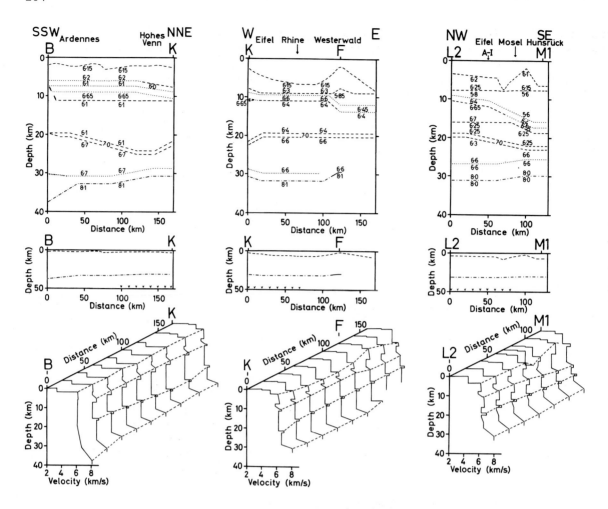

Fig. 5. Two-dimensional cross-sections (*top* with vertical exaggeration 5:1; *middle* without vertical exaggeration) and velocity-depth functions (*bottom*) for the P-wave velocity structure beneath the cross-profiles. Key to cross-sections: see Fig. 4

Table 1. Traveltime differences between observed and theoretical arrivals, and record sections where observed and theoretical critical distances are more than 10 km apart

Phase	Average absolute difference between observed and theoretical arrivals (s)	No. of observations	Record sections where observed and theoretical critical distances are more than 10 km apart
a	0.06	184	
b	0.09	107	W_{sw}, K_{ssw}, L2
b'	0.09	24	
b_1	0.07	33	F_w, M1
c	0.12	130	G_{sw}, K_{ssw}
d	0.04	7	
e	0.06	24	
f	0.06	8	
All	0.08	517	

from the 1978-1979 experiments. Maximum observation distances were in some cases around 140 km (see, e.g., record sections B_{ne}, D_{sw}, D_{ne}, G_{sw}; Fig. 2c, f, g, and i). However, in other cases the maximum observation distance was considerably smaller at around 90 km (see, e.g., record sections B_{sw}, G_{ne}, I_{sw}; Fig. 2b, k, and p), and it is uncertain whether this is a result of a difference in the vertical velocity gradient or whether it is due to lack of energy transmission at large offset distances caused by a lateral change in physical properties and geology.

First estimates of the true velocities of this phase and the depths to the top of the layer through which the phase is propagating were obtained from a time-term analysis (Willmore and Bancroft, 1960; Berry and West, 1966), the results of which are shown in Mechie et al. (1982). These first estimates of velocities and depths were then iterated manually during the ray-tracing interpretation until the final model (see Figs. 4 and 5), in which the differences between the theoretical and observed traveltimes were acceptably small, was obtained.

With strong amplitude arrivals out to distances of more than 100 km, there must exist a vertical velocity gradient in the layer which propagates this phase. However, there is a restriction on the maximum size of the vertical velocity gradient, because of the need to fit the intracrustal phases at 10-15 km depth. Thus, for example, on the main profile northeast of shot-point B and on the cross-profiles B-K and L2-M1, a value of 0.013 km s^{-1} km was settled for in the ray-tracing analysis and the calculation of synthetic seismograms (see Figs. 4, 5, and 6b-c), while between shot-points A and B a value of 0.027 km s^{-1} km was chosen (see Figs. 4 and 6a). However, it is possible that, northeast of shot-point G, the value is higher than 0.013 km s^{-1} km, because the maximum observation distance here seems to be rather less than 140 km and only about 100 km.

Figures 4 and 5 show the depths to the top surface of the layer propagating the P_g phase. Southwest of the Rhenish Massif, towards the Paris Basin, these depths range from 1.5-4 km, within the Massif itself they range from 2-7.5 km, and northeast of the Massif, in the Hessian Depression, they range from 3.5-6.5 km. Along profile L2-M1, two major structures, the Hunsrück anticline and the Mosel syncline, recognized in the surface geology, can also be deduced from the seismic data (see Fig. 5). In considering what the top surface of the layer propagating the P_g phase represents, it can be stated that it can represent the minimum depth to a crystalline (igneous/metamorphic) basement. The velocities, deduced here, for the P_g phase in the Rhenish Massif and surrounding areas are similar to the velocities found by Sapin and Prodehl (1973) in the Variscan domain of France for crystalline basement rocks and by Bayerly and Brooks (1980) and Mechie (1980) in the Variscan fold belt of South Wales, British Isles, for Precambrian crystalline basement rocks. However, where the sedimentary sequence is known to exceed the depths to the layer propagating the P_g phase, the top of this layer must represent a high-velocity sedimentary rock.

The Intracrustal Phases, b, b', and b,

The other intracrustal phases, b, b', and b, (see Figs. 2a-q and 3a-j) are of varying lateral extent and occur between 9 and 24 km depth (see Figs. 4 and 5). b is a reflected phase, often from a strong transition zone, and phase b' is its occasionally observed associated refraction (see, e.g., Fig. 2a and h). When two intracrustal reflected phases have been recognized on a record section, as in the case of the cross-profiles (see Fig. 3a-d and g-j), the later one is referred to as b,. Apart from the unincorporated record sections H_{sw} and H_{ne}, all record sections from the main profile, except that for shot-point C_{sw}, show the presence of one intracrustal reflected phase although, after positioning the reflecting zones at their correct offset distances along the profile, it appears that

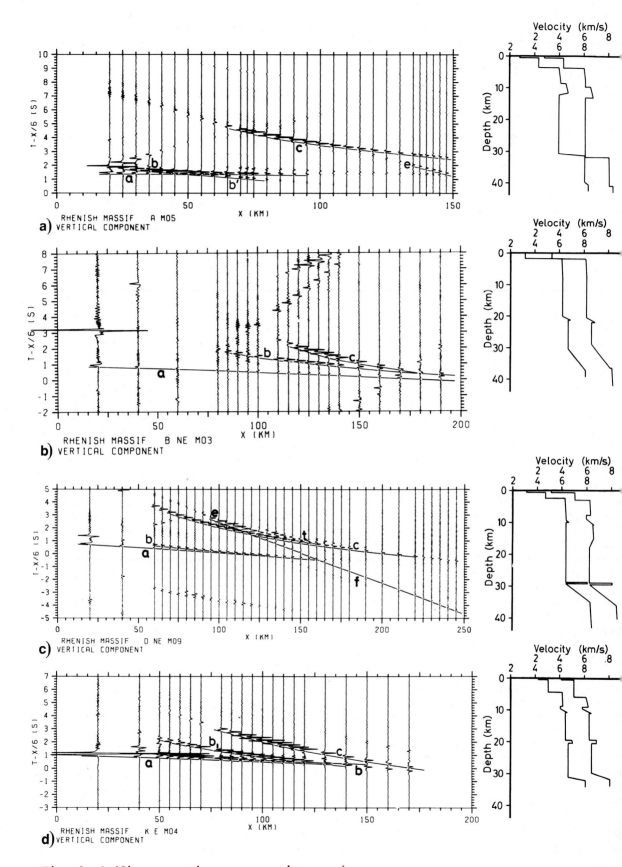

Fig. 6a-d (figure caption see opposite page)

beneath some parts there exist two depth ranges, e.g., 10.5-12 km and 18-21 km in the distance range 190-220 km (Fig. 4) giving rise to intracrustal reflected phases. Record sections A and F_{sw} from the main profile and record sections B_{nne}, L1, L2, M1, and M2 from the cross-profiles show the refracted phase b' associated with the reflected phase b (see Figs. 2a and h and 3a and g-j). All record sections from the cross-profiles, where recordings were made in the distance range 20-100 km, show the presence of two intracrustal reflected phases (see Fig. 3a-d and g-j).

Figures 4 and 5 show that the crustal structure beneath the base of the layer propagating the P_g phase is a series of alternating low- and high-velocity layers. For example, in the case of the cross-profiles there are typically three low-velocity layers separated by two high-velocity "teeth", while at the southwestern end of the main profile there is only one low-velocity layer between the layer propagating the b' refracted phase and the crust-mantle boundary. The low-velocity layers result mainly from fitting the critical point of the reflected phase from the high-velocity layer immediately below and, in some cases (e.g., on the main profile, the low-velocity layer at 20-29 km depth in the distance range 170-210 km; see Fig. 4), from the observation that the high-velocity layer immediately above does not propagate a refracted phase. The low-velocity layers are also the main regions of non-uniqueness in the model because the velocity distribution in them cannot be uniquely determined (see e.g., Giese, 1976) and because the depth to their top surface is uncertain when the high-velocity layer immediately above does not propagate a refracted phase. In all cases, the minimum possible thickness of the high-velocity layer immediately above has been assumed, and the highest possible average velocity compatible with the observations has been chosen to represent the velocity distribution within the low-velocity layers.

It would appear (see Figs. 4 and 5) that, under the profiles, there exist up to two intracrustal high-velocity zones below the layer propagating the P_g phase. The shallower, of greater lateral extent, exists at about 10-15 km depth, while the deeper, of lesser lateral extent, exists at about 19-24 km depth. This intracrustal picture of layers with alternating high and low velocities has already been recognized in Germany and is extensively discussed by Mueller (1977). In the Rhenish Massif, the wide-angle reflection, common-depth point profile, 240-LO-060 (see Fig. 1), which of the pre-1978 profiles has the highest density of recordings along it, has also been interpreted in terms of alternating intracrustal low- (3) and high- (2) velocity layers (Meissner et al., 1976; Mooney and Prodehl, 1978). Mueller (1977, Fig. 13) proposes the possibility of interpreting this sequence of alternating high- and low-velocity layers in terms of changes in rock-type with different chemical composition and, especially in the lower crust, progressive metamorphism. On the other hand, Giese (1983) basing his results on profile L2-M1 (see Fig. 3i), other geophysical evidence (see Meissner et al., 1983, and Jödicke et al., 1983) and the geological setting of the Rhenish Massif at the front of the Variscan thrust and fold belt, relates the sequence of alternating velocities to tectonic thrusting during the Variscan orogeny in which, for example, high-velocity lower crust was thrust on top of low-velocity upper crust.

In looking at the high-velocity layers in more detail, it can be said that where such a layer propagates a refracted

Fig. 6. Synthetic seismogram sections and velocity-depth (V-Z) functions for a) shot-point A, b) shot-point B_{ne}, c) shot-point D_{ne}, d) shot-point K_e. Left-hand V-Z function is that used in calculating the synthetic seismogram section. Right-hand V-Z function is taken from two-dimensional cross-section (Figs. 4 and 5) for comparison

phase to a far offset distance, the layer requires to have a considerable thickness (e.g., the layer at 11-18 km depth on profile L2-M1, Fig. 5). If a refracted phase cannot be observed, then a thin layer (e.g., the layer at 19.5-20.5 km depth on profile K-F, Fig. 5 and synthetic seismogram section and V-Z function, Fig. 6d) or a nose-like structure (e.g., the layer at 9.5-11 km depth on profile K-F, Figs. 5 and 6d) may be considered. In some cases, e.g., phase b from record section D_{ne} (see Fig. 2g, synthetic seismogram section and V-Z function, Fig. 6c, and amplitude-ratio versus distance plots, Fig. 7), a nose-like structure has been found to be the best representation of the high-velocity layer.

The Uppermost Mantle Phases, c, d, e, f, and g

Phase c (P_MP), the reflected phase from the crust-mantle boundary or the strong transition zone immediately above, is observed with varying quality. At the two extremes, the phase can be seen clearly and correlated easily on record sections D_{ne}, F_{ne}, C_{sw} (Fig. 2g, i, and d) while, on record section D_{sw} (Fig. 2f), although some energy can be observed between 120 and 150 km any correlation is extremely uncertain. A further variation shown by this phase along the main profile is the difference in its critical distance, i.e., the offset distance at which it first appears with large amplitudes. For example, for record sections B_{ne}, C_{sw}, C_{ne}, and D_{sw} (Fig. 2c-f) the critical distance is at 110-120 km, while for all other record sections from the main profile it is at 70-80 km (Fig. 2a-b and g-q). This difference in the critical distance results in the difference in depths to the crust-mantle boundary, e.g., between shot-points B and D it is at the base of a 7.5-km-thick transition zone at 36.5-37.5 km depth while along the rest of the main profile it is at a first-order discontinuity at 28-32 km depth (Fig. 4).

Between shot-points B and D, the 7.5-km-thick transition zone, shown in Fig. 4 to exist between the crust and mantle, arises from the need, as discussed above, to have velocities somewhat less than 6.9 km s^{-1} in the lower crust beneath the high-velocity zone at 18-21 km depth in this area. Figure 6b shows a synthetic seismogram section for shot-point B_{ne}, together with very alike V-Z functions for the one- and two-dimensional models. Figure 7 shows amplitude-ratio versus distance plots which show that the theoretical and observed seismogram sections are in good agreement. The large P_MP amplitudes in comparison with the other phases' amplitudes in the theoretical seismogram section for shot-point B_{ne} also fits the observed amplitude pattern of the phases on record sections C_{sw} and C_{ne} (Fig. 2d-e), bearing in mind the poor quality of these record sections. However, it does not fit the observed pattern from the good-quality record section D_{sw} (Fig. 2f), for which a similar traveltime model, involving a thick transition zone between crust and mantle, has been derived (see Fig. 4). As large amplitude ratios for P_MP/other phases can even be expected from a first-order discontinuity or a thin transition zone between crust and mantle (see synthetic seismogram section for shot-point A, Fig. 6a, and amplitude-ratio versus distance plots, Fig. 7), and as it cannot be observed that energy is leaking through the crust-mantle boundary structure to be returned from a deeper boundary (cf. discussion below concerning record section D_{ne}), it seems possible to postulate the existence of a region of higher attenuation (lower Q) in that part of the structure through which the P_MP phase, from shot-point D_{sw} alone, is propagating, i.e., in the depth range 30-36.5 km at 210-230 km distance along the main profile (see Fig. 4).

Along the rest of the main profile, there exists a sharp crust-mantle boundary. In the ray-tracing analysis, this boundary was modelled as a first-order discontinuity (see Fig. 4), although it was found, during the calculation of synthetic seismograms for shot-point A (see Fig. 6a), that a thin (<1 km) transition zone may also be allowed in the distance range 30-60 km. Along most of the length of the cross-profiles, there exists a 2-4 km-thick transition zone between the crust and mantle, with

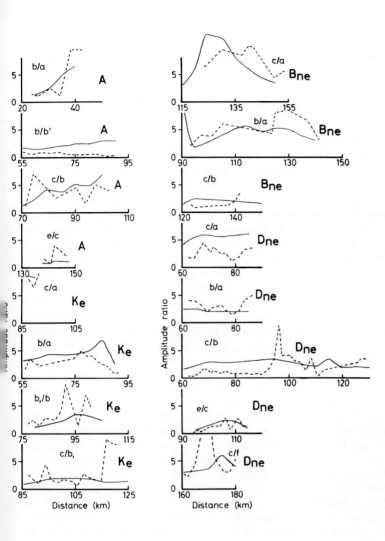

Fig. 7. Amplitude-ratios versus distance plots for synthetic seismogram sections shown in Fig. 6 (———) and observed record sections (---). In each case the shot-point and the phases used in the ratios are clearly marked

the boundary being reached at 30-33 km depth (see Fig. 5).

On only one record section, D_{sw} (see Fig. 2f), can recognition and correlation be made of phase d (P_n), the refracted phase which propagates through the uppermost mantle and the traveltime curve of which is tangential, at the critical point, to the traveltime curve of the reflected phase c (P_MP). Explanations as to why P_n is not visible on other record sections from shot-points on the main profile southwest of shot-point D and on record sections from the cross-profiles include that P_n is absent or is very weak and thus usually hidden in the background noise which in turn means that it can only be observed on the best-quality record section, D_{sw}. The reasons why P_n may be very weak are that in the uppermost mantle there exists either only a small positive velocity gradient, e.g., 0.013 km s^{-1} km as has been used in the ray-tracing model between shot-points B and D (see Fig. 4), or a zero-velocity gradient, e.g., see ray-tracing model between shot-points A and B (Fig. 4) and synthetic seismogram section for shot-point A (Fig. 6a). For P_n to be completely absent, a negative velocity gradient is required, a case which has not been used in the ray-tracing analysis of the main profile, but which may possibly exist on a local scale. On the main profile, northeast of shot-point D, the reason for the absence of P_n is thought to be that, below the crust-mantle boundary, there is only a thin (<1 km) layer of material with high velocity of 8.1 km s^{-1}, below which layer the velocity again drops to less than 7 km s^{-1} (see Fig. 4). Such a thin layer will have insufficient thickness to propagate a P_n phase with recognizable amplitudes over large offset distances (see synthetic seismogram section for shot-point D_{ne}, Fig. (6c). This type of structure for the uppermost mantle, involving a thin layer, is required to explain the low P_MP/phase e amplitude ratios (usually ~1) seen on the observed record sections D_{ne}, F_{sw}, F_{ne}, and I_{sw} (Fig. 2g, h, i, and p).

Phase e is a reflected phase from within the uppermost mantle. It is designated to the phase seen on record section A in the distance range 135-150 km at a reduced time of 1.4-1.8 s (Fig. 2a) and also to the phases seen on some of the record sections from shot-points northeast of shot-point D on the main profile in the distance range 100-130 km at a reduced time of 1.1-2.6 s. Of the latter group, the phase can be best seen on record sections D_{ne} and F_{ne}, and is also distinguishable on record sections F_{sw} and I_{sw} (Fig. 2g, h, i, and p). As mentioned above, modelling of P_MP/phase e amplitude ratios on record sections from shot-points northeast of shot-point D suggests that, along this part of the main profile, beneath the thin high-velocity layer at the top of the mantle, there exists a 3-6 km-thick transition zone from low velocities of 6.3-7.0 km s^{-1} at its top to normal upper-mantle velocities of 8.4 km s^{-1} at its base which is reached in the depth range 34-36 km (see Fig. 4 and synthetic seismogram section for shot-point D_{ne}, Fig. 6c). Interpretation of phase e on record section A, gives rise to a reflector (see Fig. 4) or, alternatively, a thin transition zone (see synthetic seismogram section for shot-point A, Fig. 6a) at about 41 km depth.

Phase f is the refracted phase, the traveltime curve of which is tangential, at the critical point, to the traveltime curve of the reflected phase e. As such, it also propagates through the uppermost mantle. The phase can be recognized on record section D_{ne} in the distance range 155-180 km and possibly again on some traces at greater offset distances of about 240 km (Fig. 2g). It may also be present on a few traces on record section F_{sw} in the distance range 155-175 km (Fig. 2h). Traveltime and amplitude modelling of this phase suggests that it propagates in the layer with a small positive velocity gradient of 0.013 km s^{-1} km below 34-36 km depth, i.e., immediately below the strong transition zone which returns the reflected phase e (see Fig. 4, and synthetic seismogram section for shot-point D_{ne}, Fig. 6c).

Figure 4 shows that the layer which propagates phase f northeast of shot-point D is a lateral continuation of the layer which propagates phase d (P_n), southwest of shot-point D. The depths to the tops of the two layers are approximately the same, the difference in the average velocity is only about 2.5%, and the velocity gradient in both is thought to be 0.013 km s^{-1} km. Thus the difference in structure, which exists at depths of less than 36 km between the two parts of the main profile northeast and southwest of shot-point D, disappears at greater depths and the upper mantle, below about 36 km depth, appears to be more homogeneous. Thus, explanations for the weakness or, in some cases, absence of phase f on the record sections northeast of shot-point D can be similar to those proposed above for the weakness or absence of phase d (P_n) on record sections southwest of shot-point D.

To emphasize that, in fact, the structure is more uniform than it may at first look, it can be seen that if the thin high-velocity layer at about 30 km depth beneath the eastern Rhenish Massif is disregarded then, in the distance range 100-500 km along the main profile (i.e., beneath the entire Massif), the crust-mantle transition would be similar and would be a linear gradient zone from crustal velocities of 6.3-7.0 km s^{-1} at around 30 km depth to upper-mantle velocities of 8.1-8.4 km s^{-1} at about 35 km depth. One possible petrological explanation for the linear gradient zone between crust and mantle, which (explanation) could still be true if, instead of a gradient zone, there existed a layer of constant velocity of about 7.4 km s^{-1}, is some kind of mixture between material with (crustal) velocities $\leq$7.0 km s^{-1} and material with mantle velocities >7.8 km s^{-1}. The preferred model with the gradient zone would involve a bias of low-velocity material

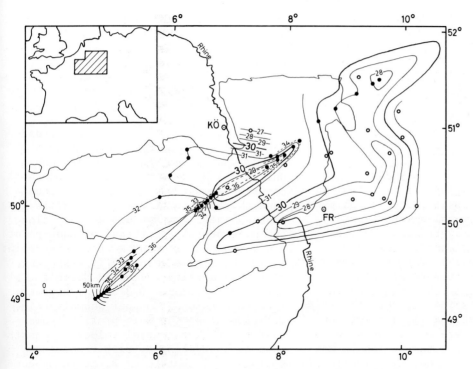

Fig. 8. Map showing depths to the Moho and the layer propagating phase f beneath the Rhenish Massif and adjacent areas. Key: ● Data points: 1978-1979 profiles. ○ Data points: pre-1978 profiles (see Mooney and Prodehl, 1978). —— Contours (in km) of depth to Moho (crustal thickness). Moho is defined as 1st-order discontinuity where velocity below the discontinuity is ≥ 7.5 km s^{-1}, *or* strong transition zone at the base of which the velocity is ≥ 7.5 km s^{-1}. --- Contours (in km) of depth to the layer propagating phase f. The outline of the Rhenish Massif is also marked. ⊙ Cities: FR Frankfurt; KÖ Cologne

at the top and high-velocity material at the bottom, whereas the model with the constant-velocity layer would involve no bias in the mixture from top to bottom.

To give an idea of crustal thicknesses beneath the Rhenish Massif and adjacent areas, a contour map of the depths to the Moho (crust-mantle boundary) has been constructed (Fig. 8) in which the pre-1978 data interpreted by Mooney and Prodehl (1978) and the 1978-1979 data presented here are both included. In addition, the depths to the top of the layer propagating phase f, beneath the northeastern part of the main 1979 profile, have been contoured (dashed lines, Fig. 8) because, as stated above, the layer propagating phase f, northeast of shot-point D, is a lateral continuation of the layer propagating phase d (P_n), southwest of shot-point D, and because the velocity structure of this layer propagating phases d and f appears to be more homogeneous than the velocity structure above it. Figure 8 shows that beneath the contoured region depths to the Moho range from 26-32 km, except between shot-points B and D along the main 1979 profile, where depths of 37.5 km are reached (see also Fig. 4). However, if one compares depths to the Moho between shot-points B and D with depths to the top of the layer propagating phase f, northeast of shot-point D, which (comparison) may be quite meaningful for the reasons stated immediately above, then it may be noted that there is only a 3.5 km change in depths over a 270 km distance range, i.e., a downdip of around 0.75° to the southwest (see also Fig. 4).

At distances greater than 200 km on record sections D_{sw} and D_{ne} (Fig. 2f-g), there exists the possibility to correlate another phase, g, which has not been modelled in the present ray-tracing analysis, but which can be interpreted as a reflected phase from a transition zone in the mantle in the depth range 44-48 km (Mechie et al., 1982).

Discussion

Interpretation of the 1978-1979 system of reversed and overlapping profiles crossing the Rhenish Massif (Fig. 1) has provided a two-dimensional P-wave velocity model, derived by ray-tracing, for the crust and uppermost mantle beneath the region (Figs. 4 and 5). This two-dimensional traveltime interpretation has been corroborated by amplitude modelling involving the calculation of synthetic seismograms (Figs. 6a-d and 7). In addition, pre-1978 and 1978-1979 data have been used to construct, for the region, a map in which depths to the Moho (solid lines) and the layer propagating phase f (dashed lines) have been contoured (Fig. 8).

To indicate the degree of uniformity of the crust and uppermost mantle beneath the Rhenish Massif, delay times were calculted for the top 40 km of the structure by ray-tracing through the model shown in Fig. 4. In the distance range 175-500 km, the variation in delay times was ±0.1 s, i.e., ±1.5%. Thus, despite the lateral variations shown in Fig. 4, the gross structure of the top 40 km, in the distance range 175-500 km, can be said to vary within ±1.5%. Delays of only ±0.1 s in the top 40 km of the structure support the hypothesis of Raikes and Bonjer (1983) that the observed teleseismic delays of up to about 1 s, encountered in the Rhenish Massif, must be attributed to an anomalous low-velocity structure in the depth range 50-150 km beneath the Massif. In addition, the smaller delay times, i.e., higher average velocities, are encountered along that part of the profile northeast of the Rhine while the larger delay times, i.e., smaller average velocities, are found southwest of the Rhine. This agrees with the result of Raikes and Bonjer (1983, Fig. 12) which shows that the top 40 km of the structure below the Rhenish Massif is characterized by high average velocities east of the Rhine and low average velocities west of the Rhine. Thus, it may be noted that the two different seismic methods have produced, independently, a similar result for the average velocity structure down to 40 km depth, beneath the Rhenish Massif.

As might be expected, however, the difference between a sharp crust-mantle boundary at 32 km depth between shotpoints A and B and the type of lowermost crust-uppermost mantle structure found beneath the rest of the main profile causes a negative Bouguer anomaly along the main profile northeast of shot-point B with respect to southwest of shot-point B, if the P-wave velocity (V_p) is converted to density (ρ) using Birch's formula $\rho = 0.252 + 0.3788\ V_p$. However, a two-dimensional Bouguer anomaly calculation along the main profile shows that the higher crustal velocities northeast of shot-point B, with respect to southwest of shot-point B, at least compensate for this negative Bouguer anomaly.

With respect to the earlier refraction data, it may be noted that the average crustal velocity and depth to the crust-mantle boundary east of shot-point D on the main profile is similar to that for the Rhenohercynian model (Mooney and Prodehl, 1978, Fig. 9), which was derived from the pre-1978 profiles located in the vicinity of the northeastern part of the main profile. On one profile, 17-240-20, which runs almost parallel to the northeastern part of the main 1979 profile and 20-40 km southeast of it, Mooney and Prodehl (1978, Fig. 4) also recognized, in addition to phase c (P_MP) a second uppermost mantle reflected phase from a depth of 33 km (cf phase e recognized here). However, because they derived a velocity of only 7.7 km s^{-1} for the layer immediately below the Moho, they did not require the existence of a low-velocity zone in the uppermost mantle.

As a result of S-waves being poorly recorded in the 1978-1979 experiments, the S-wave velocity structure is not known, and thus information about the variation in Poisson's ratio and hence the physical properties of the materials beneath the region is lacking.

In attempting to interpret the P-wave velocity structure of the crust and uppermost mantle in terms of a petrological model, discussion has already been made about the first topic to be considered, which (topic) is what the top surface of the layer propagating the P_g phase represents. In this connection, it may be added that where the main profile and the cross-profile, K-F, traverse the Rhine, at which places there is some degree of geological control, in neither case is it thought necessary for the Devonian sequence to extend to depths greater than those at which the top surface of the layer propagating the P_g phase is encountered (Meyer and Stets, 1975). Thus, at these two places the top surface of the layer propagating the P_g phase likely represents the depth to pre-Devonian (possibly Cambrian-Ordovician) sediments or crystalline (metamorphic) basement. The observed P_g velocities correspond to rocks of granitic and/or granodioritic composition (Birch, 1960).

In considering the structure of the crust below the layer propagating the P_g phase, the alternative views of Mueller (1977) and Giese (1983) have already been discussed. Mueller's hypothesis, which attributes the velocity variations, e.g., the layers with alternating high and low velocities, in the crust to changes in chemical composition and physical state, is a general one possible for all areas and thus the sequence of rock types may be regarded as that which is normally produced. Giese, on the other hand, proposes a model which is applicable only to the crust of orogenic belts where large-scale thrusting has played an important role in bringing deeper high-velocity rocks on top of shallower low-velocity rocks. In support of Giese's model for the Rhenish Massif, Hirn et al. (1982) have interpreted low-velocity layers in the upper crust in the Iberian and French parts of the Variscan orogenic belt in terms of thrust and nappe emplacement.

At middle and lower crustal depths, low velocities less than 6.9 km s^{-1} may represent gneiss or mica shists, while high velocities greater than 6.9 km s^{-1} could represent granulite (Christensen, 1965, and Christensen and Fountain, 1975). However, at depths of around 30 km, either in the lower crust or uppermost mantle, low velocities less than 6.3 km s^{-1} such as are shown in Fig. 4 along the main profile between shot-points A and B and between shot-points D and F are most likely explained in terms of rocks with free water content at normal temperatures of around 550°C at about 30 km depth. When such low velocities occur at such great depths, it is reasonable to invoke an explanation such as "high water content in pore spaces" (Gordon and Davis, 1968).

Discussion has already been made concerning the transition zone from low velocities of 6.3-7.0 km s^{-1} at 30 km depth to high velocities of 8.1-8.4 km s^{-1} at 35 km depth in the distance range 100-500 km along the main profile. However, of particular interest is the type of uppermost-mantle structure derived along the main profile in the distance range 280-500 km (see Fig. 4 and synthetic seismogram section for shot-point D_{ne}, Fig. 6c). In interpreting a record section, Florac 03, from the Massif Central, France, Fuchs and Schulz (1976, Fig. 9) show an uppermost-mantle structure similar to that shown in Fig. 6c. The type of uppermost-mantle structure shown in Fig. 6c produces tunnel waves (Fuchs and Schulz, 1976) — waves which tunnel through the thin high-velocity layer at the top of the mantle and are reflected from the transition zone below. The tunneled phase is labelled as phase t in Fig. 6c. Like the Rhenish Massif, the Massif Central contains outcrops of rocks deformed during the Variscan orogeny and Quaternary volcanics. In addition, Perrier and Ruegg (1973) and Coisy and Nicolas (1978) postulated that, beneath the Quaternary volcanics of the Massif Central, there exists in the upper mantle an anomalous structure which has given rise to 1-5 km mantle upwelling in the last 5 m.y., i.e., 1-5 mm yr^{-1}. The massif itself is thought to have risen by 0.4-1.7 km at the end of the Pliocene. Thus, there appear to be *some* similarities in the deep geological and seismic structure between the Massif Central and the Rhenish Massif.

The question must also be asked if there is a connection between the anomalous low-velocity structure in the upper mantle (Raikes and Bonjer, 1983) and the uppermost-mantle structure northeast of shot-point D or the lowermost crust structure between shot-points B and D. It is possible that the transition zone in the depth range 30-35 km (the low-velocity zone northeast of shot-point D) may be explained by magmatic intrusion, i.e., a proportion of the partial melt rising up from the anomalous low-velocity structure may be trapped at this level. Schmincke et al. (1983) postulate the need for intermediate magma reservoirs to account for the different types of Quaternary volcanic rocks found in the Eifel and favor the crust-mantle boundary for the location of these reservoirs because (1) there is a density jump at the crust-mantle boundary and (2) the volcanic rocks contain granulite (i.e., lower crust) xenoliths. Alternatively, the transition zone (low-velocity zone) may be due to metasomatism, i.e., the introduction of volatile-rich fluids which can cause a velocity decrease without a significant density change. In this connection, Seck (1983) discusses metasomatism of mantle xenoliths carried up by the Quaternary volcanic rocks and also metasomatism of the Quaternary magmas themselves. With either of these two explanations, the difference in the lowermost crust-uppermost mantle structure on either side of shot-point D may have been caused by the magmatic or metasomatic material having completely altered the uppermost mantle southwest of shot-point D but having left a thin skin of original mantle northeast of shot-point D.

Acknowledgments. The financial support of the Deutsche Forschungsgemeinschaft for this seismic-refraction project is first and foremost acknowledged. Additional financial support and organization was provided by the IPG Paris (Dr. A. Hirn), Messieurs Flick and Feidt-Mahowald at Luxembourg, the III Korps of the German Army (Mr. Goos), N.E.R.C. (Britain) and ETH Zürich. The help of the following authorities is acknowledged: Oberbergamt Clausthal-Zellerfeld; Bergämter Aachen, Goslar, Kassel, and Koblenz; the State Forestry Commission of Hessen, Niedersachsen, Nordrhein-Westfalen, and Rheinland-Pfalz. A full list of participants in the field experiments and data processing is given in Mechie et al. (1982). In particular, personnel and instruments were provided from all German geophysical university institutes and state agencies, as well as from the ETH Zürich, IPG Paris, the Universities of Birmingham, East Anglia, Leicester, Swansea, Vienna, and Uppsala, and the Serviço Meteorologico Nacional of Portugal. Special thanks are accorded to W. Kaminski for his help with the data processing and plotting of the record sections and his discussions of various problems connected with the experiment. For computing facilities, the Geophysical Institute and the Computer Centre of Karlsruhe University are thanked. R. Stangl helped with the figures, while G. Bartman typed the manuscript.

References

Bayerly, M. and Brooks, M., 1980. A seismic study of deep structure in South Wales using quarry blasts. Geophys. J.R. Astron. Soc., 60:1-20.
Berry, M.J. and West, G.F., 1966. An interpretation of the first-arrival data of the Lake Superior experiment by the time-term method. Bull. Seismol. Soc. Am., 56:141-171.
Birch, F., 1960. The velocity of compressional waves in rocks to lo kbar, 1. J. Geophys. Res., 65:1083-1102.
Červený, V., Langer, J., and Pšenčik, I., 1974. Computation of geometric spreading of seimic body waves in laterally inhomogeneous media with curved interfaces. Geophys. J.R. Astron. Soc., 38:9-19.
Christensen, N.I., 1965. Compressional wave velocities in metamorphic rocks at pressures to lo kbar. J. Geophys. Res., 70:6147-6164.
Christensen, N.I. and Fountain, D.M., 1975. Constitution of the lower continental crust based on experimental studies of seismic velocities in granulite. Geol. Soc. Am. Bull, 86: 227-236.
Coisy, R. and Nicolas, A., 1978. Regional structure and geodynamics of the upper mantle beneath the Massif Central Nature (London), 274:429-432.

Fuchs, K. and Müller, G., 1971. Computation of synthetic seismograms with the reflectivity method and comparison with observations. Geophys. J.R. Astron. Soc., 23:417-433.

Fuchs, K. and Schulz, K., 1976. Tunneling of low-frequency waves through the subcrustal lithosphere. J. Geophys., 42:175-190.

Giese, P., 1976. Problems and tasks of data generalization. In: Giese, P., Prodehl, C., and Stein, A. (eds.) Explosion seismology in Central Europe. Springer, Berlin, Heidelberg, New York, pp 137-145.

Giese, P., 1983. The evolution of the Hercynian crust - some implications to the Uplift Problem of the Rhenish Massif. This Vol.

Gordon, R. and Davis, L., 1968. Velocity and attenuation of seismic waves in imperfectly elastic rock. J. Geophys. Res., 73:3917-3935.

Hirn, A., Senos, L., Sapin, M., and Victor, L.M., 1982. High to low velocity succession in the upper crust related to tectonic emplacement: Tras os Montes-Galicia (Iberia) Brittany and Limousin (France). Geophys. J.R. Astron. Soc., 70:1-10.

Jödicke, H., Untiedt, J., Olgemann, W., Schulte, L., and Wagenitz, V., 1983. Electrical conductivity structure of the crust and upper mantle beneath the Rhenish Massif. This Vol.

Mechie, J., 1980. Seismic studies of deep structure in the Bristol Channel area. Ph.D. thesis, Univ. Wales, 362 pp.

Mechie, J., Prodehl, C., Fuchs, K., Kaminski, W., Flick, J., Hirn, A., Ansorge, J., and King, R., 1982. Progress report on Rhenish Massif seismic experiment. Tectonophysics, 90:215-230.

Meissner, R., Springer, M., Murawski, H., Bartelsen, H., Flüh, E.R., and Dürschner, H., 1983. Combined seismic reflection-refraction investigations in the Rhenish Massif and their relation to recent tectonic movements. This Vol.

Meissner, R., Bartelsen, H., Glocke, A., and Kaminski, W., 1976. An interpretation of wide-angle measurements in the Rhenish Massif. In: Giese, P., Prodehl, C., and Stein, A. (eds.) Explosion seismology in Central Europe. Springer, Berlin, Heidelberg, New York, pp 245-251.

Meyer, W. and Stets, J., 1975. Das Rheinprofil zwischen Bonn und Bingen. Z. dt. geol. Ges., 126:15-29.

Mooney, W.D. and Prodehl, C., 1978. Crustal structure of the Rhenish Massif and adjacent areas; a reinterpretation of existing seismicrefraction data. J. Geophys., 44:573-601.

Mueller, S., 1977. A new model of the continental crust. Geophysical Monograph 20, The Earth's Crust. Am. Geophys. Union, Washington, D.C., pp 289-317.

Perrier, G. and Ruegg, J.C., 1973. Structure profonde du Massif Central français. Ann. Geophys., 29:435-502.

Raikes, S. and Bonjer, K.-P., 1983. Large-scale mantle heterogeneity beneath the Rhenish Massif and its vicinity from teleseismic P-residual measurements. This Vol.

Sapin, M. and Prodehl, C., 1973. Long-range profiles in Western Europe - I. Crustal structure between the Bretagne and the Central Massif of France. Ann. Geophys., 29:127-145.

Schmincke, H.-U., Lorenz, V., and Seck, H.A., 1983. The Quaternary Eifel volcanism. This Vol.

Seck, H.A., 1983. Eocene to Recent volcanism within the Rhenish Massif and the Northern Hessian Depression - Summary. This Vol.

Willmore, P.L. and Bancroft, A.M., 1960. The time-term approach to refraction seismology. Geophys. J.R. Astron. Soc., 3:419-432.

7.2 Combined Seismic Reflection-Refraction Investigations in the Rhenish Massif and Their Relation to Recent Tectonic Movements*

R. Meissner[1], M. Springer[1], H. Murawski[2], H. Bartelsen[1], E. R. Flüh[1], and H. Dürschner[3]

Abstract

Four seismic reflection surveys were carried out in the area of the Rhenish Massif during the years 1968 to 1978. They were located in the tectonically most interesting areas and gave detailed information on the fine structure of the crust and the shape and depth range of prominent fault zones. Part of the uplift is attributed to displacements along these faults which were generated during the Hercynian orogeny and partly re-activated by a similar Tertiary stress pattern. Another, probably major part of the uplift seems to be related to a plume-like interaction between mantle and crust, as indicated by an enhanced density of reflectors at the base of the crust and by a relatively high and possibly newly formed Moho below the Rhenish Massif.

Introduction

During the years 1968, 1973, 1975, and 1978, combined seismic reflection-refraction investigations were carried out in the area of the Rhenish Shield. Three of these profiles belong to the project *Geotraverse Rhenoherzynikum*, which was later incorporated into the research programme *Vertical movements and their causes in the Rhenish Massif*. It was the objective of geoscientists working together on the Geotraverse project to integrate along a number of selected profiles all available geological, geophysical and related data. The tectonically most interesting parts of the profiles were covered by reflection surveys which are considered to be the most powerful method for the detection of the fine structure of the crust and the definition of the shape of fault zones. Following this concept, major reflection investigations were carried out in those regions where the Geotraverse crossed master faults, such as the Western fault of the Rhine Graben near Landau, the southern boundary fault of the Hunsrück (the southeastern part of the Rhenish Massif) and the Aachen thrust fault which delineates the norther: limit of the Variscan foldbelt. The results of the Rhine Graben and the Hunsrück boundary reflection profiles were previously given by Meissner et al. (1980), first results of the Aachen profile were published by Meissner et al. (1981) and the results of the 1968 reflection survey are described in Glock and Meissner (1976) and Meissner et al. (1976). Figure 1 shows the location of these profiles.

---

1 Institut für Geophysik, University of Kiel, D-2300 Kiel, Fed. Rep. of Germany

2 Geologisch-Paläontologisches Institut, University of Frankfurt, D-6000 Frankfurt, Fed. Rep. of Germany

3 Gewerkschaften Brigitta und Elwerath Betriebsführungsgesellschaft, D-3000 Hannover, Fed. Rep. of Germany

* Pub. No. 243 Institut für Geophysik, Kiel.

Plateau Uplift, ed. by K. Fuchs et al.
© Springer-Verlag Berlin Heidelberg 1983

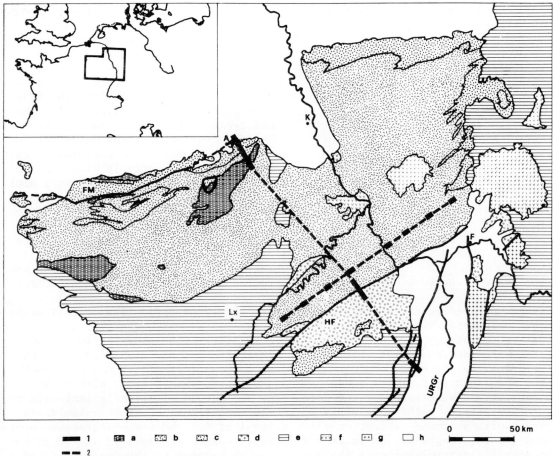

Fig. 1. Location map of seismic profiles in the Rhenish Massif. 1) reflection profiles; 2) refraction profiles. a) Cambrian and Ordovician; b) Devonian; c) Carboniferous; d) Permian; e) Triassic and Jurassic; f) Volcanic rocks; g) Crystalline basement; h) Tertiary and Quaternary; FM - Faille du Midi; HF - Hunsrück fault; URGr - Upper Rhine Graben; A - Aachen; F - Franfurt; K - Köln; Lx - Luxemburg

Field Procedure

The reflection seismic surveys of the *Geotraverse Rhenoherzynikum* were carried out by Prakla-Seismos GmbH, Hannover, Germany. Explosives were used throughout the whole investigation to permit a recording of wide-angle and refracted arrivals. In the wide-angle experiments MARS 66 equipment (Berckhemer, 1970) was used at distances up to 180 km.

In this way the shots of the reflection profiles were also used for the refraction seismic investigations on the remaining parts of the profile. Except for the northernmost reflection survey, the field procedures were those common in routine seismic prospecting but with a wider geophone spacing. Across the Hunsrück border fault zone the recording was done with 24-channel equipment. The length of the spread was 2400 m. The profile was 23 km long. The subsurface coverage was 3-4-fold. The line across the Rhine Graben fault was 12 km long. With a 3200 m geophone spread of 48 channels a coverage of 4-6-fold was obtained.

In the northern section three 48-channel recording systems were used simultaneously. With a trace spacing of 160 m each spread covered a length of 7.5 km. In between these three spreads gaps were left, each 4 km wide; these were filled with refraction stations. The effective length of the whole set-up was 30.7 km. The three reflection systems recorded all shots synchronously. The aim of this unusual set-up was to facilitate a better determination of velocities down to the Moho. A long geophone spread is necessary to arrive at move-out times for Moho reflections sufficient to calculate interval velocities.

Results

As most of the results have already been published previously, only the data relevant to significant tectonic features will be discussed in the following.

The 1968 Hunsrück Reflection Profiles

The eight short reflection segments along a ENE-WSW orientated line approximately parallel to the strike of the Hunsrück mountains, provided basic information about the quality of reflection data inside the Rhenish Massif. These steep-angle reflections were incorporated into the velocity data from the wide angle experiment with a common reflection element and a total length of 180 km providing an accurate determination of the velocity depth function in the center of the profile. Steep-angle reflections occur at major velocity jumps obtained from the wide angle survey (Glocke and Meissner, 1976, Fig. 6). These data encouraged us to record a perpendicular profile crossing the southeastern Hunsrück borderzone in 1973. Results of this line indicate that the strong dips observed in the SW part of the 1968 profile result from interferences of the Hunsrück boundary fault.

The 1973 Hunsrück Boundary Profile

The crust below the Hunsrück, which forms the southwestern part of the Rhenish Massif, is characterized by nearly horizontal reflectors, whereas the reflectors below the Saar-Nahe trough dip strongly northwards. From a migrated section, which shows the true locations of the reflecting elements, it is evident that the zones of horizontal reflectors in the north and the dipping interfaces in the south are well separated by a SSE-ward-dipping listric zone that is characterized by decreased reflection density. This zone, which reaches down to the crust-mantle boundary, is identified as the fault zone separating the Hunsrück from the Saar-Nahe trough. Further evidence for the existence of this fault zone is provided by reflected refractions which originated from both sides of the postulated fault zone. Also some arrivals with extremely high apparent velocity and strong amplitudes, visible in the records of the MARS 66 stations, give evidence of that fault zone (Meissner et al., 1980). These arrivals are explained by a double over-critical reflection in the fault region as well as at a reflector dipping to the north below the Saar-Nahe trough.

The increased reflection density between 7 and 9.5 s, which was also observed on other reflection profiles in Germany (Bartelsen et al., 1982) is explained as a zone of lamellation above the Moho (Meissner, 1967, 1973). Within the crust, a zone of slightly increased reflection frequency at the time/depth level of 4.3 and 6 s - corresponding to depth ranges of about 13 and 18 km - can be related to the Conrad and the sub-Conrad discontinuity, respectively. These occur at greater depths in the northern part of the 1973 profile and in the westernmost part of the 1968 lines than on their southern and eastern extensions respectively.

The 1975 Western Rhine Graben Profile

The profile across the western-boundary fault of the Rhine Graben is characterized by relatively poor reflection quality. No reflections are obtained from the fault plane or from the Moho. Even the wide-angle records only exhibit weak Moho reflections. This may be a consequence of a rather gradual crust-mantle transition below the graben and in its vicinity (Meissner et al., 1980). The dip of the Rhine Graben boundary fault can be inferred from the travel times of the first arrivals. These data suggest that it has a considerably shallower dip than previously assumed.

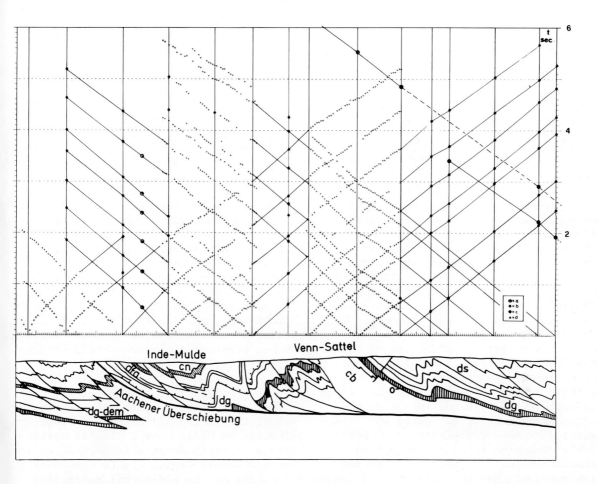

Fig. 2. Shallow reflector near Aachen in a geological cross-section and shallow refracted arrivals. *Crosses* first arrival travel times of 48-channel reflection spreads. a) travel times measured by MARS 66 equipments; b) reciprocal travel times; c) travel times computed by linear interpolation; d) travel times computed from parallel travel time curves

The refraction and wide angle observations between the Rhine Graben and the Hunsrück provide information mainly on the depth of the basement (Fig. 8). An intracrustal discontinuity (Conrad) is found at a depth of 13 to 14 km. Wide-angle reflections from the Moho are weak in this area.

The 1978 Profile across the Northern Variscan Deformation Front near Aachen

The most prominent result of the reflection survey in this region was a strong, gently southward-dipping reflector occurring at a depth of only 3-4 km. Meissner et al. (1981) describe this reflector and interpret it as corresponding to the subsurface trace of the well-known Aachen thrust fault which carries the Cambro-Ordovician strata outcropping in the Hohes Venn anticline. Extrapolation to the surface of the increased dip of the reflector near its northern end provides a reasonable correlation with the trace of the Aachen thrust fault, known from surface geological mapping.

As the publication by Meissner et al. (1981) was of preliminary nature this profile is described here in more detail. Figure 2 displays a geological cross-section along the trace of the surveyed line and plotted below it the configuration of the main reflector discussed above. These arrivals clearly observed on the three geophone spreads indicate that this shallow reflector must correspond to a thin layer. Neither is it associated with a change in refraction velocity as should be expected if this reflection originated from the upper surface of a thick higher-velocity layer, nor is there a gap and an offset in the travel time curve of the first arrivals. Refraction branches are continuous up to a distance of more than 30 km. This is not possible if the reflector corresponded to the top of a thick layer characterized by lower velocities than the overburden. Moreover, no high frequency, highly damped refracted arrivals are observed, as would be expected to originate from a thin layer of higher velocity. This reasoning corresponds to the observation that the polarity of the reflection arrival is negative (Springer, 1982); then this reflector is identified as a thin layer of lower velocity.

This interpretation is consistent with petrological observations of outcropping fault zones, always showing a reduction of seismic velocities (Etheridge and Wilkie, 1979; Kern, 1978). It may be a consequence of increased pore pressure, decreased grain size or of a hydrolytic weakening in general. Amplitude considerations and the single wavelet character of the reflected arrival impose limits on the possible range of thickness of the layer. The ratio of layer thickness to wave length within the thin layer is between 0.1 and 0.3. Since the actual velocity within such a thin layer at that depth cannot be determined by means of seismic measurements, the thickness of the layer can only be estimated. With an assumed P-wave velocity of 5.2 km s^{-1} the most probable layer thickness ranges between 20 m and 80 m (Springer, 1982).

A further interesting feature of a reflection originating from a thin layer that is apparent on the field records as shown in Fig. 3 is that the maximum amplitude of the reflected arrival does not occur at zero offset but at a shot-geophone distance of 3 to 4.5 km. This is confirmed by the results of synthetic seismograms which are computed with the reflectivity method of Fuchs and Müller (1971). This shift of the position of the maximum observed amplitude to a larger offset is the consequence of interference between the waves that are reflected at the top and at the bottom of the thin layer. Constructive interference occurs only if the distance which the wave travels within the low-velocity layer equals half the wave length within the layer. This condition requires a certain angle of incidence which seems to correspond to the offset of 3 to 4.5 km, where the maximum amplitudes are observed.

Since no continuous reflections are observed neither above nor immediately below the reflector, interpreted to originate from the thrust fault zone, it is doubtful whether this zone is underlain by basement rocks or sediments.

As mentioned above, velocity-depth functions show no discontinuity in the depth range of the shallow reflector. Therefore, it seems to be possible that sediments are present beneath the thrust plane.

Other reflection seismic surveys located along the Variscan deformation front (Faille du Midi) show a similar reflector at a depth comparable to that of the Aachen Thrust fault. Figure 4 shows the location of the additional reflection surveys and a preliminary depth contour map of the fault zone. In the Famenne region, Belgium, the depth is 4 to 6 km (Bless et al., 1980) and in northern France, near Avesnes, it increases to more than 6 km (Clément, 1963). Further to the west it has also been detected by profiles of exploration surveys (Laumondais, pers. commun.). Several wells drilled in the area south of the Faille du Midi show Devonian sediments lying above Carboniferous strata. These observations and some continuous reflectors below the thrust plane recognized on the French profiles, show the possibility of sedimentary rocks lying below the thin-skinned nappe.

Further to the west, across the English Channel, a similar overthrust fault has also been observed on seismic reflection data in Wiltshire, England (Kenolty et al., 1981). From this it is concluded that the structural style of the Variscan Externides between the Rhenish Massif and Southern England is charac-

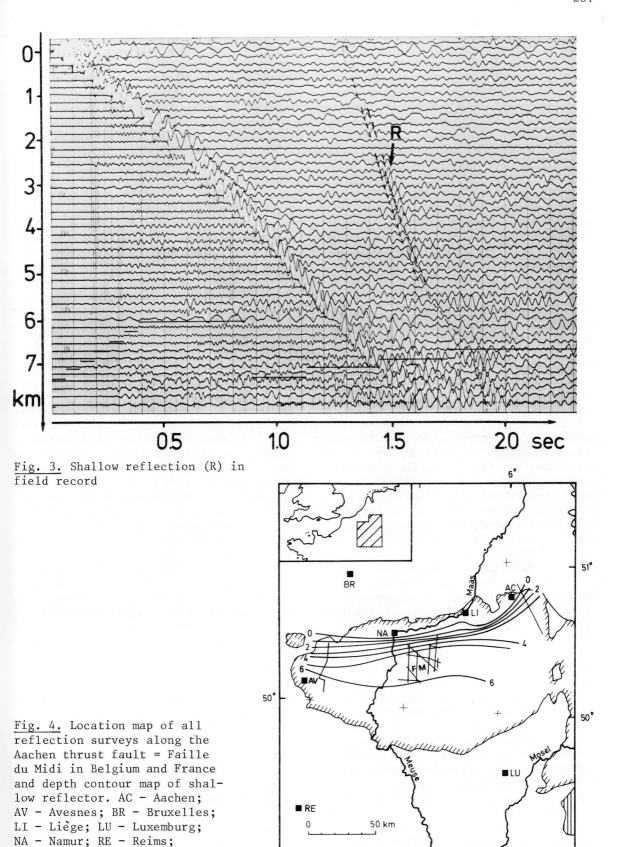

Fig. 3. Shallow reflection (R) in field record

Fig. 4. Location map of all reflection surveys along the Aachen thrust fault = Faille du Midi in Belgium and France and depth contour map of shallow reflector. AC - Aachen; AV - Avesnes; BR - Bruxelles; LI - Liège; LU - Luxemburg; NA - Namur; RE - Reims; FM - Famenne region

terized by thin-skinned thrust faulting of major proportions. This implies that the late Carboniferous Variscan diastrophism involved large scale decollement thrustings and correspondingly major compression. As these thrust faults skirt the apparently stable London-Brabant Massif, the latter may have acted as a ramp for the approaching orogenies from the south.

Although the presently available seismic reflection data permit an understanding of the structural style of the Variscan externides they do not allow us to assess the amount of horizontal shortening that was taken up in the sediments involved in them. Yet reflection data from southern Belgium indicate that the horizontal displacement along the Faille du Midi, which carries the allochthonous Dinant Basins, exceeds 30 km (Ziegler, 1982).

Presently, minor earthquakes occur along the shallow thrust fault (Ahorner, 1983). Recent vertical movements in the Hohes Venn area take place at a rate of up to 1.6 mm yr^{-1} (Mälzer, 1983). Both observations in the same region seem to be related to each other. Although the seismic slip along the fault zone is presently smaller than the uplift rate, it is probable that an additional aseismic slip is present and a conversion of horizontal to vertical movement takes place along listric faults below the Venn area.

Below the thrust fault down to 4 s two-way traveltime there are only short discontinuous reflections. Some curved events visible in the stacked but not in the migrated section indicate diffractions originating at small-scale structures. Below 6 s two-way traveltime, corresponding to about 18 km in depth, the reflectivity increases. This may be related to the nature of the ductile lower crust in which the formation of crystallization seams after cooling of intruded magmas causes an increase of impedance contrast of sufficient magnitude to observe reflections (Meissner and Strehlau, 1982). In addition, some reflectors may be interpreted as parts of listric fault zones.

Figures 5 and 6 show the stacked and migrated sections of the profile. In the southern part of the profile at about 6 s, a synclinal structure and many reflection elements that dip in a northerly direction are observed. These reflectors may possibly be related to lower crustal discontinuities that developed in conjunction with movements along the Aachen thrust plane. The different termination of two prominent deep reflectors at their northern end and its small southerly dip in this area indicates the possibility of a deep and disturbed fault-like zone at the northern boundary of the Variscan mountains. The northern part of the profile seems to exhibit a lower density of reflections as compared to the southern part. This is probably a consequence of a substantial decrease in signal-to-noise ratio in the northern part where thick Tertiary and Quaternary sediments occur near the surface.

Reflections at 6 s two-way traveltime (18 km) may be interpreted as a "Conrad" discontinuity, and even a "Subconrad" boundary between Conrad and Moho may be identified similar to observations along the 1968 Hunsrück profile (Glocke and Meissner, 1976). The Moho consists of a band of discontinuous reflectors, some of them up to 3 km long at 10 to 11.5 s two-way traveltime corresponding to a depth of about 30 to 35 km. This is in accordance with the results of refraction seismic experiments in the northern parts of the Rhenish Massif (Mechie et al., 1983). The Moho seems to be horizontal and rather undisturbed. This may indicate that the crust-mantle-boundary is younger than the dipping reflectors in the lower crust.

Figure 7 shows a comparison of the reflection density along the four reflection profiles. A common feature is the strongly increased reflectivity of the lower crust. Obvious peaks of reflection density appear to correspond to the Conrad, Subconrad and Moho discontinuities.

Figure 8 finally shows a cross-section along the NNW-SSE arranged reflection profiles summarizing the information of reflected as well as refracted data.

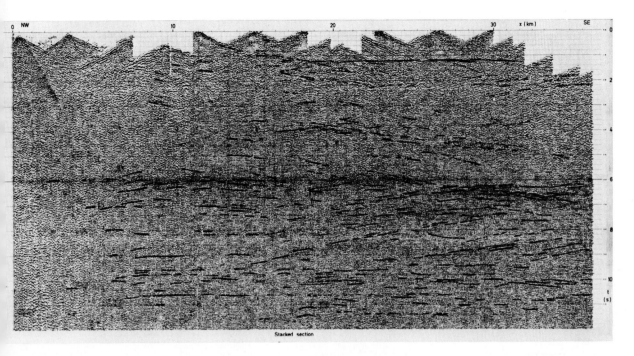

Fig. 5. Unmigrated stacked cross-section of the Aachen profile down to 12 s two-way traveltime

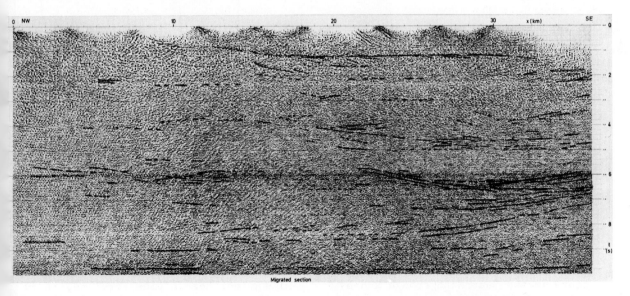

Fig. 6. Migrated stacked cross-section down to 10 s two-way traveltime

Relation of the Results to the Tectonic Development and the Present-Day Uplift

There is little doubt that the Variscan fold belts developed in response to a compressional stress regime whereby orogenic activity moved from SSE to NNW during late-Early and Late Carboniferous (Ziegler, 1978; Weber, 1981). It is uncertain whether the present-day Hunsrück boundary fault is superimposed on a major thrust fault that was active during the Hercynian diastrophism and reactivated in post-Carboniferous time, or whether it corresponds to a structural discontinuity that developed during the post-Hercynian Stephanian Autunian phase of wrench fault-

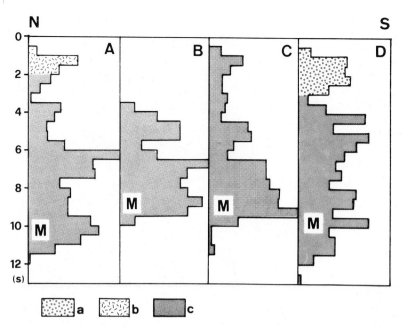

Fig. 7. Comparison of the density of reflections for the four different areas of seismic reflection surveys. *A* Hohes Venn near Aachen; *B* Hunsrück 1968; *C* part of the 1973 profile north of the Hunsrück border fault; *D* southern part of 1973 profile; *M* Moho discontinuity; a) sedimentary basin; b) thrust fault c) upper and lower crystalline crust

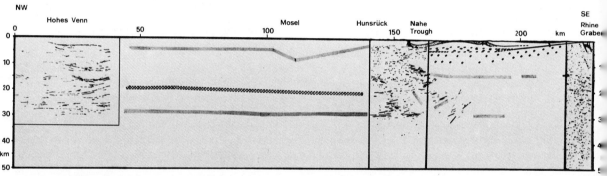

Fig. 8. Cross-section through a NNW-SSE line between Aachen and the Upper Rhine Graben, based on reflection investigations and the refraction studies of Meissner et al. (1980) and Mechie et al. (1983). Results of refraction and wide angle reflection investigations are *dotted*

ing. In any case Late Carboniferous-Early Permian activity along the Hunsrück-border fault is illustrated by the thickness and the dip of corresponding strata in the Saar-Nahe trough and by the extensive Autunian volcanism in this basin.

In general there is an orogenic and postorogenic uplift associated with compressional stress patterns and a subsidence associated with tensional tectonics, but there is also an uplift associated with plumes or other diapiric movements from the mantle in the initial stage which might develop later to the formation of grabens with raised rims or even to continental rifting if tectonic activity is strong (Meissner, 1981).

Considering present-day stress patterns in the northern Alpine foreland, it is evident that the Alpine orogeny dominate and influences the tectonic development. The direction of compressive stresses seems to be a consequence of the Alpine development, as seen in the maps of Greiner (1978) or Illies and Baumann (1982). This means that the area of the Rhenish Massif with the possible exception of the Lower Rhenish Embayment and the upper Rhine Graben, is dominated by a slightly divergent σ_1 axis directed NNW-SSE, i.e., in the same direction as during the Hercynian orogeny. It is in a most favorable condition to reactivate the old Hercynian fault zones. Such a process is indicated by the seismicity along the northern deformation front and the Hunsrück boundary fault as well as by focal plane solutions of selected earthquakes (Ahorner, 1983; Ahorner and Murawski, 1975). The increased vertical uplift in the Hohes Venn area as depicted in the map of Illies and Baumann (1982), as well as the displacement vectors, almost certainly are the consequence of a reactivation of the seismically detected overthrust fault seen in Fig. 2.

These present compressive features along the northern and southern boundaries of the Rhenish Massif, at least partly associated with strong uplifts, do not, however, seem to be in harmony with the Quaternary volcanism and with the strong tensile stresses in the middle of the Rhenish Massif. It seems that even changes in stress direction associated with major fault zones with their decreased viscosity cannot account for the widespread recent extensional features inside the Rhenish Massif. A second process has to be involved. Based on the anomalous low velocity area found from traveltime residuals (Raikes, 1980) and based also on the increased reflectivity of the lower crust especially along the Hunsrück profiles of 1968 and 1973 (Glocke and Meissner, 1976; Meissner et al., 1980) an autochthonous, possibly rift-related, interaction between crust and mantle has to be assumed.

According to Ziegler (1982), the development of the Rhine Graben rift system started in the Late Eocene and has continued with little interruption to the present. Updoming of the Rhenish Massif began during the early Miocene causing the severance of the oligocene marine connection through the Rhine-Leine Graben. This doming, coupled with continued crustal extension may have caused a reactivation of the Hunsrück border fault. At the same time an upper mantle anomaly developed under the Rhine-Ruhr-Leine Graben triple junction. This was accompanied by extensive volcanic activity that finds a late Pleistocene peak in the development of the Eifel Maars.

As mentioned before, the beginning of extensional developments is associated with an uplift or a doming. Plume activity below the Rhenish Massif would indeed explain the general uplift and the extensional features at its center. The increased reflectivity in the lower crust is assumed to originate from intruding plume material which can easily enter the low-viscosity lower crust of warm continental areas and spread out laterally. The different reflectors and even a new Moho may be formed in fractional crystallization of the intruded material. This process of formation of a new crust-mantle boundary is basically identical with that below oceanic ridges and other active rifts such as the East African rift system and seems to represent a common feature of extensional regimes. The application of such processes to the Rhenish Massif seems to be justified by a decreased Moho depth of only 27 to 28 km in the center of the 1968 reflection profile, while depths of 30 to 31 km were recorded at the western side of that profile and also on the northern and southern reflection profiles. It might be added that also in the Urach area, where a very detailed reflection profile was observed (Bartelsen et al., 1982), a huge low velocity body of intruded mantle material in the crust is associated with a drastic decrease of crustal depth.

Conclusions

Based on seismic reflection measurements in the Rhenish Massif and its boundaries, the shape and the depth pattern of prominent fault zones and the detailed structure of the crustal sections could be determined. A huge thin-skinned nappe forms the northern Variscan deformation front. The recent uplift in the Hohes Venn area could possibly be related to nearly horizontal movements along the reactivated Aachen overthrust fault. For the Hunsrück area only small movements along the Hunsrück Border Fault may have taken place, both kind of movements being compatible with a NNW-SSE directed compressional stress pattern similar to the one during Carboniferous times when these faults originated. For the bulk of the uplift of the Rhenish Massif a plume-supported rift related interaction between mantle and crust is indicated.

Acknowledgments. We thank E. Lüschen for critical comments and our colleagues from the Geologic-Paleontologic Institute of the RWTH Aachen for providing the geologic profile of the Hohes Venn area. The reflection investigations were supported by various grants from the Deutsche Forschungsgemeinschaft (German Research Foundation). The manuscript benefited greatly from critical reviews by P.A. Ziegler and C. Prodehl.

References

Ahorner, L., 1983. Historical seismicity and present-day microearthquake activity of the Rhenish Massif. Central Europe. This Vol.

Ahorner, L. and Murawski, H., 1975. Erdbebentätigkeit und geologischer Werdegang der Hunsrück-Südrand-Störung. Z. Dtsch. Geol. Ges., 126:63-82.

Bartelsen, H., Lueschen, E., Krey, Th., Meissner, R., Schmoll, H., and Walter, Ch., 1982. The combined seismic reflection-refraction investigation of the Urach geothermal anomaly. In: Haenel, R. (ed.) The Urach geothermal project. Schweizerbart, Stuttgart, pp 247-262.

Berckhemer, H., 1970. MARS 66; eine Magnetbandapparatur für seismische Tiefensondierung. Z. Geophys., 36:501-518.

Bless, M.J.M., Bouckaert, J., and Paproth, E., 1980. Environmental aspects of some Pre-Permian deposits in NW Europe. Meded. Rijks. Geol. Dienst, 32:3-13.

Clément, J., 1963. Résultats préliminaires des campagnes géophysiques de reconnaissance dans les permis de recherches "Arras et Avesnes" de l'Association Shell Francaise - P.C.R.B. - SAFREP - Objectifs du forage profond Jeumont-Marpent No. 1. Ann. Soc. Géol. Nord, 83:237-241.

Etheridge, M.A. and Wilkie, J.C., 1979. Grainsize reduction, grain boundary sliding and the flow strength of mylonites. Tectonophysics, 58:159-178.

Fuchs, K. and Müller, G., 1971. Computation of synthetic seismograms with the reflectivity method and comparison with observations. Geophys. J.R. Astron. Soc., 23:417-433.

Glocke, A. and Meissner, R., 1976. Near-vertical reflections recorded at the wide angle profile in the Rhenish Massif. In: Giese, P., Prodehl, C., and Stein, A. (eds.) Explosion seismology in Central Europe, data and results. Springer, Berlin, Heidelberg, New York, pp 252-256.

Greiner, G., 1978. Spannungen in der Erdkruste - Bestimmung und Interpretation am Beispiel von in situ - Messungen im süddeutschen Raum. Dissertation, Univ. Karlsruhe, 192 pp.

Illies, J.H. and Baumann, H., 1982. Crustal dynamics and morphodynamics of the Western European Rift System. Z. Geomorphol. N.F., Suppl., 42: 135-165.

Kenolty, N., Chadwick, R.A., Blundell, D.J., and Bacon, M., 1981. Deep seismic reflection survey across the Variscan Front of southern England. Nature (London), 293:451-453.

Kern, H., 1978. The effect of high temperature and high confining pressure on compressional wave velocities in quartz-bearing and quartz-free igneous and metamorphic rocks. Tectonophysics, 44:185-203.

Mälzer, H., Hein, H., and Zippelt, G., 1983. Height changes in the Rhenish Massif: determination and analysis. This Vol.

Mechie, J., Prodehl, C., and Fuchs, K., 1983. The long-range seismic refraction experiment in the Rhenish Massif. This Vol.

Meissner, R., 1967. Exploring deep inter faces by seismic wide angle measurements. Geophys. Prospect., 15:598-617.

Meissner, R., 1973. The "Moho" as a transition zone. Geophys. Surv., 1: 195-216.

Meissner, R., 1981. Passive margin development. A consequence of specific convection patterns in a variable viscosity upper mantle. Oceanol. Acta, pp 115-121.

Meissner, R. and Strehlau, J., 1982. Limits of stresses in continental crusts and their relation to the depth frequency distribution of shallow earthquakes. Tectonics, 1:73-89.

Meissner, R., Bartelsen, H., Glocke, A., and Kaminski, W., 1976. An interpretation of wide-angle measurements in the Rhenish-Massif. In: Giese, P., Prodehl, C., and Stein, A. (eds.) Explosion seismology in Central Europe, data and results. Springer, Berlin, Heidelberg, New York, pp 245-251.

Meissner, R., Bartelsen, H., and Murawski, H., 1980. Seismic reflection and refraction studies for investigating fault zones along the Geotraverse Rhenoherzynikum. Tectonophysics, 64:59-84.

Meissner, R., Bartelsen, H., and Murawski, H., 1981. Thin-skinned tectonics in the northern Rhenish Massif, Germany. Nature (London), 290:399-401.

Raikes, S., 1980. Teleseismic evidence for velocity heterogeneity beneath the Rhenish Massif. J. Geophys., 48:80-83.

Springer, M., 1982. Auswertung reflexionsseismischer Messungen im Hohen Venn. Diplomarbeit, Univ. Kiel, 139 pp.

Weber, R., 1981. The structural development of the Rhenische Schiefergebirge. Geol. Mijnbouw, 60:149-159.

Ziegler, P.A., 1978. North-West Europe: tectonics and basin development. Geol. Mijnbouw, 57:589-626.

Ziegler, P.A., 1982. Geological Atlas of Western and Central Europe. Shell Int. Petrol. Maatschappij B.V., 130 pp.

Zoback, M.D. and Zoback, M.L., 1981. State of stress and intraplate earthquakes in the United States. Science, 213:96-104.

7.3 Electrical Conductivity Structure of the Crust and Upper Mantle Beneath the Rhenish Massif

H. Jödicke[1], J. Untiedt[1], W. Olgemann[2], L. Schulte[1], and V. Wagenitz[2]

Abstract

A magnetotelluric survey was undertaken in the Rhenish Massif and adjacent areas. Observations were made in the period range 5-3000 s at more than 70 sites arranged along four profiles. The profiles Eifel-Palatinate, Sauerland-Taunus, and Münsterland-Spessart run perpendicular to the main trend of the Variscan fold belt. The profile Eifel-Rothaargebirge follows the Variscan trend. Results from a 1D inversion of observed rotationally invariant resistivity and phase values show mainly a southward dipping intracrustal conductor, in striking correspondence to a subfluence model or to large-scale over thrusting as geodynamic models of the Rhenish Massif. Preliminary 2D-modeling indicates that there is no major dislocation of this conductor at the southern margin of the Rhenish Massif (Hunsrück-Südrand-Störung). In some areas - especially outside the Rhenish Massif proper - an additional increase of conductivity at shallower depths was found. According to the present stage of data analysis and interpretation, there are no distinct regional differences at upper mantle depths within the Rhenish Massif.

Introduction

In the years 1977-1981 a magnetotelluric survey was undertaken in the Rhenish Massif and adjacent areas to investigate the distribution of the electrical conductivity at depth. Observations were carried out at more than 70 sites. As shown in Fig. 1 the magnetotelluric stations were arranged along four profiles, three of which running roughly NNW to SSE, i.e., perpendicular to the main trend of the Variscan fold belt (cf. Murawski, 1983). The fourth profile from the Eifel to the Rothaargebirge follows the main trend and coincides with section D-F-G of the long seismic refraction line (Mechie et al., 1983).

The westernmost profile is the only one situated west of the Rhine. It runs from the eastern part of the Eifel in the north to the Saar-Nahe basin (Palatinate) covered by younger sediments in the south. The fault forming the southern margin of the Rhenish Massif (Hunsrück-Südrand-Störung; Ahorner and Murawski, 1975) is crossed between the stations ELLE and SPON. Continuation of this profile to the north, as well as realization of a profile through the western part of the Eifel, which both would have been desirable, were not possible due to the presence of strong electric noise of regional extent. Similarly, noise problems in the Dortmund and Frankfurt areas prevented continuation of the central profile Sauerland-Taunus beyond the margins of the Paleozoic. The easternmost profile starts in the Münsterland basin with its Cretaceous sediments and crosses, from north to south, the Ostsauerland anticline (stations HELM-HEBO

---

1 Institut für Geophysik der Westfälischen Wilhelms-Universität, Corrensstaße 24, D-4400 Münster, Fed. Rep. of Germany

2 DEMINEX, Dorotheenstraße 1, D-4300 Essen 1, Fed. Rep. of Germany.

Plateau Uplift, ed. by K. Fuchs et al.
© Springer-Verlag Berlin Heidelberg 198

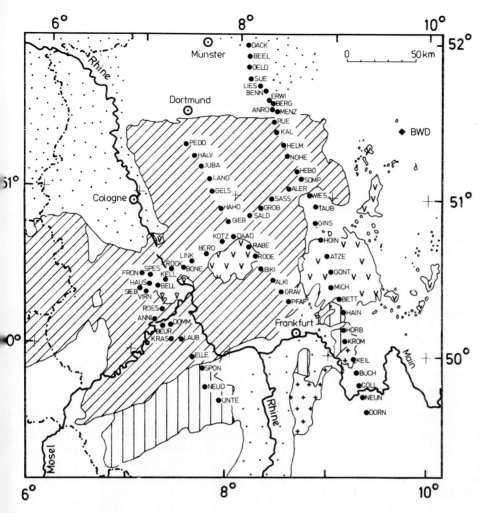

Fig. 1. Map showing sites of magnetotelluric soundings. For legend regarding geology, see Mechie et al. (1983). BWD site of magnetotelluric sounding by Richards et al. (1981)

the Hessian depression (stations WIES-HORB) partly covered by the Tertiary lava flows of the Vogelsberg, and the crystalline part of the Spessart (stations KROM and KEIL). This profile finally penetrates the Mesozoic predominating the area south of the river Main.

The Magnetotelluric Method

This method (for details see, e.g., Patra and Mallick, 1980) is based on simultaneous observations of the time-varying part of the geomagnetic field and the corresponding time-varying electric field induced in the Earth. The geomagnetic variations as measured in components B_x (towards geographic north), B_y (towards east), and B_z (downwards), are caused by both ionospheric currents and currents induced within the Earth. At the Earth's surface the induced currents are known as telluric currents. They are connected to the time-varying horizontal electric field components E_x and E_y by Ohm's law. Two examples of field records are shown in Fig. 2. The desired information on the electrical conductivity at depth is contained in the observed amplitude and phase relationships between the components of the electric and magnetic fields at different periods. Generally, observations made at longer periods give information at greater depths. However, good conductors have a shielding effect so that in the presence of well-conducting layers, even with rather long period variations, it may be impossible to resolve conductivity structures at greater depths.

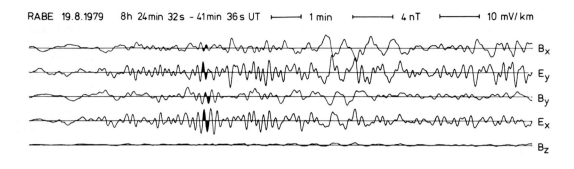

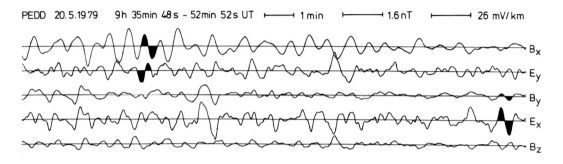

Fig. 2. Example of band-pass (5-130 s) filtered magnetotelluric records from sites RABE and PEDD. UT - Universal Time (Greenwich Mean Time)

If E_x, B_x etc. denote complex numbers characterizing both amplitude and phase of the respective field components in the frequency domain (after, e.g., Fourier transformation or filtering), the following linear relations are valid for a specific period T:

$$E_x = Z_{xx} B_x + Z_{xy} B_y$$
$$E_y = Z_{yx} B_x + Z_{yy} B_y \quad (1)$$

Z_{xx} etc. are complex numbers, dependent on both period T and site. They are called transfer functions and contain the full information on the conductivity distribution.

For a horizontally layered Earth (one-dimensional or 1D-case), $Z_{xx} = Z_{yy} = 0$ and $Z_{xy} = -Z_{yx}$. This implies that the pairs (E_x, B_y) and $(-E_y, B_x)$ show equal amplitude ratios and phase differences. As an example, this holds approximately for the shaded portions on the RABE record in Fig. 2. Theoretically, there exists a unique relation (Weidelt, 1972) between $Z_{xy}(T) = -Z_{yx}(T)$ and the resistivity ρ (which is the reciprocal value of the conductivity) as a function of depth z. Accordingly $\rho(z)$ may be determined from measured transfer functions $Z_{xy}(T)$ or $Z_{yx}(T)$ by some inversion technique, with limitations due to imperfect data.

Usually, in order to obtain a first idea of the conductivity distribution, an apparent resistivity ρ_a defined by

$$\rho_a = \mu_o T |Z_{xy}|^2/2\pi = \mu_o T |Z_{yx}|^2/2\pi \quad (2)$$

is displayed as a function of period T. $\rho_a(T)$ has a meaning similar to that of $\rho_a(L/2)$ (L = current electrode separation) in Schlumberger geoelectrics.

If, in addition to its dependence on depth, the conductivity changes in one arbitrary horizontal direction (2D-case) then generally $Z_{xx} \neq 0$, $Z_{yy} \neq 0$, and $Z_{xy} \neq -Z_{yx}$. However, by appropriate rotation of the x- and y-axes a new coordinate system may be found in which $Z_{xx} = Z_{yy} = 0$. One of the rotated coordinate axes will coincide with the strike of the conductivity distribution. Note that in the new system Z_{xy} and $-Z_{yx}$ in general remain unequal [as is clearly demonstrated by comparing the shaded portions of the (E_x, B_y) and $(-E_y, B_x)$

records of PEDD in Fig. 2]. Thus, two different apparent resistivities (ρ_{axy} and ρ_{ayx}) are defined by Eq. (2).

In the most general 3D case it will not be possible to bring Z_{xx} and Z_{yy} to zero by rotation. However, a rotated coordinate system may be found in which $|Z_{xx}|^2 + |Z_{yy}|^2$ has a minimum.

The Data and Their Analysis

The data were obtained by means of two METRONIX automatic magnetotelluric stations (Karmann, 1977; Kröger, 1977). These systems include a three-component induction coil magnetometer, calomel probes for measurement of the electric field, and microprocessor-controlled digital recording on cassette. The signals were measured separately in two period bands (about 5-130 s and 130-3000 s). Short period variations were recorded intermittently only at times of sufficient geomagnetic activity, such times being recognized by an automatic signal detection circuit. Servicing of the stations was necessary about once a week. On the average, total running time at a single site was four weeks.

The transfer functions Z_{xx} etc. were determined from the data using a single-event technique (Jödicke, 1978; see also Hermance, 1973). In this technique, time intervals with enhanced magnetic activity in several period bands are selected in order to reduce the influence of noise.

Examples of $\rho_a(T)$ curves derived from such transfer functions using Eq. (2) are shown in the upper part of Fig. 3, containing data from three adjacent stations (cf. Fig. 1, central profile Sauerland-Taunus). The lower part of this figure gives the corresponding curves after rotation as outlined above. At RABE, after rotation, ρ_{axx} and ρ_{ayy} are negligibly small with respect to ρ_{axy} and ρ_{ayx}. At GIEB, all four sets of ρ_a-values are of about the same magnitude. Thus, RABE may be considered to represent a 2D case, whereas at GIEB the conductivity distribution must be assumed to be three-dimensional. DAAD seems to illustrate a case in between. None of the more than 70 stations was found to represent a true 1D case. However, many stations showed ρ_{axy}- and ρ_{ayx}-curves which were not very different from each other (RABE being an example). In approximation, such stations may be treated as 1D cases.

In order to show some first results of the data analysis $|Z_{xy}|$- and $|Z_{yx}|$-values (after rotation; period T = 100 s) are displayed in Fig. 4 for the whole set of stations. For each station the cross-bars shown indicate the directions of the rotated x- and y-axes, and their lengths represent twice the magnitudes of $|Z_{xy}|$ and $|Z_{yx}|$, respectively. Apparently, there is a tendency for the axes of the new coordinate systems to be either parallel or perpendicular to the main trend of the Variscan fold axes. Note that, since B_x and B_y may be assumed to vary spatially but little over the area, the strong differences in $|Z_{xy}|$, $|Z_{yx}|$ and rotation angle from site to site as indicated in Fig. 4 must be largely due to spatial variations in the electric field. These differences may be either of a more local (e.g., stations LINK, GRAV) or a more regional origin. For instance, in the south-west the pattern is characteristic of a transition from a bad conductor in the north (Rhenish Massif) to a better conductor in the south (Saar-Nahe basin): the component of the electric field perpendicular to the southern margin of the Rhenish Massif (and accordingly the corresponding transfer function) drops appreciably across the separating fault, whereas the parallel electric field component and the corresponding transfer function remain almost constant.

There are several groups of stations for which, according to Fig. 4, $|Z_{xy}|$ and $|Z_{yx}|$ are not very different from each other and for which $|Z_{xx}|$ and $|Z_{yy}|$ are small. These correspond approximately to 1D cases of conductivity structure. They include the eastern Eifel, the area south-east of the Westerwald, the Hessian Depression and the northern part of the Münsterland Basin. Distinct 2D cases are found especially in the northern part of the

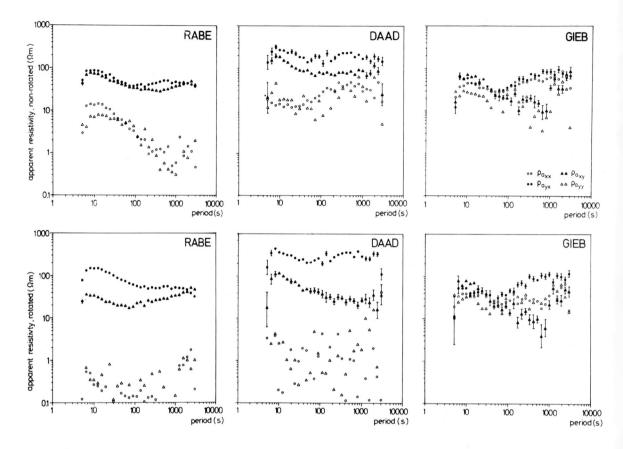

Fig. 3. Apparent resistivities ρ_a as functions of period derived from the four transfer functions Z_{xx}, Z_{xy}, Z_{yx}, and Z_{yy}, observed at three sites. *Upper line* geographically oriented coordinate system. *Lower line* coordinate systems dependent on local conditions (see text). *Error bars* drawn with ρ_{axy} and ρ_{ayx}, and only shown if larger than symbols

Rhenish Massif (Ebbe Anticline and Ostsauerland Anticline), in the southeastern part of the Münsterland Basin (Lippstädter Gewölbe, Jödicke et al., 1982), near the southern margin of the Rhenish Massif, as mentioned above, and in the area of the crystalline part of the Spessart and its southern neighborhood.

Figure 5 presents first results from the data analysis in an even more condensed form. Here, an average apparent resistivity $\bar{\rho}_a$ is displayed which was calculated from the rotationally invariant transfer function

$$\bar{Z} = (Z_{xy} - Z_{yx})/2 \qquad (3)$$

using Eq. (2). $\bar{Z}$ was introduced by Berdichevsky and Dmitriev (1976) because in many situations this quantity is less

Fig. 4. Absolute values of transfer functions Z_{xy} and Z_{yx} drawn in the locally rotated coordinate systems (see text), for period T = 100 s.

Fig. 5. Site values and isolines of apparent resistivity $\bar{\rho}_a$ derived from rotationally invariant transfer function $\bar{Z} = (Z_{xy} - Z_{yx})/2$, for period T = 100 s. Numbers in Ωm. *Crosses* Results from Fluche (unpubl.). In the north (Münsterland), results from Jödicke et al. (1982 added. *Isolines* in the south (Rhine Graben) drawn with regard to results by Scheelke (1972). Values given with isolines approximately equally spaced on a logarithmic scale

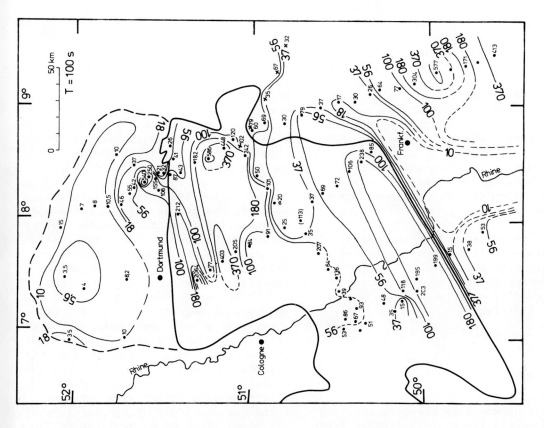

Fig. 5 (figure caption see page 292)

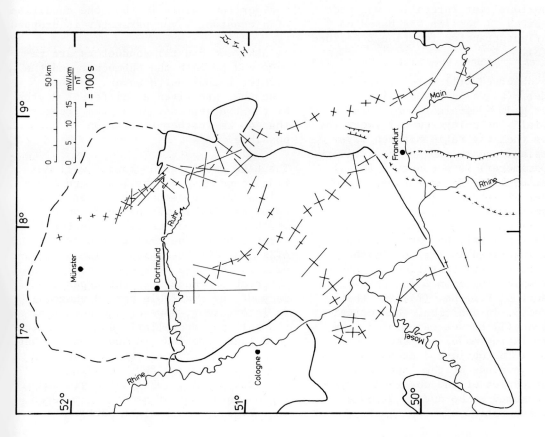

Fig. 4 (figure caption see page 292)

affected by deviation from the 1D case than is Z_{xy} or Z_{yx}. Apparently, the distribution of $\bar{\rho}_a$ corresponds rather well to the main geological structures. The main anticlines, as well as the crystalline part of the Spessart, exhibit large values, whereas the Münsterland Basin and the continuation of the Upper Rhine Graben into the Hessian Depression show low apparent resistivity values. In the central part of the Rhenish Massif a zone of decreased apparent resistivity seems to follow the main Variscan strike direction.

It should be stressed that the above discussion as well as the figures refer to one specific period of T = 100 s only. For appreciably shorter or longer periods conditions were found to be somewhat different.

1D Interpretation

As a first approach to the determination of the conductivity distribution at depth from the observed transfer functions, a one-dimensional interpretation was made for all sites. Inversion was based on the invariant transfer function $\bar{Z}$ defined by Eq. (3).

From the observed $\bar{Z}(T)$-values, a resistivity model $\rho(z)$ was derived using Schmucker's (1974) ψ-algorithm. This method yields an n-layer model under the condition that the ratio between layer thickness and square root of resistivity (layer parameter) is a constant for all n layers. Both the layer parameter and the number of layers n is optimized in the process.

For three stations, the method is partially illustrated in Fig. 6. In the upper part, for every station the observed complex function $\bar{Z}$ is represented in terms of $\bar{\rho}_a$ [cf. Eq. (2)] and the phase $\bar{\phi}$ of $\bar{Z}$. In addition, theoretical $\rho_a(T)$- and $\phi(T)$-curves calculated from the best-fitting model are shown for comparison. The obvious decrease of measured $\bar{\rho}_a$-values at the lower and upper boundaries of the observed period band is assumed to be due to a low signal-to-noise ratio.

At all three stations the most prominent and most reliable feature in the models is a good conductor in the middle or lower part of the crust. This corresponds to the minimum of the $\bar{\rho}_a$-values at a period of about 100 s. For station ATZE, the best-fitting five-layer model (with 68% confidence limits) is shown together with a four-layer model, also derived from Schmucker's ψ-algorithm. Clearly, in both models the good conductor appears at about the same depth, but with different thickness and resistivity. However, the product of thickness and conductivity (integrated conductivity) is about the same in both cases, illustrating the principle of equivalence. Below the good crustal conductor, the five-layer model indicates a second less distinct conductor at about 50-60 km depth. Here, the four-layer model shows only a half-space with a resistivity about equal to the average resistivity observed in the lower part of the five-layer model. Furthermore, the good crustal conductor is confirmed by application of a different inversion technique developed by Eichler (unpubl.; cf. model given by dashed line in Fig. 6). This method differs from Schmucker's ψ-algorithm mainly in that the condition of a constant layer parameter is dropped

At ALKI two crustal conductors are recognized by both the Schmucker and the Eichler technique. However, in this case they are given at different depths by the two methods. At depths greater than 40 km further resolution is not possible.

The LAUB model again shows a well-defined conductor in the lower crust and a less certain second conductor in the upper mantle. It may be questioned whether the upper mantle conductor indicated by the best-fitting models at ATZE and LAUB has a real existence, especially because it relies heavily on observations made near the upper boundary of the available period range. It is interesting, however, to note that the only magnetotelluric sounding which used longer periods in addition and was done in a nearby part of the Variscan fold belt (station BWD, see Fig. 1; Richards et al., 1981), also yielded a good conductor at about 50 km depth (cf. Fig. 6, lower right part).

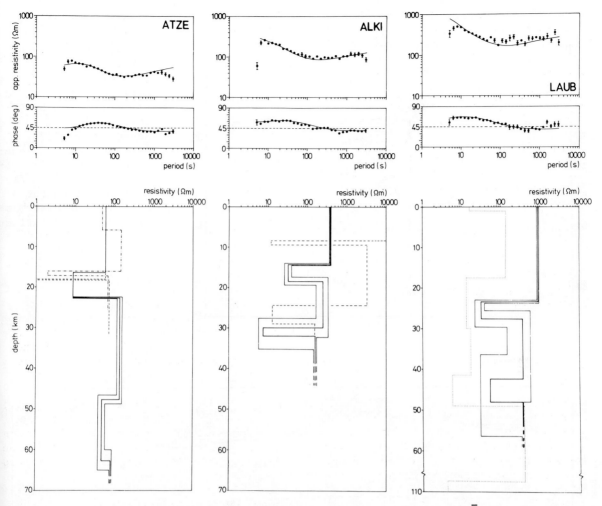

Fig. 6. *Upper part* apparent resistivities and phases derived from $\bar{Z}$ as a function of period, for three sites. *Solid lines* denote theoretical curves calculated from best-fitting models as shown in lower part. *Lower part*: *Solid lines* best-fitting models with confidence limits from 1D-inversion of data shown in upper part, using Schmucker's (1974) ψ-algorithm. *Dashed-dotted line* best-fitting Schmucker model if number of layers is assumed to be 4. *Dashed lines* best-fitting Eichler (unpubl.) models. *Dotted line* model derived for station BWD (cf. Fig. 1) by Richards et al. (1981)

Figures 7 to 10 summarize the results of 1D interpretation for all four profiles (cf. Fig. 1). Good conductors are shaded. The most remarkable result is the existence of a well-conducting zone in the middle or lower crust beneath the entire area for which the crustal thickness is about 30 km (Mechie et al., this vol.). Along the western and central profiles (Figs. 7 and 8, respectively) this zone seems to dip southward. Near the northern margin of the Rhenish Massif on the eastern profile (Fig. 10) a remarkable upwelling of this zone is observed. However, one has to keep in mind that such an effect may at least partially be due to nonvalidity of the 1D interpretation in a region governed by 2D or 3D structures (cf. also Fig. 4). This may be especially true for the southernmost part of the eastern profile (Fig. 10, Spessart region). Two groups of stations in the Rhenish Massif proper (south of the Westerwald, Fig. 8, and west of the Siegerland, Fig. 9) show a second crustal conductor at smaller

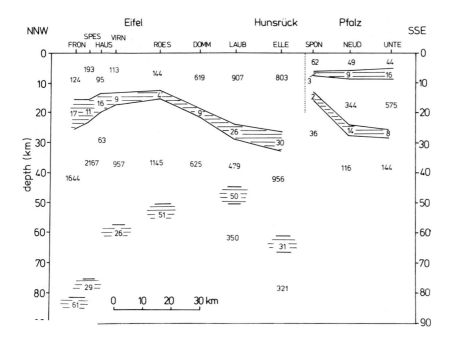

Fig. 7. Resistivity distribution (values in Ωm) in crust and upper mantle for profile Eifel - Palatinate derived from observed $\bar{Z}(T)$ functions by 1D inversion using Schmucker's (1974) ψ-algorithm. Good conductors *shaded*. *Vertical dotted line* marks position of southern margin of Rhenish Massif. Pfalz = Palatinate

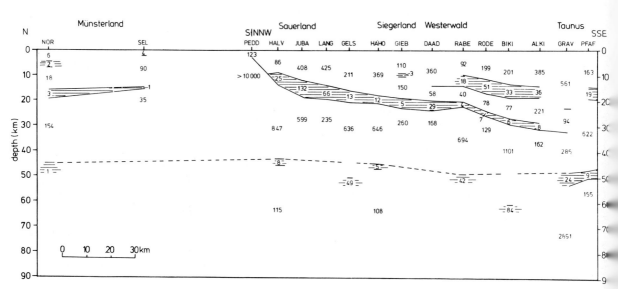

Fig. 8. As Fig. 7, but for profile Sauerland-Taunus. In the north (Münsterland), results from Jödicke (1980) added

depths (see, e.g., Fig. 6, ALKI). Well-conducting layers at even shallower depths below the Münsterland (Figs. 8 and 10) and the Palatinate (Fig. 7) are caused by sediments in the sub-Variscan foredeep and the Saar-Nahe Basin, respectively. Finally, a large number of the stations scattered over the entire area show the existence of slightly better conducting layers at depths around 50 km, similar to the cases of ATZE and LAUB discussed above (cf. Fig. 6).

2D Modeling

Two-dimensional modeling may be applied to the data in order to study to some extent the validity of a 1D interpre-

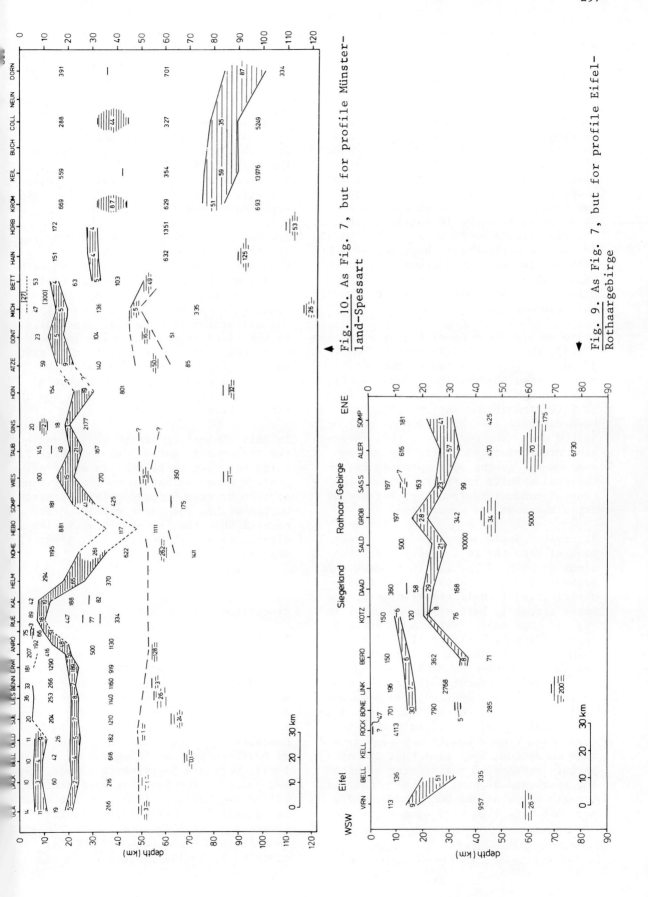

Fig. 10. As Fig. 7, but for profile Münsterland-Spessart

Fig. 9. As Fig. 7, but for profile Eifel-Rothaargebirge

tation. Such a study was undertaken for the western profile. Starting from the crustal resistivity results of the 1D analysis as shown in Fig. 7, a two-dimensional block model was designed, which is given by Fig. 11. The main crustal conductor was assumed to have a constant thickness and resistivity and to cross the southern margin of the Rhenish Massif (between ELLE and SPON) without any flexure or dislocation. For comparison, Fig. 11 also contains the position of the crustal good conductors as obtained from the 1D analysis (shaded, cf. Fig. 7). Below 28 km depth the model includes a horizontally homogeneous substratum. The well-conducting half-space beginning at 45 km depth is still unrealistic. It was chosen mainly for numerical reasons.

Using an algorithm developed by Schmucker (1971), theoretical $\bar{Z}(T)$-curves for the 2D model were calculated in the period range of the observations. Subsequently, these curves were subjected to the same 1D inversion (Schmucker, 1974) as had been used for interpretation of the observed data. The results of this inversion are summarized in Fig. 12 in the same manner as the corresponding observational results in Fig. 7. Again, the crustal conductors from Fig. 7 (shaded) are also shown for comparison.

At crustal depths there is good agreement between the resistivity distributions of Figs. 7 and 12. This means that probably the gross crustal resistivity distribution is well represented by the upper (crustal) part of the 2D model in Fig. 11. One major result appears to be that the apparent dislocation of the main crustal conductor at the southern margin of the Rhenish Massif as indicated from 1D inversion (Fig. 7) can not be taken as real. Instead, the apparent dislocation can be explained by the distortion of the electric field near the transition from a bad to a good near-surface conductor. Note also that the apparent increase of resistivity within the southward dipping crustal conductor beneath the stations ROES, DOMM, LAUB, and ELLE (cf. Fig. 7) does not have to be real.

As Fig. 12 shows, 1D inversion of $\bar{Z}$-values calculated from the 2D model in Fig. 11 gives rather poor results at greater depths (cf. broken line in Fig. 12 in comparison to full line at 45 km depth in Fig. 11). Much better results at great$\underline{er}$ depths can be obtained if instead of $\bar{Z}$ the transfer function Z_{xy} is used with the x direction (direction of the electric field) running parallel to the strike of the 2D conductivity distribution (known as E-polarization; see, e.g., Patra and Mallick, 1980).

However, a 1D inversion of observed Z_{xy}-values rotated according to the case of E-polarization was not used in the present study because of the large scatter of these values from site to site. This scatter is assumed to be due to both local distortion effects and, especially in the Eifel region, to noise in the data. Consequently, the 1D results given in Figs. 7-10 at upper mantle depths must be considered with great caution at the present stage of analysis and interpretation.

In this context it may be mentioned that the observed B_z variations (cf. Fig. 2), which have yet to be finally analyzed, contain some indication of the existence of a 2D or even 3D conductivity distribution within the upper mantle. First results show that non-Variscan strike directions probably prevail at greater depths.

Discussion

As discussed in great detail by Brace (1971), conductivity (because of its dependence on porosity and other parameters) should generally decrease with depth. Thus the middle and lower crust should be expected to be rather resistive. Examples for such behavior may be found in Southern Germany (Richards et al., 1981), Northern Scandinavia (Jones, 1982) and parts of the Canadian Appalachians (Kurtz and Garland, 1976). On the other hand, well-conducting intra-crustal layers have been detected at many places in different tectonic environments, for example, in Iceland (e.g., Beblo and

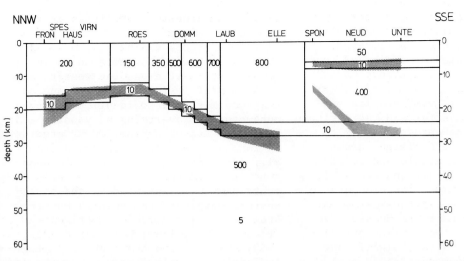

Fig. 11. Resistivity distribution (values in Ωm) for profile Eifel-Palatinate adopted in 2D modeling. Good conductors from Fig. 7 added for comparison (*shaded*). Half-space with 5 Ωm below depth of 45 km is unrealistic

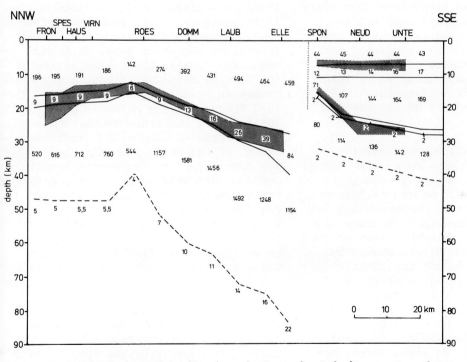

Fig. 12. Resistivity distribution (values in Ωm) in crust and upper mantle for profile Eifel-Palatinate, derived from $\bar{Z}(T)$ functions calculated from 2D model shown in Fig. 11, by 1D inversion using Schmucker's (1974) ψ-algorithm. Good conductors from Fig. 7 added for comparison (*shaded*)

Björnsson, 1980), in the Rio Grande rift valley (e.g., Hermance and Pedersen, 1980), in the Carpathians (Ádám, 1980), in Southern Africa (Van Zijl, 1977; Blohm et al., 1977), in the Adirondacks (Connerney et al., 1980), and in several regions of the U.S.S.R. (Berdichevsky et al., 1972).

Usually the following effects are considered as possible explanations of such anomalous conductivity enhancement: melts or partial melts, presence of water or other fluids, accumulation of well-conducting minerals like graphite, pyrite, magnetite, or hydrated minerals and thermally activated semiconduction.

In our case, results from xenolith studies in the eastern Eifel and the northern part of the Hessian Depression seem to preclude the presence of larger amounts of free water in the middle and lower crust (Voll, 1983; Mengel and Wedepohl, 1983). Also, no accumulation of hydrated minerals was detected in these studies. Xenoliths from the northern Hessian depression brought to the surface pyriclasites that contain in part significant amounts (0%-10%) of magnetite (Mengel and Wedepohl, 1983). Xenoliths from a depth of approximately 15-20 km beneath the Eastern Eifel also contain small amounts of magnetite, as well as ilmenite. In this depth range graphite-rich schists of low-grade metamorphic amphibolite facies were found, indicating thin layers of metamorphic black shales (Voll, pers. commun.). The existence of graphite at a depth of 15-20 km, estimated by Voll, may explain, in particular, the high integrated-conductivity at this depth (cf. Fig. 7).

According to heat flow measurements (Hänel, this Vol.) there seem to be no Indications of temperatures high enough to generate partial melt at crustal depths. Schmincke et al. (1983) suppose that hot magmatic rock bodies may be present at different depths within the crust of the volcanic part of the Eifel. Obviously, this assumption does not explain the widely observed occurrence of intra-crustal conductors in the other parts of the Rhenish Massif.

Independently, Weber (1978, 1981) and Behr (1978) as well as Giese (1983) developed geodynamic models of the Rhenish Massif which both contain southward-dipping crustal structures as essential elements. The subfluence model by Weber and Behr is based on a wealth of geological and petrological observations. Giese explains reflection and refraction seismic results from the Rhenish Massif and the region south of it in connection with the problem of crustal shortening during Variscan folding by large-scale overthrusting. The intra-crustal conductors of our present study as shown in Figs. 7 and 8 exhibit a southward dip in striking correspondence to these models, at least along the western and central profiles. Accordingly, one assumption might be to identify the good conductors as overthrusted younger sediments (cf. results from COCORP studies in the Eastern United States, Cook et al., 1979) or even as being the shear zone itself. Overthrusted younger sediments have been inferred at the northwestern margin of the Rhenish Massif from reflection seismic experiments (Bless et al., 1980; Meissner et al., 1983). East of the Rhine where geologic indications for large-scale overthrusting do not exist, reflection seismic experiments are lacking. Crack porosity generated by shear movements would enhance the conductivity due to electrolytic or surface conduction (Brace, 1971). However, such cracks may be assumed to have healed since the Carboniferous. On the other hand, it is obvious that during the recent uplift of the Rhenish Massif, Variscan shear zones became at least partly reactivated (cf. Ahorner, 1983; Meissner et al., 1983). It is uncertain whether this is also true for the middle and lower crust. If the well-conducting layers are considered to represent metasediments as discussed above (e.g., graphite bearing metamorphic black shales of Precambrian age), then they indicate structures within the crust which may result from processes like subfluence or overthrusting. Finally, the occurrence of well conducting metasediments would explain why the intra-crustal good conductors are observed not only in the Rhenish Massif proper but also beneath the sub-Variscan foredeep in the North and beneath the Saar-Nahe Basin in the South (cf. Figs. 7,8, and 10).

Raikes and Bonjer (this vol.) detected a zone of decreased seismic velocities at depths below about 50 km in a region west of the Rhine which they assume to be caused by partial melt. Our own results, at the present stage of analysis, do not show any distinct differences of conductivity at upper mantle depths between the parts of the Rhenish Massif west and east of the Rhine. On the other hand we may safely assume the conductivity at these depths to be appreciably higher beneath the Münsterland than beneath the Rhenish Massif (cf. Figs. 8 and 10).

Acknowledgements. We are most indebted to the Deutsche Forschungsgemeinschaft for enabling us to carry out this study. Special thanks are due to U. Schmucker, Göttingen, for many discussions and for permission to use his computer programs for 1D inversion and 2D-modeling. We were also supported by the Institut für Geophysik and Meteorologie and the Institut für Nachrichtentechnik of the Technische Universität, Braunschweig, who allowed us the temporary use of their magnetotelluric equipment. W. Meyer, Bonn, and H. Noll, Köln, kindly gave us geological advice. Field work would not have been possible without the assistence of many forestry officials and farmers. K. Eichler, R. Giesbert, M. Keil, and K. Lange participated in field work and data processing.

References

Ádám, A., 1980. The change of electrical structure between an orogenic and an ancient tectonic area (Carpathians and Russian Platform). J. Geomag. Geoelectr., 32:1-46.

Ahorner, L., 1983. Historical seismicity and present-day microearthquake activity of the Rhenish Massif, Central Europe. This Vol.

Ahorner, L. and Murawski, H., 1975. Erdbebentätigkeit und geologischer Werdegang der Hunsrück-Südrand-Störung. Z. Dtsch. Geol. Ges., 126:63-82.

Beblo, M. and Björnsson, A., 1980. A model of electrical resistivity beneath NE-Iceland, correlation with temperature. J. Geophys., 47:184-190.

Behr, H.-J., 1978. Subfluenzprozesse im Grundgebirgsstockwerk Mitteleuropas. Z. Dtsch. Geol. Ges., 129:283-318.

Berdichevsky, M.N. and Dmitriev, V.I., 1976. Basic principles of interpretation of magnetotelluric sounding curves. In: Ádám, A. (ed.) Geoelectric and geothermal studies (East Central Europe, Soviet Asia). Akad. Kiado, Budapest, pp 165-221.

Berdichevsky, M.N., Van'yan, L.L., Fel'dman, I.S., and Porstendorfer, G., 1972. Conducting layers in the earth's crust and upper mantle. Gerlands Beitr. Geophys., 81:187-196.

Bless, M.J.M., Bouckaert, J., and Paproth, E., 1980. Environmental aspects of some Pre-Permian deposits in NW Europe. Meded. Rijks Geol. Dienst, 32-1:3-313.

Blohm, E.-K., Worzyk, P., and Scriba, H., 1977. Geoelectrical deep sounding in Southern Africa using the Cabora Bassa power line. J. Geophys., 43:665-679.

Brace, W.F., 1971. Resistivity of saturated crustal rocks to 40 km based on laboratory measurements. In: Heacock, J. (ed.) The structure and physical properties of the earth's crust. Am. Geophys. Union, Washington, D.C. Geophys. Monogr., 14:243-255.

Connerney, J.E.P., Nekut, A., and Kuckes, A.F., 1980. Deep crustal conductivity in the Adirondacks. J. Geophys. Res., 85:2603-2614.

Cook, F.A., Albaugh, D.S., Brown, L.D., Kaufman, S., Oliver, J.E., and Hatcher, R.D., 1979. Thin-skinned tectonics in the crystalline Southern Appalachians: COCORP seismic reflection profiling of the Blue Ridge and Piedmont. Geology, 7:563-567.

Giese, P., 1983. The evolution of the Hercynian crust - some implications to the uplift problem of the Rhenish Massif. This Vol.

Hermance, J.F., 1973. Processing of magnetotelluric data. Phys. Earth Planet. Inter., 7:349-364.

Hermance, J.F. and Pederson, J., 1980. Deep structure of the Rio Grande Rift: a magnetotelluric interpretation. J. Geophys. Res., 85:3899-3912.

Jödicke, H., 1978. Auswertungsverfahren Münster. In: Haak, V. and Homilius, J. (eds.) Protokoll über das Kolloquium "Elektromagnetische Tiefenforschung" in Neustadt/W. vom 11.-13. April 1978. Inst. Geophys. Wiss. FU Berlin, NLfB Hannover, pp 147-154.

Jödicke, H., 1980. Magnetotellurik Norddeutschland - Versuch einer Interpretation. In: Haak, V. and Homilius, J. (eds.) Protokoll über das Kolloquium "Elektromagnetische Tiefenforschung" in Berlin-Lichtenrade vom 1.-3. April 1980. Inst. Geophys. Wiss FU Berlin, NLfB Hannover, pp 271-288.

Jödicke, H., Keil, M., Blohm, E.-K., and Wagenitz, V., 1982. Magnetotellurische und geoelektrische Untersuchungen im Gebiet der magnetischen Anomalie von Soest-Erwitte und ihre Bedeutung für die stratigraphische Einstufung des prädevonischen Konduktors im Untergrund Nordwestdeutschlands. Fortschr. Geol. Rheinland Westfalen, 30:363-403.

Jones, A.G., 1982. On the electrical crust-mantle structure in Fennoscandia: no Moho, and the asthenosphere revealed? Geophys. J. R. Astron. Soc., 68:371-388.

Karmann, R., 1977. Search-coil magnetometers with optimum signal-to-noise ratio. Acta Geodaet. Geophys. Mont. Acad. Sci. Hung. Tomus, 12:353-357.

Kröger, P., 1977. An automatic recording station for magnetotelluric measurements. Acta Geodaet. Geophys. Mont. Acad. Sci. Hung. Tomus, 12:359-364.

Kurtz, R.D. and Garland, G.D., 1976. Magnetotelluric measurements in Eastern Canada. Geophys. J.R. Astron. Soc., 45:321-347.

Mechie, J., Prodehl, C., Fuchs, K., 1983. The long-range seismic refraction experiment in the Rhenish Massif. This Vol.

Mengel, K. and Wedepohl, K.H., 1983. Mantle xenoliths from the Tertiary and Quaternary volcanics of the Rhenish Massif and the Tertiary basalts of the Northern Hessian Depression. This Vol.

Meissner, R., et al., 1983. Combined seismic reflections-refractions experiment in the Rhenish Massif. This Vol.

Murawski, H., et al., 1983. Regional tectonic setting and geological structure of the Rhenish Massif. This Vol.

Patra, H.P. and Mallick, K., 1980. Geosounding principles, 2. time-varying geoelectric soundings. Elsevier, Amsterdam. 419 pp.

Raikes, S. and Bonjer, K.P., 1983. Large-scale mantle heterogeneities beneath the Rhenish Massif and its vicinity from teleseismic P-residuals measurements. This Vol.

Richards, M.L., Schmucker, U., Steveling, E., and Watermann, J., 1981. Erdmagnetische und magnetotellurische Sondierungen im Gebiet des mitteleuropäischen Riftsystems. Forschungsber. T81-111. Bundesminist. Forsch. Technol.

Scheelke, I., 1972. Magnetotellurische Messungen im Rheingraben und ihre Deutung mit zweidimensionalen Modellen. GAMMA 20, Inst. Geophys. Techn. Univ. Braunschweig, 199 pp.

Schminke, H.-U., Lorenz, V., Seck, H.A., 1983. The Quaternary Eifel volcanism. This Vol.

Schmucker, U., 1971. Neue Rechenmethoden zur Tiefensondierung. In: Weidelt, P. (ed.) Protokoll über das Kolloquium "Erdmagnetische Tiefensondierung" in Rothenberge/Westf. vom 14.-16. September 1971. Inst. Geophys. Univ. Göttingen, pp 1-39.

Schmucker, U., 1974. Erdmagnetische Tiefensondierung mit langperiodischen Variationen. In: Berktold, A. (ed.) Protokoll über das Kolloquium "Erdmagnetische Tiefensondierung" in Grafrath/Bayern vom 11.-13. März 1974. Inst. Geophys. Univ. München, pp 313-342.

Voll, G. and Wedepohl, K.H., 1983. Crustal xenoliths in the Rhenish Massif and the Hessian Depression. This Vol.

Weber, K., 1978. Das Bewegungsbild im Rhenohercynikum - Abbild einer varistischen Subfluenz. Z. Dtsch. Geol. Ges., 129:249-281.

Weber, K., 1981. The structural development of the Rheinische Schiefergebirge. Geol. Mijnbouw, 60:149-159.

Weidelt, P., 1972. The inverse problem of geomagnetic induction. Z. Geophys., 38:257-289.

Zijl, van, J.S.V., 1977. Electrical studies of the deep crust in various provinces of South Africa. In: Heacock, J. (ed.) The Earth's crust. Am. Geophys. Union Washington D.C. Geophys. Monogr., 20:470-500.

7.4 The Evolution of the Hercynian Crust – Some Implications to the Uplift Problem of the Rhenish Massif

P. Giese[1]

Abstract

The crust of the Rhenish Massif as part of the Rhenohercynian Zone as well as the Saxothuringian Zone are interpreted as being built up by sedimentary and basement slices which are detached from their underlying stratum, displaced northwards, and stacked during the Hercynian orogeny. The compressional movements started from an attenuated crust which was stretched and thinned by rifting processes. At the end of the Hercynian orogeny the crust was thickened by a factor of 2-3 (to about 30-35 km). Thus the Rhenish Massif underwent a slow uplift or even attained an isostatic equilibrium. On the other side, the northern foreland continued to subside, and the crustal thickness increased continuously. Thus it must be concluded that the position above sea level and, since Late Paleozoic the slow uplift of the Rhenish Massif, are a long-term process which may recently have been intensified by geothermal events in the lithosphere.

Introduction

The discussions about the young vertical uplift of the Rhenish Massif mainly presume that the causes are also young. But it must be kept in mind that during post-Hercynian times the Rhenish Massif largely behaved as a massif which tended to stay above sea level. This more or less continuous slow uplift is certainly controlled by the isostatic rebound of a thickened continental crust of a fold belt. Thus it seems resaonable to ask how the crust of the Rhenish Massif developed during the Hercynian orogeny. This chapter treats this problem and aims to give some new ideas on the crustal formation during Hercynian times.

Main Tectonic Zones of the Hercynian Orogene

The development of the Hercynian orogene shows a clear polarity pointing from S to N. In central Europe the Hercynian chain can be divided into four zones, each of them different in composition, structure and age. The main features and characteristics of these zones are the following ones (Jacobshagen, 1976; Weber, 1978; Behr, 1978).

1. The Moldanubian Zone is exposed in the Vosges, the Black Forest, the Bavarian Forest, and the Bohemian Forest. Here, highly metamorphosed schists are interspersed with late and post-kinematic intrusions of mainly granitic character. This zone is partly covered by Paleozoic sediments.

2. The Saxothuringian Zone can be divided into three sections. The main element, a crystalline ridge, is exposed in small outcrops in the north of NW Saxony and northern Thuringia and in the Thurin-

---

[1] Institut für Geophysikalische Wissenschaften der Freien Universität Berlin, Rheinbabenallee 49, D-1000 Berlin 33, Fed. Rep. of Germany.

Plateau Uplift, ed. by K. Fuchs et al.
© Springer-Verlag Berlin Heidelberg 1983

gian Forest. The Spessart, Odenwald, and Palatinate Forest also belong to this ridge. Acid and mafic rocks intruded into highly metamorphosed schists. This so-called Mid-German Crystalline Rise started to emerge already during the Lower Devonian. In the Odenwald a layer of 10-15 km has been eroded within 50 million years (Weber, 1981). SE of this ridge a large area of Paleozoic and even Precambrian sedimentary rocks are exposed which southeastwards gradually become metamorphosed. Finally a broad anticline with crystalline cores (e.g., Fichtelgebirge) borders the Moldanubian Zone.

The northern border of the Saxothuringian Zone is marked by a deep-reaching fracture zone, the Hunsrück-Taunus fault zone.

3. Large parts of the Rhenohercynian Zone crop out in the Rhenish Massif. This zone comprises mainly Devonian and Carboniferous sedimentary rocks reaching a maximum thickness of 5-10 km. The characteristic tectonic features of this zone are folds and imbricate (schuppen) structures showing in general a vergency towards NW. Nappe structures are known in the Ardennes and in the southern zone (Gießen Graywacke). The folded sediments prove a shortening factor of at least 0.5 (Wunderlich, 1964).

4. The northern foredeep is filled with thick Devonian and Carboniferous sediments and is called the Subvariscian Zone. Whereas in its southern part the sediments were strongly folded and partly thrust, the tectonic activity died out northwards. This zone borders on the North-European Fennoscandian-Baltic shield.

Young volcanic activity occurred during the Cenozoic time in Central Europe. The largest center of this young volcanism in Central Europe is the Vogelsberg. Further sites of volcanic activity are the Eifel, Neuwied Basin, Siebengebirge, Westerwald, and Rhön.

Models of Crustal Growth and Thickening

Figure 1 shows two models sketching crustal thickening during an orogeny (Behr et al., 1980).

1a. In the first model the crustal thickening is caused by material ascending from the upper mantle and being accumulated at the base of crust.

1b. A completely different model describes the crustal thickening by detachment of crustal slices, their horizontal displacement and subsequent stacking. This concept of thin- and thick-skinned tectonics for the evolution of mountain belts is proposed, e.g., for the southern Appalachians, a Paleozoic orogene comparable with the Hercynian chain in Europe (Cook et al., 1979). A discrimination of both models should be possible by a detailed study of intracrustal discontinuities and of the velocity distribution in the lower crust.

Area Under Study

The area under study comprises the Rhenish Massif of the Rhenohercynian Zone and the adjacent Saxothuringian domain. The main features of crustal structure are described by two crustal sections traversing these zones. The choice of these sections is an attempt to bypass regions of young volcanic activity. A detailed study of the P^MP-wave group in the seismic refraction data reveals an interaction of magmatic activity and structure and composition of the crust/mantle transition (Giese, 1976; Mooney and Prodehl, 1978). During the ascension of magma from the upper mantle through the crust, the crust/mantle boundary has been disrupted and to a certain extent been replaced by a new boundary at intermediate depths at 20-25 km.

Figure 2 shows the network of seismic-refraction and reflection profiles in Central Europe. The heavy lines display

305

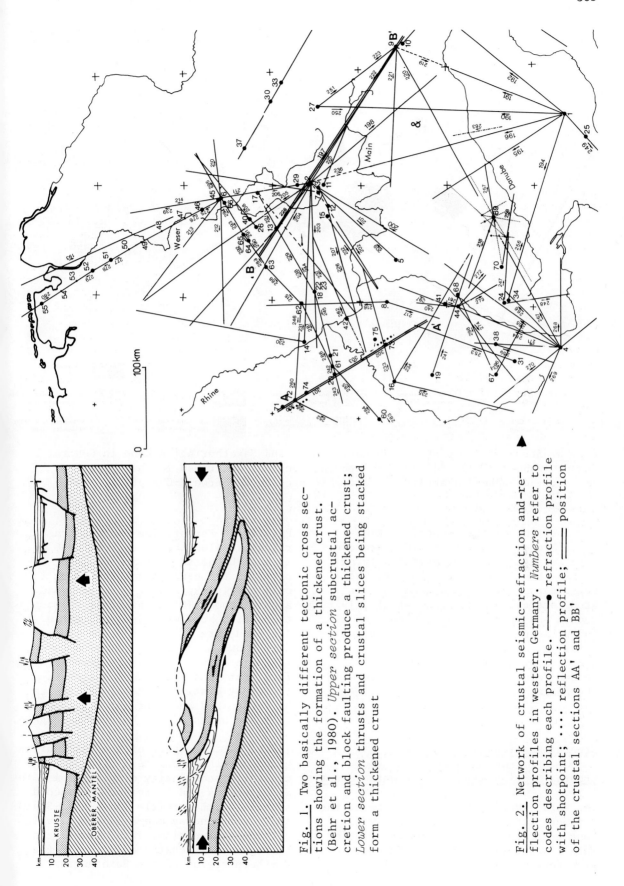

Fig. 1. Two basically different tectonic cross sections showing the formation of a thickened crust. (Behr et al., 1980). *Upper section* subcrustal accretion and block faulting produce a thickened crust; *Lower section* thrusts and crustal slices being stacked form a thickened crust

Fig. 2. Network of crustal seismic-refraction and-reflection profiles in western Germany. *Numbers* refer to codes describing each profile. ● refraction profile with shotpoint; ···· reflection profile; ═══ position of the crustal sections AA' and BB'

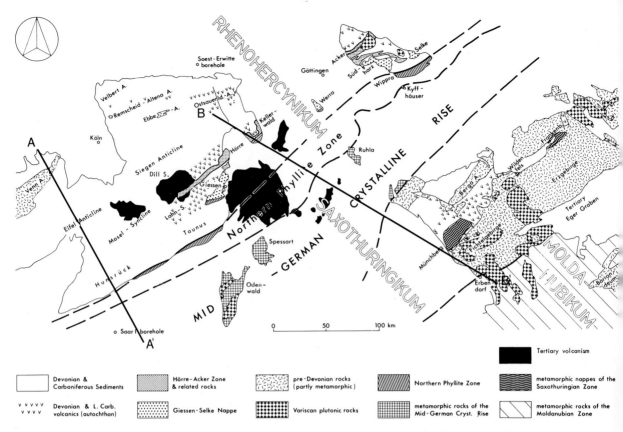

Fig. 3. Structural map of the Rhenohercynian and Saxothuringian Zone in Central Europe (Weber, 1981). *AA'* and *BB'* crustal sections described in this paper. *AA'* section Aachen-Palatinate Forest; *BB'* section Kellerwald-Oberpfalz

the position of the crustal sections described here. These sections are constructed using data of profiles lying within this line or crossing it. The seismic refraction data used in this study are taken from Giese et al., 1976, Mooney and Prodehl, 1978, Mechie et al., 1983. Figure 3 shows a simplified geological map of Central Europe with the location of the crustal cross-sections Aachen-Palatinate Forest and Kellerwald-Oberpfalz.

Crustal Sections

Crustal Section Aachen-Palatinate Forest

The first section (Fig. 3) starts at the northern margin of the Rhenohercynian Zone near Aachen, traverses the Rhenish Massif and extends into the Saxothuringian Zone where it terminates in the Palatinate Forest. The average crustal thickness along this section is 30 km. It must be noted that in the Mosel depression the crust/mantle boundary is only poorly expressed. Here, a broad zone of about 10 km thickness forms the transition between crust and mantle. The structure of the crust/mantle boundary shows no clear correlation to the tectonic structures observed at the surface. But in the internal structure of the crust distinct lateral changes can be recognized and correlated with the tectonic units of the Hercynian orogene.

In the northern part of the Saxothuringian Zone there exists a clear intracrustal discontinuity dipping from N to S. Figures 4 and 5 present the record section of the profile Taben-Rodt-S and the corresponding one-dimensional velocity-depth function. This example can be regarded as characteristic for the northern region of the Saxothuringian

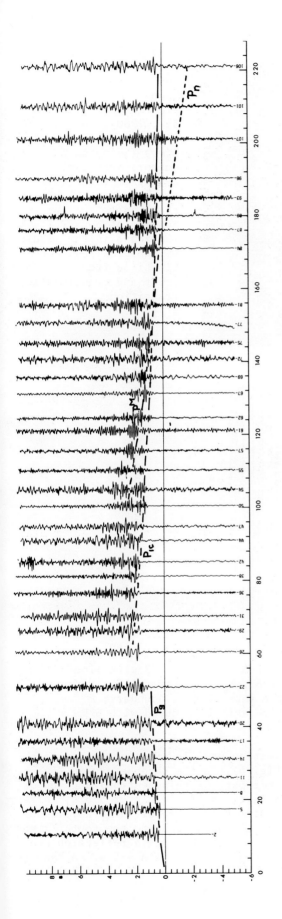

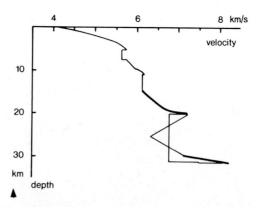

Fig. 5. One-dimensional velocity-depth function of the profile Taben-Rodt-S being typical for the northern part of the Saxothuringian Zone. The rectangular and v-shaped velocity distribution in the deep low-velocity layer presents two kinematically equivalent models

◀ Fig. 4. Record section of the seismic-refraction profile Taben-Rodt-S (shotpoint 16 with line 235 in Fig. 2). This section is typical for the northern region of the Saxothuringian Zone. The intracrustal phase P_{ic} is distinctly developed. P_g wave, travelling through the uppermost crust; P_{ic} wave, reflected at the intracrustal boundary; P^M wave, reflected at the crust/mantle boundary; P_n wave, travelling through the uppermost mantle. The time axis is reduced with v = 6 km s^{-1}

Zone. The updip of this intracrustal discontinuity is proved by a profile recorded at the northern margin of the Saxothuringian Zone. It could be detected here at 12-14 km depth. Between this discontinuity and the crust/mantle boundary a low-velocity layer is intercalated forming the lower crust.

Examples of record sections observed in the Rhenohercynian Zone are presented in the following figures. Figure 7 shows a seismogram section observed in the Hunsrück region, the southern part of the Rhenish Massif. Figure 8 displays the corresponding depth-velocity function (one-dimensional). An intracrustal discontinuity is clearly visible. In contradiction to the record section in Fig. 7 record sections observed in the northern part of the Rhenish Massif show only poorly developed intracrustal wave groups.

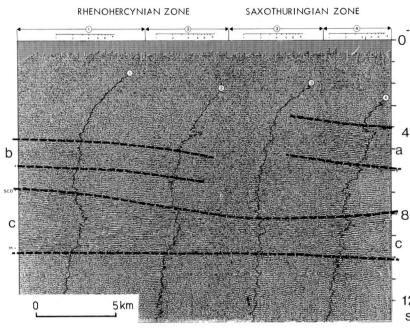

Fig. 6. Seismic reflection profile (two-way traveltime section) traversing the Hunsrück-Taunus fault zone (section after Meissner et al., 1980). The zones a,b,c bounded by *dashed lines* (being added by the author) are distinguished by higher reflectivity. a,b intracrustal reflections, note the interruption of both reflecting zones in the middle of the profile); c reflections from the lower crust and crust/mantle transition

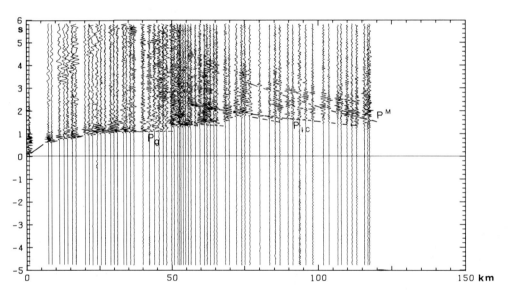

Fig. 7. Record section of the profile Baumholder-N (shotpoint 73 with line 302 in Fig. 2), record section after Mechie et al., 1983. P_g wave, travelling through the uppermost crust; P_{ic} wave, reflected at the intracrustal boundary; P^M wave, reflected at the crust/mantle boundary. The time axis is reduced with 6 km s^{-1}

The crustal section in Fig. 9a summarizes the results of a number of velocity-depth functions. In the Rhenohercynian as well as in the Saxothuringian Zone, an intracrustal boundary and an underlying low-velocity layer are present. The interruption and displacement of these boundaries at the Hunsrück-Taunus fault zone are proved by a seismic-reflection profile crossing this zone (Fig. 6). In the Rhenohercynian Zone (left part of this section) a group of reflections is visible at 5-6 s, corresponding to a depth of 15-18 km. In the Saxothuringian Zone to the south (right half of the section) a reflection can be recognized between 4-5 s, corresponding to a depth range of 11-14 km. The

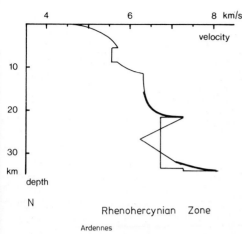

Fig. 8. One-dimensional velocity-depth function of the profile Baumholder-N. It represents the velocity distribution in the southern region of the Rhenish Massif (Hunsrück)

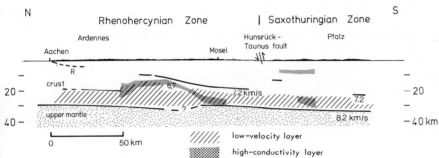

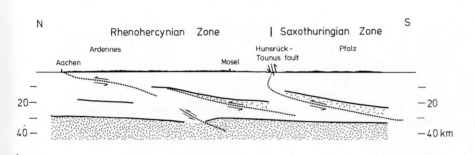

Fig. 9. a. Crustal cross-section Aachen-Palatinate Forest being composed of the results of inline and crossing profiles (see Fig. 2). Boundaries shown in this section are derived from overcritical reflections. *Hatched zone* in the lower crust indicates the existence of a low-velocity layer. *Densely dashed signature* shows zones where magnetotelluric soundings detected high electrical conductivity (Jödicke et al., 1983). Note the agreement of the low-velocity layer with the high-conductivity layer in the central part of the profile. b. Simplified tectonic interpretation of the section Aachen-Palatinate Forest. The concept is based on two items: a strong shortening of a thinned pre-Hercynian crust and the nature of intracrustal discontinuities (in the southern Rhenohercynian Zone and the Saxothuringian Zone) which are seen as ancient lower crust and/or fragments of the crust/mantle transition

broad band of reflected waves between 8 and 10 s is correlated with the crust/mantle transition between 25 and 30 km depth.

The updip of the intracrustal boundary in the southern and central part of the Rhenish Massif is proved by a crossing longitudinal profile, showing this interface at about 12-13 km depth at the crossing point.

In the northern region of the Rhenish Massif, a deeper intracrustal boundary exists, but it seems to be separated from the previous one and may belong to the thin high-velocity layers within the velocity inversion zone of the lower

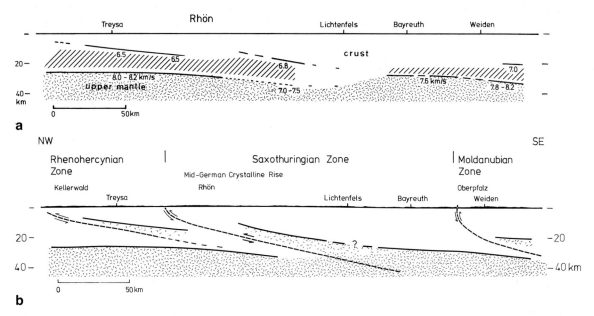

Fig. 10a. Crustal section Kellerwald-Oberpfalz. Several in-line and crossing profiles have been used for the compilation of this section. b. Simplified tectonic interpretation of the crustal section Kellerwald-Oberpfalz. The interpretation is based on the same concept as described in Fig. 9b

crust (Mechie et al., this vol.). A seismic reflection profile across the Stavelot-Venn (northern margin of the Ardennes) shows a strong reflector at 3-4 km depth with a slight dip to the SSE. This reflector is interpreted as thrust fault along which a huge horizontal nappe displacement took place (Meissner et al., 1981 and this vol.).

Crustal Section Kellerwald-Oberpfalz

This section starts in the eastern Rhenish Massif, runs in SE direction, passes the Saxothuringian Zone, and ends in the Moldanubian Zone near Weiden. A number of coinciding and crossing seismic refraction lines were used for the compilation of this crustal section shown in Fig. 10a. Here again the intracrustal boundaries are interrupted and dip down southeastwards. A low-velocity layer forms the lower crust. Beneath the central part of the Saxothuringian Zone the lower crust is distinctly thickened, and in the records the crust/mantle boundary is not so clearly defined as in other regions.

Tectonic Interpretation

The tectonic interpretation of the crustal sections described above is based on the following geological and geophysical constraints.

1. The tectonic development of the Hercynian orogene proceeded from S to N, it started in the Moldanubian Zone and terminated in the Subvariscian foredeep.

2. The faulted, folded and even partly thrust sediments indicate a crustal shortening of at least 50% or in some regions even more (Wunderlich, 1964). This means, for example, that the present width of the Rhenish Massif must be doubled in order to obtain the initial extension of the Rhenohercynian belt. The distance between Aachen, the northern border of the Rhenohercynian Zone, and the Hunsrück-Taunus fault measures 150 km. A reconstruction of the Rhenohercynian belt requires its southern border to be located near the present position of Stuttgart.

3. There is no strong evidence for the existence of a wide oceanic basin during the Hercynian orogeny. The facies of the Devonian and Carboniferous sediments suggest a deposition in a continental environment (Weber, 1981). The development of sedimentary basins must be caused by some kind of continental rifting or oceanization of crustal material. During rifting phases the continent is stretched and thinned in response to regional extension of the lithosphere. Besides pure mechanical crustal stretching thermally-induced physico-chemical processes can affect the lower crust and cause an upward displacement of the crust/mantle boundary thus also contributing to crustal thinning. McKenzie (1978) gives some quantitative considerations on the development of sedimentary basins. A subsidence of 5000 m requires a crustal stretching by the factor of 2. Thus it can be assumed that the Hercynian orogeny started with a thinned continental crust. A thickness of 10-15 km may be speculated for this pre-compressional Hercynian crust.

4. Some allochthonous sedimentary series, for example, the Gießen Graywacke, cannot be restored to a known and presently visible basement. The necessary sedimentary belt must have been situated N of the Mid-German Crystalline-Rise. The same is true for the nappe system in the Ardennes. The original basement of these thrust series must be suspected somewhere beneath the central or southern part of the Rhenish Massif. When tracing back the system of overthrusts to the internal zones of the orogene, overriding of continental crustal slices must be postulated.

5. Listric overthrusts combined with folds are the dominant structural features in the upper crust (Weber, 1978, 1981). These overthrusts arise at different depths from horizontal thrust planes. After Weber (1978) major structures can be related to shear planes at 6-8 km depth being parallel to the subhorizontal layering in the infrastructure below. Shear planes cannot only develop in the sedimentary cover and the uppermost basement but also in the deeper crust. Partial melting of sialic components in the lower crust reduces strongly the shear strength. Temperatures near the melting point of sialic rocks (~700°C) can be expected at the base of the crust if rifting processes are active and crustal thinning takes place. Meissner and Strehlau (1982) calculated for the lower crust a minimum of stress and viscosity.

These constraints and considerations are the base for the tectonic interpretation of the crustal cross-section Aachen-Pfalz, shown in Figs. 9b and 10b.

The intracrustal discontinuities in the southern Rhenohercynian and northern Saxothuringian Zone are interpreted as lower crust and parts of the crust/mantle transition of a thinned pre-Hercynian crust. During the compressional movements sedimentary as well as basement complexes were sheared off and horizontally displaced over distances of several tens of kilometers, perhaps even up to 100 km. Following this concept the present crust of the Hercynian Europe has been formed by stacking of sedimentary and basement slices. The principle of this model is sketched in Fig. 12. The size of the slices tends to decrease from internal to external, that means from S to N. Each slice is of wedge-like form and thins out in a northern direction.

In the northern Saxothuringian Zone the present intracrustal boundary is seen as the base of the crust of the Mid German Crystalline-Rise, which has overridden the northern adjoining crustal segment. It was previously mentioned that in front of the Mid German Crystalline-Rise a crustal segment must have been subducted as depositional domain of the Gießen Graywacke. Thus the present lower crust beneath the northern part of the Saxothuringian Zone can be regarded as subducted continental segment.

The Mid-German Crystalline Rise itself developed by uplifting since Lower Devonian times. The upheaval which continued up to Lower Carboniferous may have been initiated by crustal overthrusting and thickening and a subsequent isostatic rise. In the Odenwald area, within a span of 50 m.y., 10-15 km of the upper crust must have been eroded (Weber, 1981).

Similarly, the crustal structure in the southern part of the Rhenish Massif (Rhenohercynian Zone) is interpreted. The upper crust is seen as allochthonous complex which is built up by numerous subunits which are allochthonous with respect to the underlying rocks. The subunits must originate from an attenuated continental crust of the southern part of the Rhenohercynian belt. The horizontal shear plane at 6-10 km depth postulated by Weber (1981) may be regarded as an internal shear horizon within this large allochtonous complex.

Within this concept the lower crust of the present southern Rhenish Massif belongs to the northern half of the ancient Rhenohercynian belt. The sedimentary cover of this overridden or subducted basement was completely or partly detached and displaced northwards. The nappe system of the Ardennes may be explained in such a way.

Conclusions

Geological and geophysical data suggest that by thrusting and stacking of sedimentary and basement slices the Hercynian crust evolved from a stretched and thinned continental crust into a thickened crust (Fig. 11). It can be suspected that the compressional movements produced a crustal thickening by a factor of 2-3. A discussion of the uplift of the Rhenish Massif should consider its relation to the adjacent zone in the N, the Hercynian foreland. During the Late Carboniferous the tectonic activity died out in the Subvariscian Zone, but the subsidence of the foreland continued and the North German Basin was filled by Permian Molasse-type sediments. Extensive Late Carboniferous to Early Permian volcanics have been encountered by exploration wells in northern Germany and Poland. These sedimentary and volcanic rocks indicate the existence of a thinned foreland crust during the Permian time which was under a tensional regime. This idea is supported by data from explosion seismology. The present crustal thickness beneath the North German Basin measures about 30 km (Giese, 1976). The sedimentary fill between Late Permian to Quaternary can reach 8 km (Ziegler, 1982). Provided that since Permian time the crust/mantle boundary has not been changed by subcrustal erosion or thermally induced physicochemical processes, a crustal thickness of about 20 km can be assumed for the Variscian foreland crust at Late Carboniferous to Permian times. Thus geological as well as geophysical data emphasize the existence of a thinned Variscian foreland crust at that time.

The crustal situation at Late Paleozoic time is sketched in Fig. 12. In the Hercynian mountain belt the crust was distinctly thickened. From an isostatic point of view the crust of the Rhenish Massif had achieved an equilibrium or it underwent a very slow uplift accompanied by little erosion. The paleogeographic maps published by Ziegler (1982) show that since Permian time the Rhenish Massif, together with the London-Brabant Massif, were mostly situated above sea level. Contrarily, the North German Basin continued to subside during the Mesozoic and Cenozoic accumulating a thick sedimentary cover. Consequently the crustal thickness increased from 20 to 30 km. Thus it can be stated that the different vertical movement of the Rhenish Massif and the North German Basin was mainly governed by different crustal thickness. In the course of Earth history the crustal thicknesses of both regions became about equal and today the crust/mantle boundary lies at an average depth of 30 km for both areas.

It was the main goal of this priority program to study the "young" uplift of the Rhenish Massif and to elaborate geodynamic models which explain this feature. This chapter tries to outline how the crustal evolution of the Rhenish Massif and adjacent regions may affect and govern vertical movements of crustal blocks. On the other hand it cannot be excluded that any "young" geothermal events in the deeper lithosphere may have additionally intensified the uplift of the Rhenish Massif during the Late Cenozoic.

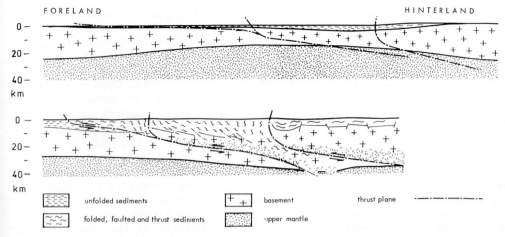

Fig. 11. Simplified model showing how the crustal thickening is produced by stacking of crustal fragments and slices. *Upper section* crustal configuration before the compressional movements started. Note the thinned crust. *Lower section* situation after the orogenetic movements cease. Note the units being detached from their substratum get thinner from the hinterland to the foreland

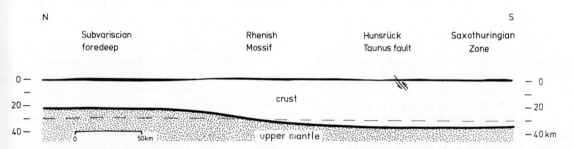

Fig. 12. N-S cross-section between the subvariscian foredeep and the Saxothuringian Zone showing the crustal thickness at the end of the Hercynian orogeny (Late Carboniferous). Whereas the crust of the Rhenish Massif has been distinctly thickened, the crust of the northern foreland remained as thinned layer. *Dashed line* displays the present crustal thickness. During Mesozoic and Cenozoic the foreland crust became thicker by continuous subsidence and sedimentation simultaneously the Rhenish Massif underwent a slow uplift and erosion, thus causing a weak crustal thinning

References

Behr, H.J., 1978. Subfluenzprozesse im Grundgebirgsstockwerk Mitteleuropas. Z. Dtsch. Geol. Ges., 129:283-318.

Behr, H.J., Engel, W., and Franke, W., 1980. Guide to Exkursion Münchberger Gneismasse und Bayerischer Wald, April 12-15, 1980. Geol. Paläontol. Inst. Mus. Göttingen, 100 pp.

Cook, F.A., Albaugh, D.S., Brown, L.D., Kaufman, S., Oliver, J.E., and Hatcher, R.D., 1979. Thin-skinned tectonics in the crystalline Southern Appalachians: COCORP seismic reflection profiling of the Blue Ridge and Piedmont. Geology, 7:563-567.

Giese, P., 1976. The basic features of crustal structure in relation to the main geological units. In: Giese, P., Prodehl, C., and Stein, A. (eds.) Explosion seismology in central Europe-data and results. Springer, Berlin, Heidelberg, New York, pp 221-242.

Giese, P., Prodehl, C., and Stein, A. (eds.) 1976. Explosion seismology in central Europe-data and results.

Springer, Berlin, Heidelberg, New York, 429 pp.

Jacobshagen, V., 1976. Main geologic features of the Federal Republic of Germany. In: Giese, P., Prodehl, C., and Stein. A. (eds.) Explosion seismology in Central Europe. Springer, Berlin, Heidelberg, New York, pp 3-18.

Jödicke, H., Untiedt, J., Olgemann, W., Schulte, L., and Wagenitz, V., 1983. Electrical conductivity structure of the crust and upper mantle beneath the Rhenish Massif. This Vol.

McKenzie, D.P., 1978. Some remarks on the development of sedimentary basins. Earth Planet. Sci. Lett., 40:25-32.

Meissner, R. and Strehlau, J., 1982. Limits of stresses in continental crusts and their relation to the depth-frequency distribution of shallow earthquakes. Tectonophysics, 1:73-89.

Meissner, R., Bartelsen, H., Glocke, A., and Kaminski, W., 1976. An interpretation of wide angle measurements in the Rhenish Massif. In: Giese, P., Prodehl, C. and Stein, A. (eds.) Explosion seismology in Central Europe. Springer, Berlin, Heidelberg, New York, pp 245-251.

Meissner, R., Bartelsen, H., and Murawski, H., 1981. Thin-skinned tectonics in the northern Rhenish Massif. Nature (London), 290:399-401.

Meissner, R. and Bartelsen, H., 1980. Seismic reflection and refraction studies for investigating fault zones along the geotraverse Rhenoherzynikum. Tectonophysics, 64:59-84.

Meissner, R., Springer, M., Murawski, H., Bartelsen, H., Flueh, E.R., and Duerschner, H., 1983. Combined seismic reflection-refraction investigations in the Rhenish Massif and their relation to recent tectonic movements. This Vol.

Mechie, J., Prodehl, C., and Fuchs, K., 1983. Crust and mantle structure and composition.- The long range seismic refraction experiment in the Rhenish Massif. This Vol.

Mooney, W.D. and Prodehl, C., 1978. Crustal structure of the Rhenish Massif and adjacent areas; a reinterpretation of existing seismic-refraction data. J. Geophys., 44:573-601.

Weber, K., 1978. Das Bewegungsbild im Rhenohercynikum - Abbild einer varistischen Subfluenz. Z. Dtsch. Geol. Ges., 129:249-281.

Weber, K., 1981. The structural development of the Rheinische Schiefergebirge. Geol. Mijubouw, 60:149-159.

Wunderlich, H.G., 1964. Maß, Ablauf und Ursachen orogener Einengung am Beispiel des Rheinischen Schiefergebirges, Ruhrkarbons und Harzes. Geol. Rundsch., 54:861-882.

Ziegler, P., 1982. Geological atlas of Western and Central Europe. Shell Int. Petrol. Maatschappij B.V. 130 pp.

7.5 Large-Scale Mantle Heterogeneity Beneath the Rhenish Massif and Its Vicinity from Teleseismic P-Residuals Measurements

S. Raikes[1,2] and K.-P. Bonjer[1]

Abstract

The azimuthal variation of teleseismic P-residuals observed at 63 stations in the vicinity of the Rhenish Massif for 350 events world-wide provides unequivocal evidence for large scale mantle heterogeneity beneath this area. Models for the velocity anomalies causing this variation, which is up to ca. 1 s for a given station and also for different stations for the same source region, have been derived by ray tracing and manual inversion and also by automatic three-dimensional inversion. All the models are extremely similar and produce a variance improvement of at least 74%. Their mean features include low velocity structures associated with the regions of volcanism and high velocities beneath the southern Rhine Graben-Vosges-Black Forest area. The regions of decreased velocity associated with the Tertiary volcanic fields of the Westerwald and Vogelsberg are relatively shallow (<60 km). The major anomaly lies beneath the West Eifel, where there is a velocity decrease of 3 to 5% extending from about 50 to 200 km depth. This structure is consistent with models for the uplift mechanism which involve lithospheric thinning or expansion of the mantle due to metasomatism and partial melting. The partial melt hypothesis is favored on the basis of gravity and seismic refraction data.

Introduction

The Rhenish Massif consists of a series of predominantly Devonian shales, sandstones and limestones which were strongly folded during the Hercynian orogeny and have subsequently undergone some 300 m of plateau uplift since the Pliocene. The nature and causes of this uplift, which continues today, have been the subject of an intensive multidisciplinary research programme sponsored by the German Research Association. A description of the uplift and its tectonic setting is given in Illies et al. (1979).

Since most models proposed to explain plateau uplift (e.g., McGetchin et al., 1979) involve processes within the upper mantle, it is clearly desirable to obtain information on the structure and properties of this region. Such information may be obtained by deep electrical sounding, modelling gravity data, analysis of mantle xenoliths found in volcanic rocks and by analysis of travel times of seismic waves from distant earthquakes to recording stations in the region of interest. This paper is concerned with the analysis of the variation of P-wave travel time residuals; this technique has proved very useful for delineating anomalous mantle structures beneath areas such as NORSAR (Aki

---

1 Geophysical Institute, University of Karlsruhe, D-7500 Karlsruhe, Fed. Rep. of Germany

2 Now with B.P. London, UK.

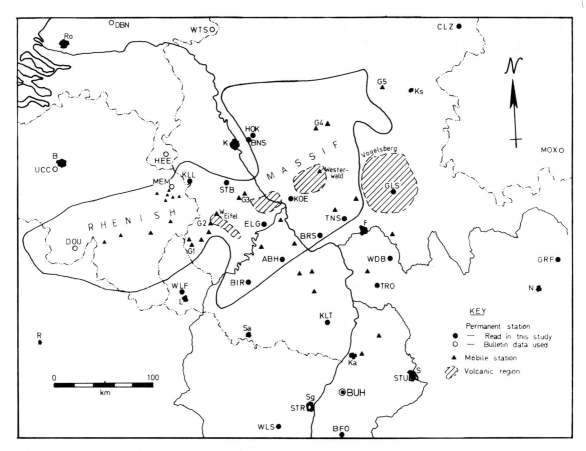

Fig. 1. Distribution of permanent and mobile stations used in this study. *Solid line* contour of the Rhenish Massif

et al., 1977), Northern California (Husebye et al., 1976) and Kilanea Volcano, Hawaii (Ellsworth and Koyanagi, 1977). Velocity models may be constructed from the observed travel time variations by ray tracing and manual inversion (Raikes, 1980b) or by automatic three-dimensional inversion using the method pioneered by Aki et al. (1977).

Determination of Residuals

Figure 1 shows the distribution of seismograph stations in the vicinity of the Rhenish Massif. The permanent stations are located largely in the active seismic belt which approximately follows the River Rhine (Ahorner, 1975). This station configuration is clearly not ideal for determining structure beneath the Rhenish Massif, and so the permanent network was augmented by mobile stations during three survey periods: June-July 1978 (Hohes Venn), April-July 1979 (throughout the German part of the Rhenish Massif) and November 1980 (the Belgian part of the Massif).

A total of 350 well-recorded events was investigated: 65 of these occurred at distances greater than 125°, and Fig. 2 shows the distribution of the remainder (although not all events are plotted). The azimuthal coverage is good, ensuring a good cross-fire of rays through the structure, which is essential for construction of stable models, although there are many more events in the northeast quadrant and very few in the southeast.

First arrivals were read with a precision of 0.1 s at 55 stations. This data set was supplemented by bulletin data from a further 8 stations.

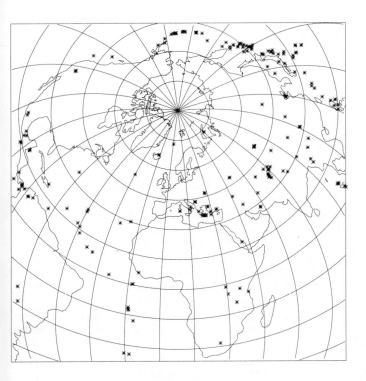

Fig. 2. Distribution of teleseisms studied

Delays were calculated with respect to the Jeffreys-Bullen (1940) arrival times (hereafter referred to as JB), corrected for the Earth's ellipticity and station elevation, using the USGS hypocentral parameters. For the core phases (PKIKP, PKHKP) Bolt's (1968) times were used because they gave a better fit to the observed ray parameters ($dT/d\Delta$). In order to minimize effects other than those arising from structure beneath the receiver network, some form of normalization is required. In this study two types of normalization were used: firstly the delays at a single station (BUH) were subtracted from those at the other stations, and secondly the mean residual over all stations for each event was subtracted. The former method has the advantage that the relative residuals are independent of the number and distribution of the stations, but any velocity models derived will be relative to the structure beneath BUH. Furthermore, any structural variation in the neighborhood of BUH will bias the residuals at all stations: care must be taken to correct for this when deriving the velocity models. The mean residual was used for normalisation of residuals input to the Aki inversion routine. Since most of the stations used are in, or on the borders of, the Rhenish Massif the mean structure is biassed to that beneath the Massif. For a given source region the scatter in residuals relative to BUH was ±0.05 to ±0.15 s for a single station except where bulletin data were used where it was ±0.15 to ±0.4 s.

A detailed discussion of the possible sources of error in relative residuals is given by Raikes (1978, 1980b) and Engdahl et al. (1977). Random errors arise largely from mislocation of the events, reading errors and heterogeneity along the ray path, and are expected to be less than ±0.2 s, which is substantiated by the observed scatter. Systematic errors due to large (~100 km) mislocations caused by structure in the source region, specific structures along the ray path or the choice of travel time tables are harder to estimate. However, observed travel times and ray parameters suggest that the use of JB as standard should not significantly bias the results. Figure 3 shows an array mislocation diagram for the stations of Fig. 1. Each event is represented by a vector whose head gives the azimuth of approach (polar angle) and ray parameter (distance from the centre) predicted by the USGS location and JB, and whose tail gives the observed azimuth and ray parameter (determined by fitting

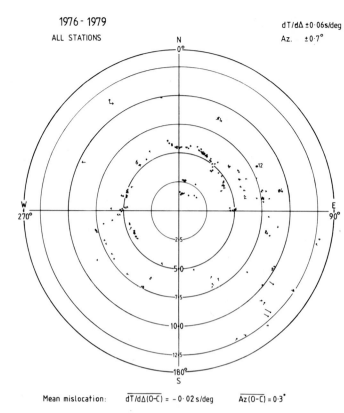

Fig. 3. Array mislocation diagram: Each *vector* represents a single event; *head of the arrow* gives the theoretical angle of approach and $dT/d\Delta$, *tail* the observed values. *Circles* are drawn at $dT/d\Delta$ = 2.5, 5.0, 7.5, 10.0 and 12.5 s deg^{-1}

a plane wave to the observed arrival times). For events in the distance range 30° to 100° the mislocations are extremely small: the mean (mean absolute) differences in ray parameter and azimuth are 0.02 (0.06)s deg^{-1} and 0.03° (1.2°). The effects of systematic mislocation are thus likely to be small. Figure 3 also indicates that there is no major structure - such as a uniformly dipping Moho - affecting the whole array.

The Observations and Their Implications

Figure 4 shows the mean residuals (relative to BUH) for each source region for each of the permanent stations. The residuals are plotted on polar diagrams in which the polar angle is the azimuth of approach and the radius vector the residual. The events are divided into 5 distance ranges: the mean residual for core phases ($\Delta > 125°$) which are essentially vertically incident, is given in the centre of each plot.

The first noticeable feature is that residuals for stations such as BFO, STU, CLZ and WTS at some distance from the Rhenish Massif show very little azimuthal variation. One possible effect associated with the choice of BUH as the reference station is that structure beneath the Rhine Graben (e.g., Wenderoth, 1978) could cause azimuthal variation in the delays at BUH and thus bias the relative residuals at other stations. However, there seems to be little evidence of this: in particular the normalized delays at CLZ are essentially independent of azimuth which would not be the case if BUH were strongly affected by local velocity changes. The main exception to this is the station GRF which has extremely late arrivals for events to the ENE. This will be discussed in more detail in the next section.

Stations within, or on the border of the Rhenish Massif exhibit very different behavior, with large azimuthal variation of residuals (e.g., BNS -0.1 to +0.9 s; STB 0.02 to 0.92 s; ELG -0.09 to +0.81 s. The late arrivals are always associated with rays that have passed beneath the Rhenish Massif west of the Rhine, in particular the Eifel volcanic region.

The large mean delays (≥ 0.5 s) at WTS, and HEE (and also at STR and KRL which are not included in Fig. 4a,b) are probably the result of low velocity sediments immediately beneath the stations.

Figure 4 provides clear evidence for lateral velocity variations in the vicinity of the Rhenish Massif, but the distribution of these variations is not easy to visualize. PKP phases are essentially vertically incident, and so delays for these phases represent structure vertically below the receiving station. A contour map of residuals for core phases (Fig. 5) thus defines the horizontal location of any velocity anomalies. There are two main regions of late arrivals: the Vogelsberg and the Rhenish Massif west of the Rhine. Delays

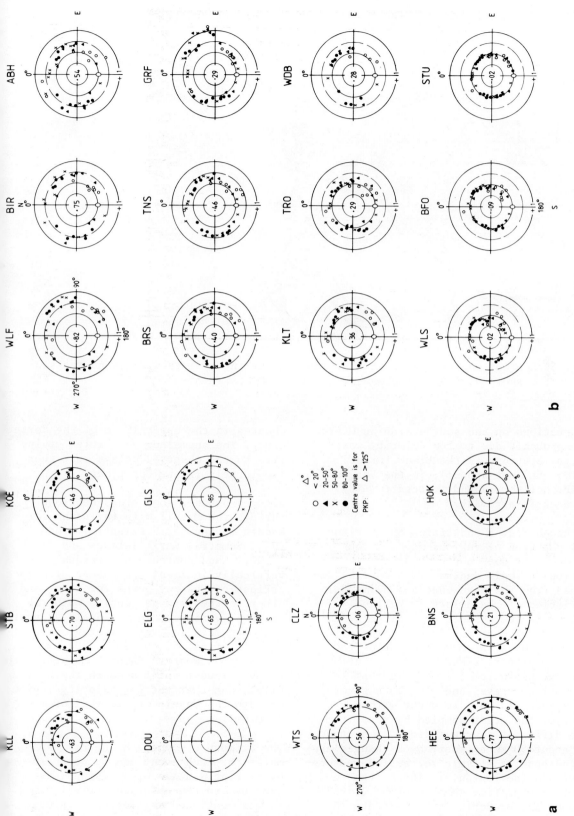

Fig. 4a,b. Azimuthal variation of residuals for the permanent stations studied

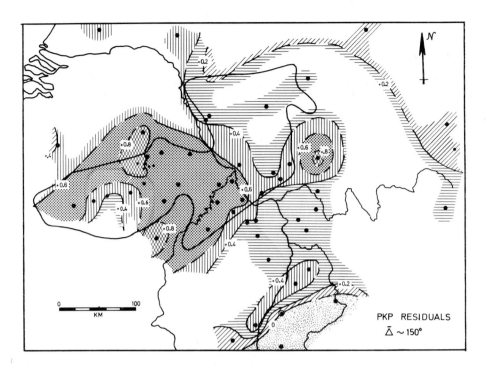

Fig. 5. Contour map of PKP residuals. The contour interval is 0.2 s, and the stations are shown as *black circles*

at stations in the southern Rhine Graben are probably due to low velocity sediments. Stations in the Black Forest (BFO; FEL (not shown)), in the Vosges (BAF, not shown) and Stuttgart (STU) have early arrivals.

Although Fig. 5 constrains the horizontal position of regions of anomalous velocity it does not provide any information on their depths. This must be deduced from the magnitude and azimuthal variation of the delays. A residual of 0.5 s, in the absence of sediments, implies a 10% decrease in mean crustal velocity or a 25 km thickening of the crust. Further, since a single station can exhibit up to 1 s of azimuthal variation, these (and even more severe) changes would have to occur within that part of the crust sampled by rays from the different source regions, a distance of about 50 km at most. Such changes are unreasonable: seismic refraction measurements in the region, including the recent long profile (Mechie et al., 1983; Mooney and Prodehl, 1978) indicate that a maximum of 0.3 s variation in residuals across the network, or about 0.15 s at a single station, can be caused by

changes in the upper 10 km of the Earth crust. The anomalous velocities giving rise to the observed delays must thus lie largely in the upper mantle.

Figures 6 and 7 show cross-sections through the Massif with ray paths for events in the North Atlantic, Hindu Kush, Sumatra, Kuril Islands and South Atlantic and contours of residuals for vertically incident PKP phases. The locations of the ray paths for these and other source regions coupled with the observed variation in residuals suggest that the main low velocity region is at a depth of ca. 50 to 150 or 200 km beneath the Western Rhenish Massif, with the shallowest point beneath the West Eifel, and that the low velocity region beneath the Vogelsberg is shallower ($\lesssim 60$ km).

The azimuth of approach and ray parameter for each event can be determined by fitting a plane wave, using least squares, to the observed arrival times. As was mentioned earlier, the values determined for the whole array show no systematic variations and are very close to those predicted by JB and the

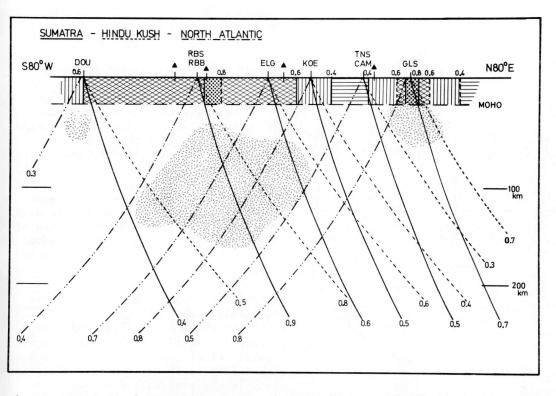

Fig. 6. Cross section through the Rhenish Massif showing ray paths for events in Sumatra, the Hindu Kush and the North Atlantic. The contours of residuals for vertically incident PKP phases are superimposed on the crust, and the residuals for other sources are given at the ends of the rays in seconds. Additional stations whose ray paths are not shown are denoted by *triangles*, and the area of inferred velocity decrease is *hatched*

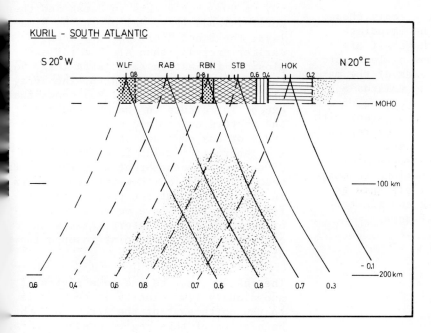

Fig. 7. Cross-section through the Rhenish Massif showing ray paths from events in the Kuril Islands and South Atlantic contours of PKP residuals

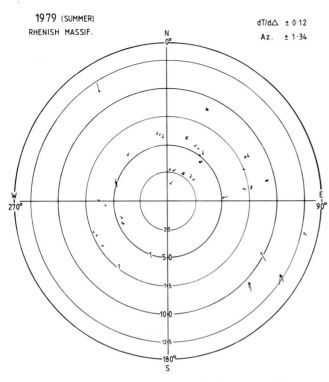

Fig. 8. Array mislocation diagram for stations within the Rhenish Massif. Values of the standard error in azimuth and dT/dΔ are given at the top and the mean mislocations at the bottom

USGS locations. However, if only those stations within, or on the immediate periphery of, the Rhenish Massif are used the mislocations show a clear pattern (Fig. 8) with the heads of the majority of vectors pointing towards a point slightly left of the centre of diagram. This pattern is consistent with the existence of a low velocity region at a depth of ca. 100 km beneath the Western Rhenish Massif. Unfortunately there are insufficient stations to merit a detailed sub-array analysis such as that carried out by Walck and Minster (1982) for Southern California.

The Anomaly at Gräfenberg (GRF)

Outside the Rhenish Massif the only station to show a marked azimuthal dependence of residuals was GRF, which had anomalously late arrivals to the ENE. GRF lies on the edge of the station network, which means that the exact location of the velocity anomaly giving rise to the delays cannot be determined. However, Gräfenberg is an array station (Fig. 9a) - the delays in Fig. 3 are for station A1 - and the residuals at the other substations and the values of azimuth of approach and dT/dΔ measured at the array can be used to place some constraints on the anomalous region.

The mean residuals for each array station (relative to A1) vary from -0.32 s at C5 to 0.02 s at A3 (Fig. 9b). This pattern is consistent with the higher crustal velocities found in the vicinity of C (Aichele, 1976). The azimuthal variation in the residuals for events at greater than 30° (Fig. 9b) indicates that there are also marked variations in mantle velocities in the surrounding region. Mean residuals for events to the north, west or south (azimuths 120° to 360°) show a pattern similar to that for the mean residuals, as do the residuals for core phases. The azimuthal variation arises largely from events at 40° to 90°. For events in Japan and the Kuril Islands (Fig. 9c) or in the Himalaya, the B and C arrays have much earlier arrivals than A1, but for explosions in Eastern Kazakhstan (Fig. 9d) the latest arrivals are at B. The array mislocation diagram (Fig. 9e) shows that except for events in the NE quadrant the vectors determined using the whole array or any combination of sub-arrays are generally similar in magnitude and direction and consistent with the crustal velocity variation. However, for events in the NE quadrant the vectors for different subarrays have very different orientations: for Eastern Kazakhstan explosions they even point in opposite directions.

The changes must be due to a region of anomalously low velocity in the mantle to the north-east of the array, and the boundaries of this region must be quite

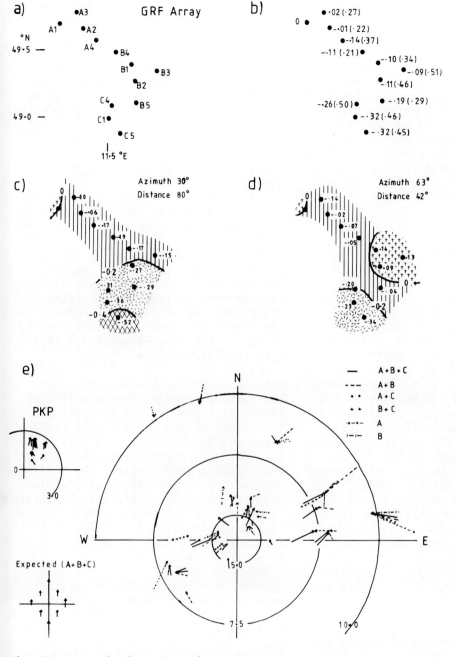

Fig. 9a-e. Variation of residuals at the Gräfenberg array. a) Location of the array sub-stations. b) Mean residual, relative to A1, for all azimuths. The numbers in parentheses are the azimuthal variation for the residuals in seconds. c) Residuals for events in Japan and the Kuril Islands. d) Residuals relative to A1 for explosions in East Kazakhstan. e) Array mislocation diagram for GRF

sharp to produce such large differences between the subarrays. It seems likely that the region of low velocity is associated with the Tertiary and Quaternary volcanics - for example the mountains of the Fichtelgebirge and the Dupp Gebirge - at the northern margin of the Bohemian Massif. If this is the case the structure must extend to depths of 200 km or more, and may be similar to that beneath the Eifel.

Modelling the Observed Delays

Two methods were used to model the observed variations. The first involves the use of ray tracing to delineate the location and magnitude of the velocity anomalies. It has been used by a number of authors, and is described in detail in Raikes (1978). The second method used was the automatic three dimensional inversion scheme pioneered by Aki and co-workers (e.g., Aki et al., 1977; Ellsworth and Koyanagi, 1977).

Models Derived by Ray Tracing and Manual Inversion

The cross-sections of Figs. 6 and 7 suggest that the velocity variations occur mainly in the depth range 50 to 150 or 200 km. Three models were constructed in which it was assumed that the velocity changes were confined to a single layer extending from 50 to 150 km, 100 to 200 km and 40 to 100 km depth, respectively. Ray paths were calculated from a representative event in each source region to each station using the Jeffreys (1939) P velocity distribution which was derived from the JB tables. The path length in the anomalous region was determined and used together with the observed delay for each station to calculate the percentage of velocity change required. This assumes that any change in raypath due to the change in velocity is neglegible: provided the velocity changes are gradual and do not exceed about 6% this is a reasonable assumption.

The model where the anomalous region lay between 50 and 150 km produced the most consistent results. The inferred velocity changes were then plotted on a map at the places where the appropriate rays pass through the midpoint of the layer (100 km) and a contour of apparent velocity change constructed. This map is shown in Fig. 10: this simple model produces a 74% improvement in the variance of the residuals which may be increased to 76% by the inclusion of a shallow low velocity zone (inset, Fig. 10) beneath the Vogelsberg.

The main features of Model 1 are the low velocities (up to 6% decrease) beneath the West Eifel and ENE of GRF. It must be emphasized that these variations are relative to the structure vertically below BUH; adjustments have been made for the slight variation of delays at BUH due to structure in the Rhine Graben area.

Models Obtained by Automatic 3-D Inversion

Although variations in crustal structure have little effect on the azimuthal variation of residuals, it is unreasonable to suppose they have no effect at all on the observed delays. Manual inversion using the method described in the previous section is impractical other than for single layer models, so the remaining models have been derived using the Aki three-dimensional velocity inversion scheme.

In this technique the region beneath the array is divided into a series of rectangular blocks, and velocity perturbations within these blocks are determined. The resolution of this method is limited by the distribution of events and stations which control the vertical and horizontal dimensions of the blocks used. The horizontal block size dictated by the size of the array (Fig. 1) is a minimum of 40 km, and the resulting structures will thus be smoothed at about this wavelength.

A total of 160 events and 53 stations were used in the inversion procedure. Details of the starting models are given in Table 1.

Initially a two layer model was generated for comparison with the ray-tracing Model 1 (Fig. 10). This model, which resulted in a 74% variance improvement, is shown in Fig. 11. There are indeed crustal velocity variations, principally in a belt extending through the centre of the Rhenish Massif. The change from low to high crustal velocities along a profile from southwest to northeast across the Massif, with the "zero" level in the neighborhood of the Rhine, is consistent with the structure deter-

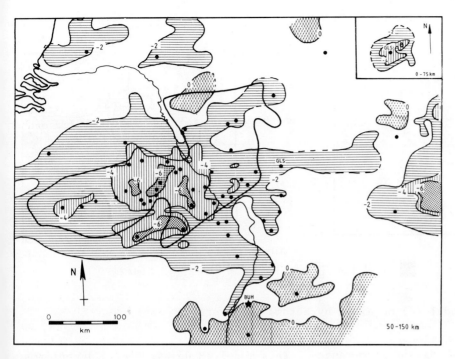

Fig. 10. Model 1, obtained by ray tracing and manual inversion. Contours of velocity change in a region from 50 to 150 km depth: negative values indicate a velocity decrease. *Inset* shows the additional structure at depths of 0 to 75 km beneath the station GLS in the Vogelsberg

Table 1. Starting models for the inversion routine

Number of layers		Model 3 2	Model 4 3	Model 5 4	Model 6 3
Layer 1: "Crust"	Block size, km Depth, km velocity, km s^{-1}	40 0–40 6.1	50 0–40 6.1	50 0–40 6.1	30 0–40 6.1
Layer 2:	Block size, km Depth, km velocity, km s^{-1}	50 40–150 8.2	50 40–100 8.2	50 40–100 8.2	35 40–100 8.1
Layer 3:	Block size, km Depth, km velocity, km s^{-1}		50 100–200 8.5	50 100–200 8.5	50 100–200 8.5
Layer 4:	Block size, km Depth, km velocity km s^{-1}			60 200–300 8.7	

Damping parameter = 100 s^2. Rays sampled 6 times per layer. Minimum number of rays per modelled block = 10.

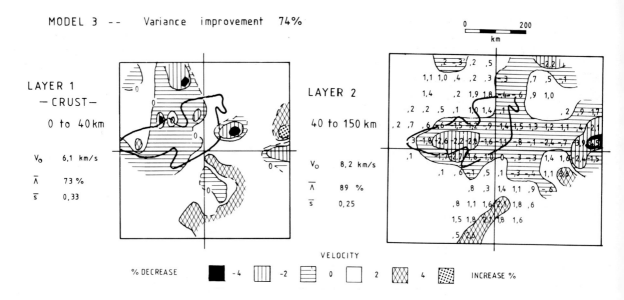

Fig. 11. Two-layer model (3) obtained by automatic inversion

mined from the Rhenish Massif long-range refraction profile (Mechie et al., 1983). There are also marked velocity decreases beneath the Vogelsberg and at those stations where sediment corrections are necessary but have not been included. The velocity variations in Layer 2 (40 to 150 km) are similar to Model 1, but the pattern is somewhat smoother. The decrease under the West Eifel is only 2.7%, but in the inversion scheme the mean residual over all the recording stations was used for normalization rather than the residual at BUH. The structure of Fig. 11 thus represents departures from the mean which, because the stations are concentrated in the Rhenish Massif, tends to a lower velocity than that found beneath BUH. In Fig. 11 the velocity beneath BUH is about 2% above average, and so the velocities beneath the West Eifel are 4 to 5% lower than those beneath BUH. This is in good agreement with Model 1 (Fig. 10).

The next series of models generated had three layers and various block configurations. All had similar velocity perturbations and variance improvement; Fig. 12 shows a composite model based on three different orientations of the block axes. The variance improvement had risen slightly to 75%, and the mean diagonal element of the resolution matrix $\bar{\lambda}$ and standard error in the velocity change $\bar{s}$ have changed by a small amount.

This model (Model 4, Fig. 12) confirms that the Vogelsberg anomaly is confined to the crust and depths less than 100 km. There is a marked velocity decrease (up to 3% or 4.5% relative to the structure beneath BUH) beneath the West Eifel in Layer 2 which appears to follow the trend of the volcanics. The 3% velocity decrease persists into Layer 3 (100-200 km) where it is somewhat broader and trends east-west.

Extending the structure to 300 km depth by the addition of an extra layer produces little change in the velocity perturbations to 200 km depth and only increases the variance improvement to 77%. The velocity changes in the deepest layer (200 to 300 km) are mostly less than ±1%. The three-layer model of Fig. 12 is thus to be considered representative of the structure beneath the array.

Since all the inversion models, whether 2, 3 or 4 layers, had similar velocity perturbations and variance improvement, it is likely that the limiting factor in determining the model is the block size. The layer thickness is dictated largely by the source distribution, and results in vertical smoothing and linkage between layers. The horizontal block size is determined by the size of the array, the station distribution and the storage available on the computer used. It would not be practical to use smaller

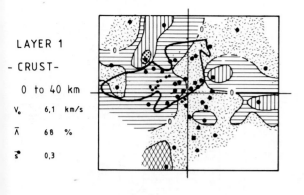
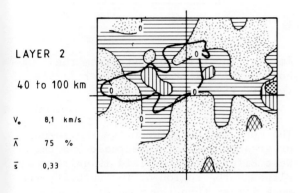
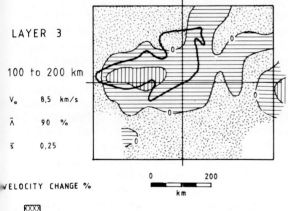

Fig. 12. Three-layer model (4) obtained by automatic inversion. The stations used are denoted by *black circles* in layer *1*; *square* is BUH

blocks for the whole array, but confining the model to the array centre by omitting the stations DOU, UCC, DBN, MOX and GRF, permits the use of smaller blocks, and thus reduces the horizontal smoothing. Figure 13 shows such a model: the blocks in Layer 1 are 30 × 30 km, in Layer 2 35 × 35 km and in Layer 3 40 × 40 km. The variance improvement has risen to 94%, although the large velocity variations, particularly in Layers 1 and 2, suggest that the damping may have been too low. The pattern of residuals is similar to Fig. 12, although the main low velocity zone beneath the Western Rhenish Massif now appears in Layer 3, and has been displaced slightly to the west relative to Fig. 12. It is probable that this shift has been caused by the absence of stations off the western margin of the anomaly. Unfortunately, although the data obtained in 1980 for mobile stations between G1 and DOU have been incorporated into the ray tracing models and support the models of Figs. 10 to 12, there were not sufficient arrivals to be included into the inversion procedure.

Discussion

Analysis of teleseismic P-residuals at stations in the vicinity of the Rhenish Massif provides strong evidence for upper mantle heterogeneity. All models derived to explain the observed variation of P-delays, whether by ray tracing and manual inversion or by automatic inversion, include a marked low velocity zone with a volume of ca. 5×10^5 km^3 beneath the Rhenish Massif west of the Rhine at depths of 50 to 200 km. The strongest velocity decrease and (shallowest) region of low velocities is beneath the West Eifel volcanic field. All models also include low velocities at depths of 100 km or less beneath the Vogelsberg, and a zone of strong velocity decrease to 200 km or more ENE of GRF. The latter is poorly constrained because it lies on the edge of the array. Data available for station WWF in the Westerwald are limited, but suggest there may also be a shallow

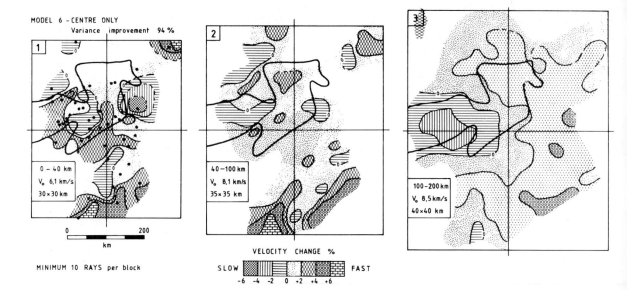

Fig. 13. Three-layer velocity model (6) obtained by inversion of data from the centre of the array. The stations used are shown by *black circles* in layer *1*; the block size and starting velocity for each layer are given

low velocity zone associated with the Westerwald volcanics. The velocity decrease probably does not exceed 4%, but this zone is otherwise similar to the one beneath the Vogelsberg.

In addition to the low velocity areas described above, the models also exhibit high velocities beneath the southern Rhine Graben, Vosges and Black Forest, particularly above 100 km. This is consistent with seismic refraction data (Ansorge et al., 1979) which show velocities of about 8.5 km s^{-1} at depths of 40 to 50 km beneath this region compared with 8.2 km s^{-1} found elsewhere in western and central Europe.

All models show a marked difference between the structure east and west of the Rhine. In view of the uplift extending over the whole region this is perhaps surprising, although the highest current uplift rates (1.6 mm yr^{-1}) are in fact found in the north Eifel just south of the station KLL (Mälzer and Zippelt, 1979; and Mälzer et al., 1983) at the northern edge of the main low velocity region. Similar asymmetry of structure about the Rhine is also found in the seismic refraction and gravity data.

The correlation between the shallowest or greatest velocity decrease and the West Eifel volcanic field is quite good, although the region of low velocity is much larger than the zone of anomalous mantle inferred from petrological and geochemical data (Mertes and Schmincke, 1981). There does not appear to be a comparable anomaly associated with the smaller East Eifel field. Both are Quaternary, but the West Eifel volcanics are younger, the most recent eruption having occured some 8000 years ago (Lorenz and Büchel, 1978), and there is also evidence from $^{87/86}$Sr-ratios that the mantle sources may be slightly different. Both areas differ from the Tertiary volcanic fields which have much lower K- and Rb-contents and higher $^{143/144}$Nd-ratios (Schmincke et al., 1978; Staudigel and Zindler, 1978). Since the mantle sources are different for the Tertiary and Quaternary lavas, differences in mantle structure in addition to those associated with thermal relaxation may be expected.

Both the Westerwald and Vogelsberg volcanics appear to be associated with shallow, rather than deep, velocity anomalies. It is worth noting that the crustal structures beneath the Vogelsberg and Westerwald are similar — the Moho is not clearly defined in either region — and both differ from that beneath the Eifel where the Moho is a good reflector (Mooney and Prodehl, 1978).

Deep electrical sounding measurements do provide evidence for a high conductivity region at a depth of about 50 to 80 km beneath the West Eifel (Jödicke et al., 1979). This layer is not marked beneath the Rhenish Massif east of the Rhine and it is tempting to correlate it with the velocity anomaly. However, the measurements are contaminated by cultural noise, and crustal effects tend to mask the deeper structure. Further measurements (Jödicke, 1981) indicate that this high conductivity zone is not unique to the West Eifel, as similar structures have been found elsewhere in Germany outside the Rhenish Massif.

The correlation between volcanism and low velocities suggests that the anomalous region in the mantle is either hotter than, or chemically different from, the surrounding area, or perhaps both. Petrological evidence from mantle xenoliths supports the idea of an abnormally hot mantle (Sachtleben, 1980) which has also undergone chemical alteration and metasomatism (Lloyd and Bailey, 1975). The exact nature of the anomalous region is, however, hard to define.

Gravity measurements show no marked anomaly associated with the West Eifel, although the Bouguer anomaly is -10 to -20 mgal compared with ~+10 mgal east of the Rhine (Jacoby and Drisler, 1981). Any model of the mantle anomaly must conform to these observations: either the velocity decrease must be accompanied by practically no density change, or the structure must be compensated by shallower changes, for example in Moho depth. The first can be achieved by partial melting: about 1% of partial melt could lower the velocity from 8.1 to 7.8 km s^{-1} (a 4% change) (Anderson and Spetzler, 1970) and only decrease the density by about 0.03 g cm^{-3}. Alternatively, velocity-density systematics (e.g., Anderson et al., 1971) indicate that by increasing the mean atomic weight it is possible to lower the velocity without significantly affecting the density. Such a change would be equivalent, for example, to mantle enrichment in Ca or Fe relative to Mg.

Jacoby and Drisler (1981) favor the second model. They propose that the ca. -60 mgal anomaly caused by the -0.1 g cm^{-3} density contrast associated with a 3% velocity decrease according to Birch's law is compensated by a Moho uplift of ca. 3 km. Such a change would have a negligible effect on the P-delays, but should be seen in the refraction data. Further, if the velocity decrease is actually around 5% the Moho uplift would have to be larger. Interpretations of the currently available refraction data provide little evidence in favor of this model.

Metasomatism accompanied by partial melting has been proposed as a mechanism for the uplift of the Rhenish Massif by a number of authors including Lloyd and Bailey (1975) and Sachtleben (1980). Their arguments were based on petrological and geochemical evidence derived from analysis both of the Eifel lavas and of mantle xenoliths.

The light element infiltration model of Lloyd and Bailey includes mantle enrichment in Fe and Ca together with a number of other elements including Ti, Mn and K plus H_2O and CO_2, together with dilution and/or abstraction of some Mg, Si, Cr and Ni. This model proposed that the uplift of the Rhenish Massif was associated with the production of phlogopite by metasomatism of enstatite and forsterite, which results in an anomalously low mantle density. However, as mentioned earlier, mantle enrichment in Fe and Ca relative to Mg can cause a velocity decrease without a large density change. Further, the presence of volatiles (such as H_2O, CO_2) can eventually lead to partial melting which will also contribute to the uplift; 1% partial melt can produce 140 m of uplift (Fuchs, 1981), and the highest observed uplift of 300 m could be achieved by increasing the degree of partial melting.

Acknowledgments. This research was performed within the DFG-priority program *Vertikalbewegungen und ihre Ursachen am Beispiel des Rheinischen Schildes.* The authors are indebted to Professor Dr. Fuchs for initiating this work, and to him, Drs. Mechie and Prodehl for many ideas and interesting discussions. The following people kindly made data of permanent stations available: Drs. Ahorner, Aichele, Baier, Flick, van Gils, Keller, Rouland, and Schneider. The following people operated the mobile stations during several surveys: K.-P. Bonjer, H. Durst, Y. Godard, W. Haberecht, H. Hoffmann, P. Jung, R. King, M. Koch, K. Küper, T. Lehnert, P. Marschall, J. Mechie, T. Moldoveanu, E. Räkers, S. Raikes, M. Rittershofer, A. Ruthard, K. Scheule, H. Siedler, G. Volk, P. Wägerle, D. Wagner, and M. Wagner.

The Geostores were provided by the N.E.R.C. and operated in cooperation with Birmingham University, and three portacorders were provided by I.S.P.H.-Bucharest. Mrs. Welzenbach and Mrs. Gadau typed the manuscript. Prof. Fuchs, Dr. Mechie, and Dr. Prodehl read the manuscript.

The computations have been carried out at computer centres of the universities Karlsruhe and Heidelberg.

References

Ahorner, L., 1975. Present day stress field and seismotectonic block movement along major fault zones in Central Europe. In: Pavoni, N. and Green, R. (eds.) Recent crustal movements. Tectonophysics, 29:233-248.

Aichele, H., 1976. Interpretation refraktionsseismischer Messungen im Gebiet des Fränkisch-Schwäbischen Jura. PhD. thesis, Univ. Stuttgart, 105 pp.

Aki, K., Christofferson, A., and Husebye, E.S., 1977. Determination of the three dimensional seismic structure of the lithosphere. J. Geophys. Res., 82:277-296.

Anderson, D.L. and Spetzler, H., 1970. Partial Melting and the low velocity zone. Phys. Earth. Planet. Inter., 4:62-64.

Anderson, D.L., Sammis, C., and Jordan, T., 1971. Composition of the mantle and core. Science, 171:1103-1112.

Ansorge, J., Bonjer, K.-P., and Emter, D., 1979. Structure of the uppermost mantle from long range seismic observations in Southern Germany and the Rhine Graben area. In: Fuchs, K. and Bott, M.H.P. (eds.) Structure and compositional variations of the lithosphere and astenosphere. Tectonophysics, 56:31-48.

Bolt, B.A., 1968. Estimation of PKP travel times. Bull. Seismolog. Soc. Am. 58:1305-1324.

Ellsworth, W.L. and Koyanagi, R.Y., 1977. Three dimensional crust and mantle structure beneath Kilanea volcano, Hawaii. J. Geophys. Res., 82:5379-5394.

Engdahl, E.R., Sinndorf, J.G., and Eppley, R.A., 1977. Interpretation of relative teleseismic P-wave residuals. J. Geophys. Res., 82:5671-5682.

Fuchs, K., 1981. Zur Diskussion zwischen Geophysik und Petrologie. Protokoll über das 6. Kolloquium im Schwerpunktprogramm "Vertikalbewegungen und ihre Ursachen am Beispiel des Rheinischen Schildes". Neustadt/Weinstr., pp 186-188.

Husebye, E.S., Christofferson, A., Aki, K., and Powell, C., 1976. Preliminary results on the three dimensional seismic structure of the lithosphere under the USGS central California seismic array. Geophys. J.R. Astron. Soc., 46:319-340.

Illies, J.H., Prodehl, C., Schmincke, H.U., and Semmel, A., 1979. The quaternary uplift of the Rhenish Shield in Germany. In: McGetchin, T.R. and Merril, R.B. (eds.) Plateau uplift: mode and mechanism. Tectonophysics, 61:197-225.

Jacoby, W.R. and Drisler, J., 1981. Schwerefeld und Massenverteilung unter dem Rheinischen Schild. Protokoll über das 6. Kolloquium im Schwerpunktprogramm "Vertikalbewegungen und ihre Ursachen am Beispiel des Rheinischen Schildes" Neustadt/Weinstr., pp 218-221.

Jeffreys, H., 1939. The times of P, S, and SKS and the velocity of P and S. Mon. Not. R. Astron. Soc. Geophys. Suppl., 4:498-533.

Jeffreys, H. and Bullen, K.E., 1940. Seismological Tables. Br. Assoc. Adv. Sci., Gray-Milne Trust, London, pp 55.

Jödicke, H., 1981. Magnetotellurik im Rheinischen Schiefergebirge. Protokoll über das 6. Kolloquium im Schwerpunktprogramm "Vertikalbewegungen und ihre Ursachen am Beispiel des Rheinischen Schildes". Neustadt/Weinstr., pp 193-196.

Jödicke, H., Schulte, L., Olgemann, W., and Untiedt, J., 1979. Magnetotellurik Rheinischer Schild. Protokoll über das 4. Kolloquium im Schwerpunktprogramm "Vertikalbewegungen und ihre Ursachen am Beispiel des Rheinischen Schildes". Neustadt/Weinstr., pp 91-92.

Lloyd, F.E. and Bailey, D.K., 1975. Light element metasomatism of the continental mantle: the evidence and the consequences. Phys. Chem. Earth, 9:389-416.

Lorenz, V. and Büchel, G., 1978. Phreatomagmatische Vulkane in der südlichen Westeifel; ihr Alter und ihre Beziehung zum Talnetz. Nachr. Dtsch. Geol. Ges., 19:30.

Mälzer, H. and Zippelt, K., 1979. Local heigh changes in the Rhenish Massif area. Allg. Vermessungsnachr., 86:402-405.

Mälzer, H., Hein, H., and Zippelt, G., 1983. Height changes in the Rhenish Massif: determination and analysis. This Vol.

McGetchin, T.R., Burke, K., Thompson, G., and Young, R., 1979. Plateau uplift: mode and mechanism. EOS Trans. Am. Geophys. Union, 60:64-67.

Mechie, J., Prodehl, C., and Fuchs, K., 1983. The long-range seismic refraction experiment in the Rhenish Massif. This Vol.

Mertes, H. and Schmincke, H.-U., 1981. Untersuchungen im Vulkanfeld der Westeifel. Protokoll über das 6. Kolloquium im Schwerpunktprogramm "Vertikalbewegungen und ihre Ursachen am Beispiel des Rheinischen Schildes". Neustadt/Weinstr., pp 197-198.

Mooney, W.D. and Prodehl, C., 1978. Crustal structure of the Rhenish Massif and adjacent areas: a reinterpretation of existing seismic refraction data. J. Geophys., 44:573-602.

Raikes, S.A., 1978. Regional variations in upper mantle compressional velocities beneath Southern California, Part I. PhD thesis, California Inst. Technol., 307 pp.

Raikes, S.A., 1980a. Teleseismic evidence for velocity heterogeneity beneath the Rhenish Massif. J. Geophys., 48:80-83.

Raikes, S.A., 1980b. Regional variations in upper mantle structure beneath Southern California. Geophys. J.R. Astron. Soc., 63:187-216.

Sachtleben, R., 1980. Petrologie ultrabasischer Auswürflinge aus der Westeifel. PhD thesis, Univ. Köln, 180 pp.

Schmincke, H.U., Mertes, H., Staudigel, H., Zindler, A., Gijbels, R., and Bowman, H., 1978. Geochemische und tektonische Kontraste zwischen den quartären Vulkanfelsen der E- und W-Eifel. Fortschr. Miner., 56, 1:124-125 (Abstr.).

Staudigel, H. and Zindler, A., 1978. Nd and Sr isotope compositions of potassic volcanics from the East Eifel, Germany: Implications for mantle source regions. Geol. Soc. Am., Ann. Meet. Toronto, p 135.

Walck, M.C. and Minster, J.B., 1982. Relative array analysis of upper mantle lateral velocity variations in Southern California. J. Geophys. Res., 87:1757-1772.

Wenderoth, R., 1978. Laufzeitanomalien teleseismischer P-Wellen im Gebiet des Oberrheingrabens. Diplomarbeit, Geophys. Inst., Univ. Karlsruhe, 170 pp.

7.6 Crustal Xenoliths in Tertiary Volcanics from the Northern Hessian Depression

K. Mengel and K. H. Wedepohl[1]

Abstract

A suite of 46 crustal xenoliths from Tertiary basaltic volcanics comprises sillimanite bearing quartzites and pegmatoids, biotite rich schists, pyribolites, pyriclasites and garnet-free granulites. The chemical composition of 11 pyriclasites and 5 garnet-free granulites resembles (within-plate) tholeiites. Equilibrium temperatures and pressures as calibrated on the base of experimental reactions, range from 700° to 900°C and from >5 to <10 kb. Present (and Upper Tertiary) temperatures at Moho depth (10 kb) have been estimated at about 550°C. Therefore granulites must have equilibrated at a former thermal event.

Petrographic and chemical analyses of xenoliths sampled from basaltic pyroclastics and basalts of the northern Hessian Depression inform about the deeper crust in the area between the Rhenish Massif and the Harz Mountains. The abundance of rock types in a suite of 46 xenoliths of metamorphic rocks and some plutonic species is presented in Table 1.

Due to the relatively small number of specimens which have been sampled by the uprising magma in a not well known mechanical process it is unlikely that this set of inclusions represents a complete crustal profile. The most abundant rock types of the Variscan mobile belt are not detected as inclusions, samples of the Mesozoic sedimentary cover are rare. The maximum thickness of the Devonian to Lower Carboniferous sedimentary cover in the Rhenohercynian basin is estimated at 6 to 8 km (Schulz-Dobrick, 1975). Two-kilometer Tertiary and Mesozoic sediments have to be added to the 5 or 6 km Carboniferous and Devonian rocks.

Compared to the probably more complete crustal profile sampled in the Eifel area (Voll, 1983) the suite of inclusions from the northern Hessian Depression contains different rock types as well as similar materials. Xenoliths of typical greenschist and amphibolite facies rocks have not yet been observed in the Hessian volcanics. More than a third of the rocks of Table 1 is represented by species as sillimanite-bearing granitic pegmatoids and quartzites. The only rock types occurring in either site are granulite facies metamorphics: pyribolites, pyriclasites, and granulites. The five granulites of our area are free of garnet, in contrast to those observed by Voll (1983) and Okrusch et al. (1979) in the series of xenoliths from the Eifel. Therefore some differences between the two crustal profiles from regions about 180 km apart are possible.

The occurrence of sillimanite in quartzites and granitic pegmatoids is a product of regional metamorphism. According to the stability conditions as reported by Holdaway (1971) and Richardson et al. (1969) these rocks have been metamorphosed at temperatures above 500° and 630°C, respectively.

---

[1] Geochemisches Institut, Goldschmidtstraße 1, D-3400 Göttingen, Fed. Rep. of Germany.

Plateau Uplift, ed. by K. Fuchs et al.
© Springer-Verlag Berlin Heidelberg 1983

Table 1. Composition of crustal xenoliths from basaltic volcanics in the northern Hessian Depression

Number of samples	Rock type	Phase assemblages
4	Sillimanite-bearing quartzites	Sillimanite, quartz ± spinel ± rutile
15	Granitic pegmatoids	Alkali-feldspar, plagioclase, quartz ± sillimanite + rutile ± spinel ± apatite, zircon
2	Biotite schists	Phlogopite, clinopyroxene ± apatite
2	Pyribolites	Plagioclase (an 52-57), pargasitic amphibole, orthopyroxene (en 73), magnetite
15	Pyriclasites	Clinopyroxene (augite), plagioclase (an 26-56), >scapolite (me 70-83) ± garnet (gross + and $\overline{84\text{-}88}$) ± alkali-feldspar (or 78), magnetite, apatite, titanite
8	Granulites (garnet free)	Orthopyroxene (en 65-78), clinopyroxene (diopsidic), plagioclase (an 38, an 74-98), magnetite

Intensively folded biotite rich schists do not contain appropriate parageneses for information about their depth of origin. According to field experience such rocks cannot be common.

All pyriclasites and pyribolites are strongly banded on a centimeter to millimeter scale. They contain plagioclase (and scapolite) rich and pyroxene rich layers. Pyriclasites and granitic pegmatoids are the most abundant rock types in our collection of xenoliths. Metamorfic rocks with a similar mineral composition are abundant in the uplifted lower crustal units which are exposed in the Ivrea Zone (Mehnert, 1975). The chemical composition of pyriclasites from both occurrences is listed in Table 2. The analysed pyriclasite inclusions are small (4-10 cm in diameter). This may explain the scattering of the major element data. The patterns of the REE distribution in these inclusions are close to those from within-plate-tholeiites (Wedepohl, 1975). Considering their widespread occurrence in the Ivrea zone and their relative abundance in our collection of xenoliths, pyriclasites probably represent a major unit of the lower crust.

The mineral composition of the granulite fragments allows some assumptions about their metamorphic history. Evaluation of microprobe data on orthopyroxene-clinopyroxene parageneses according to the Wells geothermometer indicates former equilibration in a range from $700°$ to $900°C$. These temperatures are close to the lower limit of the method. They either represent temperatures in the lower crust during Miocene time or are quenching products from a system which has not reequilibrated due to the low rates of diffusion. According to estimates of geothermal gradients in continental crustal and upper mantle sections 200 Ma after an orogenic event, reported by Pollack and Chapman (1977), the present (and Miocene) temperature at the Moho depth is probably as low as $530°C$. The lack of garnet in our granulites can indicate a rather steep temperature gradient in the lower crust introduced during the former high temperature event. According to the experimentally determined stability fields of garnet bearing and garnet free granulites (Green and Ringwood, 1967) the metamorphic pressures could not have exceeded 5 kbar at $700°C$, 7.5 kbar at $800°C$, 10 kbar at $900°C$ and so on. Otherwise garnet is expected to

Table 2. Major element chemistry of crustal xenoliths

	Mean of 11 xenoliths, Hessian Depression	[a] Melanocratic pyriclasites, Ivrea (Mehnert, 1975)	Mean of 5 xenoliths, Hessian Depression	[b] Tholeiitic basalt used in the experiment by Green and Ringwood (1967a)
SiO_2	43.8 (± 1.2)	44.8	47.6	52.2
TiO_2	3.21(± 1.03)	2.09	1.8	1.9
Al_2O_3	12.5 (± 1.5)	15.6	16.8	14.6
Fe_2O_3	8.9 (± 2.8)	1.95	n.d.	2.5
FeO	4.7 (± 1.4)	9.0	n.d.	8.6
MnO	0.20(± 0.04)	0.16	0.2	0.1
MgO	5.5 (± 2.0)	11.6	8.7	7.4
CaO	15.4 (± 4.1)	12.2	12.0	9.4
Na_2O	2.51(± 1.2)	2.15	1.8	2.7
K_2O	0.29(± 0.2)	0.36	0.6	0.7
P_2O_5	0.42(± 0.1)	0.03	n.d.	0.2
Total iron (Fe_2O_3)	14.2 (± 2.7)	12.0	12.1	12.1

[a]Pyriclasites, [b]Granulites (in brackets: standard deviation).

occur, because the chemical composition of the granulite xenoliths is close to the material used by Green and Ringwood (1967). P-T estimates about the equilibration of granulite inclusions from the Eifel area are in the same range (Voll, 1983), but the rocks from the Eifel contain garnet. This difference to our observations indicates a slightly less steep p-T gradient for the Eifel area.

In a suite of spinel harzburgite and spinel lherzolite xenoliths in alkali basalts from the northern Hessian Depression the sample with the lowest temperature has been equilibrated at 880°C (Oehm, 1980). This temperature is in the same range as that of equilibration of crustal granulites. The lack of regular partition of several elements between "paragenetic" minerals of low temperature peridotite xenoliths indicates not achieved equilibria (Oehm et al., 1983).

Apart from some rare peridotitic xenoliths sampled from pyroclastics, the great majority of harzburgites and lherzolites is characterized by a coarse grained structure without indications of major shearing. Therefore, widespread diapiric uprise of hot mantle material into the upper mantle underneath the Hessian Depression during late Tertiary time is rather unlikely. According to these observations on upper mantle rocks the temperatures of 700° to 900°C at pressures of 5 to 10 kbar, estimated for the granulites, cannot represent the p-T conditions of the lower crust during Miocene time. It is more likely that the granulites and the pyriclasites have been formed during an older metamorphic event.

References

Clark, S.P. and Ringwood, A.E., 1964. Density distribution and constitution of the mantle. Rev. Geophysics, 2: 35-88.
Den Tex, E., 1965. Metamorphic lineages of orogenic plutonism. Geol. en Mijnbouw, 44:105-132
Green, D.H. and Ringwood, A.E., 1967a. An experimental investigation of the gabbro to eclogite transformation and its petrological applications. Geochim. Cosmochim. Acta, 31:767-833.
Green, D.H. and Ringwood, A.E., 1967b. The stability fields of aluminous pyroxene peridotite and garnet peridotite and their relevance in upper

mantle structure. Earth Planet. Sci. Lett., 3:151-160.
Holdaway, M.J., 1971. Stability of andalusite and the aluminium silicate phase diagram. Amer. J. Sci., 271: 97-131.
Mehnert, K.R., 1975. The Ivrea Zone. N.Jb. Miner. Abh., 125:156-199.
Oehm, J., 1980. Untersuchungen zu Equilibrierungsbedingungen von Spinell-Peridotit-Einschlüssen aus Basalten der Hessischen Senke. Dissertation, Göttingen.
Oehm, J., Schneider, A., and Wedepohl, K.H., 1983. Upper mantle rocks from basalts of the Northern Hessian Depression. TMPM (in press)
Okrusch, M., Schröder, B., and Schnüttgen, A., 1979. Granulite-facies metabasic ejecta in the Laacher See area, Eifel, West Germany. Lithos, 12:251-270.
Pollack, H.N. and Chapman, D.S., 1977. On the regional variation of heat flow, geotherms and lithospheric thickness. Tectonophysics, 38:279-296.
Richardson, S.W., Gilbert, M.C., and Bell, P.M., 1969. Experimental determination of cyanite-andalusite and andalusite-sillimanite equilibria: the aluminium silicate triple point. Amer. J. Sci., 269:259-272.
Schulz-Dobrick, B., 1975. Chemischer Stoffbestand variskischer Geosynklinalablagerungen im Rhenoherzynikum. Dissertation, Göttingen.
Voll, G., 1983. Crustal xenoliths and their evidence for crustal structure underneath the Eifel volcanic district. This Vol.
Wedepohl, K.H., 1975. The contribution of chemical data to assumptions about the origin of magmas from the mantle. Fortschr. Miner., 52:141-172.
Wells, P.R.A., 1977. Pyroxene thermometry in simple and complex systems. Contrib. Miner. Petrol., 62:129-139.

7.7 Crustal Xenoliths and Their Evidence for Crustal Structure Underneath the Eifel Volcanic District

G. Voll[1]

Abstract

Volcanoes of the Eifel district have sampled the total crust. From 1000 Xenoliths the following profile is deduced (downwards): 5 km devonian; 10 km greenschist facies phyllites; 5 km low amphibolite facies micaschists; 5-10 km medium to high grade amphibolite facies gneisses. A thin layer of granulite facies metabasites between these and the Moho are regarded as remainder of an older series, the other rocks have suffered Variscan Barrow-regional metamorphism. Many xenoliths display young contact metamorphism at deep chambers and short pyrometamorphism superimposed. The crust has suffered rotational deformation increasing downwards in amount and complication. Heating has not surpassed $300°C$ 15 km below the recent surface since Variscan times.

Introduction

More than 1000 xenoliths have been studied from Siebengebirge (Finkenberg, Alkali-olivine-basalt), Laacher See (gray tuff, tephrite); Gleeser tuff and Hüttenberg tuff (Wehr volcano-tectonic depression, phonolite and trachyte), Kappiger Ley (Wehr depression, phonolite), from a tuff of the W-side of the Kyller Kopf volcano (W-Eifel, tephrite), from Kempenich (nephelinite - tuff), the Karmelenberg - volcano (nr. Ochtendung, basanite), the upper Mendig - lava (tephrite) and Thelenberg (basanite).

Together these volcanoes sample the total crust. Some draw from restricted levels (Kempenich: pyribolites, pyriclasites and gneisses only; Karmelenberg: migmatic gneisses only; Kyller Kopf; garnet-staurolite - micaschists only). Others have sampled all crustal levels: Laacher See, Hüttenberg, Kappiger Ley. The xenoliths may be arranged in order of increasing regional metamorphism. I take this to be a series of increasing depth of derivation, as none of the xenoliths show signs of tectonic restacking such as diaphthoresis. The rock types and levels are (Fig. 1):

Rhenohercynian Devonian, (sandstones, siltstones, slates) and lower triassic sandstones.

Greenschist Facies Rocks, phyllites, greenschists, calcareous and/or micaceous sand- and siltstones. All transitions exist from the devonian with its beginning metamorphism to amphibolite facies rocks: clastic quartzes gradually recrystallize, grain sizes of recrystallized quartz and micas increase, actinolite, epidote-clinozoisite, biotite, chloritoid, garnet + albite + chlorite make their appearance.

Lower Amphibolite Facies Rocks, first appearance of albite + oligoclase, of at first tiny staurolites can be traced. Types with chlorite + clinozoisite + oligoclase, with chlorite + garnet + biotite + clinozoisite + oligoclase lead to types with garnet + clinozoisite

[1] Mineralogisch-Petrographisches Institut, University of Köln, Zülpicherstraße 49, D-5000 Köln, Fed. Rep. of Germany.

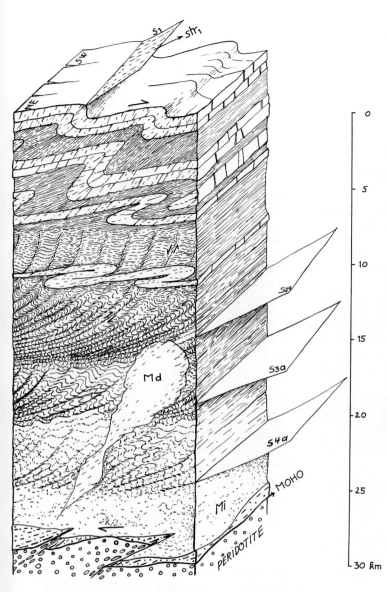

Fig. 1. Structural development within the Eifel crust. Surface = recent surface. From 0-4.5 km: Rhenic Devonian; 4.5-15 km: greenschist facies rocks; 15-20 km: lower amphibolite facies rocks. Below 20 km: medium and high grade amphibolite facies and migmatites (Mi). Md = Prevariscan metadiorites. Peridotite = mantle rocks. On top of "peridotite" granulite facies rocks: 1.5-0 km above the Moho. These rocks and Moho penetrated by shear zones. s_1, str_1 = first cleavage and stretching direction. s_{2a}, s_{3a}, s_{4a} = antithetic planes of second, third and fourth cleavage. General rotation: clockwise looking SW. *Small arrows* indicate sense of movement for antithetic (higher) and synthetic (lower) s_2-cleavage planes. Earlier cleavages are omitted where a new one takes over. *Large arrows* indicate sense of shearing within the total crust

within garnet but none outside, kyanite + staurolite + almandite + oligoclase. Sizes of garnet, staurolite, biotite, muscovite, oligoclase, quartz increase accordingly. There are transitions into calc silicate rocks, hornblende schists, epidote amphibolites and amphibolites. These rocks change steadily into:

<u>Medium Amphibolite Facies Gneisses</u>, with andesine-oligoclase + biotite + muscovite ± garnet. And from there to:

<u>Higher Amphibolite Facies Gneisses</u>, with fibrolite + biotite + andesine-oligoclase ± K-feldspar. Then cordierite-sillimanite gneisses and, finally, migmatites are reached.

Within medium and higher amphibolite facies rocks calcsilicate-rocks and amphibolites continue to occur.
Granites (Kappiger Ley) are a rare exception. They are magmatic rocks with zircone and apatite strongly adsorbed to biotite. A last step downwards leads to

<u>Granulite Facies Rocks</u>, pyriclasites and pyribolites ± scapolite.
They are very similar to rocks from the Ivrea Zone, South of the Alps: unmixed ortho- and clinopyroxenes, plagioclase

with Huttenlocher unmixing, brown hornblende, garnet (with unmixed rutile and 20-25 mol % pyrope). These rocks have also been studied by Okrusch et al., 1979).

Small Scale Structure

Increasing depth changes the style and intensity of deformation: Devonian and uppermost phyllites always display bedding and one penetrative cleavage and stretching direction only. Within the lower greenschist facies a second cleavage starts to appear: at first with high angles s_1/s_2. It causes crumples (B_2) always normal to the first stretching lineation (Fig. 1). Downwards - towards higher grade of regional metamorphism - this second cleavage is rotated: the angle s_1/s_2 becomes smaller, s_2-distances and angles between the limbs of B_2-foldlets decrease. A synthetic set of s_2 joins the first-formed, antithetic one. Pressure-exsolution of quartz from s_{2a} and s_{2s} forms a distinct metamorphic banding with hinges of B_2-microfolds enriched in quartz and chlorite or biotite. This development is complete shortly before the amphibolite facies is reached (Fig. 1).

At the very beginning of the amphibolite facies the same situation is found. Then, downwards, a third cleavage, a second refolding start, repeating the second cleavage and rotating downwards in the same way. This second refolding is completed at the boundary of lower to medium amphibolite facies.

Within the lower amphibolite facies a late retrogressive kinking near $300^{\circ}C$ must occur frequently, undulating quartz grains. Within lower levels it is impossible to count acts of refolding with certainty. Quartz-dioritic augengneisses with orthite and plagioclase phenocrysts (partly recrystallized) often display one cleavage and stretching only. They have been harder rocks and, therefore, less deformed.

The lowermost granulite facies returns to simple deformation: with one penetrative planar + linear (c of hornblendes and pyroxenes) deformation only (drag by mantle flow?). Static annealing in granulite facies and a wide spread but not omnipresent change to amphibolite facies under mainly static conditions follow. This again is a similar situation to that in the Ivrea Zone.

Depth of Facies Boundaries

I favor the following depth of facies boundaries (Voll, 1978, 1979, 1981) (Fig. 1):

Lower boundary of the devonian: 4-5 km below the recent surface (Meyer and Stets, 1975). 4 km have been eroded after the Variscan orogeny, most of these until the permian.

Lower boundary of the greenschist facies: at ca. $520^{\circ}C$ (Variscan orogeny): 15 km under the recent surface.

Lower boundary of the low-T-amphibolite facies: at ca. $600^{\circ}C$: 20 km under the recent surface.

Lower boundary of granulite facies rocks: 28-30 km under the recent surface = Moho (these rocks may be missing locally and have a thickness of 0-5 km).

Age of Metamorphism and Mountain Building

There is a continuous structural development including symmetry-constant folding and refolding from the Devonian downwards to the lowermost gneisses and migmatites. In the same direction there is a steady increase of regional metamorphism. Therefore, all this must be Variscan. Consequently the sediments below the Devonian must, in parts, be Precambrian. During the Variscan orogeny this crust has been 30 + 4 (now eroded) km thick. Nevertheless this crust of normal, continental thickness showed deformation and regional metamorphism typical for orogenic events throughout. The rotational deformation increasing steadily downward is best explained by increasing softening downward by increasing T. The whole crust was a huge shear zone with rigid higher parts (cool, thrusting) and increasing plastic deformation by mantle drag downwards. Most likely it abutted against a Variscan mantle more rigid than the lowermost crust.

T increased downwards to 650°C and partial melting within the lowermost gneisses only a few km above the Moho. This crust must have cooled pretty quickly as low-T kinkfolds formed at a depth of 15-18 km underneath the recent surface. These folds are found in micaschists which reached a maximum T from 520°-580°C at the Variscan regional metamorphism. This folding strained quartz grains heavily and these quartz grains have never recrystallized since, i.e., a T of barely 300°C has never been surpassed thereafter. However, where the same micaschists are heated by young intrusions, the same quartz may recrystallize under elimination of this residual strain from Variscan times.

As partial melting started only directly above the Moho, there was not sufficient volume of partial melts to rise as Variscan granites. Consequently only one xenolith of a true granite was found. However, there are many xenoliths derived from meta-diorites and quartz-diorites. They have suffered the same Variscan deformation as meta-psammopelites from the same level: 1-2 Variscan cleavages. Their plagioclase phenocrysts recrystallized (i.e., was deformed above 500°C) but coexisted with epidote-clinozoisite during regional deformation, i.e., these intrusions must be localized within lower amphibolite facies-micaschists. They were intruded before the Variscan orogeny and deformed together with the sediments they intruded by this orogeny. They cannot be responsible for any of the observed contact metamorphism.

This situation is different from more southerly regions of the variscan mountain range. Within Black Forest and Massif Central the migmatites (and granites) are now at the surface and the Moho is still 30 km below the surface. There the Variscan crust must have been twice as thick and thickening started from the Taunus- and Hunsrück-hills towards the S.

The metamorphic series leading into lower crustal levels has been a straightforward Variscan Barrow facies series. The thermal gradient was ca. 27°C km^{-1}. This is quite different from the Venn-Stavelot-Massiv to the N where Kramm (1982) found a local Variscan gradient up to 50°-55°C km^{-1} and low-pressure andalusite formation.

The granulite facies rocks (Okrusch et al., 1979 and own work) have yielded T and p for garnet- and pyroxene-equilibration up to 850°C and p up to 12 kbar. If the lowermost rocks of the continuous Barrow series - the Variscan migmatites - reached partial melting at 650°-670°C and 7-8 kbar there must be a gap in p-T conditions between these migmatites and the granulite facies rocks. Therefore I regard the granulite facies series as the largely eroded remainder of an older, Precaledonian orogenic series which was eroded down to a near-Moho-level (similar again to the Ivrea Zone where, however, the granulite facies metamorphism proved Caledonian). In my opinion the granulite facies rocks were, after erosion, covered by the thick series of sediments now represented as high-amphibolite facies to Devonian xenoliths. Both granulite facies underground and this sedimentary cover were (re)deformed at Variscan times. This Variscan metamorphism and deformation caused high-amphibolite facies retrogression within the granulite facies underground. This was not penetrated by deformation at this Variscan retrogression - deformation was more one of shear-zone formation and may well have penetrated as such into the mantle, imbricating the Moho, causing unsharp reflections.

If up to 850°C and 12 kbar reflect the conditions of a last equilibration for the granulite facies rocks, this cannot be a Variscan event. Therefore I think it impossible that the granulite facies metamorphism reflects a local variation from wet amphibolite to dry granulite facies conditions during Variscan times. To reach a pressure of 12 kbar there was not enough space between the lowermost gneisses and migmatites and the Moho. For the same reason I cannot imagine this last T-equilibration of the granulite facies rocks to be an even younger event connected to Eifel magmatism.

Woerner et al. (1982) assume that a
thrust underneath the Wehr depression
has cut out nearly all the greenschist
facies rocks. The thrust is an assumed
South continuation of the Aachen thrust
outcropping 90 km further N. They assume
that it passes underneath the Wehr depression at 5 km depth and envisage the
low-amphibolite facies micaschists to
start directly underneath. This I reject
as: (1) A thrust should place higher
grade rocks on lower grade ones while it
moves N and cuts up section. (2) Cut
out greenschist facies rocks of many km
thickness should be found somewhere
further N, which is not the case.
(3) Amongst all the xenoliths I have
found not one which could be derived
from a thrust horizon (sharp deformation
and preferred orientation, characteristic quartzfabrics). (4) The continuous
development of deformation, folding
and refolding is contrary to a break in
the sequence, caused by a thrust, moving after regional metamorphism.
(5) Greenschist facies rocks are not
extremely rare (1%) as quoted by these
authors. They reach 29% of the xenoliths
from the Hüttenberg-tuff (Wehr depression) and are not much less in tuffs
from Laacher See and Kappiger Ley.

Geophysical Consequences

The crustal model suggested here implicates a small and steady increase in
density and wave velocity within rocks
from the Rhenic Devonian to 15-20 km
underneath the recent surface. There
may be a slightly more rapid increase at
a depth of 15 km (a kind of Conrad-discontinuity) where garnet, staurolite
and kyanite appear in larger amounts.
The density should then remain the same
until the granulite facies rocks are
reached. They may be interlayered or imbricated with peridotites near the Moho.
Though locally higher greenschist facies
and lower amphibolite facies metapelites
may contain up to more than 10 vol% magnetite and though there are graphite-bearing rocks within the same levels
this will hardly explain a significant
increase in electrical conductivity.

Contact Heating

Rocks from all levels have suffered
contact heating. Though there are transitions at the margins of deep-seated
magma chambers, usually a slow contact
heating further down and a fast one after
immersion into the melts and following
ejection may be distinguished.

There must be magma chambers of Tertiary
to Recent age at all levels of the
crust. Contact heating at the higher
of these chambers produces hornfelses
ranging from cordierite-andalusite-spotted ones to others with high amounts
of partial melting. All transitions are
found. At lower levels hornfelses from
outer aureole zones are missing. But
partial melting is very incomplete even
within xenoliths from cordierite-sillimanite gneisses: there is no indication
of a Subrecent overall T-rise to the
gneiss-level above the Moho causing young
regional melting. At higher levels contact heating may be repeated: underneath
the Wehr depression and Laacher See
higher grade phyllites, mica schists
and gneisses have suffered high grade
contact metamorphism and partial fenitization twice followed by cooling and unmixing of KNa feldspars until low albite was formed. Then a third immersion
and fast heating occurred in connection
with a final ejection. Dealkalization
and desilicification are widespread
within outer to inner aureoles. Andalusite, regional metamorphic kyanite,
sillimanite may be transferred into
spinel or corundum, biotite into orthopyroxene, staurolite and contact metamorphic corundum into spinel. A fluid
phase must carry SiO_2 and alkalies
away - outwards. Partial melting occurs
near the chamber margins and, again,
may be carried outwards. In front of it
realkalization is widespread, causing
cordierite, sillimanite, corundum, spinel or garnets to be changed into feldspars and biotite. This alkalizing front
moves outwards behind the dealkalizing
and desilicifying one.

A last fast pyrometamorphic heating may
often be seen superposed upon the contac
metamorphism of outer and medium contact

zones. This fast heating is due to a last immersion into the melt during and just before ejection. It causes highly metastable partial melting to widely differing degrees. It produces thin rims of melt between suitable mineral contacts (above all quartz/feldspar), then broader rims spreading to other mineral contacts, finally pockets and masses of melt. Staurolites and zoned andalusites, even muscovites may be in contact with this quickly produced and quenched melt. Using experimental results from Mehnert et al. (1973) it was derived that mica schist xenoliths from the Kyller Kopf have been heated within a tephrite melt from less than 1 h to 1 day before the melt was chilled. Many xenoliths of suitable composition reach the surface without any pyrometamorphic effects. This is true even for part of the granulite facies rocks from Kempenich. High grade gneisses always display at least small amounts of partial melting, but xenoliths from all higher levels may have missed it altogether. This is best displayed for part of the Hüttenberg xenoliths. Everywhere, however, biotite has lost pleochroic haloes, which to develop was ample time from the Variscan orogeny. Very often the melts release vapor during pressure release, producing foamy xenoliths. There is a wide range of devolatilization-phenomena.

Fenitization occurs at many localities (f.i. underneath the Laacher See, Hüttenberg-, Kappiger Ley- and Kyller Kopf-tuffs). It affects contact rocks from all zones, mainly from the inner ones, gradually replacing biotite by aegirine. The final end products are fenites with course aegirine ± fluotaramite + sanidine (often unmixed, cooled to low-albite-formation). Apparently the fenitiazation is zoned.

Information on Situation of Magma Chambers and Mode of Eruption

The xenoliths provide information about formation, halt, modification and rise of volcanic melts and gases. Where an eruption is gas-rich and yields many xenoliths from the total crust and many of them without much contact and pyrometamorphic heating, the explosion must have occurred at the level of the xenoliths of lowermost derivation. Therefore, I assume the trachyte from Hüttenberg, the phonolite of Kappiger Ley and the nephelinites from Kempenich to be exploded from the crust-mantle-boundary. The Laacher See basanites, tephrites and phonolites have moved up more slowly, necking and leaving drops of melt behind as chambers at various levels. A last eruption, however, yielding the youngest gray tuffs takes even hardly altered gneisses to the surface. Here I assume that the system exploded by seathing progressing rapidly downward and taking xenoliths from all levels to the surface. I cannot agree with Woerner et al. (1982) that this volcanoe caused contact metamorphism of Devonian slates. In my opinion the highest, fenitizing, chamber occurs ca. 7 km below the surface, causing fenites and contact rocks within higher phyllites. At the Kyller Kopf a fenitizing chamber must occur within 15 km below the surface, most likely connected to a carbonatite. According to the crystal sizes of contact minerals this chamber must be the smallest of the ones mentioned. The basanites of the Karmelenberg are drawn from a chamber within the very low migmatites all of which are heated to the same degree for 1 to a few days.

These deep seated explosions have blasted solidified plutonic rocks from chambers above the explosion locality and contact hornfelses around them to the surface. I take it that the whole crust underneath the Eifel district is riddled by such chambers and aureoles. Many of the melts have not reached the surface but contributed their heat to the crust. The fluid phase causing such explosions near the Moho must be CO_2-rich. Occurrence of carbonatites also testifies a high amount of CO_2 within the gaseous phase.

References

Kramm, U., 1982. Die Metamorphose des Venn-Stavelot-Massivs nordwestliches

Rheinisches Schiefergebirge: Grad, Alter und Ursache. Decheniana, 135: 121-178.

Mehnert, K.R., Büsch, W., and Schneider, G., 1973. Initial melting at grain boundaries of quartz and feldspar in gneisses and granulites. Neues Jahrb. Miner. Monath., 165-183.

Meyer, W. and Stets, J., 1975. Das Rheinprofil zwischen Bonn und Bingen. Z. Dtsch. Geol. Ges., 126:15-29.

Okrusch, M., Schröder, B., and Schnüttgen, A., 1979. Granulite facies metabasic ejecta in the Laacher See area, Eifel, West Germany. Lithos, 12:251-270.

Voll, G., 1978. Die Erdkruste unter der Eifel, zusammengesetzt aus Auswürflingen von Vulkanen. Vortrags-Kurzfassung 130. Hauptversammlung Dtsch. Geol. Ges. Aachen, Okt. 1978.

Voll, G., 1979. Untersuchungen von vulkanischen Auswürflingen aus Kruste und Mantel im Raum der Eifel. Protokoll 4. Kolloquium Schwerpunkt Vertikalbewegungen und ihre Ursachen am Beispiel des Rheinischen Schildes, Neustadt a. d. Weinstr., Nov. 1979.

Voll, G., 1981. Die Erdkruste unter der Eifel, ermittelt aus Vulkan-Auswürflingen. Protokoll 5. Kolloquium Schwerpunkt Vertikalbewegungen und ihre Ursachen am Beispiel des Rheinischen Schildes, Neustadt a.d. Weinstr., Nov. 1981.

Wörner, G., Schmincke, H.-U., and Schreyer, W., 1982. Crustal xenoliths from the Quarternary Wehr volcano (East Eifel). Neues Jahrb. Miner. Abh., 144:29-55.

7.8 Mantle Xenoliths in the Rhenish Massif and the Northern Hessian Depression

H. A. Seck[1] and K. H. Wedepohl[2]

Abstract

Mantle xenoliths from the Tertiary volcanics of the Rhenish Massif are strongly deformed rocks with porphyroclastic textures. On the basis of their temperature histories and their deformation, they are assumed to be related to updoming of hot mantle material. According to preliminary data, all xenoliths were affected by LIL element enrichment. In addition to porphyroclastic xenoliths similar to those from Tertiary volcanics, amphibole free xenoliths with coarse-grained textures and amphibole-bearing xenoliths with mosaic textures occur in the Quaternary West Eifel volcanic field. Both groups are equilibrated in a narrow temperature range each, at about 1150°C and 950°C, respectively. Amphibole free xenoliths are LIL element depleted to very slightly enriched, with $^{87}Sr/^{86}Sr$- and $^{143}Nd/^{144}Nd$-isotope ratios characteristic for MORB. Amphibole-bearing xenoliths were affected by strong LIL element enrichment and a concomitant change in Sr- and Nd-isotope ratios towards those of CHUR. It is inferred that the xenoliths are related to a region of P-wave attenuation which has its maximum elevation at 50 km depth beneath the West Eifel.

---

[1] Mineralogisch-Petrographisches Institut, Zülpicher Str. 47, D-5000 Köln, Fed. Rep. of Germany

[2] Geochemisches Institut, Goldschmidtstr. 1, D-3400 Göttingen, Fed. Rep. of Germany.

The average modal composition of lherzolite and harzburgite xenoliths from the Hessian Depression is comparable to that of the common depleted mantle inclusions from the Eifel volcanics (74 vol % olivine, 18 vol % o-pyroxene, 6.7 vol % cl.-pyroxene). They are predominantly equilibrated in the temperature range from 1000° to 1100°C and have a coarse-grained structure. The absence of strongly deformed and reequilibrated xenoliths from the majority of the Tertiary volcanics indicates static conditions in the upper mantle underneath the Hessian volcanic region. An exception is the Habichtswald area, where pyroclastics contain recrystallized and metasomatically altered mantle peridotites.

Introduction

The young alkali basaltic volcanism in western Germany has raised mantle rocks to the Earth's surface. Their investigation contributes to our knowledge about the composition and properties of an upper layer of the continental lithosphere. A deeper layer, which acted as a source of the alkali basaltic magmas, is indirectly known from the composition of the basaltic rocks. The most abundant mantle xenoliths of our area are spinel lherzolites and spinel harzburgites. Dunites, websterites, wehrlites, pyroxenites and eclogites are rare. Amphibole and phlogopite bearing peridotites from volcanics in the Eifel and the Hessian Depression have been identified as products of reaction of metasomatic fluids with pre-existing ultramafic rocks. It is still unknown which pro-

---

Plateau Uplift, ed. by K. Fuchs et al.
© Springer-Verlag Berlin Heidelberg 1983

cesses have caused the migration of such fluids. Balances between metasomatically altered and unaltered peridotites on the base of data reported by Sachtleben (1980) demonstrate an introduction of K, Na, Ca, Ti, Al etc. by mantle fluids. The rate of the potassium accumulation is 500% of the original concentration whereas metasomatic peridotite has on average gained not more than 6% Al. The addition of some incompatible and volatile trace elements to the preexisting peridotite exceeds a factor of 5.

Our investigation has revealed resemblances and characteristic differences between xenoliths from the two major regions of sampling within and outside the Rhenish Mass. Detailed discussions on the regions will be presented by the two authors separately (Rhenish Massif by H.A.S.; Hessian Depression by K.H.W.). There is no systematic difference in the average mineral composition between peridotite xenoliths from the two areas (see Table 3 in Fuchs and Wedepohl, 1983). The sampled lithospheric layer mainly consists of depleted spinel peridotite. It contains only slightly more than half the clinopyroxene proportion of pyrolite as reported by Ringwood (1975). Due to quite different intensities of tectonism in the upper mantle under the Rhenish Massif and under the Hessian Depression, abundant xenoliths from volcanics of these two regions contain either textures from porphyroclastic recrystallization or the original coarse grained structures, respectively.

Peridotite xenoliths from the Eifel and the Hessian Depression have been used in early investigations about the origin of these inclusions in basaltic rocks (Ernst, 1935; Frechen, 1948). Ernst had already discovered rock fabrics in these materials which cause anisotopics properties (Fuchs, 1983).

Mantle Xenoliths from the Tertiary Volcanics of the Rhenish Massif

Perioditic mantle xenoliths occur as inclusions in Tertiary basaltic lavas of Eifel, Siebengebirge, and Westerwald as is shown in Fig. 1. Strongly deformed rocks with porphyroclastic textures are predominant. They are characterized by large elongated crystals of orthopyroxene and olivine up to 8×2.5 mm in size in a matrix of recrystallized olivine, orthopyroxene, clinopyroxene, and spinel as neoblasts. The temperature histories of xenoliths from all locations as deduced from orthopyroxene compositions (Seck and Witt, 1982) is very similar. The cores of the orthopyroxene porphyroclasts have exsolved clinopyroxene and spinel whereas the rims of the grains are free of exsolution lamellae. Core compositions as obtained by reintegrating the unmixed phases by scanning microprobe analysis yield minimum temperatures of $1050°$ to $1100°C$ before unmixing. The rims of the porphyroclasts are depleted in CaO, Al_2O_3, and Cr_2O_3 relative to core bulk compositions and are similar in composition to neoblasts. Neoblasts reveal grain-to-grain chemical variation and obviously recrystallized in a temperature range from about $750°$ to $900°C$. In xenoliths of both Eifel and Siebengebirge, there is evidence of incompatible element metasomatism from the presence of phlogopite and/or potassic melt inclusions believed to have formed from the breakdwon of amphibole and/or phlogopite (Seck and Reese, 1979). In addition, xenoliths from the Eifel reveal relative enrichment of LREE with La/Yb-ratios between 25 and 40.

Mantle Xenoliths from the Quaternary West Eifel

Peridotitic mantle xenoliths from the Quaternary West Eifel volcanic field were mostly brought to the surface by Maar-type explosions. Apart from xenoliths with porphyroclastic textures and temperature histories similar to those from the Tertiary lavas, the West Eifel field is characterized by high-temperature coarse-grained and amphibole-bearing tabular mosaic type nodules. These later two groups are identical with the Ib- and Ia-suites, respectively, distinguished by Frechen (1963, pers. commun.) on the basis of their different mineralogies. Ia- and Ib-xenoliths were ejected from two maar type volcanoes (Dreiser Weiher, Meerfelder Maar) in the center of the West Eifel field whereas xenoliths with porphyroclastic textures are confined to the NW end.

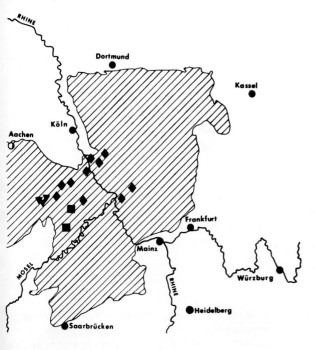

Fig. 1. Spatial distribution of mantle xenoliths (studied in this paper) within the Rhenish Massif. Tertiary: *diamonds* porphyroclastic. Westerwald (v. Gehlen, unpubl.), Hocheifel (Huckenholz, unpubl.). Quaternary: *squares* coarse-grained, mosaic; *triangles* porphyroclastic. For a more detailed record of xenolith locations of the West Eifel see Mertes (1982) and Schmincke et al. (1983)

All three types of nodules comprise a complete suite of spinel peridotites ranging from lherzolites to dunites. However, the Ia-suite alone contains amphibole and/or its break down products that is melt (glass), clinopyroxene, olivine, and spinel (Sachtleben, 1980).

Minerals of the coarse-grained Ib-suite are perfectly homogeneous and are found to have been equilibrated in a narrow temperature range from $1150°$ to $1170°C$. Temperatures reported in this paper were calculated from the CaO-solubilities of orthopyroxene (Sachtleben and Seck, 1981).

The amphibole-bearing Ia-xenoliths also equilibrated in a narrow temperature interval, however, at distinctly lower temperatures around $950°C$. Orthopyroxenes in these rocks are usually free of unmixing lamellae. Rare orthopyroxenes with exsolved clinopyroxene and spinel have bulk compositions much higher in CaO, Al_2O_3, and Cr_2O_3 and are believed to be relics of a former high-temperature state.

Pressures reported in this paper are maximum pressures for Cr-poor spinel peridotites as obtained experimentally from natural and synthetic systems.

Apart from the equilibration temperatures, the amphibole bearing Ia-suite and the amphibole free Ib-suite differ markedly by their incompatible element geochemistries. This is evident from a comparison of selected REE-patterns of clinopyroxenes (Fig. 2) which are the main carriers of the REE. Ia-clinopyroxenes (e.g., Ia/171) are characterized by a strong relative enrichment of LREE which is at variance with the geochemical experience that the upper mantle is depleted in LREE due to basalt magma extraction. REE patterns of Ib-clinopyroxenes are more variable. They range from chondritic for primitive lherzolite Ib/8, to relatively LREE-depleted for lherzolite Ib/58. However, clinopyroxenes from Ib-harzburgites (like Ib/3) may show slight relative enrichment of LREE. Stosch and Seck (1980) concluded that Ia-xenoliths come from a part of the mantle which - originally dry - has been affected by metasomatism. Besides a partial hydration (amphibole formation), this led to the addition of incompatible elements and a concomitant change in the $^{87}Sr/^{86}Sr-$ and $^{143}Nd/^{144}Nd$-isotope ratios (Stosch et al., 1980). $^{87}Sr/^{86}Sr-$ and $^{143}Nd/^{144}Nd$-ratios of dry Ib-xenoliths not affected by the metasomatism are equal to those of present day oceanic tholeiites whereas $^{87}Sr/^{86}Sr-$ and $^{143}Nd/^{144}Nd$-ratios of peridotites which have undergone metasomatism approach values found for the Quaternary volcanics of the West Eifel (e.g., Kramers et al., 1981). The fact that the enrichment of the LREE in Ib-olivines is related to the abundance of fluid inclusions in them (Stosch, 1982) lends support to the idea of Frey and Green (1974) that mantle metasomatism is caused by fluid phases migrating upwards from deeper levels of the mantle.

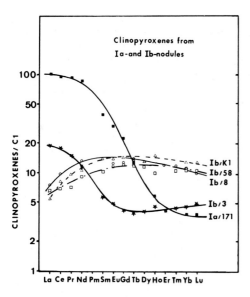

Fig. 2. Representative C1-normalized REE distribution patterns of clinopyroxenes from amphibole bearing Ia- and amphibole-free Ib-xenoliths. Ia-cpxs, generally, are strongly LREE enriched, Ib-cpxs are LREE-depleted to slightly enriched (Stosch, 1980)

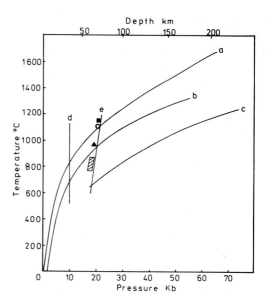

Fig. 3. PT-diagram showing equilibrium temperatures of Ia- (*triangles*) and Ib-xenoliths (*squares*). For porphyroclastic xenoliths, temperatures of porphyroclasts before unmixing (*circles*) and those of neoblasts (*shaded*) are given. Pressures assumed are maximum pressures for spinel lherzolites. *Curves* shown in the diagram are: a) High-temperature oceanic pyroxene geotherm (Mercier and Carter, 1975); b) Low-temperature oceanic pyroxene geotherm (Mercier and Carter, 1975); c) Continental geotherm (Clark and Ringwood, 1964); d) Moho; e) Spinel peridotite/garnet peridotite transition (Obata, 1976)

Discussion

According to a reconnaissance study of Panza et al. (1980) on the lithosphere structure of Europe, a zone of thin and anomalous lithosphere extends from the Upper Rhine Graben beneath the Rhenish Massif to the Lower Rhine Basin lending support to the assumption of Illies (e.g., Illies et al., 1981) that the Rhenish Massif forms an integral part of the Rhine Rift System. This zone finds its surface expression in a NW-SE trending fault system which has served as pathways for magma ascent since the Miocene. Tensional forces perpendicular to this direction are still active today (Ahorner, 1983; Baumann and Illies, 1983). Properties of mantle xenoliths occurring in the Tertiary volcanics of the Rhenish Massif fit into this picture in that they provide evidence for upwelling of hot mantle material. Temperatures derived for pyroxene pairs are plotted in Fig. 3. Temperatures of the reconstructed cores of porphyroclasts plot considerably above the continental shield geotherm of Clark and Ringwood (1964) and even above the high-temperature geotherms of Mercier and Carter (1975) characteristic for active oceanic rift systems.

The spatial distribution of xenoliths shows that anomalously high temperatures were reached in wide parts of the upper mantle underneath the Rhenish Massif during the Tertiary. Further, the temperatur history and the deformation of the xenoliths preclude heating by conductive heat transfer, but require convective heat flow by diapiric upwelling of hot mantle material. From the fact that the neoblasts recrystallized at temperatures much closer to a reasonable steady-study conductive shield geotherm, it is concluded that hot mantle material intruded in a "colder" upper mantle. It is important to note that core temperatures and recrystallization temperatures do not plot on the same geotherm, even if due allowance is made for the pressure uncertainty. The basic problem whether the uplift of the Rhenish Massif was triggered by active asthenospheric upwelling or by crustal compression related to the Alpine orogeny cannot be solved unless the age relation between mantle updoming and crustal uplift is known. The xenoliths themselves do not provide direct information related to this problem, because it is impossible to date the age of their deformation.

Apart from porphyroclastic xenoliths occurring in the NW, mantle peridotites from the center of the Quaternary West Eifel volcanic field are markedly different from those observed elsewhere in the Rhenish Massif. Both their geochemical and textural properties, as well as the equilibrium temperatures calculated, are indicative of unusual mantle conditions beneath this part of the shield. According to experimental evidence, Ia- and Ib-xenoliths come from rather shallow depth, presumably no deeper than 50 km. Temperatures of 1150°C (Ib-suite) in such a shallow depth are at variance with the low surface heat flow. Values of 70 to 80 mW m^{-2} measured (Haenel, 1983) are much lower than those of at least 100 mW m^{-2} to be expected under the condition of steady-state conductive heat transfer (Pollack and Chapman, 1977). This points to a young thermal perturbation in the upper mantle beneath the West Eifel, probably due to convective upwelling of material from greater depth. Supporting evidence for this comes from the temperature histories of deformed xenoliths and the existence of a region of P-wave attenuation in a depth range of 50 to 150 km centered underneath the West Eifel volcanic field (Raikes, 1980). The assumption of Raikes (1980) that the delay of P-waves is caused by the presence of small amounts of melt, requires temperatures in the order of 1200°-1300°C at a depth of 50 to 60 km. According to the maximum depth range and equilibrium temperatures of 1150° derived for Ib-peridotites, these xenoliths could be constituent parts of the margin of the low-velocity body. However, if this body is conceived as the source region of the Quaternary West Eifel magmas which are characterized by high enrichment of LIL elements, then it is impossible for geochemical reasons to look at the Ib-xenoliths as an integral part of it. It is widely believed today (Frey et al., 1978) that formation of such magmas requires enrichment of incompatible elements in the source region prior to partial melting. Further, the time integrated $^{87}Sr/^{86}Sr$- and $^{143}Nd/^{144}Nd$-isotope ratios of the West Eifel magmas are close to those of the "Chondritic Uniform Reservoir" (CHUR) (DePaolo and Wasserburg, 1977). Thus, there are constraints to the geochemical properties of the "low-velocity" body from both incompatible element abundances and radiogenic isotopes. As was shown above, Ib-xenoliths do not confine to these constraints and, consequently, are geochemically not related to the "low-velocity" body. From both textures and phase equilibria observed, the porphyroclastic xenoliths of the West Eifel could be strongly deformed and cooled down Ib-peridotites, providing additional evidence that the Ib-rocks come from deeper parts of the mantle. It cannot be precluded, however, that the West Eifel porphyroclasts are of the same age as the Tertiary ones.

Amphibole-bearing Ia-xenoliths are believed to occupy a shallower level in the upper mantle. From relics of orthopyroxene with unmixing lamellae of clinopyroxene and spinel, Sachtleben (1980) inferred that Ia-type peridotites, before their recrystallization, had a deformation and temperature history similar to the xenoliths with porphyroclastic textures. A concomitant metasomatism led to the formation of amphibole, a strong enrichment of incompatible elements, and a change of $^{87}Sr/^{86}Sr$- and $^{143}Nd/^{144}Nd$-ratios towards those of the Quaternary magmas. This indicates a relation between magma genesis and mantle metasomatism which is still poorly understood. Stosch et al. (1980) have placed an upper age limit of 200 m.y. for the metasomatic event, which shows that it may be linked to young mantle diapirism. It is hoped that a more precise age of the metasomatism may be derived from isotope studies which are under way now.

To account for the fact that Ia-xenoliths are enriched in LIL elements, whereas Ib-xenoliths are hardly affected by metasomatism, it is implied that the amphibole-bearing peridotites may have undergone metasomatism and recrystallization in an uplift event before the emplacement of the young diapir. The Ib-xenoliths are thought to be constituents of the deepest lithosphere depleted in LIL elements, which were uplifted on top of the rising diapir, the porphyroclastic xenolith being representatives of the strongly deformed margins of the Ib-rocks. This model offers a possible explanation for both the strongly contrasting incompatible element and isotope geochemistries of the Ia- and Ib-suites, and the fact that both suites equilibrated in a narrow temperature interval each at 950° and 1150°C, respectively, which is difficult to understand in a static stratified mantle.

Xenoliths from Tertiary Volcanics of the Northern Hessian Depression

A large proportion of the Tertiary olivine nephelinites and nepheline basanites but not more than a quarter of the alkali olivine basalt necks and flows contain peridotitic xenoliths. Pyroclastic deposits (which are mainly of alkali olivine basaltic composition) also include xenoliths from the upper mantle.

The depth of origin of the peridotite inclusions can be estimated from the temperature of equilibration of pyroxenes. Petrologic experiments on solidus temperatures of their host basalts inform about maximum temperatures attained in the upper mantle of our region. Plagioclase and garnet containing peridotites have not been observed. The Moho depth in this area is about 30 km. The average composition of spinel lherzolites and spinel harzburgites from the Hessian Depression is listed in column one of Table 3 reported by Fuchs and Wedepohl (1983). The stability field of Fe-Cr-containing spinel mainly depends on the Cr composition. Some xenoliths of this area contain 40% of the $MgCr_2O_4$ molecule in spinel expanding its stability field to pressures of about 30 kb (~90 km depth) (O'Neill, 1981). The REE patterns of all the basalt species exposed in the northern Hessian Depression are almost identical in heavy REE composition. According to partition data published by Harrison (1981), the concentration of heavy REE in these basalts has more than twice the abundance to be expected in melts which have been equilibrated with garnet. Therefore even the olivine nephelinites and melilite-containing olivine nephelinites of our area have probably been formed within the mantle volume containing spinel peridotite.

Oehm (1980) analyzed paragenetic pyroxenes in 21 xenoliths from different types of basalts by microprobe. Her results have been evaluated on the base of different "geothermometers" depending on the pyroxene solvus or partition of elements between phases etc. If we select her data according to the evaluation suggested by Wells (1977) there exists a range from 874° to 1112°C for the final equilibration. Only two samples are equilibrated below 1000°C. The majority of temperatures ranges from 1000 to 1075°C. Xenoliths from alkali olivine basalts tend to the lower and those from nephelinites and limburgites to the higher temperatures within this range. The residual or depleted character of peridotites caused by former events of partial melting is indicated by relatively small proportions of clinopyroxene and by relatively high concentrations of Cr in the pyroxenes and in chromite. The degree of depletion of our xenoliths is not correlated with temperature of final equilibration. This can only be explained by random distribution of layers with more residual peridotite and not by a regular vertical sequence of advanced depletion. For correlation of temperatures with pressure (or depth) we have used the geotherms reported by Pollack and Chapman (1977). An interpolation between two continental geotherms characteristic for a mantle which was affected by a recent orogeny and another one affected by an early Paleozoic orogeny is needed because the last thermal event for the crust and mantle of the Hessian Depression was the Devonian Carboniferous orogeny. A tentative profile through crust and mantle of this area has been reported by Fuchs and Wedepohl (1983) as Fig. 3 of their contribution. The application of the mentioned geotherm for a p-T-correlation gives a consistant picture with the source layers of the xenoliths and of the basaltic melts at increasing depth. Two garnet and pyroxene bearing xenoliths (eclogite and garnet websterite), for which pressures and temperatures could be estimated by Mengel (1981) on the base of microprobe investigations, also plot at a reasonable depth in this diagram.

Some spinel lherzolites predominantly from pyroclastics of the Hessian Depression contain phlogopite or amphibole (Mengel, 1981, Oehm, 1980; Vinx and Jung, 1977). Mengel (1981) presented structural evidence for a secondary formation of phlogopite during reequilibration of the host peridotite. Wedepohl (1983) on the base of high concentrations of incompatible and volatile elements in nepheline-bearing basalts demonstrated by modelling partition data that the formation of alkali olivine basalts, nepheline basanites and olivine nephelinites cannot be explained by partial melting of the depleted peridotites (represented by common xenoliths). They require mantle material containing a few percent phlogopite and a small fraction of carbonate and phosphate. A correlation of F and K in basalts can be explained as caused by melting of phlogopite. The average $^{87}Sr/^{86}Sr$ ratio of peridotite xenoliths and all basaltic species from the Hessian Depression is close to 0.7036 (Mengel, unpubl.). This ratio is appreciably higher than that of ocean ridge tholeiites which are typical partial melting products of depleted upper mantle rocks.

References

Ahorner, L., 1983. Historical seismicity and present-day microearthquake activity of the Rhenish Massif, Central Europe. This Vol.

Baumann, H. and Illies, H., 1983. Stress field and strain release in the Rhenish Massif. This Vol.

Clark, S.P. and Ringwood, A.E., 1964. Density distribution and constitution of the mantle. Rev. Geophys., 2:35-88.

DePaolo, D.J. and Wasserburg, G.J., 1977. The sources of island arcs as indicated by Nd and Sr isotopic studies. Geophys. Res. Lett., 4: 465-468.

Ernst, T., 1935. Olivinknollen der Basalte als Bruchstücke alter Olivinfelse. Nachr. Ges. Wiss., Göttingen. Math. Phys. Kl. Fachgr. IV N.F. 1, 13:147-154.

Frechen, J., 1963. Kristallisation, Mineralbestand, Mineralchemismus und Förderfolge der Mafitite vom Dreiser Weiher in der Eifel. Neues Jahrb. Miner. Monath.:205-224.

Frechen, J., 1948. Die Genese der Olivinausscheidungen vom Dreiser Weiher (Eifel) und Finkenberg (Siebengebirge). Neues Jahrb. Miner. Abh., 79A:317-406.

Frey, F.A. and Green, D.H., 1974. The mineralogy, geochemistry and origin of lherzolite inclusions in Victorian basanites. Geochim. Cosmochim. Acta, 38:1023-1059.

Frey, F.A. and Prinz, M., 1978. Ultramafic inclusions from San Carlos, Arizona: petrologic and geochemical data bearing on their petrogenesis. Earth Planet. Sci. Lett., 38:129-176.

Frey, F.A., Green, H.D., and Roy, S.D., 1978. Integrated models of basalt petrogenesis: A study of quartz tholeiites to olivine melilitites from South Eastern Australia utilizing experimental and geochemical data. J. Petrol., 19:463-513.

Fuchs, K., 1983. Recently formed elastic anisotropy and petrological models for the continental subcrustal lithosphere in southern Germany. Phys. Earth Planet. Int., 31:93-118.

Fuchs, K., and Wedepohl, K.H., 1983. Relation of geophysical and petrological models of upper mantle structure of the Rhenish Massif. This Vol.

Haenel, R., 1983. Geothermal investigations in the Rhenish Massif. This Vol.

Harrison, W.J., 1981. Partitioning of REE between minerals and coexisting melts during partitial melting of a garnet lherzolite. Am. Mineral., 66: 242-259.

Hutchison, R., Chambers, A.L., Paul, D.K., and Harris, P.G., 1975. Chemical variation among French ultramafic xenoliths - evidence for a heterogeneous upper mantle. Mineral. Mag., 40: 153-170.

Illies, H., Baumann, H., and Hoffers, B., 1981. Stress pattern and strain release in the Alpine foreland. Tectonophysics, 71:157-172.

Jaques, A.L. and Green, D.H., 1980. Anhydrous melting of peridotite at 0-15 kb pressure and the genesis of tholeiitic basalts. Contrib. Mineral. Petrol., 73:287-310.

Kramers, J.D., Betton, P.J., Cliff, R.A., Seck, H.A., and Sachtleben, T., 1981. Sr and Nd isotopic variations in volcanic rocks from the West Eifel and their significance. Fortschr. Mineral., 59:246-247.

Mengel, K., 1981. Petrographische und geochemische Untersuchungen an Tuffen des Habichtwaldes und seiner Umgebung und an deren Einschlüssen aus der tieferen Kruste und dem oberen Mantel. Dissertation, Univ. Göttingen.

Mercier, J.C.C., Carter, N.L., 1975. Pyroxene geotherms. J. Geophys. Res., 80:3349-3362.

Mertes, H., 1982. Aufbau und Genese des Westeifeler Vulkanfeldes. Dissertation, Ruhr-Univ. Bochum, pp 1-415.

Obata, M., 1976. The solubility of Al_2O_3 in orthopyroxenes in spinel and plagioclase peridotites and spinel pyroxenite. Am. Mineral., 61:804-816.

Oehm, J., 1980. Untersuchungen zu Equilibrierungsbedingungen von Spinell-Peridotit-Einschlüssen aus Basalten der Hessischen Senke. Dissertation, Univ. Göttingen.

Panza, G.F., Müller, St., and Calcagnile, G., 1980. The gross features of the lithosphere-astenosphere system in Europe from seismic surface waves. Pageoph., 118:1209-1213.

O'Neill, H.S.C., 1981. The transition between spinel lherzolite and garnet lherzolite, and its use as a geobarometer. Contrib. Mineral. Petrol., 77: 185-194.

Okrusch, M., Schröder, B., and Schnüttgen, A., 1979. Granulitefacies metabasite ejecta in the Laacher See area, Eifel, West Germany. Lithos, 12: 251-270.

Pollack, H.N. and Chapman, D.S., 1977. On the regional variation of heat flow, geotherms and lithospheric thickness. Tectonophysics, 38:279-296.

Raikes, S., 1980. Teleseismic evidence for velocity heterogeneity beneath the Rhenish Massif. J. Geophys., 48: 80-83.

Ringwood, A.E., 1975. Composition and Petrology of the Earth's Mantle. McGraw-Hill Inc., New York.

Sachtleben, T., 1980. Petrologie ultrabasischer Auswürflinge aus der Westeifel. Dissertation, Univ. Köln.

Sachtleben, Th. and Seck, H.A., 1980. Diapiric uprise in the upper mantle below the West-Eifel volcanic field? EOS, 61:413.

Sachtleben, T. and Seck, H.A., 1981. Chemical control of Al-solubility in orthopyroxene and its implications on pyroxene geothermometry. Contrib. Mineral. Petrol., 78:157-165.

Seck, H.A. and Reese, D., 1979. Entstehung und Zusammensetzung von Gläsern in Peridotiten der Westeifel. Fortschr. Mineral., 57, Beih. 1: 224-225.

Seck, H.A. and Witt, G., 1982. Thermal history of porphyroclasts as evidence for mantle diapirism underneath the West Eifel (West Germany). Terra Cognita, 2:240.

Stosch, H.G., 1980. Zur Geochemie der ultrabasischen Auswürflinge des Dreiser Weihers in der Westeifel: Hinweise auf die Evolution des kontinentalen oberen Erdmantels. Dissertation, Univ. Köln.

Stosch, H.G., 1982. Rare earth element partitioning between minerals from anhydrous spinel peridotite xenoliths. Geochim. Cosmochim. Acta., 46:793-811.

Stosch, H.G. and Seck, H.A., 1980. Geochemistry and mineralogy of two spinel peridotite suites from Dreiser Weiher, West Germany. Geochim. Cosmochim. Acta, 44:457-470.

Stosch, H.G., Carlson, R.W., and Lugmair, G.W., 1980. Episodic mantle differentiation: Nd and Sr isotopic evidence. Earth Planet. Sci. Lett., 47:263-271.

Vinx, R. and Jung, D., 1977. Pargasitic-kaersutitic amphibole from a basanitic diatreme at the Rosenberg, North of Kassel (North Germany). Contrib. Mineral. Petrol., 65:135-142.

Wedepohl, K.H., 1983. The Late Tertiary basaltic volcanism of the northern Hessian Depression (NW Germany) and its genesis. (in prep.)

Wells, P.R.A., 1977. Pyroxene thermometry in simple and complex systems. Contrib. Mineral. Petrol., 62:129-139.

7.9 Relation of Geophysical and Petrological Models of Upper Mantle Structure of the Rhenish Massif*

K. Fuchs[1] and K. H. Wedepohl[2]

Abstract

The multi-disciplinary investigations of the plateau uplift of the Rhenish Massif provided an excellent opportunity to compare geophysical and petrological models of structure, composition and dynamics of the upper mantle. A low-velocity volume immediately below the crust-mantle boundary is observed east of the river Rhine whereas an absence of a sharp Moho is characteristic for the area west of this river. A low-velocity volume at 50 to 150 km depth in the Westeifel region is explained by the presence of 1% partial melt. This abnormal mantle could have caused 200-300 m uplift. The occurrence of anisotropy of P-velocities in the upper mantle can be correlated with depleted mantle compositions observed in peridotite xenoliths. The abundant depleted spinel-peridotite xenoliths sampled from Tertiary and Quaternary alkalic basalts in the Eifel, Westerwald, Hessian Depression areas contain on average about 70% olivine. These lherzolites and harzburgites cannot explain the chemical composition of nepheline bearing basalts except by assuming very low degrees of partial melting of the former rocks ($\leq 1\%$). A few percent (or less) partial melt in peridotites has a very low flow velocity in the porous space of peridotite and probably cannot form magma reservoirs of reasonable size in the upper mantle to feed volcanism. But a small proportion of immobile partial melt can explain low velocity volumes in the upper mantle without effecting the density in the area of the Rhenish Massif significantly. The cooling and solidification of this anomalous mantle needs time in excess of 10 m.y. Peridotite xenoliths are mainly equilibrated in the range from 900° to 1150°C related to a depth range from 60 to 80 km. Coarse grained xenoliths without recrystallization after heavy shearing are abundant in the Hessian Depression whereas porphyroclastic peridotites predominate in the volcanic areas of the Rhenish Massif. The latter probably indicate shearing in the border zones of uprising diapirs in the upper mantle. Metasomatically altered peridotites (with about 4% amphibole or phlogopite) occur as xenoliths in basalts from the sampled areas. They are probably the source rocks for the magmas of the nepheline containing basalts. Geothermal conditions as derived from petrological models for the upper mantle outside the Rhenish Massif have been presented in a tentative profile.

1 Geophysikalisches Institut, Universität Karlsruhe, Hertzstr. 16, D-7500 Karlsruhe 21, Fed. Rep. of Germany.

2 Geochemisches Institut, Universität Göttingen, Goldschmidt.Str. 1, D-3400 Göttingen, Fed. Rep. of Germany

* Contribution No. 256 Geophysical Institute Karlsruhe.

Geophysical Models (K. Fuchs)

Mantle Composition at the Moho

The crust-mantle boundary is quite differently developed west and east of the river Rhine (Mechie et al., 1983). On the eastern part of the seismic profile the Moho forms a complicated zone with a thin high velocity layer at a depth of about 29 km on top of a low velocity layer which merges gradually into upper mantle velocities at a depth of about 37 km. The low velocity zone disappears further to the east whereby a shallow "normal" crust mantle boundary is formed at a depth of 30 km. The low velocity zone is also not found at the western most part of the seismic profile (Luxembourg, France). In the region west of the Rhine the low-velocity zone is observed, the thin high-velocity layer, however, has not been detected here. Therefore, the Moho forms a wide gradual transition from crustal to mantle velocities. This transition feature is in the neighborhood of the mantle heterogeneity of low velocity material (Raikes, 1980; Raikes and Bonjer, 1983). Quaternary volcanism of the West-Eifel also appears in this area of the transitional Moho.

The shallow low velocity zones in the upper mantle on both sides of the river Rhine can be explained tentatively as due to intrusions of basaltic magmas connected with the volcanic activities in the Eifel and Westerwald areas. The youngest volcanic events in both regions are of Pleistocene or post-Pleistocene age (Lippolt, 1983). In quantity only the Eifel volcanism of this young age is important. The top layer of the upper mantle underneath the Moho in the area east of the Rhine consists of a thin veneer of material with high seismic velocities. If a basaltic intrusion starts to consolidate at pressures from about 11 to 15 kb and a temperature of 900°C the product is expected to be of garnet granulite composition (Ringwood, 1975, Figs. 1-1, 1-2, 1-6). At pressures higher than ~15 kb eclogite should form. If the temperature above a volume of melt is lower than 900°C the stability conditions for garnet granulite and eclogite will shift to lower pressures. The present steady-state temperature close to the Moho is expected to be as low as 550°C (Pollack and Chapman, 1977). At this temperature eclogite can occur at Moho depth (Ringwood, 1975, Fig. 1-6). The density of garnet granulite ranges from 3.2 to 3.4 g cm^{-3}. The higher value is identical to the density of abundant peridotites. Some eclogites have a slightly higher density than peridotite. The observed seismic velocities do not contradict the assumption of a basaltic melt capped by garnet granulite (or/and eclogite) occurring at the Moho in the Westerwald region. West of the river Rhine a smooth transition in seismic velocities from the upper mantle to the lower crust has been observed. This transition zone can also be explained as an intrusion of melt. Because of younger age or larger size and observable consolidation in the form of granulite or eclogite is lacking in this area.

Low Velocity Volume in the Mantle of the West Eifel

Another asymmetry of the upper mantle west and east of the Rhine is established by the body of low velocity material in the upper mantle of the West-Eifel deduced from an inversion of teleseismic traveltime anomalies (Raikes, 1980; Raikes and Bonjer, 1983) in a depth range between 50 and 150 km. The anomaly has a volume of about 5×10^5 km^3. Application of the velocity-density relation (Birch, 1958) would lead to a deficit of Bouguer gravity of about -150 mgal which is not seen in the observed gravity (Gerke, 1957; Drisler and Jacoby, 1983). In spite of the uncertainties of the gravity values in the Rhenish Massif such a deficit should be recognizable, but in fact it is not.

Therefore, Raikes (1980) proposed the presence of partial melt of about 1% in the average in the anomalous volume. This partial melt would reduce the P-wave velocities sufficiently to explain the delay of traveltimes while keeping the density almost unaltered. A possible alternative mechanism of compensation for the mass deficit in the upper mantle would be an updoming of the Moho above the anomalous body. This, however, has not been observed during the seismic refraction survey Mechie et al. (1983).

The origin of the anomalous body of low velocity and its composition is of significance for the history and mechanism of uplift and of volcanism of the Rhenish Massif. Lloyd and Bailey (1975) claimed that the metasomatism observed in certain mantle xenoliths in the West Eifel could explain the uplift of this part of the plateau. Metasomatized mantle rocks should be almost equal in density to "normal" peridotites (see Wedepohl, this contribution). Applying the Clausius-Clapeyron equation the transition from solid to liquid with 2-3% partial melt would result in a volume augmentation which could roughly explain the observed uplift of 200-300 m (e.g., see Part 4, this vol.).

A strong corroboration of the presence of partial melt could be its traces in xenoliths. The partial melt is very difficult to detect because it would occur only in very fine drops. The observed metasomatism is the prerequisite of this partial melt. The dating of the metasomatism would be important for an understanding of the history of the formation of the anomalous body and the uplift.

The thermal history of the Rhenish Massif is of great importance to explain the history of formation of the anomalous body in the upper mantle. If the volume as a whole has been rising like a diapir reaching the solidus and being affected by partial melt, then the temperature increase would have been a relatively slow process. If, however, fluid and volatile phases penetrated the subcrustal lithosphere from the asthenosphere rather quickly, then the rise in temperature would have occurred much more suddenly. Which tectonic process could have opened the paths for the fluid and volatile phases west of the Rhine? What prevented them from rising east of the Rhine?

The asymmetry of the mantle anomaly with regard to the Rhine is paralleled by the occurrence of Quaternary volcanism only west of the Rhine. Is there any direct or indirect connection? The young volcanism follows a lineament which is striking parallel to the present direction of maximum horizontal compression and therefore is opened by dilatational stress. Is the same stress field also responsible for the opening of paths and eruption of the melt in the upper mantle? How could such openings occur in a medium which, under the high temperatures near solidus, would only deform plasticly?

Is it possible that the stress field led to a thinning of the lithosphere which allowed a diapir to rise from the asthenosphere at this specific location?

Many additional unanswered questions remain at the end of this research program. If the anomalous volume of low velocity in the subcrustal lithosphere is responsible for the uplift of the West Eifel region, what has caused the uplift of the eastern part of the plateau? Are we looking only at the rest of a formerly much larger volume of partial melt which extended over much of the Rhenish Massif, and since has cooled well below solidus over its major part? If so, why is there no Quaternary volcanism connected with it in the eastern part? Has it not come to the surface? Why do we not observe subsidence in the eastern part during freezing of the partial melt? This could indicate that the partial melt is generated by supply of fluid and volatile phases from below rather than from within a rising diapir forced near the solidus temperature. These are, of course, rather speculative ideas.

The occurrence of partial melt in the zone of low velocities is challenged by magneto-telluric measurements in the Rhenish Massif. No zone of anomalously high electrical conductivity is observed in this depth range (Jödicke et al., 1983) which is not also present at other places of the Rhenish Massif and even outside. The depth range 50-150 km is very close to what could be resolved with the magneto-telluric survey, and the signal-noise ratio deterriorated very much due to industrial activities in the region of the West Eifel. Nevertheless, an attempt should be made with longer period instrumentation to verify whether this anomalous body can be seen also as an electrical conductivity anomaly or not. This would be a crucial test on its physical and petrological state.

Composition and Seismic Anisotropy of the Upper Mantle

The occurrence of mantle xenoliths in the West Eifel and in the Hessian Depression offered an unique possibility to compare

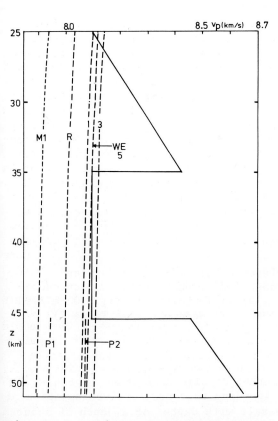

Fig. 1. Comparison of observed and predicted P-velocities; *solid line* observations on long-range profile in Southern Germany; *dashed lines* predictions for petrological models in Table 1 (Fuchs, 1983)

P-velocities observed in explosion seismic experiments and those predicted from the composition of the xenoliths (Fuchs, 1983). Although the energy of the seismic waves generated in the Rhenish Massif experiment (Mechie et al., 1983) was not sufficient for the observation of rays penetrating into the subcrustal lithosphere, a comparison with seismic observations in Southern Germany (Ansorge et al., 1979) was worthwhile since the composition of xenoliths in northwestern and northeastern Bavaria (Huckenholz and Noussinanos, 1977; Huckenholz and Schröder, 1981) is almost the same as of those from the Rhenish Massif.

The observed P-wave velocity distribution in Southern Germany is shown in Fig. 1 (thick line). It is characterized by an anisotropy at Moho level (Bamford, 1973) and a strong increase of velocity with depth of about 30×10^{-3} s^{-1}. It has been shown (Fuchs, 1983) that it is impossible to explain this observed depth distribution of seismic velocities with isotropic petrological models. The models are listed in Table 1, and the isotropic velocities predicted along a geotherm from their compositions are depicted in Fig. 1, as well. These predicted velocities all fall below the observed velocities. Only in the low velocity zone does the model 5 from the West Eifel come very close to the observed velocities. Typically all predicted velocity distributions possess negative gradients, they all decrease in the depth range from 25-50 km. Increase of the olivine content to 73% (model 5) raises the predicted isotropic velocity only to 8.1 km s^{-1}, i.e., the velocity in the low-velocity zone of the observed model.

It is therefore impossible to predict the observed strong increase of P-velocity with depth by a change in composition. Since phase changes in the depth range from Moho to 50 km are unlikely, the most obvious explanation of this unusual velocity gradient is an increase of anisotropy with depth. Although a true anisotropy experiment, i.e., a common depth point experiment on at least three crossing profiles, is still lacking for the subcrustal lithosphere in Southern Germany the hypothesis of an anisotropy increasing with depth is supported by two observations. (1) The topmost mantle is definitely anisotropic as deduced from a time-term analysis of Pn phases in Southern Germany (Bamford, 1973). (2) The amplitude distribution of P-waves in the subcrustal lithosphere is compatible with an increase of anisotropy with depth: large amplitudes in the direction of fast velocities, small amplitudes in that of slow velocities (Fuchs, 1983). Figure 2 gives those compositions which are compatible with the observed seismic Pn-anisotropy according to Bamford (1973). These compatible compositions signify a rather fertile topmost mantle. The compositions from the West Eifel, Hessian Depression and Bavaria fall well outside the range of compatibility. This means that these xenoliths do not originate from the topmost mantle immediately below the crust-mantle transition. It is an unsolved problem why this part of the upper mantle has not been sampled by xenoliths.

Table 1. Modal compositions (Vol%) of mantle xenoliths

Model/Location	Symbol	OL	OPx	CPx	SPi	GA	References
Spinel-Pyrolite	R	52	28	16	4	–	1
Garnet-Pyrolite	GA	57	17	12	–	14	2
PHN-1569	P1	55.2	39.7	2	–	3.1	3
PHN-1611	P2	63.4	9.5	16.7	–	10.4	3
NE-Bavaria	1	65	23	9	3	–	4
Heldburg	2	69	25	5	1	–	4
Dreiser Weiher	3	75	17.5	8.4	0.9	–	5-7
North Hessian Depression	4	73.4	18.6	6.7	1.3	–	8-11
Western Eifel	5 WE	73	19	7	1	–	5-7,11
	M1	50	45	4	1	–	this chapter

References: 1 Green and Liebermann (1976); 2 Leven et al. (1981); 3 Boyd and McCallister (1976); 4 Huckenholz and Noussinanos (1977), Huckenholz and Schröder (1981); 5 Sachtleben (1980); 6 Seck and Reese (1979); 7 Stosch (1980); 8 Oehm (1980) 9 Mengel (1981); 10 Wedepohl and Ritzkowski (1980); 11 Wedepohl (see Table 2).

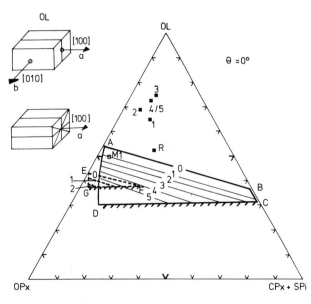

Fig. 2. OL-OPX-CPX triangle. The area ABCD is the range of compatibility for orthorhombic fabric with b-axis and a-axis horizontal for various percentages of SPi(0-5). All compositions in the area ABCD match observed $v_{max} = 8.31$ and $v_{min} = 7.73$ km s^{-1} at a depth of $z = 25$ km (Fuchs, 1983)

To fit the observed increase of velocity with depth it is not only required that the amount of preferred orientation of olivine increases to 100%, but in addition an augmentation of the relative content of olivine becomes necessary corresponding to a progressive depletion of basalt with depth. At a depth of about 35-45 km the correspondence of measured properties and observed depleted compositions of mantle xenoliths from Germany was reached.

For Southern Germany, Fuchs (1983) obtained a fit of the orientation of olivine crystals to the observed velocities which is apparently related to the orientation of the present crustal stress field derived from fault plane solutions of earthquakes (Ahorner, 1970) The a-axis of olivine and the strike of its vertical b-plane both coincide with the plane of maximum shear stress in the crust. It is likely that the crustal stress field leaking into the upper mantle is responsible for the preferred orientation of olivine in the topmost mantle.

Petrological Models

Relation Between Magma Species and Mantle Xenoliths

Petrological and geochemical investigations on composition and structure of the upper mantle are based on xenoliths in alkali basalts and kimberlites and on magmatic rocks produced in the upper mantle. A direct mantle origin for certain magmas can be inferred from a specific composition of their Sr, Nd and Pb isotopes or, more directly, the occurrence of peridotitic xenoliths. The chemical character of volcanic rocks allows certain conclusions on mantle conditions and composition based on melting experiments of peridotitic rocks (Jaques and Green, 1980) and on partition of trace elements between magma and source rock during partial melting (cf. Allègre and Minster, 1978). The direct sampling of peridotitic rocks by magmas is restricted to certain mantle layers above the sources of the melts. Some of these layers are expected to be altered by reactions with uprising solutions of unknown source into a composition with lower solidus temperature (mantle metasomatism). Only melts with high speed of uprise have overcompensated the settling speed of mantle xenoliths caused by their higher density and have transported these rocks to the Earth's surface. A residence of peridotitic inclusions in melts for more than a few tens of hours causes their disintegration (Scarfe et al., 1980). Fast uprising magmas are those containing relatively high concentrations of volatiles which might reach the state of oversaturation on the way to the Earth's surface. A slow uprise of their magmas or/and the residence in magma chambers are possible reasons for the worldwide absence and very rare occurrence of mantle xenoliths in tholeiitic and andesitic rocks, respectively. Because of their radiogenic isotope composition, it is known that tholeiitic and andesitic magma species are derived from mantle rocks. We do not know the reasons why magmas apparently do not sample mantle and crustal layers randomly. Because of non-random sampling it is dangerous to reconstruct stratigraphic columns of upper mantle and crustal sections according to abundances of xenolith species.

Thermal Conditions of the Upper Mantle

Petrologists have reliable thermometry for two-pyroxene rocks but no reliable barometry for spinel peridotites. Therefore, the depth of origin of spinel lherzolites and spinel harzburgites can only be investigated on the basis of computed geotherms. We have used geotherms, computed by Pollack and Chapman (1977), appropriate for a continental mantle which has experienced a Lower Paleozoic orogeny and possibly a more recent thermal event. For our conditions of a mantle we have interpolated between two reported geotherms and use temperature values consistent with a 65 mW m^{-2} surface heat flow density (Fig. 1). Such a heat flow density (without the effect of Pleistocene glaciation) is typical for our area (Haenel, 1983). Mantle xenoliths from the Eifel and the Northern Hessian Depression (Sachtleben, 1980; Oehm, 1980; Seck and Wedepohl, 1983) are equilibrated in the range from 800° to 1150°C (thermometry according to Wells, 1977). If we relate temperatures to pressures according to the above mentioned geotherm these peridotites originate from 50 to 80 km depth.

Depleted Upper Mantle Rocks

Compared to a pristine mantle composition, defined in this sense as pyrolite reported by Ringwood (1975), the abundant spinel lherzolites and spinel harzburgites from the Eifel and the Northern Hessian Depression are depleted in clinopyroxene (and Al, Ca, Fe, alkaline incompatible and volatile elements) (see Table 2). The average mineral compositions from both areas have a remarkable resemblance.

Even the composition of a comparable layer from the mantle underneath the French Massif Central does not deviate far from the depleted mantle typical for our area (Hutchison et al., 1975). The depletion of the mantle must result from former partial melting events. The degree of depletion (higher or lower Cr/Al ratios of pyroxenes and spinels in harzburgites or lherzolites) and temperatures of the equilibration are not correlated. This indicates that the sampled mantle volume is not layered with respect to a vertical gradient of depletion. Former melting events have left pockets of more depleted peridotite surrounded by less depleted peridotite in a random distribution.

Table 2. Mineral composition (weight percent) of abundant continental upper mantle rocks

Minerals	Northern Hessian Depression ($\leq$100 km depth) (N = 30)	Westeifel[a] (N = 28)	Westeifel[a] (metasom. altered) (N = 19)	Massif Central[b] (France) (N = 83)	Pyrolite (Ringwood)
Olivine	73.5	75.8±8.6	70.4±7.0	66.8	57
Orthopyroxene	18.0	16.5±5.9	19.8±7.6	23.8	17
Clinopyroxene	6.7	6.8±5.0	6.1±4.9	7.6	12
Spinel	$\leq$1.8	0.9±0.4	0.9±0.7	1.9	14 Garnet
Amphibole			3.9±2.9		

[a] Sachtleben (1980), Seck (unpubl.) 3 addit. samples contain phlogopite.
[b] Computed after Hutchison et al. (1975).

The maximum depth of origin of spinel lherzolites is controlled by the spinel garnet reaction $MgAl_2O_4 + 4MgSiO_3 \underset{T}{\overset{P}{\rightleftharpoons}} Mg_3Al_2(SiO_4)_3 + Mg_2SiO_4$ and the influence of Cr and Fe on this reaction. The importance of the latter elements has been evaluated by O'Neill (1981). Because of a wide range of Cr_2O_3 proportions in spinels from xenoliths of the area of investigation (10-45%) we have to expect the existence of a mantle layer at 70 to 90 km depth in which both spinel and garnet peridotite occur. Garnet peridotite has not been observed as inclusion in the Tertiary and Quaternary volcanics of West Germany. Areas with basalts containing garnet peridotite xenoliths are rare in a worldwide compilation. It is to be assumed that partial melting is mainly restricted to that part of the upper mantle which contains spinel peridotite and its marginal zone to the garnet peridotite layer.

Mantle Metasomatism

It is not possible to explain the formation of olivine nephelinites, nepheline basanites and alkali olivine basalts by partial melting of depleted lherzolites or harzburgites. The relatively high concentrations of incompatible and volatile elements in the mentioned rocks (see Wedepohl et al., 1983) require very low degrees of partial melting of less than 1% for depleted peridotites. Such small proportions of partial melt have very low flow velocities in porous rocks with low permeability at a grain size of 2 mm and at reasonable pressure gradients in the tectonic stress field (for instance 1 bar per 10 km). According to estimates of Spera (1980) flow rates under these conditions are typically of 2×10^{-11} cm s^{-1} or 7 km per 10^9 years. The available geologic time for partial melts to migrate through the interconnected porous space of peridotites into magma pockets is $\sim 10^7$ years. This time allows migration of 10^4 t melt from 10^6 t rocks over 70 m distance under the above mentioned conditions. Magma reservoirs for a single small volcano erupted in one event must have the size of more than 10^6 t, however. These data demonstrate that flow velocities of 1% melt in peridotitic rocks probably cannot feed volcanism. If the proportion of partial melt is two percent the flow rate increases by a factor of 20. Several percent of partial melting can explain the chemical composition and a necessary size of magma reservoirs of our alkali basalts. For the given degree of partial melting and an appropriate composition in volatile and incompatible elements depleted peridotites must have undergone a metasomatic alteration observable in the formation of phlogopite and/or amphibole. Extensive proof for the occurrence of metasomatic mantle fluids uprising from yet unknown sources has been reported in the literature (see for instance: Proceedings of the First and Second International Kimberlite Conference: Ahrens et al., 1975; Boyd and Meyer, 1979). These fluids, which probably consist of H_2O and CO_2 as major components (see fluid inclusions in certain peridotites; Roedder, 1965), apparently transport dissolved K, Na, Ca, P, F, LREE etc. and react in the upper layers of the mantle with the preexisting minerals to form phlogopite and/or amphibole. Metasomatically altered xenoliths with proportions of 4% amphibole have been described from the West Eifel by Sachtleben (1980) (see Table 2) and those with 2% to 5% phlogopite have been mentioned from the Habichtswald area in the Hessian Depression by Mengel (1981). Beside transporting dissolved components the uprising fluids also transport heat. Spera (1981) has estimated that they can cause an increase of temperature of 50° to 70°C in certain layers of the upper mantle. If these mantle layers at 70 to 90 km depth are close to solidus temperatures (above 1200°C), partial melting as described above can be initiated by the introduction of metasomatical fluids at about 1150°C (Modreski and Boettcher, 1973).

Further evidence for a chemical alteration of certain mantle layers is contributed by data on the Sr isotopic composition of xenoliths and basaltic rocks of the Eifel and the northern Hessian Depression (Stosch et al., 1980; Stosch and Seck, 1980; Mengel, pers. commun.; Seck and Wedepohl, 1983). In both areas metasomatically altered peridotites and basaltic rocks have an average $^{87}Sr/^{86}Sr$ ratio of 0.7036 in contrast to pyroxenes from unaltered xenoliths from the Eifel, which are in the range of ocean ridge basalts (MORB) below 0.7030. In the northern Hessian Depression all investigated peridotites, even those without visible phlogopite or amphibole, have a $(^{87}Sr/^{86}Sr)_o$ ratio close to 0.7036. This might be explained by a relatively small spatial distance of the source layer of xenoliths and the layer with extensive metasomatical alteration, from which the alkalibasaltic magma must originate.

Peridotite with four percent amphibole (density ~3.0 g cm^{-3}) or phlogopite (density ~2.89 g cm^{-3}) has a density which still plots in the range of common spinel lherzolites and harzburgites (density of olivine, orthopyroxene, clinopyroxene ~3.23 g cm^{-3}). Therefore, metasomatically altered and unaltered mantle layers can probably not be discriminated by seismic observation. But peridotite containing one percent melt already attenuates the velocity of seismic waves effectively. The volume of 5×10^5 km^{-3} anomalous mantle matter at 50 to 150 km depth under the West Eifel reported by Raikes (1980) might be explained by 1% melt containing peridotite (see Fuchs, this chap.). As described above, such small proportions of melt will be very immobile. The more than 5×10^{13} t melt could be the immobile fraction of partial melt formed during a Quaternary event of mantle metasomatism. A smaller proportion of melt from this event might have segregated into magma pockets to feed the Quaternary basaltic volcanism of the Eifel region. The decrease in density of the 100 km thick anomalous mantle body caused by the advection of heat through metasomatic fluids and an increase of temperature of about 5% could have produced an uplift of the surface of the crust of 200 to 300 m through expansion (see Fuchs, this contribution). Time required for solidification of the melt fraction and subsequent cooling of the anomalous mantle volume down to its normal geothermal state is well in excess of 10 million years (Lachenbruch et al., 1976).

Indications of Tectonism and Uplift in the Upper Mantle

The effect of solid state mobilization of mantle rocks can be recognized in the structure of the affected peridotite and in unmixing during reequilibration of its pyroxenes. Mercier and Nicolas (1975) have classified the typical structures of peridotites which have experienced tectonic shear stress and/or mobilization. Different classes of grain size due to partial recrystallization of minerals form a cataclastic or porphyroclastic structure. In extreme cases the minerals are completely recrystallized. If severe shear stress were absent, a coarse-grained, equigranular structure would be preserved. Peridotites with porphyroclastic structures are abundant in the Eifel, Siebengebirge and Westerwald volcanics (Seck and Wedepohl, 1983) but very rare in those of the Hessian Depression (exception Habichtswald Mts.). The observation of porphyroclastic xenoliths in Oligocene basalts constrains the tectonic disturbance of certain mantle layers to be at least of Mid Tertiary age in the East Eifel-Siebengebirge-Westerwald region. Seck (in: Seck and Wedepohl, 1983) in the Eifel and Mengel (1981) in the Habichtswald mountains have observed unmixing and zonation of pyroxene crystals which must be caused by a decrease of temperature in a gradient from $1050°/1100°C$ down to $700°/900°C$. This gradient can be explained by an almost adiabatic ascent of the host mantle volume. Only marginal zones of upwelling diapirs experience shear stress from differential movement of neighboring volumes. The abundance of porphyroclastically structured xenoliths in Eifel volcanics indicates a large proportion of marginal zones in uplifting volumes, and therefore relatively small cross sections of individual diapirs. If diapirs with large cross-sections (more than tens of kilometers) rise up close to the Moho, large volumes of upper mantle rocks have to be replaced. The horizontal mass movement of the preexisting peridotites in convecting systems must cause effective horizontal displacement of crustal segments. But this tectonic phenomenon (large scale rifting) has not been observed in the Eifel-Westerwald-Hessian Depression centres of volcanic activity. There is also no evidence for an updoming of the Moho caused by an uplift of a large mantle volume in the area of the Rhenish Massif. If diapirs rise up from the depth of garnet stability, garnet peridotite will be quenched in the marginal zones and garnet plus olivine will react to form spinel plus orthopyroxene in the slowly cooling core of the diapir. Such processes have been interpreted from studies of the Serrania de Ronda peridotite in Spain (Den Tex, 1982).

Local diapiric uplift is needed to explain partial melting of metasomatically unaltered or slightly altered depleted mantle material at shallow depth to form tholeiitic magmas under two relatively small areas of the Hessian Depression. According to experimental evidence pressures smaller than 5 kb or 15 kb are needed to form quartz tholeiite or olivine tholeiite magma respectively from dry depleted lherzolite by **partial** melting (Jaques and Green, 1980). On this base quartz tholeiite can only be explained as a differentiation product of olivine tholeiite. But olivine tholeiite occurs only very rarely in the northern Hessian Depression. The melting condition to form tholeiite at shallow depth by partial melting of depleted mantle rocks with a low proportion of water can only be attained by adiabatic uprise of peridotite from a deeper level (80-90 km) up to about 50 km depth and melting due to decompression. Convective mobilization of mantle material under the area of tholeiite volcanism is indicated by almost contemporaneous local rifting in the Tertiary sediments of the Reinhardswald region (Backhaus et al., 1980). Because of the absence of xenoliths in this magma species there is no possibility to confirm mantle ascent by observation of cataclastic rocks.

Profile Through Crust and Upper Mantle

Figure 3 presents a tentative cross-section through the crust and upper mantle under the northern Hessian Depression. Correlation of temperature and depth is according to a geotherm previously mentioned (Pollack and Chapman, 1977) and should be taken with about 10% uncertainty. We have confirmation for this geotherm at about 80 km depth by geothermometry and geobarometry of an eclogite and a garnet websterite inclusion from basaltic tuff in the northwestern Hessian Depression (Mengel, 1981). The probable origin of both rocks plots correctly above the layer of magma production and within the source region of depleted peridotite xenoliths. Depleted spinel peridotite is the common xenolith in the basaltic volcanics from the Eifel and Hessian Depression (Table 2). At Moho depth this geotherm has a temperature as low as $530°C$. A low temperature is in line with the relations between seismic properties and upper mantle composition

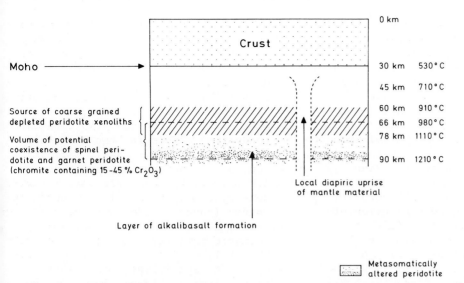

Fig. 3. Tentative profile through the crust and upper mantle in the area of the northern Hessian Depression

as suggested by Fuchs (1983). The lower crust under the Hessian Depression and the Eifel contains granulites which are equilibrated in the range from 700° to 900°C (Mengel, 1981, 1983; Mengel and Wedepohl, 1983). On the base of structural investigations of medium and high grade metamorphic xenoliths from the crust under the Eifel Voll (1983) assumes that the high temperature of granulites is a relict of an older orogenic event. In other words: these rocks are not reequilibrated at the present temperature conditions. Figure 3 contains a layer from 530° to 900°C from which we apparently have no xenoliths of peridotites (with the exception of reequilibrated metasomatically altered ultramafic inclusions in tuffs from the Habichtswald area). A possible explanation for the absence of coarse grained peridotites with temperatures lower than 900°C is the lack of equilibration due to the very low rate of diffusion in solid minerals at this range of temperatures (Hofmann and Hart, 1978).

References

Ahorner, L., 1970. Seismo-tectonic relations between the Graben zones of the upper and lower Rhine valley. In: Illies, J.H., Müller, St. (eds.) Graben Problems. Schweizerbart, Stuttgart, pp 155-166.

Ahrens, L.H., Dawson, J.B., Duncan, A.R., and Erlank, A.J. (eds.), 1975. Physics and chemistry of the earth, Vol. 9. Pergamon Press, Oxford, New York, 940 pp.

Allègre, C.J. and Minster, J.F., 1978. Quantitative models of trace element behavior in magmatic processes. Earth Planet. Sci. Lett., 38:1-25.

Ansorge, J., Bonjer, K.-P., and Emter, D., 1979. Structure of the uppermost mantle from long-range seismic observations in Southern Germany and the Rhinegraben area. Tectonophysics, 56: 31-48.

Backhaus, E., Gramann, F., Kaever, M., Lepper, J., Lohmann, H.H., Meiburg, P., Preuss, H., Rambow, D., and Ritzkowski, S., 1980. Erläuterungen zur Geologischen Karte des Reinhardswaldes 1:50 000. Hess. Landesamt Bodenforsch., Wiesbaden, 32 pp.

Bamford, D., 1973. Refraction data in Western Germany - a time-term interpretation. Z. Geophys., 39:907-927.

Birch, F., 1958. Interpretation of seismic structure of the crust in the light of experimental studies of wave velocities in rocks. Contrib. Geophys., 6:158-170.

Boyd, F.R. and McCallister, R.H., 1976. Densities of fertile and sterile garnet peridotites. Geophys. Res. Lett., 3:509-512.

Boyd, F.R. and Meyer, H.O.A. (eds.), 1979. The mantle sample: inclusions in kimberlites and other volcanics. Proc. 2nd Int. Kimberl. Conf. Vol. 2, Am. Geophys. Union, Washington D.C., 424 pp.

Den Tex, E., 1982. Dynamothermal metamorphism across the continental crust/mantle interface. Fortschr. Miner., 60:57-80.

Drisler, J. and Jacoby, 1983. Gravity anomaly and density distribution of the Rhenish Massif. This Vol.

Fuchs, K., 1983. Recently formed elastic anisotropy and petrological models for the continental subcrustal lithosphere in Southern Germany. Phys. Earth Planet. Inter., 31:93-118.

Gerke, K., 1957. Die Karte der Bouguer-Isanomalen 1:1000000 von Westdeutschland. Dtsch. Geodät. Kommiss., Reihe B, Heft 46: Teil I, Frankfurt a.M., 13 pp.

Green, D.H. and Liebermann, R.C., 1976. Phase equilibria and elastic properties of a pyrolite model for the oceanic upper mantle. Tectonophysics, 32:61-82.

Haenel, R., 1983. Geothermal investigations in the Rhenish Massif. This Vol.

Hofmann, A.W. and Hart, S.R., 1978. An assessment of local and regional isotopic equilibrium in the mantle. Earth Planet. Sci. Lett., 38:44-62.

Huckenholz, H.G. and Noussinanos, Th., 1977. Evaluation of temperature and pressure conditions in alkali-basalts and their peridotite xenoliths in NE Bavaria, Western Germany. Neues Jahrb. Mineral. Abh., 129, 2:139-159.

Huckenholz, H.-G. and Schröder, B., 1981. Die Alkalibasaltassoziation der Heldburger Gangschar (Exkursion I am 25. April 1981. Jahresber. Mitt. Oberrhein. Geol. Ver., N.F., 63:97-110.

Hutchison, R., Chambers, A.L., Paul, D.K., and Harris, P.G., 1975. Chemical variation among French ultramafic xenoliths - evidence for a heterogeneous upper mantle. Miner. Mag., 40:153-170.

Jaques, A.L. and Green, D.H., 1980. Anhydrous melting of peridotite at 0-15 kb pressure and the genesis of tholeiitic basalts. Contrib. Mineral. Petrol., 73:287-310.

Jödicke, H., Untiedt, J., Olgemann, W., Schulte, L., and Wagenitz, V., 1983. Electrical conductivity structure of the crust and upper mantle beneath the Rhenish Massif. This Vol.

Lachenbruch, A.H., Sass, J.H., Munroe, R.J., and Moses, Jr., T.H., 1976. Geothermal setting and simple heat conduction models for the Long Valley Caldera. J. Geophys. Res., 81:769-784.

Leven, J.H., Jackson, I., and Ringwood, A.E., 1981. Upper mantle seismic anisotropy and lithospheric decoupling. Nature (London), 289:234-239.

Lippolt, H.J., 1983. Distribution of volcanic activity in space and time. This Vol.

Lloyd, F.E. and Bailey, D.K., 1975. Light element metasomatism of the continental mantle: the evidence and the consequences. Phys. Chem. Earth, 9:389-416.

Mechie, J., Prodehl, C., and Fuchs, K., 1983. The long-range seismic experiment in the Rhenish Massif. This Vol.

Mengel, K., 1981. Petrographische und geochemische Untersuchungen an Tuffen des Habichtswaldes und seiner Umgebung und an deren Einschlüssen aus der tieferen Kruste und dem oberen Mantel. Dissertation, Univ. Göttingen, 104 pp.

Mengel, K. and Wedepohl, K.H., 1983. Crustal xenoliths in Tertiary volcanics from the northern Hessian Depression. This Vol.

Mercier, J.C.C. and Nicolas, A., 1975. Textures and fabrics of upper mantle peridotites as illustrated by xenoliths from basalts. J. Petrol., 16:454-487.

Modreski, P.J. and Boettcher, A.L., 1973. Phase relations of phlogopite in the system $K_2O-MgO-CaO-Al_2O_3-SiO_2-H_2O$ to 35 kbars: a better model for micas in the interior of the earth. Am. J. Sci., 273:385-415.

Oehm, J., 1980. Untersuchungen zu Equilibrierungsbedingungen von Spinell-Peridotit-Einschlüssen aus Basalten der Hessischen Senke. Dissertation, Univ. Göttingen, 78 pp.

O'Neill, H.St.C., 1981. The transition between spinel lherzolite and garnet lherzolite and its use as a geobarometer. Contrib. Mineral. Petrol., 77:185-194.

Pollack, H.N. and Chapman, D.S., 1977. On the regional variation of heat flow, geotherms and lithospheric

thickness. Tectonophysics, 38:279-296.
Raikes, S., 1980. Teleseismic evidence for velocity heterogeneity beneath the Rhenish Massif. J. Geophys., 48:80-83.
Raikes, S. and Bonjer, K.-P., 1983. Large-scale mantle heterogeneities beneath the Rhenish Massif and its vicinity from teleseismic P-residual measurements. This Vol.
Ringwood, A.E., 1975. Composition and petrology of the Earth's mantle. McGraw-Hill, New York, 618 pp.
Roedder, E., 1965. Liquid CO_2 inclusions in olivine bearing nodules and phenocrysts from basalts. Am. Mineral., 50:1746-1782.
Sachtleben, T., 1980. Petrologie ultrabasischer Auswürflinge aus der Westeifel. Dissertation, Univ. Köln, 160 pp.
Scarfe, C.M., Takahashi, E., and Yoder, H.S., 1980. Rates of dissolution of upper mantle minerals in alkali olivine basalt melt at high pressures. Carnegie Inst. Washington, Yearb., 79:290-296.
Seck, H.A. and Reese, D., 1979. Entstehung und Zusammensetzung von Gläsern in Peridotiten der Westeifel. Fortschr. Mineral., 57, 1:224-225.
Seck, H.A. and Wedepohl, K.H., 1983. Mantle xenoliths from the Tertiary and Quaternary volcanics of the Rhenish Massif and the Tertiary basalts of the northern Hessian Depression. This Vol.
Spera, F.J., 1980. Aspects of magma transport. In: Hargraves, R.B. (ed.) Physics of magmatic processes. Princeton Univ. Press, Princeton N.J., 585 pp.
Spera, F.J., 1981. Carbon dioxide in igneous petrogenesis: II Fluid dynamics of mantle metasomatism. Contrib. Mineral. Petrol., 77:56-65.
Stosch, H.-G., 1980. Zur Geochemie der ultrabasaltischen Auswürflinge des Dreiser Weihers in der Westeifel: Hinweise auf die Evolution des kontinentalen oberen Mantels. Ph.D. thesis, Univ. Köln, 233 pp.
Stosch, H.G. and Seck, H.A., 1980. Geochemistry and mineralogy of two-spinel peridotite suites from Dreiser Weiher, West Germany. Geochim. Cosmochim. Acta, 44:457-470.
Stosch, H.G., Carlson, R.W., and Lugmair, G.W., 1980. Episodic mantle differentiation: Nd and Sr isotopic evidence. Earth Planet. Sci. Lett., 47:263-271.
Voll, G., 1983. Crustal xenoliths and their evidence for crustal structure underneath the Eifel volcanic district. This Vol.
Wedepohl, K.H. and Ritzkowski, S., 1980. Die nördliche Hessische Senke. Fortschr. Mineral., 58, 2:4-31.
Wedepohl, K.H., Mengel, K., and Oehm, J., 1983. Depleted mantle rocks and metasomatically altered peridotite inclusions in Tertiary basalts from the Hessian Depression (NW Germany). Proc. 3rd Int. Kimberl. Conf. Terra Cognita (in press).
Wells, R.A., 1977. Pyroxene thermometry in simple and complex systems. Contrib. Mineral. Petrol., 62:129-139.

8 Attempts to Model Plateau Uplift

8.1 Gravity Anomaly and Density Distribution of the Rhenish Massif

J. Drisler and W. R. Jacoby[1]

Abstract

The principal aim of this study is to explain the Bouguer anomaly by a geometrical model of deep Rhenish Massif structure constructed on the basis of available seismological, geological and other data. The gravity effect of the three-dimensional model is computed and least-squares fitted to the observed anomaly to render density contrasts for the model layers. This procedure successfully reproduces the main features of the Bouguer anomaly with the prominent density contrast found for the seismic 7 km s^{-1} interface, more than for the M-discontinuity, and for the lithosphere-astenosphere transition. Well reproduced are the gravity decrease toward the Alps, the regional Rhine Graben anomaly, the positive values of the eastern Massif, with highs in the Odenwald/Spessart, Bergisches Land/Sauerland, and the Harz regions, and the negative values of the western Massif; not reproduced is the gravity high of the Palatine, calling for high density mass in the crustal layers. The Eifel low-velocity body in the subcrustal lithosphere has no conspicious gravity effect, at most -20 mgal; hence, on the basis of our model, the density contrast is of the order of only -0.01 g cm^{-3} rather than -0.1 g cm^{-3} which would be predicted by the Birch relationship. Alternatively, the anomalous body might have a stronger density contrast if positive density anomalies are hidden nearby, but this is considered less likely. In contrast, the mantle anomaly below the southern Rhine Graben shows little velocity decrease but a clear density contrast of the order of -0.05 g cm^{-3}. In both regions the computed densities are in excellent agreement with isostatic estimates. The discrepancy between the velocity-density characteristics of both anomalies suggests that beside heating also fluid phases affect uplift and volcanism. Fluid phases appear to be particularly important in the initial stages as in the Eifel region; the southern Rhine Graben anomaly is at a more mature stage, degassed, but at its top still several hundred degrees hotter than the surrounding mantle.

Introduction

In this context of summing up a large-scale interdisciplinary study on the nature and the causes of crustal uplift, we present a discussion of the deep density structure of the Rhenish Massif region. It is based on model computations to fit the observed gravity anomaly with constraints from additional, particularly seismological, data. Gravity has been presented already in Chapter 6.7 (Jacoby et al., 1983) together with a discussion of qualitative aspects of interpretation.

The ultimate aim of our study has been to provide insight into the dynamics of uplift under the force of gravity, in line with the objective "Vertical motions and their causes as exemplified by the

---

1 Institut für Meteorologie und Geophysik, Universität Frankfurt, Feldbergstraße 47, D-6000 Frankfurt am Main 1, Fed. Rep. of Germany.

Rhenish Massif"; of course, knowledge of the density distribution is a necessary precondition for quantitative calculations of the body forces and their effects (see Neugebauer et al., 1983). Here we wish to emphasise our results in terms of density structure and present their geodynamic implications in a qualitative fashion. Some fundamental insights have been gained during this project on the nature and evolution of mantle anomalies which may plausibly be considered to cause surface deformations.

Since the given data define meaning and limitation of our models, we must discuss the results in the light of the data, but a detailed discussion is not necessary since many new observations are presented elsewhere in this volume. The computational methods applied in gravity modelling are here only briefly described, enough to enable the reader to appreciate our handling of the non-uniqueness problem of gravity interpretation; a more detailed treatment of the methodical aspects is given by Drisler and Jacoby (1983).

Gravity Field and Geology of the Region

The Bouguer anomaly map of the Rhenish Massif is presented in Fig. 1. It is computer plotted from digitized values read from the Bouguer anomaly map on a 21 km square grid (see Fig. 1 in Chap. 6.7; Jacoby et al., 1983); these values are used in the quantitative modelling (see below). The main features related to the Massif are (1) the generally positive values over the eastern part extending north (Bramsche Massif), northeast (Harz - Fläming), and south (Palatine - Spessart - Odenwald); (2) the generally negative values over the western Massif (Eifel, Ardennes) radiating northwestward (lower Rhine Graben) and southwestward (Lorraine); (3) the narrow gravity lows related to the graben sediments; (4) the decrease to the south (Alps) and north (North German Basin). Apart from the conspicious radiating pattern mentioned (lower Rhine Graben, Lorraine, Vlanderen west of Brussels, Ardennes, Rhine Graben - all radiating to the western quadrants), we also recognize a ring pattern of uncertain significance: a ring of lows around the Massif (Münsterland - Kassel area - Rhön - northern Rhine Graben - Ardennes - lower Rhine Graben) surrounded by a ring of highs (Bramsche Massif - Harz - Würzburg/Kraichgau - Alsace - north of Reims - Brabant Massif).

Most of the above features correspond to identifiable tectonic units; not, however, the division into eastern and western Massif. Its history from the Tertiary to the present is one of uplift and denudation, volcanism and rifting. The recent impulse largely responsible for today's morphology is only the latest stage of a process lasting for 40 Ma (Chaps. 3 and 4, this Vol.) and forming the peculiar patterns described; the rift pattern cuts the Massif somewhat east of its present center in the East Eifel, Neuwied basin region of recent volcanism and subsidence (center of the ring pattern); the volcanic regions form an approximately equidistant sequence (Rhön - Vogelsberg - Westerwald - Siebengebirge, East Eifel - West Eifel) with no clear age succession (Chap. 5, this Vol.).

Outline of the Data Base

Here we briefly summarise the data used for constructing the models.

Geologically, the Massif consists of folded Devonian shales, sandstones and limestones, perhaps forming one or more Hercynian nappes thrust from the southern crystalline terrain (Meissner et al., 1983; and Chap. 2, this Vol.). Depth to crystalline basement has been estimated from several sources such as, e.g., geological (Meyer and Stets, 1975; W. Meyer, pers. commun., 1980; R. Teichmüller, pers. commun., 1980; H. Grabert, pers. commun., 1980), magnetic anomalies (W. Bosum, pers. commun., 1980), and seismic reflection (Meissner et al., 1983) and refraction (E. Plein, pers. commun., 1980; references quoted below) with the basement corresponding to P velocity $v_p \approx 6$ km s^{-1}. Post-Hercynian sediments are generally thin

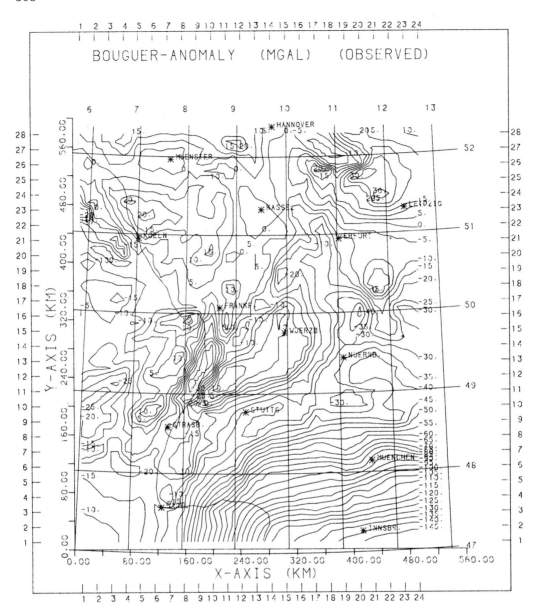

Fig. 1. Simplified Bouguer anomaly contour map (Jacoby et al., 1983) of the Rhenish Massif and its vicinity (reference: Potsdam gravity system; International Gravity Formula of 1930; Bouguer density 2.67 g cm^{-3}), replotted from digitized values

(<1 km) around the Massif except in the Rhine Graben, the Saar-Nahe trough and in the basins of North Germany, Paris and the Molasse (Chap. 2, this Vol.)

Deeper crustal structure has been extensively studied with seismic refraction and reflection surveys, summarised by Giese et al. (1976), Mooney and Prodehl (1978), and Mechie et al. (1983). Our initial models were constructed from Giese's (1976) depth contour maps of the 6 and 7 km s^{-1} interfaces and the M-discontinuity and were updated during the project on the basis of new results from seismic refraction and, to a minor degree, from magneto-tellurics (H. Jödicke, pers. commun., 1981; Jödicke et al., 1983).

The structure of the lithosphere-asthenosphere transition has been studied by an isostatic model (Grohmann, 1981), by surface wave dispersion (Panza et al.,

1980; review in: Müller and Lowrie, 1980), and by teleseismic travel time residuals (Raikes, 1980; Raikes and Bonjer, 1983). The latter work has demonstrated a low-velocity body in the depth range of the subcrustal lithosphere below the western Massif, the Eifel anomaly, which will lead us to a discussion in connection with the gravity anomaly. The Eifel anomaly is of crucial importance for an understanding of the uplift dynamics.

Modelling Method

Ultimately we wish to find the density distribution below the Rhenish Massif. Our principal question in the modelling procedure is whether or not we can explain the Bouguer anomaly by a model of layers or bodies of uniform, but differing densities, constructed from the above data, largely seismological, on deep structure. The velocity information is not immediately translated into densities; our aim has rather been to compute the layer density constrasts by a least-squares fit of the model gravity effect to the observed Bouguer anomaly on a grid of stations (see Jacoby, 1973; Drisler and Jacoby, 1983). The gravity calculation is based on a division of the model space into cubes or parallelepipeds which at great distance from an "observation" point can be combined to larger blocks and/or be approximated by point masses at the element centers of gravity; in this study the basic elements had the size $4.2 \times 4.2 \times 4.2$ km^3; they were combined to blocks of $5 \times 5 \times 5$ elements (in some models the vertical element size was smaller than 4.2 km).

The procedure we followed consisted of the following steps:

1. construction of the density contrast surfaces on the basis of seismic profiles and depth contours (see above) by digital sampling at a sufficiently dense but non-uniform grid of points and subsequent least-squares surface fitting (two-dimensional splines of various degrees or similar polynomials);

2. calculation of the weighted mean densities of the elements if intersected by a model surface ($\bar{\rho} = \sum_j \rho_j \Delta V_j/V$ where $\sum_j \Delta V_j/V = 1$, V, ΔV_j = total, partial element volumes, respectively), otherwise allocation of the layer densities to the elements;

3) computation and storage of the gravity effect at a grid of stations (here usually 21 km by 21 km) by exploiting the symmetries of the square division with stations only at block centers (for a more detailed description, see Drisler and Jacoby, 1983);

4) storage of the gravity effects e_{ik} of each surface k at stations i for $\rho = 1$, followed by least-squares calculation of the density contrasts $\Delta\rho_k$ by fitting the model effect to observed gravity in the sense of a linear regression analysis; thus the linear inversion problem is solved;

5) if desired, "manual" modification of the geometrical surfaces on the basis of new information or simply to obtain a closer fit of gravity and repetition of steps (1) to (4).

The method, as outlined, was developed during the project and applied to a number of models of the Rhenish Massif. It has, however, not been our aim to achieve an absolute fit of the observed gravity field for three reasons: (1) the fact that a particular model fits the observations perfectly does not prove its correctness or uniqueness, for one can construct many such models, (2) the gravity data in some areas are of limited accuracy (see Jacoby et al., 1983), and (3) the model of only a few density contrast surfaces is anyway far too simple.

Results

In this section we describe the model surfaces (based on the references quoted above and subject to future improvements) and present the computed gravity effects, first for the M-discontinuity alone, and then for the whole model after the least-squares adjustment. In view of the general model uncertainties, the contour maps of the surfaces have no

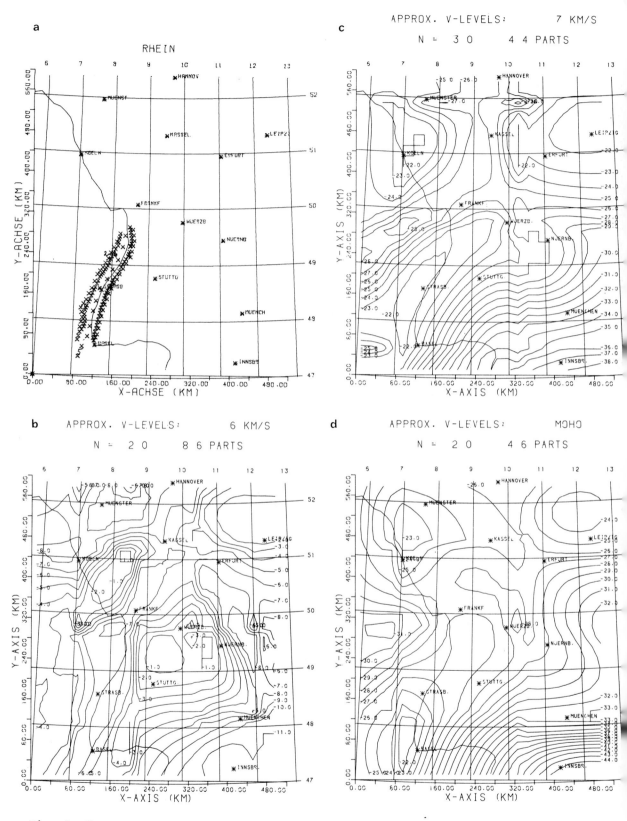

Fig. 2a-f

371

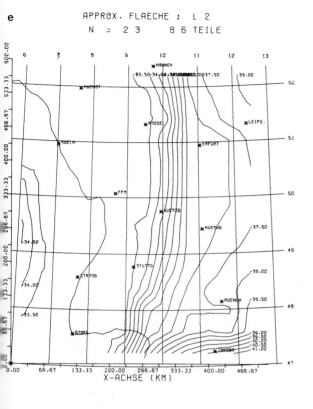

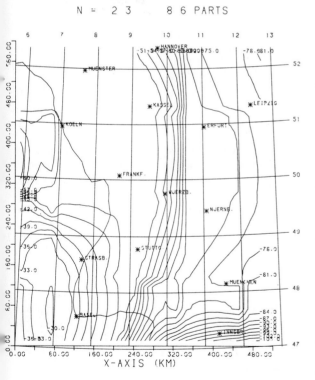

Fig. 2a-f. The density contrast interfaces (in km below NN or mean sea level) of the final three-dimensional model for the Rhenish Massif. a) Bottom of the Tertiary Rhine Graben sediments as sampled from Doebl and Olbrecht (1974); simplified, b) Crystalline basement or 6 km s^{-1} surface after Giese (1976) and updated with new data (see text); contours plotted after sampling and least-squares surface fitting. c) 7 km s^{-1} surface after Giese (1976), updated (see text); contoured after surface fitting. d) M-discontinuity or approx. 8 km s^{-1} interface after Giese (1976) and updated with data from C. Prodehl (pers. commun., 1980) and others (see text); contouring after surface fitting. e) "L2" surface below M-discontinuity, assumed to be similar in shape to lithosphere-asthenosphere transition (less relief, shallower) except below southern Rhine Graben where "L2" does not rise proportionally to top of asthenosphere (see text). f) Lithosphere-asthenosphere transition mostly after Müller and Lowrie (1980); below southern Rhine Graben the interface rises to nearly the M-discontinuity (see text)

other purpose than to display the gross features of the model. To avoid edge effects in the gravity computation the model surfaces are extended beyond the actual study area (shown in Fig. 2) by 200 km and more from where on they assume constant depths equal to their average depths within the model space.

The model surfaces as sampled or as plotted after sampling and surface fitting are shown in Fig. 2a-f.

a) The bottom of the Tertiary Rhine Graben sediments (shown in Fig. 2a) as sampled) forms a narrow trough, up to 3 km deep (Doebl and Olbrecht, 1974); outside, the surface is set to zero and the lower Rhine Graben has been neglected.

b) The 6 km s^{-1} surface (shown as plotted after surface fitting) correlated with the top of the crystalline basement, closely follows the exposed geology with narrow troughs (Rhine Graben - Hessian Depression - Saar-Nahe Trough, lower Rhine Graben, NE margin of the South German Triangle). The Rhenish Massif and the northern part of the Triangle are relative basement highs. Away from the Massif the basement dips towards the North German Basin and towards the Alps.

c) The 7 km s^{-1} surface (again plotted) generally describes the lower crustal region of strong velocity increase with depth. The relief partly follows that of the 6 km s^{-1} surface (a drop towards the Alps and a trough along the NE margin of the South German Triangle beyond Nuremberg and Würzburg towards Kassel), and partly has the opposite behavior ("ridges" beneath the Upper and Lower Rhine Graben and a depression below the Rhenish Massif),

d) The M-discontinuity with V_p approximately 8 km s^{-1} (see Giese, 1976) is rather similar to the 7 km s^{-1} surface, though more complicated, probably because of more detailed mapping: swells below the Upper and the Lower Rhine Graben, about 30 km deep below the Rhenish Massif, depressed beneath the Saar-Nahe Trough. Generally the crust thickens towards the Alps and thins towards North Germany.

e) The "L2" surface within the subcrustal lithosphere has been assumed to allow for subcrustal density variations and, with low-amplitude relief, is similar to the lithosphere-astenosphere transition (see below) in shape except beneath the southern Rhine Graben where it has no special uplift. L2 is positioned shortly below the M-discontinuity.

f) The lithosphere-asthenosphere transition shown as sampled in Fig. 2f (from which L2 has been constructed) resembles the M-discontinuity in a much smoothed fashion: uplift approximately along the River Rhine, depression towards the Alps. Our model deviates from that of Müller and Lowrie (1980) by a more pronounced rise of the surface beneath the southern Rhine Graben very close to the M-discontinuity. The teleseismically discovered mantle low-velocity anomaly ("Eifel anomaly") (Raikes, 1980; Raikes and Bonjer, 1983) below the western Massif near the map periphery has not been correlated with the asthenosphere-lithosphere transition, partly because of its different depth and partly because of its incompatible density contrast as discussed below. The body is, indeed, not required to explain gravity.

Each model surface was first treated separately; surface fitting required careful testing. The gravity effects were computed with both assumed density contrasts and by least-squares fitting. The individual surfaces are of differing importance. Clearly, the graben sediments and the basement relief can only explain smaller features of the field and the lithosphere bottom has only a very smooth effect. Since crustal thickness is widely believed to dominantly control the Bouguer anomaly we present in Fig. 3 the gravity effect of the model M-discontinuity with an assumed $\Delta\rho = 0.25$ g cm^{-3}. Some gross features of the Bouguer anomaly (Fig. 1) are well reproduced, others not at all. The amplitude of the gravity variation is of the right order; the drop towards the Alps, the relative high over the eastern Rhenish Massif (particularly in the mountains of Bergisches Land/Sauerland, Harz, Odenwald, and the low over the western Massif are reasonably explained by crustal thickness variations alone. But there are also striking discrepancies, particularly the model gravity high following the Rhine Graben without the narrow gravity troughs. It may be added here that the 7 km s^{-1} surface has a very similar gravity effect as the M-discontinuity.

To evaluate the whole model we can either combine all surfaces with presupposed density contrasts and then compute the gravity effect or we can first store the effects of the individual surfaces and then perform the least-squares density adjustment. If the densities are strongly constrained, the adjustments of, in our case six, density contrasts by trial and error is unnecessary; but in fact, the seismic velocities (as may be taken from the refraction data on which our models are based) do not tightly constrain the densities (Jacoby, 1975). We have therefore chosen the least-squares density calculation as our principal method as already stated.

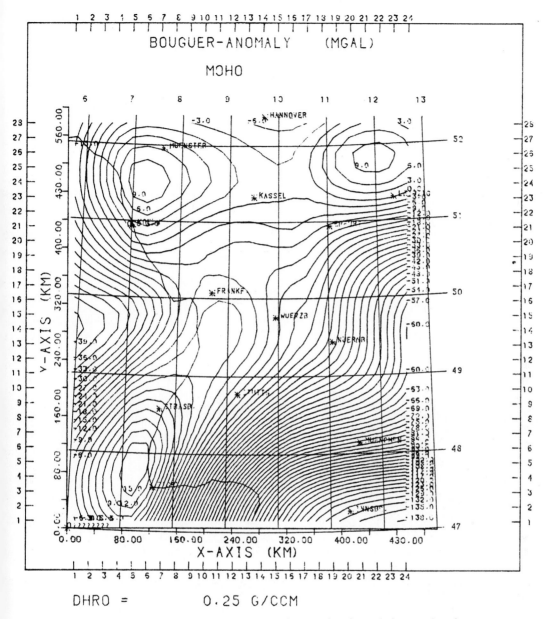

Fig. 3. Gravity effect computed for M-discontinuity (Fig. 2d) with an assumed density contrast of +0.25 g cm^{-3}

Our final result as based on the surfaces of Fig. 2 is presented in the form of the computed model gravity effect in Fig. 4. When new information becomes available, the model will need corrections. We shall first discuss the gravity map and then the computed densities.

A comparison of Figs. 3 and 4 shows that the complete (least-squares) model fits the Bouguer anomaly in more detail than the M-discontinuity alone (with an assumed density contrast) does while the gross features are similar. The gravity field over the western and eastern Massif (Eifel low, Bergisches Land/Sauerland, Harz, and Odenwald/Spessart highs) and the fall-off towards the Alps are again well duplicated, now also crudely the broad-scale anomaly of the Rhine Graben region. Not well represented are the narrow gravity troughs of the Rhine Graben – Hessian Depression and Lower Rhine Graben and the gravity high of the Palatine. These features are considered peripheral to the present study and the

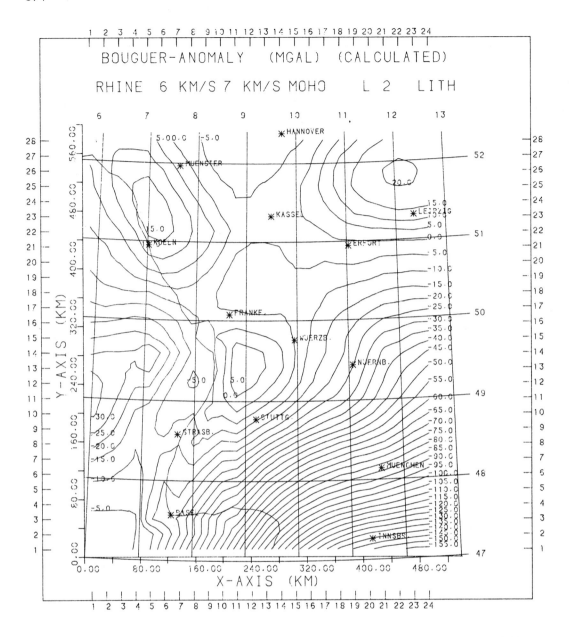

Fig. 4. Gravity effect (contoured at 5 mgal intervals) computed for model consisting of all interfaces of Fig. 2; density contrasts calculated by least-squares fit of model effect to observed Bouguer anomaly. Downward density contrasts (in g cm^{-3}) from top to bottom listed below map

discrepancies are related to model deficiencies such as the neglect of the sedimentary cover, partly to contouring problems in the coarse grid. In the case of the Palatine the high Bouguer anomaly requires additional high-density mass in the crust, not evident in the seismic data, for attempts to modify the model within the given constraints failed. In this case the assumption of homogeneous layers derived from seismic cross-sections seems to break down, as also to a lesser degree beneath the Black Forest and the Vosges.

The similarity of the main features between the total model effect, the crustal thickness effect and the observed Bouguer anomaly might, at first glance, be taken as proof that the crustal thickness variation is the dominant factor except in the Alsace, Lorraine, and the Palatine. Inspection of the individual interface effects after least-squares adjustment demonstrates, however, that some surfaces cause a comparable or even larger gravity variation than the M-discontinuity in this solution. In particular, the 7 km s^{-1} surface contributes about 100 mgal variation of the 150 mgal observed throughout the whole map (Figs. 1, 3 and 4), which is the effect of the M and 7 km s^{-1} surfaces together. The lithosphere bottom relief contributes no less than 70 mgal, though partly cancelled by the effect of the subcrustal "L2" interface. On the other hand, the basement relief contributes very little, about 10 mgal. The effect of the graben sediments seems to be underestimated; obviously the narrow graben anomaly is not sufficiently reproduced in Fig. 4 and we attribute this to the coarse grid and to contouring difficulties.

Modifications of the model could change some of the relative contributions of the surfaces. We believe, however, that no larger changes will occur except between the mutually very similar surfaces; but even the 7 km s^{-1} and M-discontinuity or the lithosphere bottom and "L2" are dissimilar enough to give stable solutions. The errors of the Bouguer anomaly have little effect on the gross feature of the inversion; we have tested this by two-dimensional models for regions where we have new and old data (see Jacoby et al., 1983). For these and other reasons already mentioned we have abstained from further model refinements. The gross fit is satisfactory. A ±1 km undulation superimposed on the surfaces with <100 km wavelength has no appreciable effect and much larger deviations would be in conflict with the constraints in most regions. The "Eifel anomaly" of subcrustal low velocity will be discussed separately.

We proceed to a discussion of the computed density contrasts at the model surfaces. The standard errors were generally in the order of ±0.01 g cm^{-3}; we consider this a reliable estimate of the uncertainties except where two surfaces are very similar to each other as in the case of 7 km s^{-1} and M or "L2" and lithosphere bottom. Note further that a purely vertical density increase can be superimposed on the computed density model; it would not alter the gravity field (Jacoby, 1973).

a) The density contrast between the Rhine Graben sediments and their basement has been computed to be -0.09 g cm^{-3}. This is probably too low and may be related to the problems discussed above. A stronger contrast of -0.2 to -0.3 g cm^{-3} would fit the observed gravity anomaly better.

b) The average density increase across the top of the crystalline basement or 6 km s^{-1} surface is very small, of the order of +0.01 g cm^{-3}. This somewhat surprising result is confirmed by other models (see Jacoby et al., 1983).

c) The density increase of +0.2 g cm^{-3} computed for the 7 km s^{-1} surface is also unexpected, especially in view of the +0.04 g cm^{-3} at the M-discontinuity.

d) As mentioned the computed density contrast at the M-discontinuity (+0.04 g cm^{-3}) is rather low, but the two surfaces (7 km s^{-1} and M) must be seen in combination. It appears realistic to imagine the lower crust as a region of rapid density increase with depth including gradient layers involving variable and interrupted discontinuities. The M-discontinuity is but the general bottom of that region. If the two surfaces depict the gross features of the region well the least-squares inversion should rather correctly determine the combined density increase, even if their relative importance may be misjudged.

e), f) "L2" and the lithosphere bottom are discussed together since they are very similar (except below the southern Rhine Graben) with different depths and amplitudes. The least-squares fit suggests a strong density increase (+ 0.4 g cm^{-3}) at 35 to 40 km depth and a decrease from the lithosphere to the asthenosphere (-0.05 g cm^{-3}) in the 45 to 75 km depth range. Actually it is mainly the asthenospheric rise under the southern Rhine Graben that is responsible for this result; elsewhere the effects of the two surfaces nearly cancel each other. It is remarkable that this result is strongly required by gravity, although not evident in the seismic data as travel time residuals (Raikes and Bonjer, 1983) and surface wave dispersion (Müller and Lowrie, 1980), although it may have been missed there.

The southern Rhine Graben density anomaly in the upper mantle is to be contrasted with the "Eifel anomaly" of low velocity, not incorporated into our model of Figs. 2 and 4 because it did not improve the gravity fit. Accordingly its gravity effect cannot be stronger than -10 or at most -20 mgal and thus its density contrast no more than -0.01 g cm^{-3}. This is an order of magnitude less than that estimated from the velocity anomaly of about -0.5 km s^{-1} on the basis of the Birch relationship $\Delta\rho \approx b \cdot \Delta v_p$ with $b \approx 0.3$ g cm^{-3}/km s^{-1} (Birch, 1960, 1961; Jacoby, 1973, 1975).

In summary, the seismic and gravity data have led to the discovery of two very different upper mantle anomalies which are related to the evolution of Central Europe. The Eifel anomaly has a strong negative seismic velocity contrast (~ -0.5 km s^{-1}) but a very small density contrast (at most -0.01 g cm^{-3}); the southern Rhine Graben anomaly has just the opposite characteristics (-0.05 g cm^{-3} and no significant velocity anomaly). In the following discussion we shall concentrate on this peculiar observation and its geodynamic implications and shall add some speculations on the evolution of the Rhenish Massif.

Discussion

Let us first restate that the density model is probably reliable: much of the Bouguer anomaly is explained by crustal thickness variations and a low-density body below the southern Rhine Graben; the seismically distinct Eifel anomaly has, however, only a very small density contrast.

Could this conclusion be erroneous? The negative density contrast might be stronger if some of the surface uplift had already been eroded and the M-discontinuity uplift had grown with time, thus isostatically compensating the mantle mass deficiency, but this explanation would require initial uplift >1 km and 3-4 km denudation leading to $\overline{M}$oho uplift >4 km and about zero Bouguer anomaly (for a more detailed quantitative analysis see Jacoby and Drisler, 1983). Such an uplift, however, has not been observed seismically (Mechie et al., 1983), nor has there been so much denudation during Pleistocene. Alternatively, the former position of the M-discontinuity previous to the uplift must then have been regionally low, with a corresponding high density somewhere in the crust and/or upper mantle (again required by isostasy). This would imply a rather complex, seismically and gravimetrically "invisible" compensation of a mass of low density (evident only in low velocity) below the western Massif. Such an explanation cannot be excluded, but in view of the unrealistic amount of denudation still required it appears unlikely. Thus the very small density contrast of the subcrustal Eifel anomaly is accepted as the most likely interpretation. The Birch relationship does not seem to hold in this case.

The above interpretation is supported by a very simple isostatic argument. Assume that the western Massif has been raised in geologically short time with respect to its surroundings by 0.3 km as the result of the expansion of an upper-mantle body of 100 km vertical extent; its density would have decreased by $\Delta\rho \approx -3$ (g cm^{-3}) $\cdot 0.3$ (km)/100 (km) ≈ -0.01 (g cm^{-3}), in excellent agreement with the gravimetric result.

Concerning the southern Rhine Graben anomaly we can convince ourselves with the same arguments that it is unlikely that the density anomaly here would have a much smaller magnitude than -0.05 g cm^{-3}, in spite of the missing velocity anomaly. Again we can support this interpretation with an isostatic argument. Assume the regional uplift of the M-discontinuity under the southern Rhine Graben to be about 5 km with a density contrast of $+0.25$ g cm^{-3} (7 km s^{-1} *and* M surfaces) and the extra uplift of the asthenosphere top to be about 20 km (see Fig. 2f); then $\Delta\rho \approx -0.25$ (g cm^{-3}) $\cdot 5$ (km)/20 (km) ≈ -0.06 g cm^{-3}, again in good agreement with the gravimetric estimate. Note, however, that this is a maximum $\Delta\rho$ estimate: if the extra anomalous mass includes deeper asthenosphere and thus has a greater vertical extent its density contrast is correspondingly smaller; if, e.g., $\Delta h = 100$ km, $\Delta\rho \approx -0.01$ g cm^{-3}; had we included such a body in the gravity inversion, the computed density contrast would also have been of this order.

How can we explain the very different upper mantle anomalies? The elastic moduli, density, and hence the seismic velocities are affected by (1) physical state as temperature and pressure; (2) geometrical structure as the presence of voids and cracks and crystal orientation; (3) phase changes as partial melt; (4) chemical composition in bulk or as migrating volatiles. Although these parameters will usually not act in isolation we can perhaps identify the most likely candidates explaining the observed velocity and density anomalies. Pressure is disregarded since we are concerned with lateral heterogeneities, but temperature must be considered; hence thermal expansivity α and the temperature derivative of velocity $\partial v_p/\partial T$ are relevant. Preferred crystal orientation, causing seismic anisotropy, may occur, but we lack the information for a meaningful discussion. We must consider possible voids and cracks which would be filled by melt or volatiles as H_2O and CO_2. If temperature alone were the cause of the anomaly, then $\Delta\rho = -\rho \cdot \alpha \cdot \Delta T$ and $\Delta v_p = \partial v_p/\partial T \cdot \Delta T$. Cracks, on the other hand, affect bulk density $\bar{\rho}$ by their volume fraction n and the density difference $\Delta\tilde{\rho} = (\rho_{rock} - \rho_{fluid})$: $\Delta\rho = \bar{\rho} - \rho_{rock} = -n \cdot \Delta\tilde{\rho}$; velocity v_p depends also strongly on crack shape, and particularly on aspect ratio β = thickness/length (O'Connell and Budianski, 1977; Mavko, 1980).

The Eifel anomaly cannot be explained by a temperature rise alone for, with $\alpha \approx 3 \times 10^{-5}$ K^{-1} and $\partial v_p/\partial T \approx -4 \times 10^{-4}$ km s^{-1}/K (Anderson, 1980), ΔT would have to be ~100 K for $\Delta\rho$ and ~2000 K for Δv_p. Very thin cracks or films, on the other hand, can explain $\Delta\rho$ and Δv_p. According to the theory of O'Connell and Budiansky (1977), Mavko (1980), and H. Schmeling (pers. commun., 1982) the parameters n and β would have to be in the range of up to several percent and 1/100 to 1/1000, respectively, for the density and velocity reduction to be as "observed": $\Delta\rho \approx -0.01$ g cm^{-3} and $\Delta V_p \approx -0.5$ km s^{-1}, if $\Delta\tilde{\rho} \approx 0.1$ to 1 g cm^{-3}, i.e., the fluid phase ranges from melt to volatile composition. Such thin cracks would have to be opened by a very high fluid or volatile pressures, perhaps driven upward by an activating temperature increase at the base of the lithosphere. A minor temperature rise could be associated with the volatile "wave" without contradicting the data. Some melt obviously rises to produce the recent volcanism of the Eifel region.

There is, however, a problem in the above model in which the upward expulsion of volatiles and the opening of very fine cracks are the most important mechanisms. The volatiles are expected to immediately react with the rock and form melt (K.H. Wedepohl, pers. commun., 1982). This would immobilize them strongly because of the relatively high melt viscosity. But would the geologically rapid Pleistocene uplift not require an agency diffusing much faster than heat? A possible reconciliation of the dilemma could be that the volatile activation is the result of the onset of convective motion in the lower lithosphere. This has been studied by Neugebauer et al. (1983) who demonstrate that the motion may accelerate rapidly after a long-lasting quasistable state. The motion may accelerate fast enough to explain the Massif uplift by itself, if the lower lithosphere viscosity is as low as 10^{21} Pa·s, which we consider unrealistically low. If, however, the beginning overturn convectively raises the isotherms in the lithosphere enough to activate the volatiles, the amount of incipient melting and the crack configuration required by the seismic and gravity data could be reached over a wide depth range nearly simultaneously. This suggestion is hypothetical at present as we lack exact knowledge of the parameters involved, as bulk and melt viscosity, temperature distribution, activation energies of volatiles, their distribution, etc.; but it seems to be a viable qualitative hypothesis explaining many of the complex observations from the Rhenish Massif. The suggestion seems also to be supported by the arguments brought forward by Lloyd and Bailey (1975). Let us finally remark that we should not imagine the anomalous upper-mantle volume as a quasi-homogeneous body; this would be only a geophysical simplification, but the bulk features may be correctly described by it.

The southern Rhine Graben anomaly, on the other hand, can probably be explained as the result of elevated temperatures (see also Werner and Kahle, 1980). If the anomalous mass has 20 km vertical extent and -0.05 g cm^{-3} density contrast, then $\Delta T \approx +500$ K; if 100 km and -0.01 g cm^{-3} are the appropriate values, then $\Delta T \approx +100$ K. An estimate of the temperature excess at the top of the anomalous mass may be obtained from heat flow, if a steady state is assumed; taking the Rhine Graben value as 100 mW m^{-2} and the reference as 70 mW m^{-2} (Čermak and Rybach, 1979) a temperature gradient of ~ 40 K km^{-1} instead of 30 K km^{-1} is implied, hence a temperature excess of $\Delta T \approx 300$ K at 30 km depth. This is indeed bracketed by the temperature contrasts estimated above. The temperature increase should also give a velocity decrease of 0.05 to 0.2 km s^{-1}, in contradiction to observation (Raikes and Bonjer, 1983). A possible explanation is that the Rhine Graben anomaly is at a mature or late and degassed stage. Such an interpretation is also supported by comparison with other rift zones (Seidler and Jacoby, 1981).

The internal consistency of the above picture should, however, not be taken as proof or as a cause to overlook the problems. In particular, the above arguments provide no explanation for the uplift of the *eastern* Rhenish Massif. No corresponding upper-mantle anomaly has been discovered there. Why then did the eastern region rise? There is not definitive answer, but we may hypothesize, on the basis of the trade-off between fluid phase and temperature effects, that the mantle in the east, while degassed by now, has been heated slowly, sufficiently (100 K on average over 100 km depth range) to explain the uplift without velocity anomaly. Unfortunately such a model can hardly be tested. The westerly drift of volcanic activity that is implied in the above speculation is not supported by the radiometric dates obtained for the volcanics of the region (Chap. 5, this Vol.) which suggest a much more scattered and complicated activity; but perhaps a simple history of volcanism cannot be expected and the model advanced here is probably too simple to make any predictions about the space-time behavior of volcanism in central Europe.

The speculative picture as envisaged here contains several dynamic aspects: migration of volatiles, heating, and bulk motion of mass. Such processes could proceed in different ways depending on the relative magnitudes of the diffusivities and the mechanical constraints. A possible and attractive speculation is sketched in Fig. 5. The main features are: (1) a temperature rise at the base of the lithosphere in a broad region, perhaps related to subduction in the Alpine arc; (2) initial volcanism and minor uplift related to the diffusion and expulsion of volatiles; (3) slower heating of the lower lithosphere (Spohn and Schubert, 1982) decrease of its viscosity ν over an increasing depth range L; as a result, the local Rayleigh number $\alpha \cdot g \cdot \Delta T \cdot L^3 / \nu \varkappa$, characterising the instability, would rise above the critical value and, first

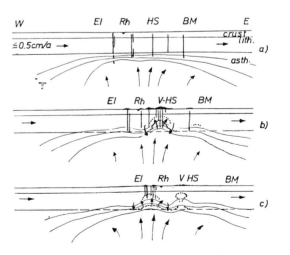

Fig. 5. Diagram of a hypothetical model for the Tertiary evolution of the Rhenish Massif. (*T* isotherms; *BM* Bohemia; *EI* Eifel; *HS* Hessian Depression; *Rh* Rhine river; *V* Vogelsberg volcano; see text for discussion)

slowly then rapidly, diapirs would develop in the asthenosphere-lithosphere region (Neugebauer et al., 1983); (4) growth of a mature mantle diapir under the Rhine Graben - Vogelsberg - Hessian Depression where the crust slightly gave way laterally; evolution of neighboring diapirs with volcanism above their crests (Westerwald, Siebengebirge, Hocheifel, Rhön etc.).

These concluding speculations are only to be taken as questions incited by the new insights into the deep Rhenish Massif gained in a great multidisciplinary project. Frankly, we do not know the causes of uplift of the Rhenish Massif, but we have extended the limits of our knowledge a little further.

Acknowledgments. We gratefully acknowledge the help we have received during this study. The interdisciplinary discussion was most valuable; it was greatly advanced by the project Vertikalbewegungen und ihre Ursachen am Beispiel des Rheinischen Schildes organized and funded by Deutsche Forschungsgemeinschaft and inspired by Henning Illies. Our acknowledgment of the help with gravity is not repeated here from Chap. 6. Other data were freely supplied by W. Bosum, Hannover; H. Grabert, Krefeld; H. Jödicke, Münster; W. Meyer, Bonn; E. Plein, Hannover; C. Prodehl, Karlsruhe; R. Teichmüller, Krefeld, and others. The computations have been done at Hochschulrechenzentrum of Frankfurt University. Deutsche Forschungsgemeinschaft funded this study (grants Ja 258/5,7,8,9,12,16-7).

References

Anderson, C., 1980. The temperature profile of the upper mantle. J. Geophys. Res., 85:7003-7010.
Birch, F., 1960. The velocity of compressional waves in rocks to 10 kbars, part 1. J. Geophys. Res., 65:1083-1102.
Birch, F., 1961. The velocity of compressional waves in rocks to 10 kbars, part 2. J. Geophys. Res., 66:2199-2224.
Čermak, V. and Rybach, L. (eds.), 1979. Terrestrial heat flow in Europe. Springer, Berlin, Heidelberg, New York

Doebl, F. and Olbrecht, W., 1974. An isobath map of the Tertiary base in the Rhinegraben. In: Illies J.H., Fuchs, K. (eds.) Approaches to taphrogenesis. Schweizerbart, Stuttgart, pp.71-72.
Drisler, J. and Jacoby, W.R., 1983. Three-dimensional gravity modelling for the Rhenish Massif: Methods. (in preparation).
Giese, P., 1976. Results of the generalized interpretation of the deep-seismic sounding data. In: Giese, P., Prodehl, C., and Stein, A. (eds.) Explosion seismology in Central Europe. Springer, Berlin, Heidelberg, New York.
Giese, P., Prodehl, C., and Stein, A. (eds.), 1976. Explosion seismology in Central Europe. Springer, Berlin, Heidelberg, New York, pp.201-214.
Grohmann, N., 1981. Die 2-scale Zellularkonvektion; Untersuchungen über die Zusammenhänge zwischen Orogenese, Kontinentaldrift und Magamatismus. Diss. Univ. München.
Jacoby, W.R., 1973. Isostasie und Dichteverteilung in Kruste und oberem Mantel. Z. Geophys., 39:79-96.
Jacoby, W.R., 1975. Velocity-density systematics from seismic and gravity data. Veröff. Zentr. Inst. Phys. Erde, 31:323-333.
Jacoby, W.R. and Drisler, J., 1983. The gravity effect of a rising dome maintaining isostasy. (in preparation).
Jacoby, W.R., Joachimi, H., and Gerstenecker, C., 1983. The gravity field of the Rhenish Massif. This Vol.
Jödicke, H., Untiedt, J., Olgemann, W., Schulte, L., and Wagenitz, V., 1983. Electrical conductivity structure of the crust and upper mantle beneath the Rhenish Massif. This Vol.
Lloyd, F.E. and Bailey, D.K., 1975. Light element metasomatism of the continental mantle: the evidence and the consequences. Phys. Chem. Earth, 9:389-416.
Mavko, G.M., 1980. Velocity and attenuation in partially molten rocks. J. Geophys. Res., 85:5173-5189.
Mechie, J., Prodehl, C., and Fuchs, K., 1983. The long-range seismic refraction experiment in the Rhenish Massif. This Vol.
Meißner, R., Springer, H., Murawski, H., Bartelsen, Flüh, E.R., and Dürschner, H., 1983. Combined seismic reflection-refraction investigations in the Rhe-

nish Massif and their relation to recent tectonic movements. This Vol.

Meyer, W. and Stets, J., 1975. Das Rheinprofil zwischen Bonn und Bingen. Z. Dtsch. Geol. Ges., 126:15-29.

Mooney, W.D. and Prodehl, C., 1978. Crustal structure of the Rhenish Massif and adjacent areas; a reinterpretation of existing seismic refraction data. J. Geophys., 44:573-601.

Müller, S. and Lowrie, W., 1980. Die geodynamische Entwicklung des westlichen Mittelmeerraumes und der Alpen. Vermess. Photogramm. Kulturtech., 12/80:470-495.

Neugebauer, H.J., Woidt, W.D., and Wallner, H., 1983. Uplift, volcanism and tectonics: evidence for mantle diapirs at the Rhenish Massif? This Vol.

O'Connell, R.J. and Budiansky, B., 1977. Viscoelastic properties of fluid saturated cracked solids. J. Geophys. Res., 82:5719-5735.

Panza, G.F., Müller, S., and Calcagnile, G., 1980. The gross features of the lithosphere-asthenosphere system in Europe from seismic surface waves and body waves. Pure Appl. Geophys., 118: 1209-1213.

Raikes, S., 1980. Teleseismic evidence for velocity heterogeneity beneath the Rhenish Massif. J. Geophys., 48:80-83.

Raikes, S. and Bonjer, K.P., 1983. Large scale mantle heterogeneity beneath the Rhenish Massif and its vicinity from teleseismic P-residual measurements. This Vol.

Seidler, E. and Jacoby, W.R., 1981. Parameterized rift development and upper mantle anomalies. Tectonophysics, 73:53-68.

Spohn, T. and Schubert, G., 1982. Convective thinning of the lithosphere: a mechanism for the initiation of continental rifting. J. Geophys. Res., 87:4669-4681.

Werner, D. and Kahle, H.G., 1980. A geophysical study of the Rhinegraben, Part I: Kinematics and geothermics. Geophys. J. R. Astronom. Soc., 62:617-629.

8.2 Uplift, Volcanism and Tectonics: Evidence for Mantle Diapirs at the Rhenish Massif

H. J. Neugebauer, W.-D. Woidt, and H. Wallner[1]

Abstract

The Cenozoic history of the Rhenish Massif is characterized by uplift, volcanism and normal faulting style tectonics. These phenomena are closely related to each other and are of a subregional and episodic nature in space and time. Anomalies associated with P-wave veloctiy in the mantle, Bouguer gravity, uplift and volcanic activity as well as an extensional crustal stress regime are located within the Rhenish Massif region and are coincident. From magma source parameters two sequences of magma source uprise during Upper Oligocene to Lower Miocene and Pliocene to Pleistocene can be deduced. The typical rates of magma source uprise are 5 and 10 km Ma, respectively.

The tectonic history of the Rhenish Massif may be associated with either the Mesozoic North Sea rifting or the Cenozoic Alpine orogeny, in which case the possible causes and mechanisms could be a deep-seated source or a lateral transmission of stress, respectively. In order to quantify these principal approaches, we investigated three numerical models: (a) thermal, convective thinning of the lithosphere; (b) diapiric thinning of the lithosphere in terms of a density instability; (c) Alpine push transmission through the lithosphere. The results of the models are discussed in the view of the complex constraints provided by the data.

The three-dimensional, elastic-plastic finite element analysis on the transmission of stresses due to the "Alpine push" explains the earthquake hypocenter distribution in the foreland very well, however, it predicts a rather negligible influence on the Rhenish Massif.

The thermal plume model on convective thinning of the lithosphere requires an increase of heat flow at the base of the lithosphere by a factor of 6-8 to meet the deduced rates, but on the other hand it exceeds the observations in terms of the calculated final thickness of the lithosphere, the surface uplift and the excess heat flow.

The diapir model represents the lithosphere-asthenosphere system by four viscous layers with an inverted density stratification for the asthenosphere. This model satisfies both, the typical lateral diameter of the thinned lithosphere of 100-200 km and the derived rates of thinning. Further, the model predicts a negative density contrasts necessary for the asthenosphere in the range of 0.0003 to 0.15 g cm^{-3} in correspondence with viscosities for the lower lithosphere between 10^{19} to 10^{23} poises.

Modeling the velocity anomaly in the mantle with a density contrast of -0.01 g cm^{-3} is in agreement with the observed Bouguer gravity anomaly. Local isostasy would provide corresponding uplift in the order of 150 m. In addition, the stress regime estimated for a diapiric structure is typically extensional and confined to the area above the diapir, features which are most appropriate to explain the tectonic and dynamic character

---

[1] Institut für Geophysik, Technische Universität Clausthal, Arnold-Sommerfeld-Str. 1, Postfach 230, D-3392 Clausthal-Zellerfeld, Fed. Rep. of Germany.

of the Rhenish Massif. Thus it is concluded that periods of uplift, volcanic activity and seismotectonics are most likely caused by diapiric developments resulting from density instabilities in the lithosphere-asthenosphere system.

Introduction

The Rhenish Massif (RM) is characterized by recent geodynamic phenomena such as crustal uplift, zonal seismotectonics and Pleistocene volcanism. However, these activities had a more obvious intensity in geological history dating back to the Middle Tertiary. This long period of activity reveals a temporal correspondence between uplift, volcanism and tectonism. In addition, an association of the geographical position of these processes with pronounced heterogeneities in the lower crust and lithosphere is obvious. The fact that volcanism refers to deep and shallow seated magma sources suggests a link between the observed processes and the determined structure.

Such or similar situation are found in many tectonic areas on continents, especially in rift zones. Further, a net low Bouguer gravity and increased heat flow are usually assigned to such places. Besides the local concentration of phenomena, their development might be affected by regional tectonic processes. Thus the RM is part of the Rhine Rift System; it is located between the Oligocene Lower Rhine region and the Eocene Upper Rhine Graben. Within an even more extended framework the Rhine Rift System itself is part of a Cenozoic active belt which represents the foreland of the Alpine orogene in northwestern Europe.

Thereafter, the specific Cenozoic development of the RM might be determined or affected by local deep-seated physical processes, pre-existing geological structures, and even by the plate underthrust and collision of the Alpine orogeny. In this context any suggestion on the origin of the total uplift of the RM should be judged by its ability to "explain" the entire association of phenomena.

On the basis of the collection of new, as well as already available data, we will discuss the interdependence of internal processes in the RM and their possible relationship to external tectonic developments. In order to find a reasonable understanding of the geodynamic development as a whole, we present numerical models on the possible development of mantle diapirs, "convective" thermal thinning of the lithosphere, as well as the influence of lateral transmission of mechanical energy resulting from the Alpine collision.

Both the new data and the numerical investigations favor the concept of a mantle diapir as the most likely mechanism for understanding the uplift of and the majority of related phenomena affecting the RM.

The Rhenish Massif in Space and Time

The subsequent compilation of data on Cenozoic features and processes for the RM is supposed to display their mutual relation in space and time. They should provide appropriate constraints in addition to the accentuated uplift to find a most reasonable explanation for the complex development by means of physical models.

Geophysical Characteristics

The RM is geologically part of the Variscan orogenic belt. However, its Cenozoic development is neither homogeneous nor congruent throughout the geological unit. Therefore, we prefer a representation by geophysical phenomena and features, which are concerned with the geodynamics of the area instead of geological units. The most striking feature concerning the deep structure of the RM is a large mantle velocity anomaly derived from teleseismic P-residuals (Raikes and Bonjer, 1983). The lateral extent of the major anomaly with a velocity decrease $\geq 4\%$ is shown in Fig. 1. The lateral diameter of the anomaly reaches 150-200 km, while regions with lesser anomaly exceed the extent of the RM. The depth range of the zone has been deduced to be between 50 and 200 km.

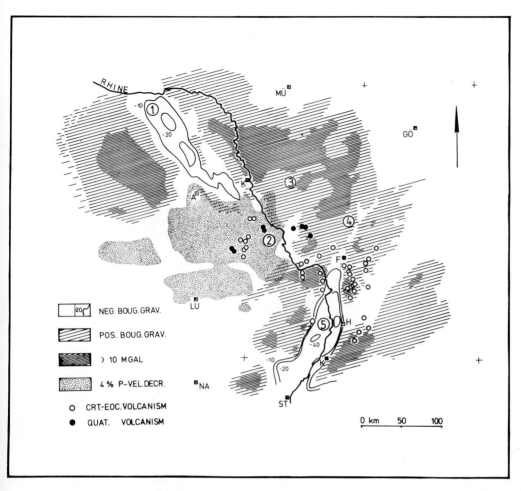

Fig. 1. Spatial compilation of geophysical data for the Rhenish Massif and adjacent areas. Sites represent: 1) Lower Rhine, Netherlands; 2) Eifel-Neuwied Basin; 3) northern rim of Westerwald; 4) Hessen Depression; 5) northern Upper Rhine Graben. Cities: A-Aachen; K-Cologne; Mü-Münster; Gö-Göttingen; F-Frankfurt; H-Heidelberg; K-Karlsruhe; St-Strassburg; Na-Nancy; LU-Luxemburg. Gravity after Jacoby (1983); P-wave velocity anomaly after Raikes and Bonjer (1983)

This upper boundary finds verification also from seismic surface wave studies (Panza et al., 1980). They suggest 90-100 km for the "normal" thickness of the lithosphere in northwestern Europe.

The detection of an unusual crustal structure west of the Rhine by long range refraction measurements concurs with the position of the anomalous mantle indicated by the P-residuals in Fig. 1. In this area, mantle velocities were eventually attained between 36 and 38 km depth, while east of the Rhine mantle velocities were encountered at shallower depths, Mechie et al. (1983). Prodehl (pers. commun.) proposed, that, above the mantle anomaly, the Moho cannot be defined with certainty by the seismic refraction data, and that it is possible that an extremely low velocity mantle is attached to the base of the crust. The lateral diameter of this zone is about 150 km in good agreement with the mantle anomaly. The Bouguer gravity field over the RM has only a small range of variation (±20 mgal), but a clear negative Bouguer gravity anomaly can be attributed to the crust and upper mantle anomalies as represented by the area of strong P-velocity decrease in Fig. 1. In accord with regions of shallow depth of the Moho, see Mechie et al. (1983, Fig. 8) we find positive Bouguer gravity anomalies.

Once proper corrections have been made for the graben zones, represented by gravity contours in Fig. 1, the area of positive Bouguer gravity anomalies has the shape of a horseshoe, surrounding the gravity low, from north-west through north to the south-west.

The open side of the "horseshoe" is closed within France by small positive anomalies north of Reims, at the southwestern tip of the Ardennes and from east along the line Strasbourg-Nancy-Bar-le-Duc towards the west (gravity map of France, BRGM 1974/75). Thus the anomalous crust and mantle congruent with a Bouguer gravity low is encircled by an indiscriminately expressed broad positive gravity anomaly which exceeds laterally the geological unit of the RM in accord with the area of weak P-wave delays.

Heat flow measurements within the RM region are most properly represented by an average of about 60 mW m^{-2} (Haenel et al., 1983). Allowing a general uncertainty of about 10%, there is still no indication of a regional increase on the basis of the data without climatic corrections, even not within the vicinity of the zones of Miocene volcanism.

Recent crustal uplift attains high values at the flanks of the Lower Rhine region in the north-west with the maximum above the anomalous mantle south of Aachen (Mälzer et al., 1983), while subsidence dominates in the northern Upper Rhine Graben south-east of the RM. The total area is characterized by normal faulting type seismotectonics and extensional horizontal stresses deduced from in situ measurements (Ahorner et al., 1983; Baumann, 1982). Seismicity and seismotectonics with predominant NW-striking B-axes are concentrated along the entire zone from the Lower Rhine region to the northern Upper Rhine Graben, sites 1 and 5 in Fig. 1, respectively. Adjacent to the northern Upper Rhine Graben, towards the Alps, strike slip faulting is the predominant mode of deformation, indicating a horizontal compressive stress regime.

While the seismicity in the RM and the northern Upper Rhine Graben is confined to a narrow belt, it widens into a triangular area towards the Alps (Bonjer and Fuchs, 1979; Schneider, 1979). This area includes the Black Forest, and is flanked by the Upper Rhine Graben in the west and the Alps in the south. The stress unconformity of the RM raises the question of the origin of such a confined extensional stress province within a collisional Alpine environment. Assuming that the Alps play a dominant role in forming the present stress field in northwestern Europe, we have to consider the influence of the Upper Rhine Graben for the specific situation in the RM.

Geological Characteristics

Detailed data on the age of volcanic activity, stratigraphic dating of Cenozoic vertical movements and fault-controlled tectonics provide a view of the geodynamic history of the RM and adjacent regions. Figure 2 shows patterns of volcanic activity, uplift, and tectonics from the Cenozoic stratigraphic record for a number of selected sites indicated in Fig. 1. However, the Mesozoic history of the RM and its geographical neighborhood might provide a better understanding of the Cenozoic development. The opening of the Atlantic and the rifting in the North Sea were the major events west and northwest of the RM in the Mesozoic. South of the RM, the Mesozoic opening of Tethys occurred, succeeded by the Late Mesozoic and Cenozoic Alpine collision between Africa and Eurasia and the Early Tertiary rifting of the Upper Rhine Graben.

The tectonic development of the North Sea Rift started in the north with Early Jurassic subsidence forming the Viking Graben (Ziegler, 1981). During the Middle Jurassic, the Central Graben was formed at its southern extension accompanied by large volcanic centers between the two grabens.

The late Middle to Late Jurassic crustal distension in the North Sea was taken up at its southern end by a system of NW-SE-striking faults in West and Central Netherlands and Lower Saxony. In the former area Late Jurassic to Early Cretaceous volcanic centers occurred. Middle Cretaceous volcanism has been dated in the RM south of the Eifel (Lippolt, 1983), and between the Black Forest and Vosges near Freiburg (Baranyi et al., 1976). Late Cretaceous and Early Cenozoic volcanism

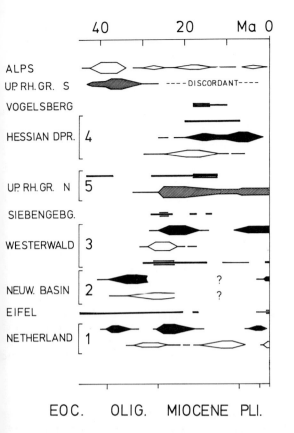

Fig. 2. Pattern of Cenozoic activity in the Rhenish Massif: Uplift (*dark areas*), tectonic activity (*light areas*), volcanism (*straight lines*). Position of sites as shown at Fig. 1 (Lippolt, 1981; Murawski et al., Chap. 2, this vol.; Meyer, 1979; Keizer and Letsch, 1963). In addition: Events of folding in the Alps (Trümpy, 1975); relative accumulation of sediments in the Upper Rhine Graben, northern and southern part (Roll, 1979)

preceding the main middle Eocene development of the structure, occurred along the entire Upper Rhine Graben.

Although the RM links two graben zones spatially, it obviously fits in with the Mesozoic north-south-trending migration of tectonic and volcanic activity in the North Sea (Fig. 2). This view is well supported by conclusions drawn from stratigraphic data from the Netherlands (Keizer and Letsch, 1963). While the first major Tertiary uplift in the Lower Oligocene occurred along a predominantly central WNW-ESE-trending zone between Nijmegen and Leiden, the second major uplift in the Lower Miocene affected mainly the northern and southwestern part of the Netherlands. This shift of activity towards the RM is consistent with the observations at the northern rim of the Westerwald (site 3, Fig. 1), and can be interpreted that the center of uplift passed through the Netherlands from the North Sea towards the RM. This development ends at the northern Upper Rhine Graben and the Hessian Depression. According to the post-Oligocene sedimentary succession of the northern Upper Rhine Graben discussed by Roll (1979) and the activities in the RM, the Miocene northern Upper Rhine Graben is most likely a consequence of the uplift in the RM. This is well expressed by the northwest-trending axis of the concordant Miocene subsidence in the graben itself and, on the other hand, by the remarkable fact that there is no significant uplift of the shoulders flanking the northern Upper Rhine Graben.

A narrow tectonically active belt is thus crossing the RM with either ends within the grabens (sites 1 and 5, Fig. 1). There is no symmetric graben within the RM. However, a major Oligocene tectonic line can be traced in the grabens and along the Mesozoic Rhine valley throughout the RM. It is marked in the asymmetric Lower Rhenish Depression by an offset of the top of the Mesozoic of about 1100 m at the major Peel-boundary fault (Ahorner, 1962; Keizer and Letsch, 1963). The Neuwied Basin (site 2, Fig. 1), has the same character, with the major Cenozoic offset of 350 to 400 m at the northeastern boundary fault (Meyer, 1979). Finally, the sedimentary wedge of the northern Upper Rhine Graben (site 5, Fig. 1), reveals a post-Oligocene northwest trending relative subsidence of the Tertiary base of 1000 m (Doebl and Olbrecht, 1974). This zone lies just at the southeastern prolongation of the Rhine valley into the graben. Along the entire zone, seismotectonic movements still repeat the style of deformation - normal faulting - which allowed the deposition of Cenozoic sedimentary successions. The Tertiary development ends locally at the eastern rim of the RM with Middle to Upper Miocene volcanism associated with uplift in the Hessian Depression and the Vogelsberg, both of which are located at the northern extension of the Upper Rhine

Graben. Quaternary rejuvenation of volcanism is grouped around site 2, Fig. 1. The westerly shift of volcanic centers is again accompanied by uplift (Fig. 2). Thereafter, the RM represents an outstanding area of extensional tectonics and uplift during Cenozoic to present times. There is obviously a close concurrence between periods of volcanic activity and uplift, which suggests a correspondence of uplift with deep-seated processes. These joint activites evidently migrate from northwest to southeast as a function of time. This means, geographically, from the Lower Rhenish Depression to the northern Upper Rhine Graben during the Oligocene and Miocene, and, finally, a last Pliocene-Pleistocene phase of uplift and volcanism within the RM.

The sequence of orogenic events in the Alps might also be important for the Cenozoic history of the RM (Fig. 2). The lower Cretaceous Austrian tectonic pulse finds little expression within the North Sea graben system, which shows continuous differential subsidence during that time (Ziegler, 1981). The main Alpine phase of orogeny at the transition between the Upper Eocene and Lower Oligocene (Trümpy, 1973) corresponds to the development of the nearby southern Upper Rhine Graben (Roll, 1979). There is only minor sedimentation in the northern Upper Rhine Graben and a conjunction with the Hessian Depression at that time. The Oligocene activity in the Lower Rhenish Depression and central RM does not seem to be correlated with either Alpine tectonics or uplift and volcanic activity east of the river Rhine. The minor Neo-Alpine period during the Upper Oligocene and Miocene is characterized by uplift of the Alps, subsidence of the Molasse Basin, and uplift and erosion of the southern Upper Rhine Graben. In the north, the Westerwald and, later, the southern Hessian Depression show uplift and volcanism. Those two areas are disconnected by the northern Upper Rhine Graben, which shows no shoulder uplift in the Odenwald and Palatinate Forest, but most intense subsidence within the graben itself along a northwestern trend during that period. This comparison suggests that there is no perceptable temporal correspondence between uplift, volcanism, and tectonics in the Rhenish Massif and the orogenic events of the Alps.

Geochemical Characteristics

The small number of dated volcanic rocks from the Mesozoic reveal a clear movement of volcanic activity from north to south and southeast (Ziegler, 1981). Early Tertiary volcanism is rather frequent and preceeds the Upper Rhine Graben formation Fig. 1 (Lippolt, 1981). At that period of time, the eruptions in the Eifel area of the RM, site 2 in Fig. 1, appear to be marginal events.

However, during the Upper Oligocene and Lower Miocene, the activity outside the Upper Rhine Graben dominates, tending towards the Hessian Depression with time. This west-to-east migration reversed during the Pliocene and Pleistocene with the most recent volcanic sites within the region of the seismic velocity mantle anomaly, Fig. 1. Detailed investigations on the depths of origin of the volcanics and their ages for the RM area shed some light on the temporal evolution of the subcrustal mantle (Neugebauer and Walter, 1983).

The main characteristics of magma source depth as a function of time are well represented for the Cenozoic volcanic regions of the RM by the curves shown in Fig. 4. The data indicate a long period of deep seated sources succeeded by a short period of accelerated uprise. The derived mean ratio of corresponding time spans is about 2:1. Petrochemically, the uprise represents the well-known differentiation sequence of magmas, where each stage requires the preceding one (Green, 1971), a behavior which demands physical transport of material as against a simple temperature upwelling. Because most of the magmas are generated at about $1100^\circ C$ the revealed uprise of magma sources outlined by their depth of origin defines also the upward encroachment of this isotherm as a function of time. Neugebauer and Walter (1983) derived two major events of magma source uprise in the RM area. Event I is characterized by a period of time of about 10 Ma for the uprise with a corresponding mean rate of 5 km/Ma. This event ends during Lower

Miocene with the main erruptions at the Westerwald and Siebengebirge. Event II shows a shorter period of about 3 Ma and a higher rate of 10 km/Ma on the contrary. It occurred during Pliocene and Pleistocene and ended with the most recent volcanic activity.

Often, both end members of the sequence of differential magmas can be observed at the same time during uprise and post-uprise in an active region. This implies the allocation of high percentage melt at the smaller and low percentage melt at the greater depth at the same time. In addition to the deduced petrogenic suite of magmas, upper mantle xenoliths tracing strong deformation and an upwelling of isotherms (Seck et al., 1983) support the view of a physical uprise of matter related to volcanic activity and thus to corresponding uplift.

The number of features and processes, superposed in their gross geographical position and in time, are considered likely to reveal an interdependence of deep seated and shallow phenomena. The presented discussion can be summarized in a number of suitable constraints (see Table 1), in addition to the accentuated Cenozoic uplift, which have to be regarded or even explained by any suggested model.

Mechanism of Uplift

A compilation of suggested mechanisms for plateau uplift by McGetchin and Merrill (1979) includes processes of thermal, geochemical and dynamical nature. Thereafter, thermal expansion of the lithosphere in response to a deep mantle plume or hot spot or even shear heating due to relative motion along the lithosphere-asthenosphere interface can be envisaged. The generation of a partial melt will be accompanied by volumetric expansion.

Metasomatism can be taken as equivalent to a thermal event; it precedes volcanic activity and yields a density decrease of up to 6% (Dawson, pers. commun., 1982). Metamorphic reactions could equally well account for changes of density. A more detailed discussion of seismic velocity inhomogeneities and their possible interpretation in terms of a density decrease is given by Aki (1982). This group of possible mechanisms can be further extended by mechanical processes such as shallow underplating, subduction or reactivation of listric thrust faults, which, however, can hardly provide appropriate explanations for the complex situation in the RM. All the preceding suggestions have the major consequence of a regional decrease of density, while changes in the thermal regime are either a cause of, or a consequence of the mechanism. However, such regions of inverted density stratification in the upper mantle will create a dynamically unstable situation which tends to stabilize by diapiric uprise which, in turn, changes the stratification.

The enumerated processes lead, in principle, to a coupled thermo-mechanical mechanism which is controlled by either the thermal or dynamical parameters of the problem. With respect to the RM, where deep seated phenomena are obviously associated with uplift and tectonics in space and time, one might expect the cause of the development to be within the upper mantle while the consequences near the surface will be strongly affected by pre-existing structures and the tectonic boundary conditions at the period of time concerned.

Possible Models

The subsequent discussion on numerical models suffers from the inability to model the thermo-mechanical problem by means of a coupled numerical approximation. Thus we must discuss both aspects on the basis of separate model approaches.

Mechanical Diapir

The concept of a mechanical diapir is based on an upper mantle structure with inverted density stratification in the asthenosphere. It has been demonstrated by means of numerical models (Woidt, 1980), that a region having lesser density relative to overlying layers can

Table 1. Cenozoic characteristics of the Rhenish Massif Region

Subject	Characteristics	Quantities	Reference
Thinned Lithosphere	1. P-wave velocity anomaly regional	Velocity decrease 4-6% - depth to top ~50 km diameter 150-200 km	Raikes and Bonjer, 1983
	2. Mantle xenoliths	Depth of origin $\leq$50 km pT estimate	Seck and Wedepohl, 1983
	3. Seismic refraction experiments	LV-mantle below 40 km depth coincident with anomalous mantle	Mechie et al., 1983
	4. Petrogenic suite: magma source pressure	Shallow sources between 30-40 km subregional diameter $\geq$50 km	Neugebauer and Walter, 1983
Rate of Thinning	5. Petrogenic suite as function of time	(5-10) km/Ma	Neugebauer and Walter, 1983
Uplift: Surface	6. Cenozoic: episodic and subregional	(400-600)m total amount; at least 3 periods	
	7. Recent: subregional, centre within mantle anomaly	Up to 2 mm/a (maximum)	Mälzer et al., 1983
Stress Field	8. Cenozoic: dominant extensive	Normal faulting tectonics	Meyer, 1979
	9. Recent: independent extensive stress province-compressive environment	Normal faulting-horizontal excess stress: tensile 1-5 MPa	Ahorner et al., 1983; Baumann, 1982
Gravity	10. Bouguer gravity-regional: negative within anomalous mantle enclosed by outer positive anomaly	-(10-20) mgal +(10-20) mgal	Jacoby et al., 1983; Gravity map of France
Heat Flow	11. No perceptible anomaly from uncorrected data	Regional average 60 mW m^{-2}	Haenel, 1983
Volcanism	12. Episodic and subregional	3 periods: center - Early Tertiary + Quaternary; eastern rim - Late Tertiary	Lippolt, 1983
	13. Petrogenic suites: subregional deep initial period, subsequent uprise	Ratio of duration 2:1 uprise period: I: 10 Ma, II: 3 Ma	Neugebauer and Walter, 1983
	14. Volcanic eruption from deep and shallow sources coincidently		Neugebauer and Walter, 1983
	15. Eifel volcanic zone can not be regarded as stationary plume		Duda and Schminke 1978; Lippolt, 1981
Tectonics	16. RM: End member of south to southeast trending North Sea Rifting	Junction with Upper Rhine Rift: Lower Miocene	This study; Ziegler, 1981

Table 2. Specification of viscous four-layer model (Woidt and Neugebauer, 1983)

Layer	Thickness (km)	Viscosity ratio	Density (g cm^{-3})
Crust	$h_c = 30$	$\mu_c/\mu_L = 10^3$	$\rho_c = 2.8$
Lower lithosphere	$h_L = 70$	$\mu_L = 1$	$\rho_L = 3.3 + \Delta\rho$
Asthenosphere	$h_A = 10-100$	$\mu_A/\mu_L = 10^{-5} - 10^0$	$\rho_A = 3.3$
Mesosphere	$h_M \to \infty$	$\mu_M/\mu_A = 3$	$\rho_M = 3.6$

lead to diapiric uprise of material. The subsequent exchange of material is characterized by a typical lateral wavelength and progressive stages which are respectively characterized by (a) an initial and sluggish growth period, (b) a period of accelerated uprise of less dense material, (c) a period of approximately constant uprise velocity, and finally (d) a stagnation period. The stagnation of the diapir is dominated by either the high viscosity of the top layer or the loss of density contrast during the ascent of the diapir.

In order to quantify parameters such as the thickness of the gravitationally unstable layer, the size of the density contrast, and the lithospheric viscosity required to match the observed mean wavelength and growth rate of lithospheric thinning, we have adopted a four-layer viscous model (for details see Table 2). An initial disturbance at the lithosphere-asthenosphere (LA) interface of wavenumber k leads to an exponential growth of the instability with time:

$$z^{LA}(t) = \sum_{j=1}^{4} c_j^{LA} \exp\gamma_j(k)t \qquad (1)$$

where z(t) is the amplitude of the sinusoidal disturbance, t is the time $\gamma_j(k)$ are characteristic eigenvalues of the problem, which determine how the deflections of the interfaces j (j = 1,2,3,4) interfere with one another. The initial amplitude of the sinusoidal disturbance at the LA boundary is given by:

$$z_0^{LA} = \sum_{j=1}^{4} c_j^{LA}(k) \qquad (2)$$

As γ_3 (γ_{LA}) and C_3 of the unstable interface are positive and very large compared with all the other eigenvalues and coefficients respectively, the development of the instability amplitude with time is well approximated by:

$$z^{LA}(t) \approx z_0^{LA} \exp\gamma_{LA}(k)t \qquad (3)$$

The growing instability is governed in space and time by a typical wavelength, $\lambda_{max} = 2\pi/k_{max}$ corresponding to the maximum growth rate, γ_{max}. The important influence of λ_{max} on the spatial development of the interface instability is in general valid even for the diapiric stage as has been shown by Woidt (1980) and Woidt and Neugebauer (1980).

Figure 3 (insets) shows the dependence of γ_{max}^* and λ_{max} on the thickness h_A of the unstable asthenospheric layer and the viscosity contrast, μ_A/μ_L. For a normalized viscosity, μ_L, the rate γ_{max}^*, varies between 1.7 (parameter 6) and 0.5 (parameter 1), while the corresponding maximum wavelength, λ_{max}, lies between 100 and 200 km. The exponential growth of the instability with dimensionless model time t^* is demonstrated for both end values and an intermediate one: $\gamma_{max}^* = 1.7, 1.0, 0.5$. The initial phase of growth is well defined by a time span $1/\gamma_{max}^*$ and exhibits a very sluggish uprise of the instability, whereas subsequent periods exhibit a very pronounced amplitude increase. Calibration of the model time can be attained by the choice of absolute viscosity, μ_L (poise), for the lithosphere and a reasonable density contrast, $\Delta\rho$ (g cm^{-3}), between the lithosphere and the unstable asthenosphere:

$$t = 5 \cdot 10^{-23} \mu_L t^*/\Delta\rho \text{ [Ma]} \qquad (4)$$

The influence of these parameters on the development of the instability is visualized in Fig. 3 for viscosities of the

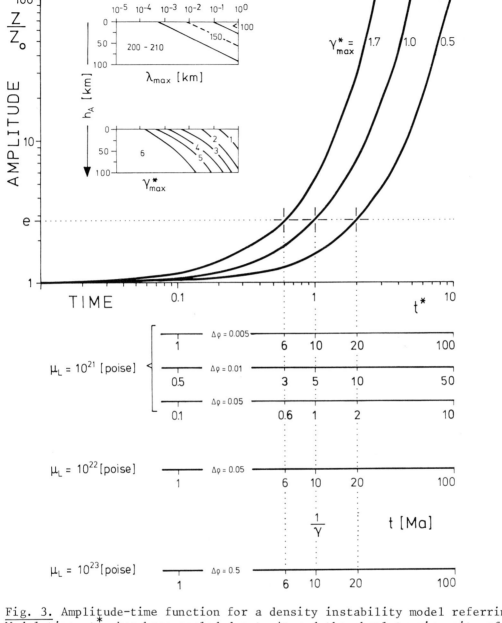

Fig. 3. Amplitude-time function for a density instability model referring to Table 2. Model time, t^*, has been scaled due to $\Delta\rho$ and the absolute viscosity of the lithosphere, μ_L. Inserts: Optimum wavelength, λ_{max}, and model time parameter γ^*_{max} as a function of the thickness of the asthenosphere (inverted density) and the ratio, μ_A/μ_L. Remember $\mu_C = 10^3 \mu_L$

lower lithosphere between $10^{21} - 10^{23}$ poise and probable density contrasts across the LA interface of $0.005 - 0.5$ g cm^{-3}.

Woidt (1980) has shown that, in general, the linear approach given here is applicable to an instability amplitude of the order of $z(t) \lesssim 0.1\lambda_{max}$. The development of a mechanical diapir can be represented by a sluggish initial phase followed by rapid growth, given by the linear model. The main phase of diapiric uprise is most likely a linear function of time while a final period of stagnation might be expected below the crustal layer, Fig. 4.

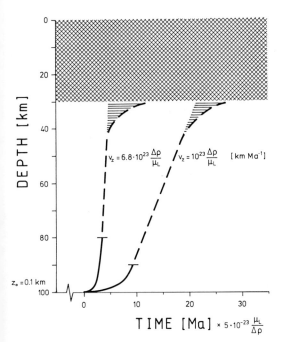

Fig. 4. Diapir uprise with time; calculated by linear model (*solid line*), predicted by linear model (*dashed line*). Stagnation of ascent (0-30 km) beneath the crust has been adopted from other numerical models and data (Woidt, 1980; Neugebauer, 1983). Initial disturbance $z_0 = 0.1$ km

stage of the uprising interface instability and an assumed decrease of uprise velocity, when the top of the diapir approaches the crust-mantle boundary at 30 km depth. The time span of the linear stage of the instability is only dependent on the initial amplitude z_0. In Fig. 4 the time $t = 0$ corresponds to a value of 0.1 km for the amplitude of the instability.

An example for the typical shape of an uprising diapir is given in Fig. 5a after Woidt and Neugebauer (1980). Two fluid layers with viscosities, μ_1 (upper), μ_2 (lower), and thicknesses, h_1 and h_2, have been described by a finite element code for incompressible flow. The model diapir develops from an initially sinusoidal perturbation of the interface between the two fluids with inverted density stratification. The model box in Fig. 5a contains a limited reservoir of low density material (layer 2) which moves upwards in the shown calculated shape. Free-slip holds for the upper boundary and fixed conditions (zero velocity) for the bottom. The shape is controlled by the viscosity ratio, $\mu_1 : \mu_2 = 100 : 1$, the uprise velocity by the absolute value of μ_1.

With this approach we can easily deduce lower and upper bounds of constant uprise velocity, v_z, for the main phase independent of the initial amplitude z_0

$$v_z = dz(t)/dt = \gamma \cdot z(t) . \quad (5)$$

For λ_{max} between 100 and 200 km the upper and lower bounds for the transition from the linear to the non-linear stage of a Rayleigh Taylor instability are bracketed by

$$20 \text{ km} \geq z(t) \geq 10 \text{ km} . \quad (6)$$

This leads to an upper and lower limit for the uprise velocity

$$6.8 \cdot 10^{23} \frac{\Delta\rho}{\mu_L} [\text{km Ma}^{-1}] \geq v_z \geq 10^{23} \frac{\Delta\rho}{\mu_L} [\text{km Ma}^{-1}] \quad (7)$$

where $\Delta\rho$ and μ_L have the same meaning as in Eq. (4). Figure 4 shows the results from Eq. (7) combined with the linear

The columns in Fig. 5a demonstrate the diapir ascent with time by the relative difference of time intervals between stages. Figure 5b shows the time development of the top (i.e., axis of symmetry in the middle of the diapiric structure) and rim syncline (i.e., interface point on the right or left hand boundary) of a single diapiric structure. The curves document the assumed transition from linear to nonlinear geometric behavior at $z(t) \leq 0.1\lambda_{max}$ and the transition from approximately constant velocity of uprise to decreasing uprise near the model surface in the late diapiric stage of development.

Following the uprise of less dense material in terms of a sequence given in Fig. 5a, a first-order approximation of a "diapiric" stress regime is attained by the numerical determination of associated stresses. For each stage the corresponding instantaneous dynamics will be determined based on the shape of the deformed interfaces, the excess body forces and a constitutive equation,

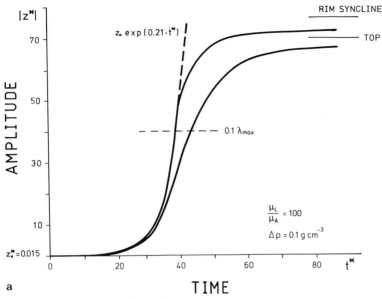
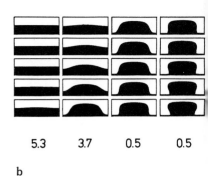

Fig. 5. a. Sequential development of a diapir derived from numerical calculations. Three periods of development can be recognized: initiation - ascent - stagnation. *Numbers* indicate relative length of the time interval between stages with reference to the scale of model time t^* in b. 5b. Continuous time development of the top (diapir) and subsidence of the rim syncline (flanks) of the diapiric structures of a

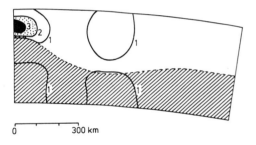

Fig. 6. Modeled stress regime for diapiric structures according to Fig. 5a. Distribution of relative effective shear stress. Zone of buoyant body forces *shaded*. *Numbers* indicate the increase of modeled effective shear stresses

without any consideration of the stress history of the specific situation. The boundaries allow free-slip and the surface is loaded due to dynamical uplift. This simplified procedure leads to a modeled shear stress concentration above the density instability for a power law (n = 3) rheology, in a plain strain two-dimensional model (Fig. 6; Ochmann, 1982; Neugebauer, 1983).

Such diapiric stress regimes are characterized by a confined zone of high effective shear above the crest of the diapir. The horizontal components of the induced excess stress are of an extensional nature which would lead to either stress release or extension.

Thermal Plume

The following numerical models are concerned with the upward movement of the lithosphere-asthenosphere (LA) boundary from an initial equilibrium position as a consequence of an increased heat flux supplied by a thermal plume. This problem has been attacked by a variety of approaches. The models can be classified as conductive or convective depending on their approach to heat transfer. An example of the former is Mareshal's (1981) description of the uplift caused by thermal expansion following the conductive heating of a restricted area of a stationary lithosphere.

Birch (1975), Gass et al. (1978) and Sandwell (1982) on the other hand restrict themselves to steady-state conductive models but incorporate a horizontal movement of the lithospheric plate as an additional parameter in the computations. All these models show that thermal conduction is a very slow process for lithospheric thinning.

In a contrasting approach, Withjack (1979) proposed convective transfer of heat into the lithosphere by magma intrusion and calculated the uprise velocity of the LA boundary by a simple energy conservation argument. Such a "convecting" model explains very rapid thinning of the lithosphere. For example, a 150-km-thick lithosphere can be thinned to 50 km in times of 2 to 15 m.y., depending on the rate of magma intrusion used. Although this penetrative magmatism provides for an efficient and rapid thinning of the lithosphere, the mechanism of the process remains somewhat arbitrary. For instance, the main parameter of this model, the rate of magma injection into the lithosphere is hard to predict. On the other hand, lithospheric thinning is critically dependent on this parameter.

An alternative approach for "convective" thinning of the lithosphere has been proposed recently by Spohn and Schubert (1982). It is based on an energy balance at the LA boundary which provides an additional explicit, non-linear differential equation for the motion of this phase boundary. This formulation is very similar to the classical Stefan-problem of the movement of an ice-water phase boundary.

Our calculations basically follow this approach. However, the results are obtained by a completely different numerical code using a finite element technique of a moving mesh system. The numerical scheme uses Hermitian approximation functions for the temperature and its gradient, which are the basic unknowns at the nodal points. It reproduces the analytically solved Stefan-problem (Ingersoll et al., 1954) with excellent accuracy. The method focuses attention on the substantial temperature-time derivative at each internal point n, given by:

$$\left.\frac{dT}{dt}\right|_n = \left.\frac{\partial T}{\partial z}\right|_n \cdot \frac{dz}{dt} + \left.\frac{\partial T}{\partial t}\right|_n \quad (8)$$

where t is the time, T is the temperature, and z is the depth. The rate of travel of each point is related to the phase boundary velocity, ds/dt by:

$$\left.\frac{dz/dt}{z}\right|_n = \frac{ds/dt}{s} \quad (9)$$

where s is the position of the phase boundary on the x-axis.

Combining these two equations with the one-dimensional heat equation including radiometric heat sources, $A(z)$, leads to:

$$\rho c \left\{ \left.\frac{dT}{dt}\right|_n - \left.\frac{\partial T}{\partial z}\right|_n \frac{z_n}{s} \frac{ds}{dt} \right\} = k \frac{\partial^2 T}{\partial z^2} + A(z) \quad (10)$$

where k is the thermal conductivity, c is the specific heat, and ρ is the density.

The phase boundary motion $s(t)$ is described by the equation:

$$\left\{ \rho L + \rho c (T_p(s) - T_s(s)) \right\} \frac{ds}{dt} = -q + k \left.\frac{\partial T}{\partial z}\right|_{z=s(t)} \quad (11)$$

where L is the latent heat of melting, $T_p(s)$ is the plume temperature, and $T_s(s)$ is the solidus. The right hand side of this equation represents the difference between the upward heat flux q, provided by a plume into the base of the lithosphere and the heat conducted upward into the lithosphere away from the LA boundary. The difference of these quantities is a heat flux available to convert lithosphere into plume material of temperature $T_p(z)$ and for the latent heat of melting. Both terms on the left hand side of the last equation are approximately equal to 840 J cm^{-3}. However, for a partial melt only a fraction of the latent heat of melting must actually be provided, and thus the latent heat term can be neglected in the calculations.

The boundary conditions are assumed to be:

$T = 10°C$ at $z = 0$

$T = T_s(z)$ at $z = s(t)$.   (12)

The temperature of the thermal plume is given by the adiabat:

$$T_p(z) = 1300 + 0.37 z \quad (13)$$

in °C, with z in km within the range suggested by Spohn and Schubert (1982).

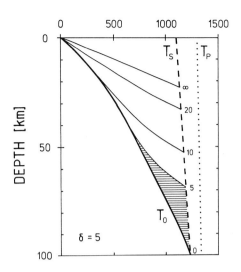

Fig. 7. Temperature-depth distribution for a disturbance of T_o by a thermal plume of strength, $q = 5\ q_i$. Undisturbed heat flux is 60 mW m^{-2} at the surface and $q_i = 31.4$ mW m^{-2} at the base of the lithospheric layer. T_o-starting geotherm; T_s-mantle solidus; T_p-plume temperature (adiabat)

Finally, the position of the LA boundary with time is defined by the intersection of the appropriate continental geotherm and a mixed volatile mantle solidus (Pollack and Chapman, 1977) given by Fig. 7:

$$T_s(z) = 1100 + 1.37\ z \quad . \tag{14}$$

The initial condition is specified by a particular geotherm appropriate for the undisturbed thermal regime of the RM. The starting geotherm, T_o, was calculated under the assumption of one dimensional steady state conductive heat transfer through a layered half space. Parameters used for the geotherm calculation have the following properties: surface heat flow of 60 mW m^{-2}; a pressure- and temperature-dependent thermal conductivity function with appropriate coefficients in the upper and lower crust and the upper mantle (Schatz and Simmons, 1972; Roy et al., 1981; Chapman, 1982); a radioactivity profile which decreases exponentially through the upper crust. The geotherm is similar to the 60 mW m^{-2} geotherm shown in Fig. 3 of Pollack and Chapman (1977), reaching a temperature of 557°C at the base of the crust (35 km) and 1237°C at the base of the lithosphere (100 km).

"Convective" thinning of the lithosphere is demonstrated in Fig. 7. $T_o(z)$ corresponds to an initial upward heat flux of $q_i = 31.4$ mW m^{-2} at 100 km depth. The assumed thermal plume provides a convective increase of heat flow at the LA boundary by a factor, $\delta = q/q_i$. The result is the upward migration of the LA boundary with time (curve parameters in Fig. 7: time in Ma) until an equilibrium thickness of the lithosphere is reached. The gradient of the final geotherm ($t \rightarrow \infty$ in Fig. 7) determines the surface heat flux while the shaded area in Fig. 7 is proportional to the surface uplift due to thermal expansion with respect to the time reached.

The rate of thinning as well as the final position of the top boundary of partially molten material is governed by the relative strength δ of the thermal plume source (see Fig. 8). While for $\delta > 5$ the thinning process is very rapid and leads to final lithospheric thicknesses smaller than 20 km, a decrease of the source strength δ provides decreasing amplitude of equilibrium position which is attained with increasing periods of time (e.g., if $\delta = 2$: $t > 100$ Ma). This mechanism of thinning is thus critically dependent on the strength of the thermal plume source.

Modeled surface uplift by thermal expansion (Fig. 9) is a function of temperature increase within the lithospheric layer. It is proportional to the shaded area in Fig. 7, and thus it is concurrent with the process of thinning as a function of time. High source strength ($\delta > 5$) leads to modeled surface uplift of more than 10^3 m within less than 25 Ma. Lower values of uplift correspond to smaller plume strength and require periods between 40 Ma and more than 100 Ma to reach their equilibrium.

The model finally allows a quantification of the corresponding excess surface heat flux (Fig. 10). As this quantity is a function of the temperature gradient of the geotherm T (Fig. 7), a relative time delay of the heat flux with respect to

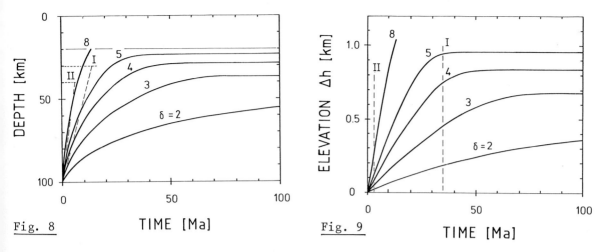

Fig. 8. Thinning of the lithosphere represented by the position of the phase boundary with time. Relative plume strength varies from $2q_i$ to $8q_i$ (*solid lines*). Thinning with time due to events I and II deduced from data (*dashed lines*). (After Neugebauer and Walter, 1983)

Fig. 9. Modeled surface uplift due to thermal expansion, $\alpha = 5 \times 10^{-5}/K$ in response to constant thermal plumes of relative strength, δ. Absolute period of time related to the onset of uprise for event I and II (*dashed lines*)

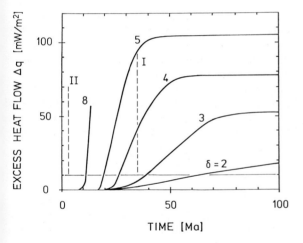

Fig. 10. Modeled excess heat flow for different plume strengths. Mean surface heat flow was assumed to be 60 mW m^{-2}. The level of accuracy of measurements was taken to be ±10 mW m^{-2}. Absolute period of time related to the onset of uprise for event I and II (*dashed lines*)

thinning and uplift of the lithosphere becomes obvious. This time delay ranges from about 10-30 Ma for the onset and 12-65 Ma for a 10 mW m^{-2} level for a thermal plume strength δ between 8 and 2 (Fig. 10). The amount of excess heat flow might be characterized by two examples: a $\delta = 8$ plume will cause 50 mW m^{-2} within 13 Ma while a $\delta = 2$ plume will not have reached the level of detection (about 10 mW m^{-2}) before 60 Ma.

Alpine Push

The stress field of the northern Alpine foreland is characterized by the dominantly extensional province of the RM and the compressive region comprising the Alps and their near vicinity. Any interaction or interdependence between these two areas would most likely be controlled or affected by the existence of the Upper Rhine Graben linking both.

In order to model the tectonic stresses in this situation, we established a three dimensional elastic-plastic four layer finite element model (Fig. 11). The ratio of dimensions is x:y:z = 7:4:1. Thus for a single layer thickness of 20 km, the dimensions of the model are 560:320:80 km. Following the relative limits on lithospheric stresses imposed by laboratory experiments (Brace and Kohlstedt, 1980), we assumed relative elastic limits for the layers A to D (Fig. 11) of 7:10:7:1 associated with the von Mises failure criterion.

The Upper Rhine Graben has been approximated by reducing the failure stresses

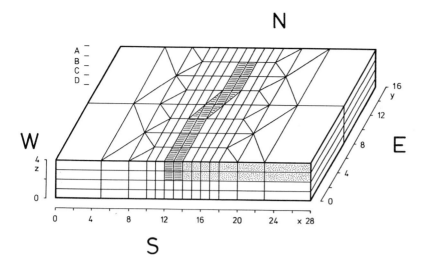

Fig. 11. Three-dimensional finite element structure approximating the northern Alpine foreland with the Upper Rhine Graben. For explanation see text

from 7 to 2 and from 10 to 3 at the marked zone within layers A and B, respectively. For simplicity, we simulated a possible action of the Alps on the foreland structure by a prescribed south to north directed displacement, at layers A and B and between $x = 14$ and $x = 28$, (Fig. 11) of 10^{-3} times the dimension of the structure along y. As a consequence of this shear motion of the graben flanks, the structure becomes coupled plastically throughout layer D to the fixed base. Shear deformation of the eastern block against the western one is most promoted by a free northern end of the structure and a N-S free-slip eastern boundary. The graben zone indicated in Fig. 11 exceeds the elastic limit along its entire length under the outlined conditions.

The modeled stress and displacement fields shown subsequently represent the elastic-plastic equilibrium in conformity with the imposed displacements.

Tectonic and seismotectonic activity refers most likely to the state of shear stresses while the mode of deformation can be adequately deduced from the stress field in terms of a principal stress representation. Figure 12 represents the relative distribution of effective shear stresses by means of cross sections, through a horizontal plane at the high strength layer B (Fig. 12a), east-west vertical plane at the front, middle, and far end of the structure (Fig. 12b), and vertical sections in the north-south direction along the graben structure and at both flanks (Fig. 12c). The contour lines depict relatively high shear stresses within the southeastern block with a steep decay towards the graben zone, and a slow decline towards northeast. Their extension to the north is limited to the southern half of the eastern block. Although the model boundaries have been chosen to enable shear deformation to occur along the entire graben zone, no tip effect in terms of a shear stress concentration at the northern boundary of the graben structure is predicted by the model calculations.

The modeled displacements (Fig. 13a-c) are representative for the maximum principal stress and confirm the stated rapid decline of influence of the Alpine action towards the north assisted by shearing along the Upper Rhine Graben. Small amplitude horizontal compression trending northwest is supposed to affect the RM area.

Hence, the action of shear, induced by the displacement of the eastern block and influenced by a graben-like structure finds a spatially rather limited response with its maximum stress effects confined to a triangular region adjacent to and east of the graben zone.

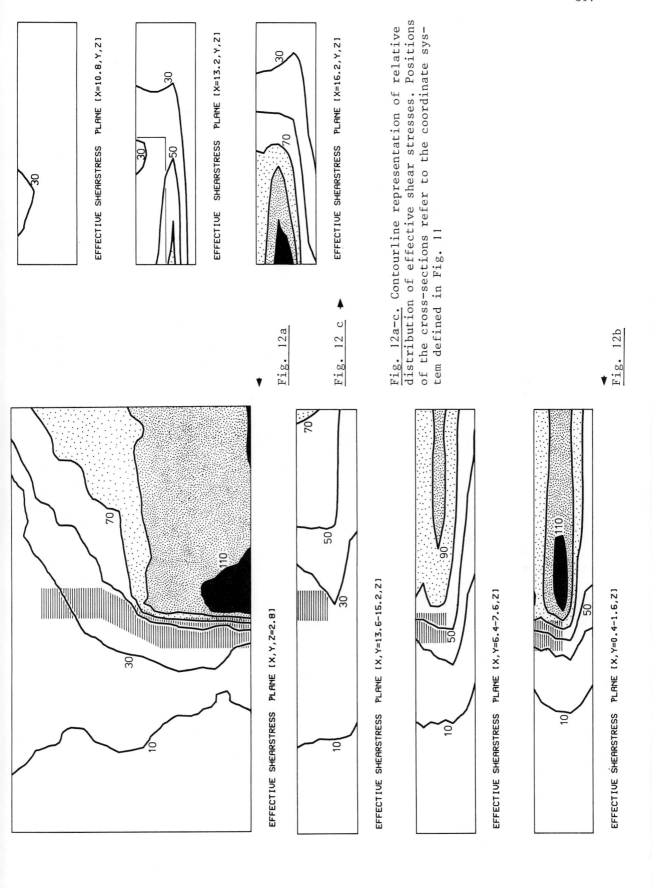

Fig. 12a-c. Contourline representation of relative distribution of effective shear stresses. Positions of the cross-sections refer to the coordinate system defined in Fig. 11

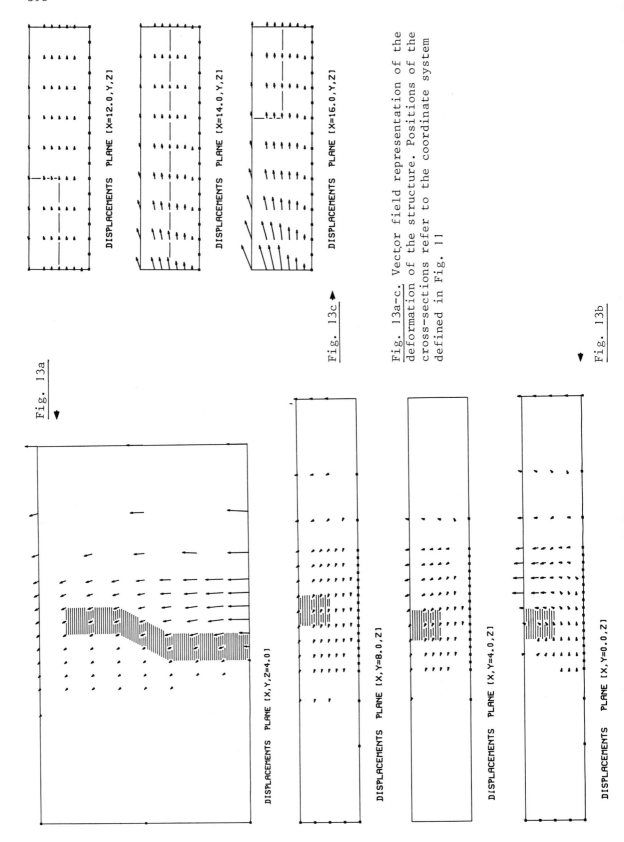

Fig. 13a-c. Vector field representation of the deformation of the structure. Positions of the cross-sections refer to the coordinate system defined in Fig. 11

Discussion and Conclusions

From the review of the data summarized in Table 1, a specific succession of phenomena linked in space and time requires explanation by a comprehensive mechanism.

On the basis of recent dynamics, including stress and uplift, the possibility of a direct influence of the present Alpine mountain belt on the RM should be considered. In this model (see Fig. 12a-c), the three-dimensional approximation of the stress field north of the Alps reproduces the seismicity of the SW end of the Upper Rhine Graben remarkably well. The eastern part of the southern and middle graben structure is a zone of high horizontal stress gradient in terms of the model and at the same time a prominent boundary of seismic activity. Asymmetry of seismicity and its decay towards the north and east, as revealed by hypocentral positions (Bonjer and Fuchs, 1979), is well reflected by the cross-sections of Fig. 12b and c. If we associate the triangular zone of maximum effective shear stress, which is obviously the result of "Alpine action" and a "shear motion" at the Upper Rhine Graben, with observed seismicity, the model would suggest a traceable effect of these sources only as far north as the middle graben zone (corresponding to the location of Karlsruhe). This zone corresponds well with the area of predominant strike-slip mode of deformation revealed by focal mechanism solutions. The northern part of the graben structure exhibits no stress concentration in the model. Here, seismicity is confined to the graben proper and normal faulting is the predominant mode of deformation. Thus, the presented three-dimensional numerical model supports the view that the northern Upper Rhine Graben belongs to a specific seismic province, the RM, which differs from the area to the south by its normal faulting style of deformation corresponding to uplift.

This stress unconformity between the RM and surrounding areas might be reasonably explained by self-reliant diapiric processes beneath the RM. Such diapiric stress regimes are characterized by a confined zone of high-value effective shear stress above the crest of the diapir. The horizontal components of the induced excess stress are of an extensional nature above the diapir which would lead to either stress release or extension (Fig. 6). Thus, the existence of diapiric structures beneath the RM could most likely provide regionally confined extensional stress provinces. Their intensity is supposed to be a function of the gravitational potential inherent in the structure, the external tectonic stress field and, finally, the tectonic and mechanical constitution of the lithosphere. This view of a negligible direct influence of the Alps on the RM is also expressed in the discussed Cenzoic history of the RM.

Both the mechanical diapir and thermal plume models represent regional or local processes with a good ability to explain the geodynamic development of the RM. In the following, we will discuss the probable predominance of either thermal or mechanical mechanisms in the light of the arguments deduced from observations. According to P-wave travel-time delays and mantle xenoliths, there is a lower bound for the thinned lithosphere under the RM of 55 km, while from seismic refraction data and depth of magma segregation with time an upper bound between 30 and 35 km is indicated.

The two petrogenic events I and II reveal a beginning of the thinning process before 35 and 3 Ma, a period of 10 and 3 Ma and a mean rate of 5 and 10 km/Ma, respectively (Neugebauer and Walter, 1983). In order to meet these rates by convective thinning, a strength for the plume in the order of $\delta = 6-8$, equivalent to $q \simeq 190-250$ mW m^{-2}, is necessary, (Fig. 8). However, such strong thermal plumes do not attain the steady state thickness of the lithosphere below a depth of 20 km, which is in contradiction to the observations.

With reference to the required rates and the absolute starting time, the predicted surface elevation and excess heat flow become substantially overestimated by the thermal plume models. Thereafter the elevation according to event I should have reached more than 10^3 m, event II would already have caused 300 m in addition (Fig. 9). The modeled excess heat flow is supposed to be about 100 mW m^{-2} with respect to event I (Fig. 10). In the contrary, the deduced Cenozoic uplift of the RM ranges between 400 to 600 m and there is no evidence for an increased heat flow according to an uncorrected mean level of 60 mW m^{-2} and an additional assumed noise level of about 10 mW m^{-2}.

A small increase of the sublithospheric heat flow $\delta < 2$, however, could serve well as a long-lasting process preceding an event of rapid thinning of the lithosphere, which is most probably controlled by mechanical parameters.

The diapir model is based on the existence of a sublithospheric layer of finite thickness and inverse density contrast. This specification implies, for either limited or infinite horizontal extent of the low density layer, the same mode of response to a small disturbance of the interface, i.e., a single or a sequence of single diapirs of characteristic final wavelength. The range of corresponding parameters of the earth could possibly limit the typical wavelengths to 10^2-10^3 km for observed thinned lithosphere. Hence, such a system implies the local exchange of only a limited volume of material where the dimensions and time constants for this exchange are defined by the mechanical and structural conditions of the system itself. Figures 4 and 5 exhibit three phases of development; initiation, ascent, and stagnation. The first and second phases have been deduced for the Rhine Rift, and the second and third phases for East Africa (Wendlandt, 1981; Neugebauer, 1982; Neugebauer and Walter, 1983).

The characteristic wavelength of lithospheric thinning in the RM is between 100 to 200 km. This quantity can be matched in the discussed model by a viscosity ratio $\mu_A/\mu_L : 10^0$-10^{-3} and $\mu_C \gg \mu_L$. The predicted layer of inverted density ranges between a minimum of 20 km to a maximum of 100 km thickness (Fig. 3, inset). The period of initiation of a possible diapir at the RM is of the order of 20 Ma or less (Table 1). For a constant amplitude z_0 of the initial disturbance, $h_A : 20$-100 km and $\mu_A/\mu_L : 10^0$-10^{-2} we attain the parameter γ^* ranging from 0.5 to 1.7 (Fig. 3, inset), which defines the onset of rapid amplitude increase after a time delay of $1/\gamma^*$. This model time transfers to absolute time by assuming a specific density inversion in the asthenospheric layer and the absolute viscosity of the lithosphere μ_L (see Fig. 3). An initiation of 10 to 20 Ma is consistent with a density contrast about or less than 0.005 g cm^{-3} for $\mu_L = 10^{21}$ poises, about 0.05 g cm^{-3} for $\mu_L = 10^{22}$ poises or even 0.5 g cm^{-3} for $\mu_L = 10^{23}$ poises. The same initiation time $1/\gamma$ can always be explained from the model results of Fig. 3 by a lower density in combination with a larger value of the growth rate γ. The allowable combinations of $\Delta\rho$ and μ_L deduced from the constraint of initiation time are bounded by the "upper bound $\Delta\rho$" line, shown in Fig. 14.

To explain uprise velocities in the diapiric stage of at least $v_z = 5$ km Ma (Table 1), we can derive a minimum density contrast directly from Eq. (7), again depending on the equivalent viscosity μ_L of the lithospheric layer. For $\mu_L = 10^{21}$ poises a lower limit of $\Delta\rho = 0.01$ g cm^{-3} is required, whereas for $\mu_L = 10^{22}$ poises, $\Delta\rho = 0.1$ g cm^{-3} is necessary. These values for $\Delta\rho$ can be taken as a lower limit (plotted as "lower bound $\Delta\rho$" in Fig. 14) to fullfil the requirement on the uprise velocity of the diapiric structure as derived from the data.

Although lower and upper bounds on the density difference as derived from the analysis of two different aspects in the dynamical behavior of a gravitational unstable asthenospheric layer do not intersect (see Fig. 14), the range of values for density differenc $\Delta\rho$ and lithospheric viscosity μ_L is in excellent agreement with data given by Aki (1982) and Murrell (1976).

Aki (1982) has shown that a seismic velocity anomaly of 5% can be explained by a decrease in density in the range of 0.0003 g cm^{-3} to 0.15 g cm^{-3} depending on different physical mechanisms. If these mechanisms are additive even higher density difference are possible. A 5% decrease in velocity is in good agreement with data from the RM (see Fig. 1). With respect to viscosity Murrell (1976) derived for the lower crust to the LA-boundary equivalent viscosities in the order of 10^{18} poises to 10^{22} poises de-

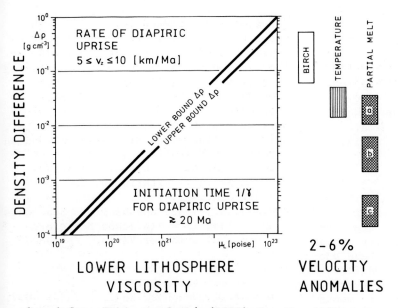

Fig. 14. Upper and lower bounds on the density difference $\Delta\rho$ dependent on the equivalent viscosity μ_L of the lithospheric layer derived from two different aspects of the diapiric uprise model (detailed discussion see text). The range of density differences $0.0001 < \Delta\rho < 0.1$ g cm^{-3} is taken from Aki (1982), who assumed different physical mechanisms to explain the measured decrease of the seismic velocities 2%-6% by a decrease in the density. *Rectangles* representing the range of possible density differences as calculated from Table 1 of Aki (1982) for the different mechanism: Birch law, temperature effect, partial melt a, b, and c. The range of equivalent viscosities for the lithosphere $10^{19} \leq \mu_L \leq 10^{23}$ poises is taken from Murrell (1976)

pendent on the assumed stress and temperature field. The crust has an equivalent viscosity many orders higher than the lower lithosphere, which is taken into account in our four layer model described in Table 2. The results of the model calculations show no dependence on the equivalent viscosity μ_C of the crustal layer if $\mu_C \geq 10^3 \mu_L$.

For the initiation phase the comparison between a convective thinning model and the diapiric uprise model shows that convective thinning results in faster rates of uprise of the LA-boundary. Therefore the convective thinning is certainly the more important process during the initial phase of diapiric uprise of a gravitational unstable asthenospheric layer. Consequently only a fraction of the initiation time of about 20 Ma or even more has to be exclusively explained by the diapir model, the larger part can be explained by convective thinning. A shorter initiation time for the diapir model results in higher density differences and shifts the upper limit on $\Delta\rho$ across the lower boundary derived from the constraint of an uprise velocity of at least 5 km/Ma (5 mm y^{-1}) for the diapiric stage.

The contradiction in Fig. 14 can easily be avoided by a contribution of the convective thinning model to the diapiric uprise in the initiation phase.

In terms of the diapir model it can be concluded that the wavelength of lithospheric thinning, $\lambda = 100-200$ km, and that the derived uplift rates of 5-10 km/Ma lead to a layer of inverted density h_A, of 20-100 km thickness and a possible density contrast, $\Delta\rho$, of 0.0003-0.15 g cm^{-3}. This range of density inversion is associated with absolute viscosities, μ_L, of the lithosphere of 10^{19}-10^{23} poise respectively. Gravity models by Fuchs (pers. commun.) for the mantle anomaly beneath the RM lead to a Bouguer anomaly of about -16 mgal for $\Delta\rho = 0.01$ g cm^{-3} without any crustal influence. However, the density contrast could be even larger if crustal compensation is allowed. On the other hand, local isostasy predicts surface uplift in the order of 0.3%-1.5% of the amount of thinned lithosphere, in accord with $\Delta\rho = 0.01-0.05$ g cm^{-3}. This yields, for example, 150-750 m uplift for a lithosphere thinned by 50 km. As shown for the thermal plume, the main periods of uplift have to be assigned to the phase of diapiric ascent. This causes surface uplift to be a more or less episodic phenomenon of variable lateral extent, a feature which can be easily explained in the close correspondence of main volcanic activity and uplift in the RM during the Cenozoic (see Fig. 2). On the basis of this data and in the context of the model, it is concluded that the sum of uplift in the RM of 400-600 m is most likely the result of subregional, episodic diapiric events as evidenced by volcanic activity during the Lower and Upper Miocene in the central area and eastern rim of the RM and, again, during the Pleistocene in the central area.

The interpretation of mantle velocity anomalies in terms of density contrasts is well accepted and a function of the process which is causing the velocity anomaly (Aki, 1982). Aki interprets these anomalies in terms of decoupling and flow in the lithosphere-asthenosphere system. On the other hand, geochemical arguments caused Anderson (1981) to claim that plumes might originate above the 220 km mantle discontinuity instead of their initially favored deep mantle origin. He suggests a continental thermal anomaly between 150-220 km depth due to possible thermal blanketing by the thick conductive continental lithosphere.

Acknowledgements. The authors wish to thank D.S. Chapman and J. Mechie for critically reviewing the manuscript. We had many helpful discussions on the models with D.S. Chapman, he provided a new geotherm for the thermal models. We are grateful to D. German, B. Stappert and R. Walter for their assistance in model calculation and data preparation, and to A. Seiz for typing the manuscript.

The financial support by the Deutsche Forschungsgemeinschaft is gratefully acknowledged.

References

Ahorner, L., 1962. Untersuchungen zur quartären Bruchtektonik der Niederrheinischen Bucht. Eiszeitalter Ggw., 13: 24-105.

Ahorner, L., Baier, B., and Bonjer, K.-P., 1983. General pattern of seismotectonic dislocations and the earthquake generating stress field in central Europe between the Alps and the North Sea. This Vol.

Aki, K., 1982. 3-D seismic inhomogeneities in the Lithosphere and Asthenosphere: Evidence for decoupling in the Lithosphere and flow in the Asthenosphere. Rev. Geophys. Space Phys., 20:161-170.

Anderson, D.L., 1981. A global geochemical model for the evolution of the mantle. Geodyn. Ser. Monogr., 5, AGU GSA, pp 6-18.

Baranyi, I., Lippolt, H.J., and Todt, W., 1976. Kalium-Argon-Alterbestimmungen an tertiären Vulkaniten des Oberrheingraben Gebietes: II. Die Alterstraverse vom Hegan nach Lothringen, Oberrhein. Geol. Abh., 25:41-62.

Baumann, H., 1982. Spannung und Spannungsumwandlung im Rheinischen Schiefergebirge. Numistischer Verlag, Koblenz, 240 pp.

Birch, F.S., 1975. Conductive heat flow anomalies over a hot spot in a moving medium. J. Geophys. Res., 80:4825-4827.

Bonjer, K.-P. and Fuchs, K., 1979. Real-time monitoring of seismic activity and earthquake mechanisms in the Rhinegraben area as a basis for prediction. Proc. ESA SP, 149:57-62.

Brace, W.F. and Kohlstedt, D.L., 1980. Limits on lithospheric stress imposed by laboratory experiments. J. Geophys. Res., 85:6248-6252.

Chapman, D.S., 1983. Lithosphere geotherms (in preparation).

Doebl, F. and Olbrecht, W., 1974. An isobath map of the tertiary base in the Rhinegraben. In: Illies, J.H. and Fuchs K. (eds.) Approaches to taphrogenesis. Schweizerbart, Stuttgart, pp 71-72.

Duda, A. and Schmincke, H.-U., 1978. Quarternary basanites, melilite nephelinite and tephrides from the Laacher See area (Germany). Neues Jahrb. Miner. Abh., 132:1-33.

Gass, I.G., Chapman, D.S., Pollack, H.N., and Thorpe, R.S., 1978. Geological and geophysical parameters of mid-plate volcanism. Philos. Trans. R. Soc. London Ser. A, 288:581-597.

Green, D.H., 1971. Composition of basaltic magmas as indicators of conditions of origin: application to oceanic volcanism. Philos. Trans. R. Soc. London Ser. A, 268:707-725.

Haenel, R., 1983. Geothermal investigations in the Rhenish Massif. This Vol.

Ingersoll, L.R., Jobel, O.J., and Ingersoll, A.C., 1954. Heat conduction. Univ. Wisconsin Press, Madison.

Jacoby, W.R., Joachimi, H., and Gerstenecker, C., 1983. The gravity field of the Rhenish Massif. This Vol.

Keizer, J. and Letsch, W.J., 1963. Geology of the Tertiary in the Netherlands. Verh. K. Ned. Geol. Mijnbouw. Genoot., Geol. Ser., 21-2:147-172.

Lippolt, H.J., 1981. K-Ar-Altersbestimmungen und zeitliche Korrelation des mitteleuropäischen Tertiärvulkanismus. Geol. Jahrb. (in press).

Lippolt, H.J., 1983. Distribution of volcanic activity in space and time. This Vol.

Mälzer, H., Hein, H., and Zippelt, G., 1983. Height changes in the Rhenish Massif: Determination and Analysis. This Vol.

Mareschal, J.C., 1981. Uplift by thermal expansion of the lithosphere. Geophys. J. R. Astron. Soc., 66:535-552.

Meyer, W., 1979. Influence of the Hercynian structures on Cainozoic movements in the Rhenish Massif. Allgem. Vermess. Nachr., 10:375-377.

McGetchin, T.R. and Merrill, R.B., 1979. Plateau uplift: mode and mechanism. Tectonophysics, 61:1-336.

Mechie, J., Prodehl, C., and Fuchs, K., 1983. The long-range seismic refraction experiment in the Rhenish Massif. This Vol.

Murrell, S.A.F., 1976. Rheology of the lithosphere - experimental indications. Tectonophysics, 36:5-24.

Neugebauer, H.J., 1983. Mechanical aspects of continental rifting. Tectonophysics, 94:91-108.

Neugebauer, H.J. and Walter, R., 1983. Petrogenic suite of magma genesis and the process of rifting at the Rhine Rift system (in preparation).

Ochmann, N., 1982. Numerische Modellrechnungen zur Dynamik von Diapirstrukturen und ihre Anwendung auf den Erdmantel. Thesis TU Clausthal, 110 pp.

Panza, G.F., Mueller, St., and Calcagnile, G., 1980. The gross features of the Lithosphere-Asthenosphere system in Europe from seismic surface waves and body waves. Pageoph., 118:1209-1213.

Pollack, H. and Chapman, D.S., 1977. On the regional variation of heat flow, geotherms and lithospheric thickness. Tectonophysics, 38:279-296.

Raikes, S. and Bonjer, K.-P., 1983. Large scale mantle heterogeneities beneath the Rhenish Massif and its vicinity from teleseismic P-residual measurements. This Vol.

Roll, A., 1979. Versuch einer Volumenbilanz des Oberrheingrabens und seiner Schultern. Geol. Jahrb. A, 52:3-82.

Roy, R.F., Beck, A.E., and Toulaukian, Y.S., 1981. In: Toulaukian, Y.S., Tudd, W.R., and Roy, R.F. (eds.) Thermophysical properties of rocks, in physical properties of rocks in minerals, vol. II-2. McGraw-Hill, New York, pp 409-502.

Sandwell, D.T., 1982. Thermal Isostasy: Response of a moving lithosphere to a distributed heat source. J. Geophys. Res., 87:1001-1014.

Schatz, J.F. and Simmons, G., 1972. Thermal conductivity of earth materials at high temperatures. J. Geophys. Res., 77:6966-6983.

Schneider, G., 1979. Seismotectonic movements inside the southern German triangle. Allgem. Vermess. Nachr., 86, 10:379-382.

Seck, H.A. and Wedepohl, K.H., 1983. Mantle xenoliths from the Tertiary and Quaternary volcanics of the Rhenish Massif and the Tertiary basalts of the Northern Hessian Depression. This Vol.

Spohn, T. and Schubert, G., 1982. Convective thinning of the Lithosphere: A mechanism for the initiation of continental rifting. J. Geophys. Res., 87:4669-4681.

Trümpy, R., 1973. The timing of orogenic events in the Central Alps. In: De Jong, K.A. and Scholten, R. (eds.) Gravity and tectonics. Wiley, New York, pp 229-251.

Wendlandt, R.F., 1981. Experimental petrology as a probe of rifting processes. Conference on the processes of planetary rifting. Lun. Planet. Inst., pp 126-133.

Withjack, M., 1979. A convective heat transfer model for lithosphere thinning and crustal uplift. J. Geophys. Res., 84:3008-3021.

Woidt, W.-D., 1980. Analytische und numerische Modellexperimente zur Physik der Salzstockbildung. GAMMA, 38, Braunschweig, 151 pp.

Woidt, W.-D. and Neugebauer, H.J., 1980. Finite element models of density instabilities by means of bicubic spline interpolation. Phys. Earth Planet. Int., 21:176-180.

Woidt, W.-D. and Neugebauer, H.J., 1983. Structural changes of the lithosphere in response to the development of mantle diapirs (in preparation).

Ziegler, P.A., 1981. Evolution of sedimentary basins in North-West Europe. Petrol. Geol. Cont. Shelf NE.-Europe, London, pp 3-39.

9 Epilogue: Mode and Mechanism of Rhenish Plateau Uplift

K. Fuchs[1], K. von Gehlen[2], H. Mälzer[3], H. Murawski[4], and A. Semmel[5]

Abstract

The main results of a 6 years' research program directed towards a better understanding of the young uplift of the Rhenish Massif as obtained from the various branches of geosciences are summarized and related to each other in this final review.

An anomalous low-velocity body discovered in the subcrustal lithosphere of the western part of the Rhenish Massif correlates with a number of other phenomena in this area: young volcanism, strongest rates of geodetic height changes, highest uplift of terraces. The generation of partial melt in the mantle body was facilitated by metasomatism which still requires dating. The discussion is focused on the properties of the subcrustal lithosphere, data on surface uplift, stress measurements, the geothermal field, and finally on numerical models of plateau uplift.
An uprise of less dense material from the lithosphere-asthenosphere boundary seems to be the most likely of the modeled mechanisms.

Finally, future research needs are indicated to further clarify the nature of the low velocity body in the mantle, and also to explain the uplift of the eastern part of the Rhenish Massif.

Introduction

Plateau uplift is one form of epeirogenic deformation of the crust. Fourteen possible mechanisms have been discussed and listed by McGetchin et al. (1980). It appears that the actual mode of uplift depends on a complicated aggregate of parameters including the geological history of the uplifted area, its mechanical behavior, the dominant stress-field, the composition of the lithosphere, and the state of temperature, energy and mass flux across the lithosphere-asthenosphere boundary. In a particular case, such as the uplift of the Rhenish Massif, it is essential that all relevant information be obtained in an effort to understand the mechanism of uplift.

---

1 Geophysikalisches Institut, Universität Karlsruhe, Hertzstraße 16, D-7500 Karlsruhe 1, Fed. Rep. of Germany

2 Institut für Geochemie, Petrologie und Lagerstättenkunde, Universität Frankfurt, Postfach 111932, D-6000 Frankfurt 11, Fed. Rep. of Germany

3 Geodätisches Institut, Universität Karlsruhe, Kaiserstraße 12, D-7500 Karlsruhe 1, Fed. Rep. of Germany

4 Geologisch-Paläontologisches Institut, Universität Frankfurt, Senckenberganlage 32-34, D-6000 Frankfurt, Fed. Rep. of Germany

5 Institut für Physische Geographie, Fachbereich Geowissenschaften, Universität Frankfurt, Senckenberganlage 36, D-6000 Frankfurt, Fed. Rep. of Germany.

Plateau Uplift, ed. by K. Fuchs et al.
© Springer-Verlag Berlin Heidelberg 1983

During the past six years the German Research Society (Deutsche Forschungsgemeinschaft) supported an interdisciplinary Priority Program the results of which are presented in this volume. In this concluding chapter, an attempt is made to summarize what has been learnt about the mechanism of plateau uplift of the Rhenish Massif. Although no definite model can be put forward, our knowledge of the influencing parameters has been extended to such a degree that alternative models can be examined and future critical experiments suggested.

Subcrustal Lithosphere

The most important finding during this research program was the discovery of a body of low P-velocity material in the western part of the Rhenish Massif in the depth range between 50 and 150 km with a volume of about 5×10^5 km^3. This body was deduced from a survey of teleseismic P-residuals in the Rhenish Massif and surrounding areas (Raikes, 1980; Raikes and Bonjer, 1983). If this low-velocity body is assumed to have risen from the asthenosphere, its depth range corroborates the depth of 50 km deduced by Panza et al. (1980) from surface wave dispersion data for the lithosphere-asthenosphere boundary beneath the Rhenish Massif. The existence of this mantle anomaly has immediate bearings on many other geoscience disciplines involved in the research program.

The geographic position of the mantle anomaly encompasses the West Eifel Quaternary volcanism (Schmincke et al., 1983), and also coincides with the region of strongest rates of present height changes (about 1.6 mm/a) obtained from geodetic measurements (Mälzer et al., 1983).

The observed Bouguer anomaly is more or less explained by crustal heterogeneities (Jacoby et al., 1983; Drisler and Jacoby, 1983). Therefore, since the mantle anomaly should possess only a negligible contrast in density relative to the surrounding subcrustal lithosphere, Raikes (1980) proposed the presence of about 1% partial melt which would reduce the velocity and practically not affect the density of the low-velocity body (see also Raikes and Bonjer, 1983).

While the crust-mantle boundary outside the Rhenish Massif is formed by a first-order discontinuity, its mode of transition below the massif is quite complicated (Mechie et al., 1983). In the eastern part a shallow low-velocity zone immediately below the Moho may be formed by instrusions of basaltic magmas. West of the Rhine on top of the deep low velocity body, the thin high-velocity lid is absent, and the crust-mantle boundary takes the form of a broad transition zone which may contain part of the source magma for the Eifel volcanism.

The occurrence of partial melt in the mantle anomaly was also discussed with regard to composition of mantle magma and xenoliths (Wedepohl, 1983; Seck, 1983; Seck and Wedepohl, 1983). The depleted lherzolites and harzburgites cannot explain the chemical composition of nepheline bearing basalts except if only a very low degree of partial melt ($\leq 1\%$) is involved. This in turn would imply very low velocities which could not lead to the formation of magma reservoirs able to feed volcanism (Fuchs and Wedepohl, 1983). However, the metasomatically altered periodites (which also occur as xenoliths in the basalts) can form the source rocks for the magmas of the nepheline containing basalts even if several (e.g., 2-3%) partial melt is allowed. Thus the presence of a metasomatism is regarded as a prerequisite in the Rhenish Massif to allow several percent partial melt to give rise to nepheline-containing basalts. The time for the cooling of the partial melt in the low-velocity body of the upper mantle is estimated to be 10 Ma. Therefore, if this body formed during the Pliocene/Quaternary, partial melt can still be expected to be present.

The presence of a magmatic source is also indicated from a C-isotope survey of CO_2. CO_2 is produced in the Eifel region at a rate of $(0.5-1) \times 10^6$ t/a amounting to a total of $(5-10) \times 10^9$ t CO_2 since the latest volcanic event 10,000 years ago (Puchelt, 1983).

There are two more findings where seismic and petrological observations with regard to the upper mantle are closely related. Seismic P_n-velocities of 8.1 km s^{-1} obtained for the topmost mantle from explosion seismic experiments are not compatible with the composition of depleted mantle xenoliths (Fuchs, 1983; Fuchs and Wedepohl, 1983). However, both P-velocities and composition become compatible at a depth of about 10 km below the crust/mantle boundary. It is not yet clear why the undepleted mantle close to the Moho is not sampled by basalts as xenoliths. The other finding is related to the observation of anisotropy in the subcrustal lithosphere of Southern Germany which was explained by a preferred orientation of olivine crystals using data on the distribution of P-velocities and on xenolith composition. The a-axis of olivine (i.e., the direction of the fastest velocity) with a vertical b-plane (i.e., the plane of best gliding) strikes N20°E, which is the same direction as that of the strike of the vertical plane of maximum crustal shear stress obtained from fault plane solutions of earthquakes. Therefore, it transpires that the crustal stress-field may leak into the upper mantle, there leading to deformation processes. This observation is of some importance for the discussion of the origin of the anomalous body in the upper mantle.

Surprisingly, the low velocity in the mantle was not detected by magneto-telluric sounding surveys; the upper mantle below the Rhenish Massif does not seem to differ in its electrical conductivity from that in the neighborhood (Jödicke et al., 1983). It is not certain whether this is due to lateral homogeneity in electrical conductivity or whether the observed maximum periods did not penetrate deep enough into the mantle to detect the anomalous body, or whether the body does not contain partial melt.

Uplift

The Quaternary uplift is not a singular event. Epeirogenic movements in the Rhenish Massif have taken place in the whole Massif or in parts of it since the Late Palaeozoic, but before the Tertiary without volcanism. Since the Tertiary the uplift has been connected with magmatic events. The young uplift started at the end of the Cretaceous/early Tertiary. The first main uplift began in the Middle/Upper Oligocene (about 30 Ma ago). At the end of the Miocene (15 Ma) the uplift accelerated and continues until present. However, the uplift appears to be interrupted by periods of stagnation and subsidence (Meyer et al., 1983; Murawski et al., 1983). The rate did not only change in time but also in space. In the Oligocene the Hohes Venn was uplifted by 460 m, the Westerwald Mountains by 200 m and the Siebengebirge Mountains remained stationary, corresponding to average rates of 0.015 mm/a, 0.006 mm/a and zero, respectively. Even the maximum average rate is smaller by orders of magnitudes than the maximum rate of the presently observed geodetic height changes (1.6 mm/a) in the Hohes Venn (Mälzer et al., 1983). The difference between the average and the present rates of uplift requires either that the general trend of uplift was interrupted by periods of stagnation or even of subsidence or that greater rates of uplift are occurring now than in the past. The lateral variations of uplift rates are also evident in the geographic distribution of geodetic height changes. There is no block uplifting of the Rhenish Massif, but rather a field of uplift changing strongly both in time and in space. This is to be observed throughout the time from Variscan to Quaternary.

The Rhine valley has existed since the Middle Miocene (about 20 Ma). Therefore, river terraces and their motion are available for the analysis of uplift of the Rhenish Massif. Despite numerous studies of terraces along the rivers Rhine, Mosel, Lahn, Rur, Sambre and Ourhte and others, it was impossible to arrive at a Massif-wide correlation of terraces from various geological times (see contributions in Chap. 4, this vol.). The heights of the major Pleistocene river terraces indicate relative motions of at least 200 m. These motions of the Rhenish Massif during the Quaternary relative to its surroundings are not homogeneous. Motions >200 m relative to the Lower Rhenish Embayment or about 70 m relative to the Mainz Basin are in contrast to minimal motions at the Eastern

Margin. According to R. Müller (1983) the uplift at the northwestern margin of the Rhenish Massif is more than 300 m. Apart from uplift, these values include the effects of subsidence in the foreland and of eustatic sea level changes.

Within the Rhenish Massif, investigations of the main river valleys (Rhine, Mosel, Lahn) revealed than in this area no major tectonic dislocations took place during the Quaternary. Exceptions include the basins of Neuwied and of Limburg.

Stress Field

According to Illies (1978; see also Illies et al., 1979) the stress field in the Alpine foreland had its horizontal compressional axis parallel to the upper Rhine Graben axis with an azimuth N20°E in the early Tertiary throwing the Upper Rhine Graben into extension. At the beginning of the Miocene an anticlockwise rotation of the stress field started producing sinistral shear deformation along the Upper Rhine Graben. This rotation continues until the present day. The horizontal compressional axis is now parallel to that segment of the European rift system running from the northern end of the Upper Rhine Graben to the Lower Rhenish Embayment. Thus this segment is subjected to extension. Geographical distributions of earthquakes and their fault-plane solutions are indicative of a buried rift structure under extension traversing the Rhenish Massif from the northern end of the Upper Rhine Graben to the Lower Rhenish Embayment (Ahorner, 1983; Ahorner et al., 1983). Extensional stresses obtained from in-situ stress-measurements are not only concordant with those derived from fault-plane solutions, but they show that the Rhenish Massif as a whole is under extensional stresses. In comparison, outside the Rhenish Massif compressional stress distributions prevail (Baumann and Illies, 1983). Most likely, the extensional stress field opened the path for the magma rise producing the Quaternary volcanism in the west Eifel (Schmincke et al., 1983). In contrast to the principal trend of the Tertiary structures volcanic fields are predominantly oriented N-S (Bussmann and Lorenz, 1983). This finding corroborates the notion of a rotation of the stress field between the Tertiary and the Quaternary. It is an important problem whether this crustal stress field opened the paths for magma rise only in the crust or whether by reaching into the upper mantle (Fuchs, 1983) the crustal stress field also affected the permeability of the subcrustal lithosphere.

Geothermal State

The heat-flow data obtained during the research program in the Rhenish Massif in shallow boreholes had to be corrected for palaeo-climatic changes. These corrected data indicate only slightly increased heat-flow in parts of the Rhenish Massif compared to the mean heat flow density of 81 mW m^{-2} for the Federal Republic of Germany (Haenel, 1983). These anomalies are not directly due to magmatic intrusions in the mantle or crust because the temperature "wave" would not have yet reached the Earth's surface. Therefore, it is assumed that the heat-flow anomalies are produced by migrating water or other shallow local differences in geothermal parameters.

Models for Mechanisms of Uplift

Most of the models proposed in this volume for the mechanism of uplift are connected with the anomalous body of low-velocity material in the upper mantle centered below the West Eifel. So far all of these models can only explain the uplift west of the river Rhine and fail to provide a mechanism for the uplift of the eastern part of the Rhenish Massif

Two to three percent partial melt in the low velocity body beneath the West-Eifel is equivalent to a volume expansion of 2-3‰ leading to an uplift of about 200-300 m (Fuchs and Wedepohl, 1983; Raikes and Bonjer, 1983). This would roughly explain the observed height changes. Uplift by thermal expansion alone would be in conflict with the observed Bouguer anomaly (Drisler and Jacoby, 1983).

While it is quite obvious that the anomalous low-velocity body in the upper mantle is responsible for the uplift in the western part of the Rhenish Massif and linked to the volcanic activity in this region, the crucial question is how this anomalous body was generated below the western Rhenish Massif.

Neugebauer et al. (1983) tested two different types of models for a diapiric uprise by finite-element calculations. For the thermal plume model with convective thinning of the lithosphere an increase of heat flow at the lithosphere-asthenosphere boundary by a factor of 6-8 is necessary to match the history of uplift. However, for such a model the final thickness of the lithosphere, the surface uplift, and the excess heat flow cannot meet the observations. The second type of model, the mechanical diapir, is due to unstable density stratification in the asthenosphere. This diapir model satisfies observed uplift rates and the lithosphere thickness of 50 km. The required density contrasts are compatible with the absence of an observable Bouguer anomaly. The model produces extensional stresses in the crust on the top of the diapir.

Such density instabilities can be supposed to occur at many places at the lithosphere-asthenosphere boundary. Why do these local instabilities develop into diapirs in some cases, why do they stabilize in most cases? Is it just a random event that placed the growing diapir beneath the Rhenish Massif? Or is there a deterministic process that triggered or enhanced the growth of the Rhenish diapir?

Illies and Greiner (1979) and Illies (in: Illies et al., 1979) advanced the concept that the tectonic activity in the Rhenish Massif, especially the activation of the buried rift, is closely linked to the rotation of the stress-field in the Alpine foreland in southern Germany. They suggested that the motion of the Rhenish Massif, induced by this field is responsible for the uplift by a rifting process and possibly by shear heating at the lithosphere-asthenosphere boundary. In particular, the blocking of the shear motion at the northern end of the Upper Rhine Graben and the consequent transformation into extension in the Rhenish Massif was regarded as a possible cause to focus uplift into the Rhenish Massif.

In a three-dimensional finite-element model Neugebauer et al. (1983) simulated the stress field in the Alpine foreland of southern Germany including the effect of the Upper Rhine Graben. The stress field is generated by an Alpine push from the south. Although this model predicts the seismicity near the southern end of the Upper Rhine Graben rather well, it only produces negligible influence on the Rhenish Massif. Preliminary as these calculations may be, they indicate that other, not yet known deterministic causes for the positioning of the mantle diapir in the Rhenish Massif have to be investigated, if one is not satisfied with a random positioning of the uprising material below the western part of the Rhenish Massif.

Conclusions

Although our knowledge on the uplift of the Rhenish Massif has expanded during the past 6 years by contributions from all geoscience disciplines, it must be admitted that the process of uplift is not yet fully understood.

A better understanding of the physical nature of the anomalous body in the mantle and of its composition is needed to constrain uplift mechanisms. Obviously the presence of mantle metasomatism has conditioned the subcrustal lithosphere of the Rhenish Massif to form partial melt. A dating of this metasomatism is urgently required to distinguish whether the generation of metasomatism is part of the young tectonic processes leading to the uplift of the Rhenish Massif or whether it is an accidental gift from the ancient past. Studies of the dissipation of P- and S-waves in this anomalous body and ultralong magneto-telluric sounding should provide important constraints on its physical properties and composition.

The present models for the uprise of material from the lithosphere-asthenosphere boundary do not take into account fast mass transport by fluid and volatile phases along grain boundaries. Such a process would drastically shorten the time for the generation of the anomalous body in the subcrustal lithosphere. Which process could widen the paths for these mobile phases? What is the role of the crustal stress field leaking into the mantle?

While the uplift of the Eifel west of the Rhine is closely linked to the presence of the low-velocity body in the mantle, the cause of the uplift in the eastern part of the Rhenish Massif is still obscure. The only deep feature in this part is a shallow low-velocity zone immediately below the crust-mantle boundary, possibly formed by a magmatic intrusion. It is found outside the Rhenish Massif. The nature of this possible intrusion should be constrained in the future by joint reflection-refraction profiles. A search for xenoliths from this zone is urgently needed. However, it should also be kept in mind that there seems to be a difference in epeirogenic activity between the eastern and western part of the Rhenish Massif ever since the end of the Variscan orogeny (Murawski et al., 1983).

Last but not least, despite numerous difficulties (e.g., eustatic sea-level changes) which became evident during this research program, efforts should continue to map the uplift history of the Rhenish Massif in space and time for the last 15 Ma from the deformation of river terraces.

References

Ahorner, L., 1983. Historical seismicity and present-day micro-earthquake activity of the Rhenish Massif, Central Europe. This Vol.
Ahorner, L., Baier, B., and Bonjer, K.-P., 1983. General pattern of seismotectonic dislocations and the earthquake generating stress field in central Europe between the Alps and the North Sea. This Vol.
Baumann, H. and Illies, H.J., 1983. Stress field and strain release in the Rhenish Massif. This Vol.
Brunnacker, K. and Boenigk, W., 1983. The Rhine valley between the Neuwied basin and the lower Rhenish Embayment. This Vol.
Bussmann, E. and Lorenz, V., 1983. Volcanism in the southern part of the Hocheifel. This Vol.
Drisler, J. and Jacoby, W.R., 1983. Gravity anomaly and density distribution of the Rhenish Massif. This Vol.
Fuchs, K., 1983. Recently formed elastic anisotropy and petrological models for the continental subcrustal lithosphere in Southern Germany. Phys. Earth Plant. Inter., 31:93-118.
Fuchs, K. and Wedepohl, K.H., 1983. Relation of geophysical and petrological models for upper mantle structure of the Rhenish Massif. This Vol.
Haenel, R., 1983. Geothermal investigations in the Rhenish Massif. This Vol.
Illies, H.J., 1978. Two stages Rhinegraben rifting. In: Ramberg, I.B. and Neumann, E.-R. (eds.) Tectonics and Geophysics of Continental Rifts. Réidel Dordrecht, 63-71.
Illies, J.H. and Greiner, G., 1979. Holocene movements and state of stress in the Rhinegraben Rift System. Tectonophysics, 52:349-359.
Illies, J.H., Prodehl, C., Schmincke, H.-U., and Semmel, A., 1979. The Quaternary uplift of the Rhenish Shield in Germany. Tectonophysics, 61:197-225.
Jacoby, W.R., Joachimi, H., and Gerstenecker, C., 1983. The gravity field of the Rhenish Massif. This Vol.
Jödicke, H., Untiedt, J., Olgemann, W., Schulte, L., and Wagenitz, V., 1983. Electrical conductivity structure of the crust and upper mantle beneath the Rhenish Massif. This Vol.
Mälzer, H., Hein, H., and Zippelt, G., 1983. Height changes in the Rhenish Massif: Determination and Analysis. This Vol.
McGetchin, T.R., Burke, K.C., Thompson, G.A., and Young, R.A., 1980. Mode and Mechanisms of Plateau Uplift. In: Bailey, A.W., Bender, P.L., Mcgetchin, T.R., and Walcott, R.I. (eds.) Dynamics of Plate Interiors. Geodynamics Series Vol. 1, AGU-GSA, Washington-Boulder, 99-100.

Mechie, J., Prodehl, C., and Fuchs, K., 1983. The long-range seismic refraction experiment in the Rhenish Massif. This Vol.

Meyer, W., Albers, H.J., Berner, H.P., Gehlen, K.v., Glatthaar, D., Löhnertz, W., Pfeffer, K.H., Schnütgens, A., Wienecke, K., and Zakosek, H., 1983. Pre-Quaternary uplift in the central part of the Rhenish Massif. This Vol.

Murawski, H., Albers, H.J., Bender, P., Berner, H.P., Dürr, St., Huckriede, R., Kauffmann, G., Kowalczyk, G., Meiburg, P., Müller, A., Müller, R., Ritzkowski, S., Schwab, K., Semmel, A., Stapf, K., Walter, R., Winter, K.-P., and Zankl, H., 1983. Regional tectonic setting and geological structure of the Rhenish Massif. This Vol.

Neugebauer, H.J., Woidt, W.-D., and Wallner, W., 1983. Uplift, volcanism and tectonics: Evidence for mantle diapirs. This Vol.

Panza, G.F., Mueller, St., and Calcagnile, G., 1980. The gross features of the lithosphere-asthenosphere system in Europe from seismic surface waves and body waves. Pageoph., 118:1209-1213.

Puchelt, H., 1983. Carbon dioxide in the Rhenish Massif. This Vol.

Raikes, S., 1980. Teleseismic evidence for velocity heterogeneities beneath the Rhenish Massif. J. Geophys., 14:80-83.

Raikes, S. and Bonjer, K.-P., 1983. Large scale mantle heterogeneities beneath the Rhenish Massif and its vicinity from teleseismic P-residual measurements. This Vol.

Schmincke, H.-U., Lorenz, V., and Seck, H.A., 1983. Quaternary Eifel Volcanisms. This Vol.

Seck, H.A., 1983. Eocene to recent volcanism within the Rhenish Massif and the Northern Hessian Depression - Summary. This Vol.

Seck, H.A. and Wedepohl, K.H., 1983. Mantle xenoliths from the Tertiary and Quaternary volcanics of the Rhenish Massif and the Tertiary basalts of the Northern Hessian Depression. This Vol.

Semmel, A., 1983. Plateau Uplift during Pleistocene time - Preface. This Vol.

Semmel, A., 1983. The early Pleistocene Terraces of the upper middle-Rhine and its southern foreland - Questions concerning their tectonic interpretation. This Vol.

Wedepohl, K.H., 1983. Tertiary volcanism in the Northern Hessian Depression. This Vol.

Final Report of Sonderforschungsbereich 48 – Göttingen
„Entwicklung, Bestand, und Eigenschaften der Erdkruste, insbesondere der Geosynklinalräume"

Intracontinental Fold Belts

Case Studies in the Variscan Belt of Europe and the Damara Belt in Namibia

Editors: **H. Martin, F. W. Eder**

1983. 300 figures, 24 plates. XIV, 946 pages
ISBN 3-540-12440-3

The book presents the main results of a 12 year interdisciplinary study conducted by a special research program at the University of Göttingen. The studies center on the Variscan Orogen in various parts of Europe and the Pan-African Damara Orogen in central Namibia. They help elucidate the sedimentological, geochemical, magmatic, tectonic, geochronological, metamorphic and geophysical features that contribute to the understanding of the geodynamic processes which governed the evolution of these intracontinental orogens. Geodynamic models are presented and discussed.

Springer-Verlag
Berlin
Heidelberg
New York
Tokyo